Springer Optimization and Its Applications

Volume 225

Aims and Scope

Optimization has continued to expand in all directions at an astonishing rate. New algorithmic and theoretical techniques are continually developing and the diffusion into other disciplines is proceeding at a rapid pace, with a spot light on machine learning, artificial intelligence, and quantum computing. Our knowledge of all aspects of the field has grown even more profound. At the same time, one of the most striking trends in optimization is the constantly increasing emphasis on the interdisciplinary nature of the field. Optimization has been a basic tool in areas not limited to applied mathematics, engineering, medicine, economics, computer science, operations research, and other sciences.

The series **Springer Optimization and Its Applications (SOIA)** aims to publish state-of-the-art expository works (monographs, contributed volumes, textbooks, handbooks) that focus on theory, methods, and applications of optimization. Topics covered include, but are not limited to, nonlinear optimization, combinatorial optimization, continuous optimization, stochastic optimization, Bayesian optimization, optimal control, discrete optimization, multi-objective optimization, and more. New to the series portfolio include Works at the intersection of optimization and machine learning, artificial intelligence, and quantum computing.

Volumes from this series are indexed by Web of Science, zbMATH, Mathematical Reviews, and SCOPUS.

Xiucui Guan • Panos M. Pardalos • Binwu Zhang

Inverse Combinatorial Optimization Problems

Springer

Xiucui Guan
School of Mathematics
Southeast University
Nanjing, Jiangsu, China

Binwu Zhang
School of Mathematics
Hohai University
Nanjing, Jiangsu, China

Panos M. Pardalos
Gainesville, FL, USA

ISSN 1931-6828 ISSN 1931-6836 (electronic)
Springer Optimization and Its Applications
ISBN 978-3-031-91174-3 ISBN 978-3-031-91175-0 (eBook)
https://doi.org/10.1007/978-3-031-91175-0

This Springer imprint is published by the registered company Springer Nature Switzerland AG
The registered company address is: Gewerbestrasse 11, 6330 Cham, Switzerland

To Professor Jason Jianzhong Zhang
We dedicate this monograph in token of affection and gratitude.

Preface

The realm of combinatorial optimization has long been a cornerstone of operations research, offering solutions to a myriad of complex decision-making problems across various domains. However, the traditional focus on forward optimization problems, which seek to optimize a given objective function subject to certain constraints, has evolved to include the burgeoning field of inverse combinatorial optimization problems (ICOPs). This monograph stands at the forefront of this paradigm shift, providing a comprehensive and scholarly exploration of ICOPs.

The concept of inverse optimization is not merely an academic curiosity; it holds profound implications for practical applications where the underlying model parameters are often unknown or obscured. By inverting the optimization process, researchers and practitioners can infer these parameters from observable outcomes, thus enabling a retrospective analysis and predictive modeling that is invaluable in fields such as network design, logistics, and strategic planning.

Our purpose in compiling this monograph is twofold. Firstly, we aim to contribute to the academic discourse by presenting a unified framework for understanding the complexities and interrelations among different ICOPs. Secondly, we seek to provide practitioners with the theoretical insights and algorithmic tools necessary to tackle real-world problems effectively.

The plan of the monograph is meticulously structured into four parts, each focusing on a distinct category of ICOPs. The first part lays the foundational concepts of ICOPs, highlighting the transition from forward to inverse optimization and its implications. The second part delves into the complexities of inverse shortest path problems, offering detailed analyses and algorithmic solutions. The third part explores inverse spanning tree problems, presenting theoretical insights and algorithmic advancements. The final part addresses inverse center location problems, with a focus on network efficiency enhancement.

Our intended audience comprises researchers, academics, and practitioners in the fields of operations research, management science, and related disciplines. We anticipate that this monograph will serve as a valuable resource for those seeking to understand and apply inverse combinatorial optimization in their work.

In conclusion, we are confident that this monograph will make a significant contribution to the field of inverse combinatorial optimization. We have endeavored to make this work accessible and informative, hoping to inspire further research and practical applications. We extend our gratitude to the readers for their interest and look forward to the advancements that this field will undoubtedly witness in the coming years.

Acknowledgments We extend our appreciation to the following authors for their specialized contributions to various chapters within the monograph:

- Junhua Jia for his work on inverse maximum capacity path problems and inverse (max + sum) spanning tree problems
- Wenjing Du for her collaboration on the chapter regarding inverse maximum capacity path problems
- Qiao Zhang for her research on shortest path interdiction problems and restricted inverse optimal value problems on shortest paths
- Yunwei Zhang for her exploration of methods to solve inverse linear programming problems
- Xiao Li for his investigation into the sum of root-leaf distance interdiction problems on trees
- Xianyue Li for his analysis of partial inverse minimum spanning tree problems
- Chen Wang for her study on inverse vertex obnoxious 1-center location problems
- Xinqiang Qian for his examination of inverse quickest 1-center location problems on trees

The collaborative efforts of these authors have not only enhanced the academic integrity of this monograph but also expanded its potential impact on the field of operations research and beyond. We are deeply thankful for their contributions and for the collective advancement of knowledge that this work represents.

Nanjing, China — Xiucui Guan
Nanjing, China — Binwu Zhang
Gainesville, FL, USA — Panos M. Pardalos
September 2024

Contents

Part I An Introduction to Inverse Combinatorial Optimization Problems

1 An Outline of Inverse Combinatorial Optimization Problems 3
1.1 A Framework of the Generalized Inverse Combinatorial Optimization Problems 3
1.1.1 Seven Classes of the Generalized Inverse Combinatorial Optimization Problems 4
1.1.2 Some Variations of the Generalized Inverse Combinatorial Optimization Problems 8
1.1.3 A Uniform Format for a Concrete *GICOP* 11
1.2 Time Complexities of Some *GICOPs* 12
1.2.1 The Forward Problem Is in $\mathcal{P}$, So Is Its Inverse Problem 12
1.2.2 The Forward Problem Is in $\mathcal{P}$, Its Inverse Problem May Be $\mathcal{NP}$-Hard 16
1.2.3 The Forward Problem Is $\mathcal{NP}$-Hard, So Is Its Inverse Problem 17
1.2.4 An Open Problem: Is there an Inverse $\mathcal{NP}$-Hard Problem Polynomially Solvable? 24
1.3 Applications of Inverse Optimization Problems 25
1.3.1 Geophysical Sciences 26
1.3.2 Healthcare Applications 26
1.3.3 Transportation and Logistics 27
1.3.4 Energy Systems 28
1.3.5 Agent Demand Models 28
1.3.6 Design Problems 29
1.3.7 Estimation Problems 30
1.3.8 Machine Learning and Data Science 31
1.3.9 Structural Estimation and Bilevel Optimization 32
1.3.10 Future Research Directions 32
1.4 Conclusions 33

2 Generalized Inverse Bottleneck Optimization Problems 37
2.1 Inverse Bottleneck Optimization Problems 37
2.1.1 (Inv, BOP, Bounded, Weighted l_1 Norm/Sum Hamming) 39
2.1.2 (Inv, BOP, Bounded, Weighted l_∞ Norm) 40
2.1.3 (Inv, BOP, Bounded, Bottleneck Hamming Distance) 42
2.2 Budget-Constrained Inverse Bottleneck Optimization Problem 43
2.3 Budget-Constrained Improvement Bottleneck Capacity Problem 46
2.3.1 (BC-Imp, BCP, Bounded, Weighted l_1 Norm) 46
2.3.2 (BC-Imp, BCP, Bounded, Weighted l_∞ Norm) 49
2.3.3 (BC-Imp, SCP, Bounded, Weighted l_1 Norm) 50
2.3.4 (BC-Imp, SCP, Bounded, Weighted l_∞ Norm) 52
2.3.5 (BC-Imp, BOP, Bounded, Weighted l_1/l_∞ Norm) 52
2.4 Budget-Constrained Interdiction Bottleneck Capacity Problem 53
2.4.1 (Int, BCP, Bounded, Weighted l_1 Norm) 53
2.4.2 (BC-Int, BCP, Bounded, Weighted l_1 Norm) 54
2.4.3 Complexities of the Problems $(BC\text{-}Int\text{-}BCP_1)$ and $(BC\text{-}Imp\text{-}BCP_1)$ 55
2.5 Budget-Constrained Interdiction Bottleneck Capacity Problem by Upgrading Nodes 56
2.5.1 A Linear Time Algorithm for Problem (*Int-Node-BCP*) on a Tree-path System 57
2.5.2 An $O(n \log n)$ Algorithm for Problem (*BC-Int-Node-BCP*) on a Tree-Path System 59
2.6 Inverse Constrained Bottleneck Problems on Networks 60
2.6.1 The Problem (ICBP_∞) Under the Proportional Case 61
2.6.2 Solve the Inverse Optimal Value Problem $(IOVCBP(p))$ 62
2.6.3 Solve the Inverse Problem $(ICBP_\infty)$ 64
2.7 Conclusions 66

3 Generalized Inverse Maximum Capacity Path Problems 69
3.1 (Int, Maximum Capacity Path, Bounded, Weighted l_1 Norm) 69
3.1.1 Introduction 69
3.1.2 Some Properties of the Problem (IntMCP) 71
3.1.3 Solve the Problem (IntMCP) 75
3.1.4 Numerical Experiments 80
3.2 (RIOV, Maximum Capacity Path, Bounded, Weighted l_1 Norm) 81

3.2.1 The Mathematical Model of the Problem ($RIOVMCP_1$) 82
3.2.2 Solve the Problem ($RIOVMCP_1$) 82
3.2.3 Numerical Experiments 86
3.3 (RIOV, Maximum Capacity Path, Bounded, Weighted l_∞ Norm) 88
3.3.1 The Capacity Value of the Given Path Remains Unchanged 89
3.3.2 Solve the Problem ($RIOVMCP_\infty$) 93
3.3.3 Numerical Experiments 95
3.4 Conclusions 97

4 Some General Methods to Solve Inverse Linear Programming Problem Under Weighted l_1 Norm 101
4.1 Introduction 101
4.2 Some General Methods to Solve the Inverse Problem (ILP) Under Weighted l_1 Norm 103
4.2.1 A Mathematical Model of (ILP) Under Weighted l_1 Norm 103
4.2.2 A Column Generation Method to Solve the Problem **(ILP1)** 107
4.2.3 A Row Generation Method to Solve the Problem **(ILP1)** 112
4.2.4 A Revised Simplex Method to Solve the Problem **(ILP1)** 115
4.3 Two Primal-Dual Algorithms to Solve the Inverse Bounded Linear Programming Problem Under Weighted l_1 Norm 119
4.3.1 A Primal-Dual Algorithm by Considering **(IBLP)** as a Primal Problem 120
4.3.2 A Primal-Dual Algorithm by Considering **(IBLP)** as a Dual Problem 124
4.4 Conclusions 135

Part II Generalized Inverse Shortest Path Problems

5 Shortest Path Improvement Problems 141
5.1 A Mathematical Model of Shortest Path Improvement Problems 141
5.2 (Imp, SP, Bounded, Weighted l_1 Norm) 142
5.2.1 Complexity of the Problem (**Imp-SP**$_{w1}$) 142
5.2.2 The Problem (**Imp-SP**$_{w1}$) with a Pair of Nodes 144
5.2.3 The Problem (**Imp-SP**$_{w1}$) on Trees with a Single Source 145
5.3 (Imp, SP, Bounded, Weighted Sum Hamming Distance) 146
5.3.1 Complexity of the Problem (**Imp-SP**$_{wsH}$) 147

5.3.2 Algorithms of **(Imp, SP, Bounded, sH-distance)** on Rooted Trees 150
5.4 Conclusions 153

6 Shortest Path Interdiction Problems on Trees 155
6.1 Introduction 155
6.2 The Budget-Constrained Shortest Path Interdiction Problem by Upgrading Edges on Trees 156
6.2.1 Some Important Definitions 157
6.2.2 (BC-Int, SP-Trees, Bounded, Weighted l_1 Norm) 158
6.2.3 ((BC-)Int, SP-Trees, Bounded, Unit l_1 Norm) 160
6.3 (Int, SP-Trees, Bounded, Weighted l_1 Norm) 163
6.4 (BC-Int, SP-Trees, Bounded, Weighted sum Hamming Distance) 164
6.4.1 Solve the Problem **(BC-Int-SPT$_{sH}$)** 165
6.4.2 (Int, SP-Trees, Bounded, Sum Hamming Distance) 169
6.5 Conclusions 170

7 Sum of Root-Leaf Distance Interdiction Problems on Trees 171
7.1 Introduction 171
7.2 The Sum of Root-Leaf Distance Interdiction Problem by Upgrading Edges/Nodes on Trees 173
7.2.1 ((BC-)Int, SRD-Trees, Bounded, Weighted Sum Hamming) 175
7.2.2 (BC-Int, SRD-Trees, Bounded, Sum Hamming Distance) 176
7.2.3 (Int, SRD-Trees, Bounded, Sum Hamming Distance) 177
7.2.4 ((BC-)Int, SRD-Trees, Bounded, Weighted l_1 Norm) 179
7.2.5 (BC-Int, SRD-Trees, Bounded, Weighted l_∞ Norm) 179
7.2.6 (Int, SRD-Trees, Bounded, Bottleneck Hamming Distance) 181
7.2.7 ((BC-)Int, SRD-Trees, Bounded, Weighted Node Cost) 183
7.2.8 (BC-Int, SRD-Trees, Bounded, Unit Node Cost) 183
7.2.9 (Int, SRD-Trees, Bounded, Unit Node Cost) 184
7.3 The Sum of Root-Leaf Distance Interdiction Problem with Cardinality Constraint by Upgrading Edges on Trees 186
7.3.1 (BC-Int-Cardinality, SRD-Trees, Bounded, Weighted l_∞ Norm) 187
7.3.2 (Int-Cardinality, SRD-Trees, Bounded, Weighted l_∞ Norm) 188

7.3.3 (BC-Int-Cardinality, SRD-Trees, Bounded, Bottleneck Hamming) 190
7.3.4 (Int-Cardinality, SRD-Trees, Bounded, Bottleneck Hamming) 191
7.4 Double Interdiction Problem on the Sum of Root-Leaf Distances on Trees 192
7.4.1 (BC-Double-Int, SRD-Trees, Bounded, wsH-l_∞) 194
7.4.2 (Double-Int, SRD-Trees, Bounded, wsH-l_∞) 200
7.5 Conclusions 201

8 Restricted Inverse Optimal Value Problem on Shortest Path Under Weighted l_1 Norm on Trees 203
8.1 (RIOV, SP-Trees, Bounded, Weighted l_1 Norm) 203
8.2 Some Important Properties of the Problem (**RIOVSPT$_1$**) 204
8.3 Construct a Slack Problem (**RIOVSPT$_1^s$**) 206
8.4 Construct and Solve the Subproblems (**RIOVSPT$_1^i$**) 207
8.5 An $O(n^2)$ Time Algorithm to Solve the Problem (**RIOVSPT$_1$**) 208
8.6 Conclusions 210

Part III Generalized Inverse Spanning Tree Problems

9 Inverse Minimum Spanning Tree Problems 215
9.1 The Inverse Minimum Spanning Tree Problem 215
9.2 The Inverse Minimum Spanning Tree Problem Under l_1/l_∞ Norm 216
9.2.1 (Inv, MST, Bounded, l_1 Norm) 216
9.2.2 (Inv, MST, Unbounded, l_1 Norm) 218
9.2.3 (Inv, MST, Unbounded, Weighted l_1 Norm) 221
9.2.4 (Inv, MST, Unbounded, l_∞ Norm) 222
9.3 The Inverse Minimum Spanning Tree Problem Under Hamming Distance 222
9.3.1 (Inv, MST, Bounded, Weighted Sum Hamming) 223
9.3.2 (Inv, MST, Bounded, Bottleneck Hamming) 225
9.4 Conclusions 228

10 Inverse Max+Sum Spanning Tree Problems 231
10.1 Inverse Max+Sum Spanning Tree Problems 231
10.2 (Inv-sum-cost, MSST, (Un)Bounded, Weighted l_∞ Norm) 232
10.2.1 (Inv-sum-cost, MSST, Unbounded, Weighted l_∞ Norm) 233
10.2.2 (Inv-sum-cost, MSST, Bounded, Weighted l_∞ Norm) 234
10.3 (Inv-sum-cost, MSST, Unbounded, Weighted l_1 Norm) 236
10.4 (Inv-sum-cost, MSST, Bounded, Hamming Distance) 239
10.5 (Inv-max-weight, MSST, Bounded, Weighted l_∞ Norm) 242
10.6 Conclusions 251

11 Restricted Inverse Optimal Value Problem on Minimum Spanning Tree ... 253
11.1 An Introduction to Restricted Inverse Optimal Value Problem on MST ... 253
11.2 (RIOV, MST, Bounded, Weighted l_∞ Norm) ... 254
11.3 (RIOV, MST, Bounded, Weighted l_1 Norm) ... 261
11.4 (RIOV, MST, Bounded, Bottleneck Hamming Distance) ... 275
11.5 Conclusions ... 282

12 Partial Inverse Minimum Spanning Tree Problems ... 283
12.1 An Introduction to Partial Inverse Minimum Spanning Tree Problems ... 283
12.1.1 A Mathematical Model ... 283
12.1.2 Preliminary Results ... 284
12.2 Computational Complexities of (*PInvMST*) ... 285
12.2.1 A Polynomial Time Algorithm for ($PInvMST^-$) ... 285
12.2.2 A Polynomial Time Algorithm for ($PInvMST^+_\infty$) ... 287
12.2.3 $\mathcal{NP}$-Hardness of ($PInvMST^+_p$) ... 289
12.2.4 $\mathcal{NP}$-Hardness of ($PInvMST^+_{wsH}$) ... 292
12.2.5 $\mathcal{NP}$-Hardness of ($PInvMST_1$) and ($PInvMST_{wsH}$) ... 293
12.2.6 Polynomial Time Algorithms for ($PInvMST_1$) and ($PInvMST_{wsH}$) with $|F| = 1$ ... 293
12.2.7 A Polynomial Time Algorithm of ($PInvMST_{bH}$) ... 297
12.2.8 A Polynomial Time Algorithm of ($PInvMST_\infty$) ... 299
12.3 Approximation Algorithms on (*PInvMST*) ... 304
12.3.1 Fitted Cut ... 304
12.3.2 Approximation Algorithms on ($PInvMST^+$) ... 305
12.3.3 Approximation Algorithms on (*PInvMST*) ... 308
12.4 Conclusions ... 309

Part IV Generalized Inverse Center Location Problems

13 Inverse Vertex Obnoxious 1-Center Location Problems ... 313
13.1 Inverse Center Location Problems ... 313
13.1.1 A Mathematical Model of Inverse Center Location Problems ... 313
13.1.2 Research Results of Inverse Center Location Problems ... 314
13.2 Inverse Median Location Problems ... 315
13.2.1 A Mathematical Model of Inverse Median Location Problem ... 315
13.2.2 Research Results of Inverse Median Location Problems ... 316
13.3 Inverse Vertex Obnoxious 1-Center Location Problems ... 317
13.3.1 An Introduction to the Problem (IVO1C) ... 317

13.3.2 A Mathematical Model of the Problem (IVO1C) 319
13.3.3 An Algorithm to Solve the Intersection Point of Two Polyline Functions 320
13.3.4 (Inv, VO1C, Bounded, Weighted l_∞ Norm) 325
13.3.5 (Inv, VO1C, Bounded, Bottleneck Hamming Distance) 332
13.3.6 Computational Experiments 337
13.4 Conclusions and Further Research 338

14 Inverse Quickest 1-Center Location Problem on Trees 339
14.1 Vertex and Absolute Quickest 1-Center Problems 339
14.2 Inverse Vertex and Absolute Quickest 1-Center Problems Under Weighted l_∞ Norm 341
14.2.1 (Inv, VQ1C, Bounded, Weighted l_∞ Norm) 341
14.2.2 (Inv, AQ1C, Bounded, Weighted l_∞ Norm) 350
14.3 Inverse Vertex and Absolute Quickest 1-Center Problems Under Weighted l_1 Norm 352
14.3.1 Maximum Transmission Time Balance Problem 352
14.3.2 (Inv, VQ1C, Bounded, Weighted l_1 Norm) 356
14.4 Conclusions 360

References 361

Acronyms

Lists of Abbreviations

Adj	Adjustment Combinatorial Optimization Problem defined as in (1.9)
AQ1C	Absolute Quickest 1-Center
BC-Imp	Budget Constrained Improvement Problem defined as in (1.16)
BC-Imp-BCP	Budget Constrained Improvement Problem on Bottleneck Capacity Problem defined as in (2.15)
BC-Imp-Node	Budget Constrained Improvement Problem by upgrading Nodes defined as in (1.24)
BC-Int	Budget Constrained Interdiction Problem defined as in (1.17)
BC-Int-Node	Budget Constrained Interdiction Problem by upgrading Nodes defined as in (1.26)
BC-Inv	Budget Constrained Inverse Problem defined as in (1.18)
BCP	Bottleneck Capacity Problem defined as in (2.14)
bH-distance	Bottleneck Hamming Distance
BLP	Bounded Linear Programming
BN	Binary-Newton
BOP	Bottleneck Optimization Problem defined as in (2.1)
BPN	Bin-Packing Number
BSM	Binary Search Method
BST	Bottleneck Spanning Tree
CBP	Constrained Bottleneck Problem defined as in (2.37)
CC-Inv	Cardinality Constrained Inverse Problem when the sum-type Hamming distance is applied on (1.18)
CL	Center Location
COP	Combinatorial Optimization Problem defined as in (1.1)
CS	Combinatorial Structure
DIBLP	Dual of Inverse Bounded Linear Programming
DP	Dynamic Programming
DRMP	Dual of Restricted Master Problem

DRP	Dual of Restricted Primal problem
ebH-distance	Expanded Bottleneck-type Hamming Distance
esH-distance	Expanded Sum-type Hamming Distance
GICOP	Generalized Inverse Combinatorial Optimization Problem
HD	Hamming Distance
IBLP	Inverse Bounded Linear Programming
IAQ1C	Inverse Absolute Quickest 1-Center
ICBP	Inverse Constrained Bottleneck Problem defined as in (2.38)
ICOP	Inverse Combinatorial Optimization Problem
ILP	Inverse Linear Programming
IMCPM	Inverse Maximum Capacity Path under Max-type distance
IMILP	Inverse Mixed Integer Linear Programming
Imp	Improvement Problem defined as in (1.2)
Imp-BCP	Improvement Problem on Bottleneck Capacity Problem defined as in (2.16)
Imp-Node	Improvement Problem by upgrading Nodes defined as in (1.23)
ImpSP	Improve Shortest Path
IMSST	Inverse Max+Sum Spanning Tree
IMST	Inverse Minimum Spanning Tree
Int	Interdiction Problem defined as in (1.3)
Int-MCP	Interdiction Problem on Maximum Capacity Path
Int-Node	Interdiction Problem by upgrading Nodes defined as in (1.25)
Int-SPT	Interdiction Problem on Shortest Path on Trees
Inv	Inverse Problem defined as in (1.7)
InvMC	Inverse Minimum Cut
InvMCF	Inverse Minimum Cost Flow
InvMF	Inverse Maximum Flow
InvML	Inverse Median Location
IOV	Inverse Optimal Value Problem defined as in (1.4)
ISP	Inverse Shortest Path
IVO1C	Inverse Vertex Obnoxious 1-Center location problems
IVQ1C	Inverse Vertex Quickest 1-Center location problems
KP	Knapsack problem
LP	Linear Programming
MC	Minimum Cut
MCF	Minimum Cost Flow
MCP	Maximum Capacity Path
MF	Maximum Flow
MILP	Mixed Integer Linear Programming
ML	Median Location
MP	Master Problem
MSST	Max + Sum Spanning Tree
MST	Minimum Spanning Tree
MTTB	Maximum Transmission Time Balance problem
PICO	Partial Inverse Combinatorial Optimization Problem

PInv	Partial Inverse Problem defined as in (1.8)
PInvMST	Partial Inverse Minimum Spanning Tree
RIOV	Restricted Inverse Optimal Value Problem defined as in (1.5)–(1.6)
RIOVMCP	Restricted Inverse Optimal Value Problem on Maximum Capacity Path
RIOVMST	Restricted Inverse Optimal Value Problem on Minimum Spanning Tree
RIOVSP	Restricted Inverse Optimal Value Problem on Shortest Path
RMP	Restricted Master Problem
RP	Restricted Primal problem
SCP	Sum Capacity Problem defined as in (2.22)
SP	Shortest Path
SPIUH-RT	Shortest Path Improvement problem under Unit Hamming distance on Rooted Trees
SRD	Sum of Root-leaf Distances
StRD	Shortest Root-leaf Distance
TSP	Traveling Salesman Problem
VQ1C	Vertex Quickest 1-Center
wsH-distance	Weighted Sum Hamming Distance

Lists of Symbols

$G = (V, E)$ A graph with a set $V = \{v_1, v_2, \cdots, v_n\}$ of vertices and a set $E = \{e_1, e_2, \cdots, e_m\}$ of edges.

$\boldsymbol{w} = (w_1, \cdots, w_m)$ A weight vector defined on E.

$\boldsymbol{c} = (c_1, \cdots, c_m)$ A cost vector defined on E.

$\boldsymbol{l} = (l_1, \cdots, l_m)$ A bound decrement vector for a modified weight vector $\bar{\boldsymbol{w}}$.

$\boldsymbol{u} = (u_1, \cdots, u_m)$ A bound increment vector for a modified weight vector $\bar{\boldsymbol{w}}$.

$\mathcal{S}$ A class of Combinatorial Structures.

$\boldsymbol{w}(S)$
- **Sum type** $w(S) = \sum_{e \in S} w(e)$;
- **Bottleneck type** $w(S) = \max_{e \in S} w(e)$;
- **Max+Sum type** $w(S) = \max_{e \in S} w(e) + \sum_{e \in S} c(e)$.

$\|\bar{w} - w\|_1 = \sum_{i=1}^{m} c_i |\bar{w}_i - w_i|$ Weighted l_1 norm.

$\|\bar{w} - w\|_\infty = \max_{i=1}^{m} c_i |\bar{w}_i - w_i|$ Weighted l_∞ norm.

$\|\bar{w} - w\|_2 = \sqrt{\sum_{i=1}^{m} c_i (\bar{w}_i - w_i)^2}$ Weighted l_2 norm.

$\|\bar{w} - w\|_{wsH} = \sum_{i=1}^{m} c_i H(\bar{w}_i, w_i)$ Weighted Sum-type Hamming distance (wsH).

$\|\bar{w} - w\|_{bH} = \max_{i=1}^{m} c_i H(\bar{w}_i, w_i)$ Bottleneck-type Hamming distance (bH).

$\|\bar{w} - w\|_{esH} = \sum_{i=1}^{m} c_i \bar{H}(\bar{w}_i, w_i)$ Expanded Sum-type Hamming distance (esH).

$\|\bar{w} - w\|_{ebH} = \max_{i=1}^{m} c_i \bar{H}(\bar{w}_i, w_i)$ Expanded Bottleneck-type Hamming distance (ebH).

$H(\bar{w}_i, w_i) = \begin{cases} 1, \text{ if } \bar{w}_i \neq w_i, \\ 0, \text{ otherwise.} \end{cases}$ Hamming distance.

$\bar{H}(\bar{w}_i, w_i) = \begin{cases} 1, \text{ if } |\bar{w}_i - w_i| > a_i, \\ 0, \text{ if } 0 \leq |\bar{w}_i - w_i| \leq a_i. \end{cases}$ Expanded Hamming distance.

List of Figures

Fig. 1.1 Relationship among the seven classes of *GICOPs* 7
Fig. 3.1 An example network G 80
Fig. 3.2 An example of problem (RIOVMCP$_1$) 86
Fig. 3.3 The Network G_L and its minimum cost cut 87
Fig. 3.4 The optimal network $\bar{G}$ 88
Fig. 3.5 An example of problem (IMCPM) 91
Fig. 3.6 The network $\hat{G}$ 92
Fig. 3.7 The network $\hat{G}_1$ 93
Fig. 3.8 The optimal network $\bar{G}$ 93
Fig. 3.9 An example of problem (RIOVMCP$_\infty$) 95
Fig. 3.10 The network G_L 96
Fig. 3.11 The optimal network $\bar{G}$ 97
Fig. 4.1 Relationship among the sets and values 130
Fig. 5.1 An example of graph G 143
Fig. 6.1 The illustration of Case 1 167
Fig. 7.1 The area labeled in blue and red are the subtrees $T_v(1)$ and $T_v(2:3)$, respectively [90] 196
Fig. 11.1 A demonstration to construct the auxiliary graph $\bar{G}^z$ 268
Fig. 11.2 An arc reversal transformation 270
Fig. 11.3 The relationships among some notations 276
Fig. 11.4 An example to show the meanings of notations 276
Fig. 12.1 An instance of ($PInvMST^+$) and its different solutions 289
Fig. 12.2 An instance of ($PInvMST$) with $|F| = 1$ under the unit l_2 norm and its different solutions 296
Fig. 12.3 An illustration of the increasing operation 301
Fig. 12.4 An instance $I = (G, w, c \equiv 1, F, l \equiv 0, u)$ of ($PInvMST_1^+$) 306
Fig. 12.5 The process of Algorithm 12.12 306

List of Tables

Table 1.1 An outline of generalized inverse combinatorial optimization problems ... 11
Table 2.1 The outline of generalized inverse bottleneck optimization problems ... 66
Table 3.1 The experimental results of Algorithm 3.1 ... 82
Table 3.2 Experimental results of Algorithm 3.2 ... 88
Table 3.3 Experimental Results of Algorithm 3.3 ... 93
Table 3.4 Experimental results of Algorithm 3.4 ... 97
Table 3.5 Research results on generalized inverse maximum capacity path (MCP) problem ... 98
Table 4.1 Research results on the problems (ILP) and (IBLP) ... 136
Table II Research results on generalized inverse shortest path problems ... 138
Table III Research results on generalized inverse MST/MSST problems ... 212
Table 9.1 The data in maximum flow problem (MWF1) ... 217
Table 12.1 The computational complexities and algorithms or ($PInvMST^-$), ($PInvMST^+$), and ($PInvMST$) ... 310
Table IV Research results on the inverse 1-center problems ... 312
Table 13.1 The experimental results of Algorithms 13.5 and 13.6 ... 338

List of Algorithms

1.1 A cutting-plane algorithm for (IMILP$_\infty$) [17, Algorithm 1] 23
2.1 A binary search algorithm to solve ($IBOP_{bH}$) [56, Algorithm 5] 43
2.2 Find a feasible solution of (BCIBOP) [49, Algorithm 1] 44
2.3 An algorithm to solve problem (BCIBOP) [49, Algorithm 2] 45
2.4 An algorithm to solve the bounded problem (BC-Imp-BCP$_1$) [145, BN-Algorithm] 48
2.5 An algorithm to solve the unbounded problem (BC-Imp-SCP$_1$) [145] 51
2.6 A subroutine to determine weight vector $\tilde{w}^i$ in Line 4 for the bounded problem (BC-Imp-SCP) [145] 51
2.7 Solve (*Int-Node-BCP*) on a tree-path system [57, Algorithm 1] 58
2.8 Solve (*BC-Int-Node-BCP*) on a tree-path system [57, Algorithm 2] 59
2.9 An algorithm to solve the problem ($IOVCBP(p)$) [58] 62
2.10 An algorithm for the problem (ICBP$_\infty$) [58] 64
3.1 An algorithm to solve (IntMCP) under weighted l_1 norm 77
3.2 An algorithm for the problem (**RIOVMCP$_1$**) 84
3.3 An algorithm to solve (**IMCPM**) under weighted l_∞ norm 90
3.4 An algorithm for solving the problem (**RIOVMCP$_\infty$**) 94
4.1 A framework of the column generation algorithm for (ILP1) 110
4.2 A row generation method to solve the problem **(ILP1)** 114
4.3 A revised simplex algorithm to solve the problem **(ILP1)** 117
4.4 A primal-dual algorithm by regarding **(IBLP)** as a primal problem 124
4.5 A primal-dual algorithm by considering **(IBLP)** as a dual problem 134
5.1 A DP Algorithm for (**Imp-SP$_{w1}$**) with a pair of nodes [142] 145

5.2 An algorithm for (**Imp-SP**$_{w1}$) on trees with a single source [142] ... 146
5.3 A greedy algorithm for Problem (**SPIUH**) on Chains [135] ... 152
5.4 A DP algorithm for problem (**SPIUH-RT**) [135] ... 152
6.1 A primal dual algorithm to (**BC-Int-SPT**$_1$) [153, Algorithm 5]. $[r^k, z^k] :=$ (**BC-Int-SPT**$_1$)$(T, Y, CD, CA, \chi, LN, w, u, c, B)$ ... 159
6.2 An algorithm to solve the problem (**Int-SPT**$_{u1}$) $r^* = OPT(L)$. [153, Algorithm 2] ... 161
6.3 An algorithm to solve (**BC-Int-SPT**$_{u1}$). [153, Algorithm 3] ... 161
6.4 The primal-dual algorithm to (**Int-SPT**$_1$).[153, Algorithm 6] $[r^k, g^k] :=$ **Int-SPT**$_1$$(T, Y, CD, CA, \chi, LN, w, u, c, D)$ ... 163
6.5 A dynamic programming algorithm [154, Algorithm 2] $[E_{v_1}(B), \bar{w}, f(T_{v_1}, B)] :=$ **BC-Int-SPT**$_{sH}$$(T, Y, w, u, B)$ ... 168
6.6 $[B^*, E^*, w^*] :=$ **Int-SPT**$_{sH}$$(T, Y, w, u, D)$ [154, Algorithm 3] ... 169
7.1 A greedy algorithm to solve the problem (**SRDIT**$_{sH}$) [151] ... 177
7.2 A greedy algorithm to solve the problem (**MCSRDIT**$_{sH}$) [151] ... 178
7.3 $[\bar{w}, C(\bar{w})] :=$ MCSRDIT$_\infty(T, n, w, u, c, D)$ [83] ... 181
7.4 $[\bar{w}, C(\bar{w})] :=$ MCSRDIT$_{bH}(T, n, w, u, c, D)$ [83] ... 182
7.5 A greedy algorithm to solve the problem (**SRDIT**$_{uN}$) [151] ... 184
7.6 A greedy algorithm to solve the problem (**MCSRDIT**$_{uN}$) [151] ... 185
7.7 $\left[\tilde{w}, \tilde{w}(T)\right] :=$ SRDITC$_\infty(T, n, w, u, c, K, N)$ [83] ... 187
7.8 $\left[\tilde{w}, C(\tilde{w})\right] :=$ MCSRDITC$_\infty(T, n, w, u, c, D, N)$ [83] ... 189
7.9 $\left[\tilde{w}, \tilde{w}(T)\right] =$ SRDITC$_{bH}(T, n, w, u, c, K, N)$ [83] ... 190
7.10 $\left[\tilde{w}, C(\tilde{w})\right] :=$ MCSRDITC$_{bH}(T, n, w, u, c, D, N)$ [83] ... 192
7.11 A greedy algorithm to solve (DIT$^1_{H\infty}$) [82] ... 195
7.12 A dynamic programming algorithm to solve ($CIT^{\lambda}_{H\infty}$) [82] ... 198
7.13 A binary search algorithm to solve (DIT$_{H\infty}$) [82] ... 199
7.14 A pseudo-polynomial time algorithm for (MCDIT$_{H\infty}$) [82] ... 200
8.1 $[\bar{w}, C(\bar{w})] :=$ **RIOVSPT**$_1$$(T, Y, w, l, u, c, D, t_0)$ [152] ... 209
9.1 An algorithm to solve problem (9.13) [62] ... 224
9.2 An algorithm to find a minimum bottleneck-weight node cover [139] ... 226
9.3 An $O(mn)$ algorithm to solve problem (9.16) [139] ... 227
9.4 An algorithm to solve problem (9.15) [139] ... 227
9.5 An $O(n^2)$ algorithm to solve problem (9.16) [37] ... 228
10.1 A Newton algorithm to solve (IMSST$^u_\infty$) [53, Algorithm 4] ... 233
10.2 A binary search method to solve (IMSST$^b_\infty$) [53, Algorithm 19] ... 235
10.3 The column generation algorithm to solve the problem (IMSST$^u_1$) [51, Algorithm 3] ... 238

10.4 A binary search method to solve (IMSST^b_{bH}) [50, Algorithm 4] 241
10.5 $(\lambda_\tau, \lambda_{\tau+1}] :=$ **BinarySearch**$(\Lambda, T^0, \boldsymbol{w}, \boldsymbol{l}, \boldsymbol{u}, \boldsymbol{c}, \boldsymbol{q})$ [67, Alg. 1] 243
10.6 $(T^0_\theta, E_u, E_Q, Q^1_\tau, Q^1_{\tau+1}) :=$ **Findsets1**$(V, E, T^0, \boldsymbol{w}, \boldsymbol{c}, \boldsymbol{q}, \boldsymbol{l}, \boldsymbol{u})$ [67, Algorithm 2] 246
10.7 $(e_0, T^0_\Delta, E_u, E_Q, Q^2_\tau, Q^2_{\tau+1}) :=$ **Findsets2**$(V, E, T^0, \boldsymbol{w}, \boldsymbol{c}, \boldsymbol{q}, \boldsymbol{l}, \boldsymbol{u})$ [67, Algorithm 3] 248
10.8 An $O(m^2 \log n)$ algorithm for $(\textbf{IMSST}_\infty)$ [67, Algorithm 4] 250
11.1 An $O(m^2 n)$ algorithm for $(\textbf{RIOVMST}_\infty)$ [65, Algorithm 3.1] 257
11.2 $(\lambda_\tau, \lambda_{\tau+1}] :=$ **BinarySearch**$(\Lambda, T^0, K, \boldsymbol{w}, \boldsymbol{l}, \boldsymbol{U}, \boldsymbol{c})$ [65, Alg. 4.1] 259
11.3 An algorithm for $(\textbf{RIOVMST}_\infty)$ when $\tilde{w}(T^0) < K$ [65, Alg. 4.2] 260
11.4 $\bar{G}^z(\bar{V}, \bar{E}, \bar{b}^z, \bar{c}) =$ **AG1**(S, z) [66, Algorithm 1] 266
11.5 $\tilde{G}^z(\tilde{V}, \tilde{E}, \tilde{b}^z, \tilde{c}, \tilde{d}^z) :=$ **AG2**$(\bar{G}^z(\bar{V}, \bar{E}, \bar{b}^z, \bar{c}), E^0_<, M^1, M^2, z)$ [66, Algorithm 2] 270
11.6 $(k_z, \psi(z)) =$ **Slope**$(S, E^0_<, \bar{z}, \underline{z}, z)$ [66, Algorithm 3] 272
11.7 $(\bar{\alpha}^*, \bar{\beta}^*) =$ $\textbf{RIOVMST}_1$$(G(V, E), T^0, w, l, u, c, K)$ [66, Alg. 4] 273
11.8 A binary search algorithm for $(\textbf{RIOVMST}_{bH})$ [128, Algorithm 3] 280
12.1 An algorithm to solve $(PInvMST^-)$ under the monotone norm 286
12.2 An algorithm to solve $(PInvMST^+_\infty)$ [81, Algorithm 1] 288
12.3 An algorithm to solve $(PInvMST^+_1)$ with $|F| = 1$ [86, Algorithm 2] 291
12.4 An algorithm for $(PInvMST^+_{wsH})$ with $|F| = 1$ [84, Algorithm 1] 292
12.5 An algorithm for $(PInvMST_{wsH})$ with $|F| = 1$ [84, Algorithm 2] 294
12.6 An algorithm to solve $(PInvMST_1)$ with $|F| = 1$ and $c \equiv 1$ [87, Algorithm 3] 296
12.7 Algorithm for the decision version of $(PInvMST_{bH})$........ 298
12.8 Algorithm for $(PInvMST_{bH})$ 298
12.9 Algorithm for the decision version of $(PInvMST_\infty)$........ 299
12.10 Algorithm for unit $(PInvMST_\infty)$ [85, Algorithm 2] 302
12.11 Algorithm for minimum fitted cut [87, Algorithm 1] 304
12.12 Approximation algorithm for $(PInvMST^+_1$ [87, Algorithm 2]) 305
12.13 Approximation algorithm for unit $(PInvMST_1)$........ 309

13.1 An algorithm for the polyline function $F_s(\rho)$ on $\rho \in [D_l(s), D_l^{max}]$. $(z_s, k_s, b_s) :=$ **Increase**$(D_l(s), D_l^{max}, A(s), l, c)$ 321
13.2 An algorithm for the polyline function $G_{v_i}(\rho)$ on $\rho \in [D_l(s), D_l^{max}]$. $(z_{v_i}, k_{v_i}, b_{v_i}) :=$ **Decrease**$(D_l(s), D_l^{max}, A(v_i), l, d)$ 323
13.3 An algorithm to find the intersection point $(\rho(v_i), C(v_i))$ of the function $F_s(\rho)$ and $G_{v_i}(\rho)$. $(\rho(v_i), C(v_i)) :=$ **Intersection**$(z_s, k_s, b_s, z_{v_i}, k_{v_i}, b_{v_i})$ 325
13.4 An algorithm to solve the transcendence point $(\rho(v_i), C(v_i))$ on $(a, b]$: $(\rho(v_i), C(v_i)) :=$ **Trans – Point**$(z_s, k_s, b_s, a, b, l^y, d, l, D_l(s), \rho_{max})$ 328
13.5 An algorithm to solve the problem (**IVO1C**$_\infty$) : $(C^*, x^*, y^*) =$ **IVO1C**$_\infty(G, l, c, d, u^x, u^y)$ 331
13.6 An algorithm to solve the problem (**IVO1C**$_{bH}$) : $(C^*, x^*, y^*) =$ **IVO1C**$_{bH}(G, l, c, d, u^x, u^y)$ 336
14.1 $[E_{11}, E_{12}] =$ **Findsets**$(T', T'', E_1, S_{E_1}, \Lambda, c_\Lambda, Q'_{c_\Lambda}, Q^{i^*}_{c_\Lambda})$ [110] 346
14.2 $[D_{\tau_1}, D_{\tau_2}] =$ **Binary Search**$(T', T'', S_E, \Lambda, c_\Lambda, Q'_{c_\Lambda}, Q^{i^*}_{c_\Lambda})$ [110] 347
14.3 $[c^*, D^*] = (IVQ1C_\infty)(T, l, c, u, d, q^+, q^-, \sigma, v^*)$ [110] 349
14.4 $[Q_k, g_k(Q_k)] =$ **MTTB**$_{e_k}(T_1, T_2, Q_l^1, Q_u^1, Q_l^2, Q_u^2)$ [109] 355
14.5 $[c^*, g(c^*)] = MTTB(T_1, T_2, E_2^{\geq}, Q_l^1, Q_u^1, Q_l^2)$ [109] 356
14.6 $[c^*, f(c^*)] =$ **IVQ1C**$_1(T, s^*)$ [109] 359

Part I
An Introduction to Inverse Combinatorial Optimization Problems

Inverse problems have been studied extensively by researchers working with geophysical data. Tarantola [122] highlighted that the essence of solving an inverse problem lies in deducing the model parameters based on observed values. Conversely, forward problems involve predicting observable parameter values given a set of model parameters. In this monograph, a typical optimization problem is classified as a forward problem because it focuses on identifying observable parameter values (such as optimal decision variables) based on model parameters (including cost coefficients, the right-hand side vector, and the constraint matrix). In contrast, an inverse optimization problem aims to determine the model parameters from the observable parameters, which may include optimal decision variables, an optimal objective value, and bounds for the optimal objective value.

The field of inverse combinatorial optimization problems (ICOP) has garnered significant attention in recent years, offering novel perspectives on traditional combinatorial optimization challenges. This introduction aims to synthesize the key findings and methodologies presented in four distinct chapters of a comprehensive study on ICOP. The chapters collectively provide a structured framework for understanding the complexities and applications of ICOPs, ranging from generalized inverse bottleneck optimization to inverse linear programming problems.

Chapter 1 lays the groundwork by outlining the broad spectrum of ICOPs, emphasizing the importance of a unified approach to understand the interrelations among various problem types. It highlights the significance of classical combinatorial optimization problems as a foundation for exploring the inverse paradigm. We introduce a standardized format for problem representation, facilitating detailed analysis and comparison across different ICOPs.

Chapter 2 delves into the intricacies of generalized inverse bottleneck optimization problems, presenting a systematic approach to formulating and solving these problems under various norms. The development of tailored algorithms and their complexity analyses underscore the chapter's contribution to the field, with practical implications extending to network design.

Chapter 3 investigates the restricted inverse optimal value problem on the maximum capacity path, proposing innovative algorithms with reduced time complexity.

The discussion on computational complexity and the interplay between forward and inverse optimization problems offers valuable insights into the potential and limitations of current computational methods.

Chapter 4 presents a comprehensive study on inverse linear programming problems, introducing innovative solution methodologies such as the column generation method, the revised simplex method, and the primal-dual algorithm. These methods transform inverse problems into solvable linear programming problems, significantly advancing the field.

In conclusion, this introduction encapsulates the essence of ICOP research, highlighting the theoretical advancements and practical applications. The chapters collectively pave the way for future research, offering a robust foundation for tackling the open challenges and complexities inherent in ICOPs. It not only lays the groundwork for the subsequent parts of the monograph but also contributes significantly to the broader field of operations research and management science.

Chapter 1
An Outline of Inverse Combinatorial Optimization Problems

Abstract In this chapter, we introduce a unified framework for Inverse Combinatorial Optimization Problems (ICOPs), encompassing a spectrum of problem types including traditional ICOPs, inverse optimal values, partial inverse problems, network improvement, and interdiction problems. This framework elucidates the intricate relationships among different problem classes and examines several polynomial-time-solvable combinatorial optimization problems such as shortest path and minimum spanning tree. We propose a standardized format to delineate the attributes of ICOPs, identifying open questions and exploring the complexity dynamics between forward and inverse problems. We question whether an inverse problem of an $\mathcal{NP}$-hard problem can be polynomially solvable, an unresolved challenge. The chapter highlights the broad applicability of inverse optimization in healthcare, transportation, logistics, and energy systems, emphasizing its role in decision-making and data analytics, with the potential to refine models and predict outcomes in complex systems.

Keywords Inverse combinatorial optimization problems · Budget constrained inverse problems · A uniform format · Time complexity · Applications of inverse optimization problems

1.1 A Framework of the Generalized Inverse Combinatorial Optimization Problems

Consider a graph $G = (V, E)$, where $V = \{v_1, v_2, \ldots, v_n\}$ represents a set of vertices, and $E = \{e_1, e_2, \ldots, e_m\}$ denotes a set of edges. Associated with each edge in E are two vectors: a weight vector $\boldsymbol{w} = (w_1, \ldots, w_m)$ and a cost vector $\boldsymbol{c} = (c_1, \ldots, c_m)$. Furthermore, we define non-negative bound vectors $\boldsymbol{l} = (l_1, \ldots, l_m)$ and $\boldsymbol{u} = (u_1, \ldots, u_m)$ that specify the allowable decrement and increment limits for adjusting the original weight vector, leading to a modified weight vector $\bar{\boldsymbol{w}}$.

Xiucui Guan and Binwu Zhang have equally contributed to this chapter.

X. Guan et al., *Inverse Combinatorial Optimization Problems*, Springer Optimization and Its Applications 225, https://doi.org/10.1007/978-3-031-91175-0_1

Let $\mathcal{S}$ be a class of *Combinatorial Structures (CS)*. In this book, $\mathcal{S}$ contains some classical *CS* including a set of paths, spanning trees, center/median locations, flows and cuts, etc.

Most *Combinatorial Optimization Problems (COPs)* formulated as (1.1) aim to find an optimal *CS* $S^* \in \mathcal{S}$ to minimize its weight $\boldsymbol{w}(S^*)$.

$$(COP) \quad \min_{S \in \mathcal{S}} \boldsymbol{w}(S). \tag{1.1}$$

The weight $\boldsymbol{w}(S^*)$ is generally defined as one of the following three formulations:

- *Sum type* $\boldsymbol{w}(S^*) = \sum_{e \in S^*} w(e)$;
- *Bottleneck type* $\boldsymbol{w}(S^*) = \max_{e \in S^*} w(e)$;
- *Max + Sum type* $\boldsymbol{w}(S^*) = \max_{e \in S^*} w(e) + \sum_{e \in S^*} c(e)$.

We mainly focus on some classical polynomial-time-solvable *COPs* including *Shortest Path (SP), Minimum Spanning Tree (MST), Bottleneck Spanning Tree (BST), Max + Sum Spanning Tree (MSST), Median/Center Location (ML/CL), Minimum Cost Flow (MCF)*, and *Minimum Cut (MC)*.

1.1.1 Seven Classes of the Generalized Inverse Combinatorial Optimization Problems

This subsection delineates seven distinct categories of Generalized Inverse Combinatorial Optimization Problems (GICOPs), outlining their mathematical formulations and interrelations.

An *Improve* problem (*Imp*) on a *COP* formulated as (1.2) is to find a new weight vector $\bar{\boldsymbol{w}}$ satisfying the *bounded constraint* $\boldsymbol{w} - \boldsymbol{l} \leq \bar{\boldsymbol{w}} \leq \boldsymbol{w} + \boldsymbol{u}$. We aim to minimize the modification $\|\bar{\boldsymbol{w}} - \boldsymbol{w}\|$ under some norm such that the weight of an optimal *CS* under $\bar{\boldsymbol{w}}$ is no greater than a given value D.

$$\begin{aligned} &\min_{\bar{\boldsymbol{w}}} \|\bar{\boldsymbol{w}} - \boldsymbol{w}\| \\ (Imp) \quad s.t. \; &\min_{S \in \mathcal{S}} \bar{\boldsymbol{w}}(S) \leq D, \\ &\boldsymbol{w} - \boldsymbol{l} \leq \bar{\boldsymbol{w}} \leq \boldsymbol{w} + \boldsymbol{u}. \end{aligned} \tag{1.2}$$

The difference between an *Interdiction* problem (*Int*) and an (*Imp*) problem is to require the weight of an optimal *CS* under $\bar{\boldsymbol{w}}$ no less than a given value D.

$$\min_{\bar{\boldsymbol{w}}} \|\bar{\boldsymbol{w}} - \boldsymbol{w}\|$$

$$(Int)\ s.t.\ \min_{S \in \mathcal{S}} \bar{\boldsymbol{w}}(S) \geq D, \tag{1.3}$$
$$\boldsymbol{w} - \boldsymbol{l} \leq \bar{\boldsymbol{w}} \leq \boldsymbol{w} + \boldsymbol{u}.$$

The intersection of an (*Int*) problem and an (*Imp*) problem defined on the same set $\mathcal{S}$ is called an *Inverse Optimal Value (IOV)* problem, which requests the weight of an optimal *CS* under $\bar{\boldsymbol{w}}$ is just equal to a given value D.

$$\begin{aligned} &\min_{\bar{\boldsymbol{w}}} \|\bar{\boldsymbol{w}} - \boldsymbol{w}\| \\ (IOV)\ s.t.\ &\min_{S \in \mathcal{S}} \bar{\boldsymbol{w}}(S) = D, \qquad (1.4) \\ &\boldsymbol{w} - \boldsymbol{l} \leq \bar{\boldsymbol{w}} \leq \boldsymbol{w} + \boldsymbol{u}. \end{aligned}$$

In the (*IOV*) problem, if we impose restriction on a given *CS* S^0 whose $\bar{\boldsymbol{w}}$-weight is also equal to the value D, then we call it a *Restricted Inverse Optimal Value (RIOV)* problem.

$$\begin{aligned} &\min_{\bar{\boldsymbol{w}}} \|\bar{\boldsymbol{w}} - \boldsymbol{w}\| \\ (RIOV)\ s.t.\ &\min_{S \in \mathcal{S}} \bar{\boldsymbol{w}}(S) = D, \qquad (1.5) \\ &\bar{\boldsymbol{w}}(S^0) = D, \qquad (1.6) \\ &\boldsymbol{w} - \boldsymbol{l} \leq \bar{\boldsymbol{w}} \leq \boldsymbol{w} + \boldsymbol{u}. \end{aligned}$$

It follows from (1.5) and (1.6) that (1.7) holds in the classical *Inverse Combinatorial Optimization (ICO)* problem (*Inv*). It minimizes the modification $\|\bar{\boldsymbol{w}} - \boldsymbol{w}\|$ under some norm such that the weight of an optimal *CS* is equal to the weight of a given *CS* S^0 under $\bar{\boldsymbol{w}}$. Hence, S^0 is an optimal *CS* under $\bar{\boldsymbol{w}}$.

$$\begin{aligned} &\min_{\bar{\boldsymbol{w}}} \|\bar{\boldsymbol{w}} - \boldsymbol{w}\| \\ (Inv)\ s.t.\ &\min_{S \in \mathcal{S}} \bar{\boldsymbol{w}}(S) = \bar{\boldsymbol{w}}(S^0), \qquad (1.7) \\ &\boldsymbol{w} - \boldsymbol{l} \leq \bar{\boldsymbol{w}} \leq \boldsymbol{w} + \boldsymbol{u}. \end{aligned}$$

The (*Inv*) problem can be be regarded as the generalization of the theory of sensitivity analysis and a supplementary framework to the theory concerning the stability of optimal solutions in optimization challenges [63]. For example, the sensitivity analysis of linear programming delves into the permissible variations of each parameter without altering the optimal solution. The inverse linear programming seeks to ascertain the minimal adjustments required under a specified norm to transform a feasible solution, denoted as x^0, into the optimal solution for the new LP problem. This approach gives us much flexibility of choices and provides us another perspective to describe the sensitivity analysis.

Furthermore, in a *Partial Inverse Combinatorial Optimization (PICO)* problem (*PInv*), a subset F^0 of a *CS* is given. The constraint condition is to find an optimal *CS* S_{F^0} under $\bar{\boldsymbol{w}}$ including F^0. Obviously, the classical *ICO* problem is a special case of its corresponding *PICO* problem when $F^0 = S_{F^0}$ is given.

$$\begin{aligned} &\min_{\bar{\boldsymbol{w}}} \|\bar{\boldsymbol{w}} - \boldsymbol{w}\| \\ (PInv)\ s.t.\ &\min_{S\in\mathcal{S}} \bar{\boldsymbol{w}}(S) = \bar{\boldsymbol{w}}(S_{F^0}),\ F^0 \subseteq S_{F^0}, \\ &\boldsymbol{w} - \boldsymbol{l} \le \bar{\boldsymbol{w}} \le \boldsymbol{w} + \boldsymbol{u}. \end{aligned} \tag{1.8}$$

In a *Subclass Inverse Combinatorial Optimization (SICO)* problem (*SInv*), a subclass $\mathcal{S}^0 \subset \mathcal{S}$ of *CSs* is given. We aim to minimize the modification $\|\bar{\boldsymbol{w}} - \boldsymbol{w}\|$ under some norm such that the minimum weight under $\bar{\boldsymbol{w}}$ of an optimal *CS* on the subclass $\mathcal{S}^0$ is equal to that on the whole class $\mathcal{S}$. Such a problem is called an adjustment problem in [88].

$$\begin{aligned} &\min_{\bar{\boldsymbol{w}}} \|\bar{\boldsymbol{w}} - \boldsymbol{w}\| \\ (SInv)\ s.t.\ &\min_{S\in\mathcal{S}} \bar{\boldsymbol{w}}(S) = \min_{S\in\mathcal{S}^0} \bar{\boldsymbol{w}}(S), \\ &\boldsymbol{w} - \boldsymbol{l} \le \bar{\boldsymbol{w}} \le \boldsymbol{w} + \boldsymbol{u}. \end{aligned} \tag{1.9}$$

Let $\mathcal{S}_{F^0} = \{S | F^0 \subseteq S\}$ be the subclass of $\mathcal{S}$ containing all the *CSs* S including the subset $F^0 \subseteq S$. Then the constraint condition (1.8) can be regarded as

$$\min_{S\in\mathcal{S}} \bar{\boldsymbol{w}}(S) = \bar{\boldsymbol{w}}(S_{F^0}) = \min_{S\in\mathcal{S}_{F^0}} \bar{\boldsymbol{w}}(S).$$

Hence, the *PICO* problem is a special case of its corresponding *SICO* problem when $\mathcal{S}^0 = \mathcal{S}_{F^0}$ is given.

Libura [88] considered the adjustment problems on *SP* and *MST*, where $\mathcal{S}^0$ is a subclass of $s - t$ paths passing through a given vertex and of spanning trees, which are also Hamiltonian paths, respectively. Observed that the adjustment problemon *SP* simplifies to a *PICO* problem when considering $\mathcal{S}^0$ as a specific subset of $s - t$ paths that traverse a predetermined edge, contributing to a partial solution of an $s - t$ path. It is worth noting that a variety of constraints can be applied to a *CS*, thereby expanding the scope of the *SICO* problem to encompass a broader spectrum of issues compared to the *PICO* problem class.

As a conclusion, we propose a framework of the generalized inverse combinatorial optimization problems, in which we use the same objective function and different constraint conditions to describe their characteristics. The relationship

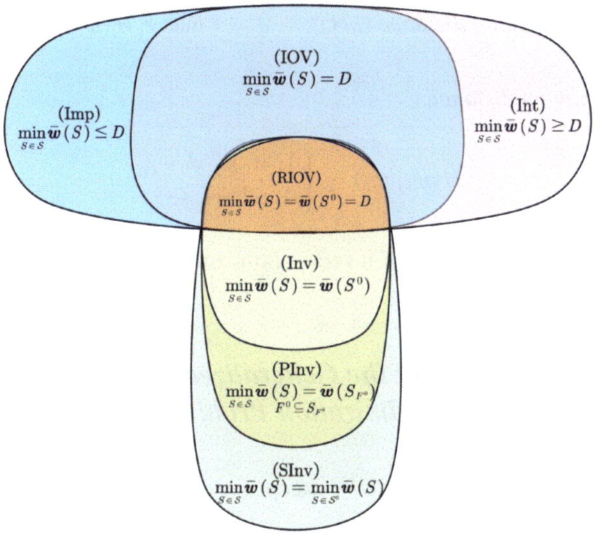

Fig. 1.1 Relationship among the seven classes of *GICOPs*

among these seven classes of *GICOPs* can be described by the following three expressions, which are shown in Fig. 1.1.

$$(Imp) \cap (Int) = (IOV),$$
$$(IOV) \cap (Inv) = (RIOV),$$
$$(RIOV) \subset (Inv) \subset (PInv) \subset (SInv).$$

Mostly used norms $\|\bar{\boldsymbol{w}} - \boldsymbol{w}\|$ in the field of *GICOP* are *weighted l_1 norm, l_∞ norm, l_2 norm, and Hamming distance* defined further. When $c_i = 1$ for $i = 1, 2, \cdots, m$, we call it a *unit norm* and often delete “unit” before the corresponding norm for simplicity.

$$\textit{Weighted } l_1 \textit{ norm}: \quad \sum_{i=1}^{m} c_i|\bar{w}_i - w_i|; \tag{1.10}$$

$$\textit{Weighted } l_\infty \textit{ norm}: \quad \max_{i=1}^{m} c_i|\bar{w}_i - w_i|; \tag{1.11}$$

$$\textit{Weighted } l_2 \textit{ norm}: \quad \sqrt{\sum_{i=1}^{m} c_i(\bar{w}_i - w_i)^2}; \tag{1.12}$$

$$\textit{Weighted sum Hamming distance } (wsH): \quad \sum_{i=1}^{m} c_i H(\bar{w}_i, w_i); \tag{1.13}$$

$$\textit{Bottleneck Hamming distance (bH)}: \quad \max_{i=1}^{m} c_i H(\bar{w}_i, w_i); \tag{1.14}$$

where the *Hamming distance*

$$H(\bar{w}_i, w_i) = \begin{cases} 1, \text{ if } \bar{w}_i \neq w_i, \\ 0, \text{ otherwise}, \end{cases} \tag{1.15}$$

was first introduced to the field of inverse optimization by He et al. [62].

1.1.2 Some Variations of the Generalized Inverse Combinatorial Optimization Problems

There are some variations of *GICOPs* except for the aforementioned seven classes of *GICOPs*.

- **The Budget Constrained Improvement Problem (BC-Imp)**

Some references such as [19, 20, 147, 150] considered a class of reverse optimization problems, where we exchange the objective function and the constraint condition in the improvement problem. Now we call it a *Budget Constrained Improvement Problem (BC-Imp)*. We aim to improve the performance of a *COP* by modifying the weight vector within a budget B. Its mathematical model can be formulated by (1.16).

$$\begin{aligned} &\min_{\bar{\boldsymbol{w}}} \min_{S \in \mathcal{S}} \bar{\boldsymbol{w}}(S) \\ (\textit{BC-Imp}) \quad s.t. \ &\|\bar{\boldsymbol{w}} - \boldsymbol{w}\| \leq B, \\ &\boldsymbol{w} - \boldsymbol{l} \leq \bar{\boldsymbol{w}} \leq \boldsymbol{w} + \boldsymbol{u}. \end{aligned} \tag{1.16}$$

It is worthy noting that the problem (*BC-Imp*) is often solved as a subproblem of the problem (*Imp*).

- **The Budget Constrained Interdiction Problem (BC-Int)**

Similarly, there is a class of *Budget Constrained Interdiction Problem (BC-Int)* [153, 154]. We aim to interdict the performance of a *COP* by modifying the weight vector within a budget B. Its mathematical model can be formulated by (1.17).

$$\begin{aligned} &\max_{\bar{\boldsymbol{w}}} \min_{S \in \mathcal{S}} \bar{\boldsymbol{w}}(S) \\ (\textit{BC-Int}) \quad s.t. \ &\|\bar{\boldsymbol{w}} - \boldsymbol{w}\| \leq B, \\ &\boldsymbol{w} - \boldsymbol{l} \leq \bar{\boldsymbol{w}} \leq \boldsymbol{w} + \boldsymbol{u}. \end{aligned} \tag{1.17}$$

Similarly, the problem *(BC-Int)* is often solved as a subproblem of the problem *(Int)*, as in the case of the shortest path interdiction problem on trees under unit Hamming distance [154]. Moreover, the problem *(BC-Int)* can be solved by a similar method for the problem *(Int)*. For example, for the shortest path interdiction problem on trees under weighted l_1 norm, both the problems *(BC-Int)* and *(Int)* were solved by a primal-dual algorithm in $O(n^2)$ time, where n is the number of vertices in a tree [153].

- **The Budget Constrained Inverse Problem (BC-Inv)**

The Budget Constrained Inverse Problem (BC-Inv) [8, 61, 68] minimizes the modification $\|\bar{\boldsymbol{w}} - \boldsymbol{w}\|$ under some norm while imposing a budget constraint on modification $\|\bar{\boldsymbol{w}} - \boldsymbol{w}\|$ under another norm. Its mathematical model can be formulated as follows.

$$\begin{aligned} &\min_{\bar{\boldsymbol{w}}} \|\bar{\boldsymbol{w}} - \boldsymbol{w}\| \\ (BC\text{-}Inv)\ s.t.\ &\min_{S \in \mathcal{S}} \bar{\boldsymbol{w}}(S) = \bar{\boldsymbol{w}}(S^0), \\ &\|\bar{\boldsymbol{w}} - \boldsymbol{w}\| \le B, \\ &\boldsymbol{w} - \boldsymbol{l} \le \bar{\boldsymbol{w}} \le \boldsymbol{w} + \boldsymbol{u}. \end{aligned} \tag{1.18}$$

We usually choose the norm in the constraint condition (1.18) by the sum-type norm (1.10) and (1.13), while in the objective function by the bottleneck-type norm (1.11) and (1.14). The reason is that the inverse problem under bottleneck-type norm always has multiple optimal solutions. Then we can cancel some unnecessary modifications by imposing the budget constraint on a sum-type norm. Notice that we restrict the number of modified elements when the unit sum Hamming distance is applied on (1.18), and we call the corresponding problem *(BC-Inv)* as the *Cardinality Constrained Inverse problem (CC-Inv)*. Hasanzadeh et al. [61] defined an *expanded Hamming distance* to ignore some "small" modifications within a given vector $\boldsymbol{a} = (a_1, \cdots, a_m)$.

$$\bar{H}(\bar{w}_i, w_i) = \begin{cases} 1, \text{ if } |\bar{w}_i - w_i| > a_i, \\ 0, \text{ if } 0 \le |\bar{w}_i - w_i| \le a_i. \end{cases} \tag{1.19}$$

Obviously, $\bar{H}(\bar{w}_i, w_i)$ is a generalization of the Hamming distance (1.15). Then we can correspondingly define

$$\textit{Expanded sum Hamming distance } (esH): \sum_{i=1}^{m} c_i \bar{H}(\bar{w}_i, w_i); \tag{1.20}$$

$$\textit{Expanded bottleneck Hamming distance } (ebH): \max_{i=1}^{m} c_i \bar{H}(\bar{w}_i, w_i). \tag{1.21}$$

Hasanzadeh et al. [61] considered the (*CC-Inv*) problem of center location on trees. They proposed an $O(n^2)$ algorithm under weighted l_∞ norm and an $O(n\log^2 n)$ algorithm under ebH-distance. They proved the $\mathcal{NP}$-hardness of the problem under weighted esH-distance and weighted l_1 norm by a polynomial time reduction from the *Partition problem*. Furthermore, they devised an $O(n^2\log n)$ algorithm under unit esH-distance and an $O(n\log n)$ algorithm under unit l_1 norm.

It is worthy noting that bottleneck Hamming distance and (weighted) l_∞ norm are not suitable in the budget constraint (1.18) since they can all be transformed into upper bound constraints on $\boldsymbol{l}$ and $\boldsymbol{u}$.

- **The Improvement/Interdiction Problem by upgrading nodes (Imp/Int-Node)**

We can also modify the weight vector by upgrading nodes [54, 57, 59, 75–77, 79]. In a *node based upgrading model*, we are given three nonnegative values $w_0(e) \geq w_1(e) \geq w_2(e) \geq 0$ for each edge $e \in E$. The value $w_i(e)$ represents the weight on the edge if exactly $i (i = 0, 1, 2)$ of its endpoints are upgraded. Thus, upgrading a node v reduces the weight on each edge incident to v with a upgrading cost $c(v)$. The cost of upgrading a subset Ω of V is $c(\Omega) = \sum_{v\in\Omega} c(v)$. The edge weight vector w_Ω resulting from upgrading the node set Ω is calculated by (1.22). Therefore, the corresponding improvement problem (*Imp-Node*) and the Budget-Constrained improvement problem (*BC-Imp-Node*) by upgrading nodes can be formulated by (1.23) and (1.24), respectively.

$$w_\Omega(e) = w_i(e), \qquad \text{for } e = (u, v) \text{ and } |\{u, v\} \cap \Omega| = i, i = 0, 1, 2. \tag{1.22}$$

$$\begin{aligned} &\min_{\Omega\subseteq V} \sum_{v\in\Omega} c(v) \\ (\textit{Imp-Node}) \quad s.t. \ &\min_{S\in\mathcal{S}} \boldsymbol{w}_\Omega(S) \leq D. \end{aligned} \tag{1.23}$$

$$\begin{aligned} &\min_{\Omega\subseteq V} \min_{S\in\mathcal{S}} \boldsymbol{w}_\Omega(S) \\ (\textit{BC-Imp-Node}) \quad s.t. \ &\sum_{v\in\Omega} c(v) \leq B. \end{aligned} \tag{1.24}$$

Similarly, we can consider the interdiction problem (*Int-Node*) and the Budget-Constrained interdiction problem (*BC-Int-Node*) by upgrading nodes, which can be formulated by (1.25) and (1.26), respectively.

$$\begin{aligned} &\min_{\Omega\subseteq V} \sum_{v\in\Omega} c(v) \\ (\textit{Int-Node}) \quad s.t. \ &\min_{S\in\mathcal{S}} \boldsymbol{w}_\Omega(S) \geq D. \end{aligned} \tag{1.25}$$

$$\max_{\Omega \subseteq V} \min_{S \in \mathcal{S}} w_\Omega(S)$$
$$(BC\text{-}Int\text{-}Node) \ s.t. \sum_{v \in \Omega} c(v) \leq B. \tag{1.26}$$

1.1.3 A Uniform Format for a Concrete GICOP

We propose a uniform format for a concrete *GICOP* as *(class of GICOP, COP, Bounded or not, type of norm)*. For example, *(RIOV, MST, Bounded,* l_∞*)* means the bounded restricted inverse optimal value problem on minimum spanning tree under l_∞ norm and *(Inv, SP, Unbounded,* l_1*)* means the inverse shortest path problem under l_1 norm.

In the big family of *GICOPs*, we can obtain a concrete *GICOP* by selecting any item in the first, third, and fifth columns in Table 1.1. For example, the wavily underlined items compose *(RIOV, MST, Bounded/Unbounded,* l_∞*)*. Notice that the *GICOPs* under weighted l_2 norm [22, 69, 142] were seldom studied since the objective functions are quadratic, which cannot be formulated as integer LP problems. Then there are total $7 \cdot 11 \cdot 4 = 308$ concrete *GICOPs* for the listed seven classes of *GICOPs*, 11 classical *COPs*, four types of norms including (weighted) l_1, l_∞ norm, wsH-distance, and bH-distance. Moreover, for the class of *BC/CC-Inv* problems, there are two sum-type norms in the constraint condition and five types of norms in the objective function including weighted l_∞ norm, bH-distance, ebH-distance, esH-distance, and wsH-distance/l_1 norm. Therefore, considering two classes of problems *Imp/Int-Node*, there are total $7 \cdot 11 \cdot 4 + 2 \cdot 11 \cdot 5 + 11 \cdot 2 = 440$

Table 1.1 An outline of generalized inverse combinatorial optimization problems

Class of GICOPs	*CS*	*COP*	*Type of w(S)*	*Norm*
Imp	*Spanning tree*	*MST*	*Sum*	
Int		*BST*	*Bottleneck*	
IOV		*MSST*	*Max + Sum*	*(Weighted)* l_1
RIOV	*Path*	*SP*	*Sum*	*(Weighted)* l_∞
Inv		*BSP*	*Bottleneck*	*(Weighted)* l_2
PInv	*Location*	*ML*	*Sum*	*(Weighted) Sum-Ham*
SInv		*CL*	*Bottleneck*	*Bottleneck-Ham*
BC-Inv		*Q1C*	*Max + Sum*	*(Expanded) Sum-Ham*
BC-Imp/Int	*Flow*	*MF*	*Bottleneck*	*(Expanded) Bot-Ham*
Imp/Int-Node		*MCF*	*Sum*	
BC-Imp/Int-Node	*Cut*	*MC*	*Sum*	

concrete *GICOPs* for the listed 11 classical *COPs*. It is worthy noting that we do not include the budget-constrained problems *BC-Imp/Inv* and *BC-Imp/Inv-Node* since they have related solution methods for the problems *Imp/Inv* and *Imp/Inv-Node*, respectively.

In this book, we only cover a part of research work in the big family of *GICOPs* listed in Table 1.1. In the end of each chapter, we will summarize the research results in tables for the corresponding *GICOPs*. Therefore, Table 1.1 also provides *a list of open problems* in the field of *GICOPs*.

Furthermore, almost every *COP* has its corresponding *GICOP*. For those who are interested in combinatorial optimization problems, they can find a comprehensive way to study an inverse *COP* through this book.

1.2 Time Complexities of Some *GICOPs*

We call an original optimization problem as *a forward problem* compared to its associate generalized inverse problem [5]. We will discuss the relationship of time complexities between the forward optimization problems and their inverse problems. Most (*Imp*), (*Int*), and (*PInv*) problems are $\mathcal{NP}$-hard [86, 140, 154]. Ahuja and Orlin [5] mentioned that an inverse linear programming problem under the l_1 as well as l_∞ norm is also a linear programming problem. Furthermore, there is a polynomially solved center location problem whose inverse problem is $\mathcal{NP}$-Hard. As an interesting *open problem*, we still do not know if there is an $\mathcal{NP}$-hard optimization problem whose inverse problem is polynomially solvable [63].

1.2.1 The Forward Problem Is in $\mathcal{P}$, So Is Its Inverse Problem

Theorem 1.1 ([5, Theorem 2]) *If an optimization problem is polynomially solvable for each linear cost function, then the corresponding inverse problems under weighted l_1, l_∞ norms are polynomially solvable.*

Obviously, Theorem 1.1 holds for inverse linear programming problems. Ahuja and Orlin [5] also proved the following results.

(i) If the forward problem is a shortest path, assignment, or minimum cut problem, then its inverse problem under unit l_1 norm can be solved by solving a problem of the same kind. Under the weighted l_1 norm, the inverse problem reduces to solving a minimum cost flow problem.
(ii) If the forward problem is a minimum cost flow problem, then its inverse problem under unit l_1 (l_∞, resp.) norm weights reduce to solving a unit-capacity minimum cost flow problem (minimum mean cycle problem, resp.) Under weighted l_1 (l_∞, resp.) norm, the inverse problem reduces to solving a minimum cost flow problem (minimum cost-to-time ratio cycle problem, resp.).

1.2.1.1 (Inv, Submodular Maximization Problem, Bounded, Weighted l_1 Norm)

It is well known that most combinatorial optimization problems, including shortest path, minimum spanning tree, weighted bipartite matching, minimum arborescence, and weighted matroid intersection, are special cases of submodular functions on digraphs [38]. Cai et al. [24] formulated *Inverse Submodular Maximization Problem (ISMP)* as a combinatorial linear program and further transformed it into a minimum cost circulation problem, which can be solved in strongly polynomial time.

Definition 1.1 ([24]) Let V be a finite ground set; the subsets $A, B \subset V$ are *intersecting* if none of $A \cap B$, $A - B$, $B - A$ is empty. If in addition, $A \cup B \neq V$, then A and B are *crossing*. A family $\mathscr{F}$ of subsets of V is intersecting [crossing] if $A \cap B, A \cup B \in \mathscr{F}$ for all intersecting [crossing] members A, B of $\mathscr{F}$. A set function f defined on an intersecting [crossing] family $\mathscr{F}$ is called *submodular* if

$$f(A) + f(B) \geq f(A \cap B) + f(A \cup B),$$

for any intersecting [crossing] members $A, B \in \mathscr{F}$.

Let $G = (V, E)$ be a digraph, let $\mathscr{F}'$ be a crossing family of subsets of V, and let f' be a submodular function defined on $\mathscr{F}'$. Without loss of generality, we can assume that $V, \varnothing \notin \mathscr{F}'$. For $S \subseteq V$, we denote by $\delta^-(S)$ the set of arcs entering S, and $\delta^+(S) = \delta^-(\bar{S}) = \delta^-(V - S)$. We define A' to be a $(0, \pm 1)$ matrix as follows:

$$A'_{S,e} = \begin{cases} -1, & e \in \delta^+(S), \\ +1, & e \in \delta^-(S), \\ 0, & \text{otherwise} \end{cases}$$

Introduce the vectors $a, d, c \in (R \cup \{\pm\infty\})^{|E|}$ with $d \leq a$, then the *Submodular Maximization Problem (SMP)* on digraphs is expressed as the linear program as follows:

$$\begin{aligned} &\max\ cx \\ &\text{s.t. } A'x \leq f', \\ &\qquad d \leq x \leq a. \end{aligned} \tag{1.27}$$

Let $b^+ \geq 0, b^- \geq 0$ and $w^+ \geq 0, w^- \geq 0$ be given bound and cost functions on E, respectively; let x^0 be a feasible solution of (1.27). The (ISMP) on digraphs is to find c^* satisfying the problem as follows:

$$\begin{aligned} &\min \sum_{e \in E} \left\{ w^+(e) \max \left\{0, c^*(e) - c(e)\right\} + w^-(e) \max \left\{0, c(e) - c^*(e)\right\} \right\} \\ &s.t.\ x^0 \text{ is an optimal solution of (SMP) under } c^*, \\ &c - b^- \leq c^* \leq c + b^+. \end{aligned} \tag{1.28}$$

Lemma 1.1 ([91]) *Let $\mathscr{F}'$ and f' be given as earlier. Define*

$$\mathscr{F} = \left\{X : X \neq \varnothing, X = \bigcap X_i, X_i \in \mathscr{F}', \bar{X}_i \cap \bar{X}_j = \varnothing\right\} \cup \{V\},$$

$$f(X) = \min\left\{\sum f'(X_i) : X = \bigcap X_i, \bar{X}_i \cap \bar{X}_j = \varnothing\right\}$$

and $f(V) = 0$. Then, $\mathscr{F}$ is an intersecting family and f is submodular on $\mathscr{F}$.

Lemma 1.2 ([42]) $\left\{x : A'x \leq f'; d \leq x \leq a\right\} = \{x : Ax \leq f; d \leq x \leq a\}$, *where A is the counterpart of A' with respect to $\mathscr{F}$.*

Definition 1.2 ([24]) A member S of $\mathscr{F}$ is called *f-critical* with respect to x^0 if $A_S, x^0 = f(S)$. We denote by $P(v)$ the intersection of all f-critical sets containing a vertex v of G; $P(v)$ depends on x^0.

We decompose E into four subsets:

$$E_1 = \left\{e \in E \mid d(e) < x^0(e) < a(e)\right\},$$

$$E_2 = \left\{e \in E \mid x^0(e) = a(e) > d(e)\right\},$$

$$E_3 = \left\{e \in E \mid x^0(e) = d(e) < a(e)\right\},$$

$$E_4 = \{e \in E \mid a(e) = d(e)\}.$$

By the complementary slackness theorem of linear programming, we have the following optimality condition.

Theorem 1.2 ([24, Theorem 2.2]) *A solution x^0 is optimal to the problem (SMP) under c^* if and only if there exists a potential p satisfying*

$$\begin{array}{ll} p(v) - p(u) = c^*(e), & \forall e = (u, v) \in E_1, \\ p(v) - p(u) \leq c^*(e), & \forall e = (u, v) \in E_2, \\ p(v) - p(u) \geq c^*(e), & \forall e = (u, v) \in E_3, \\ p(u) \geq p(v), & \forall u \in P(v), \forall v \in V. \end{array} \tag{1.29}$$

Theorem 1.3 ([24, Theorem 2.3]) *The problem (ISMP) can be formulated as the following linear program:*

$$
\begin{aligned}
\max - \sum_{e\in E_1\cup E_2} w(e)^+\theta^+(e) - \sum_{e\in E_1\cup E_3} w(e)^-\theta^-(e),\\
s.t.\quad p(v) - p(u) - \theta^+(e) + \theta^-(e) = c(e), \quad & \forall e = (u, v) \in E_1,\\
p(v) - p(u) - \theta^+(e) \le c(e), \quad & \forall e = (u, v) \in E_2,\\
p(v) - p(u) + \theta^-(e) \ge c(e), \quad & \forall e = (u, v) \in E_3,\\
p(u) \ge p(v), \quad & \forall u \in P(v), v \in V,\\
0 \le \theta^+(e) \le b^+(e), \quad & \forall e \in E_1 \cup E_2,\\
0 \le \theta^-(e) \le b^-(e), \quad & \forall e \in E_1 \cup E_3.
\end{aligned}
\tag{1.30}
$$

The number of constraints of (1.30) is not more than $|V|^2+3|E|$, and the number of variables is not more than $|V| + 2|E|$. As the elements of the constraint matrix are 0, 1, -1, (1.30) is in fact a combinatorial linear program, which can be solved by a strongly polynomial algorithm [123].

Theorem 1.4 ([24, Theorem 3.1]) *The dual linear program of (1.30) is equivalent to a mum cost circulation problem (MCCP) on network $N = (V, E^*, k, l)$ as follows.*

(i) The vertex set V is the same as G.
(ii) The arc set is defined as

$$
\begin{aligned}
E^* = &\left\{e^1 = (u, v), e^2 = (u, v), e^3 = (v, u), e^4 = (v, u) \mid e = (u, v) \in E_1\right\}\\
&\bigcup \left\{e^1 = (u, v), e^2 = (u, v) \mid e = (u, v) \in E_2\right\}\\
&\bigcup \left\{e^1 = (v, u), e^2 = (v, u) \mid e = (u, v) \in E_3\right\}\\
&\bigcup \left\{e^* = (u, v) \mid u \in P(v), v \in V\right\}.
\end{aligned}
$$

(iii) The capacity on each arc is

$$
k(u, v) = \begin{cases} w^+(e), & \textit{if } (u, v) = e^1, e = (u, v) \in E_1 \cup E_2,\\ w^-(e), & \textit{if } (u, v) = e^3, e = (v, u) \in E_1,\\ w^-(e), & \textit{if } (v, u) = e^1, e = (v, u) \in E_3,\\ +\infty, & \textit{otherwise.} \end{cases}
$$

(iv) The cost on each arc is

$$l(u,v)=\begin{cases} c(e), & \text{if } (u,v)=e^1, e=(u,v)\in E_1\cup E_2,\\ b^+(e)+c(e), & \text{if } (u,v)=e^2, e=(u,v)\in E_1\cup E_2,\\ -c(e), & \text{if } (u,v)=e^3, e=(v,u)\in E_1,\\ b^-(e)-c(e), & \text{if } (u,v)=e^4, e=(v,u)\in E_1,\\ -c(e), & \text{if } (u,v)=e^1, e=(v,u)\in E_3,\\ b^-(e)-c(e), & \text{if } (u,v)=e^2, e=(v,u)\in E_3,\\ 0, & \text{if } (u,v)=e^*, u\in P(v), v\in V. \end{cases}$$

We can use the repeated capacity scaling algorithm [4] to solve the problem (MCCP). This algorithm is a strongly polynomial algorithm, which produces first a dual-optimal solution $\left(p,\theta^+,\theta^-\right)$ of (1.30).

Furthermore, they also gave a necessary and sufficient condition for the feasibility of the inverse problem.

Theorem 1.5 ([24, Theorem 3.2]) *The problem (ISMP) (1.30) is feasible if and only if there are no negative cycles in* $N^*=N-\left\{e^1\mid e\in E_1\cup E_2\cup E_3\right\}-\left\{e^3\mid e\in E_1\right\}$.

1.2.2 The Forward Problem Is in $\mathcal{P}$, Its Inverse Problem May Be $\mathcal{NP}$-Hard

Let $G=(V,E,w)$ be a connected edge-weighted graph with $|V|=n$, $|E|=m$ and $w\geq 0$. The classic *Vertex 1-Center (V1C)* problem seeks a vertex $v^*\in V$, named a vertex 1 center, to minimize the longest distance from it to all the other vertices [27]. Let P_{v,v^*} be the shortest path between v and v^* and $w(P_{v,v^*})=\sum_{e\in P_{v,v^*}}w(e)$. Then the problem *(V1C)* can be formulated as

$$(V1C):\quad \min_{v^*\in V}\max_{v\in V} w(P_{v,v^*}). \tag{1.31}$$

The problem *(V1C)* is strongly polynomially solvable, and the complete computation is dominated by finding all-pairs shortest paths in G, which can be done using Fredman and Tarjan's $O\left(mn+n^2\log n\right)$-time algorithm [43], or $O\left(m^*n+n^2\log n\right)$ time algorithm by Karger et al. [70] in which m^* is the number of edges included by shortest paths, or Pettie's $O\left(mn+n^2\log\log n\right)$-time algorithm [107].

Given a specific vertex $v^0\in V$, the inverse vertex 1-center location problem *(Inv,V1C,Unbounded,l_1)* aims to modify w to $\bar{w}$ such that v^0 becomes a V1C of G

under $\bar{w}$ under unit l_1 norm. Its mathematical model can be formulated as

$$\min \sum_{e \in E} |\bar{w}(e) - w(e)|$$
$$s.t. \quad \max_{v \in V} w(P_{v,v^0}) = \min_{v' \in V} \max_{v \in V} w(P_{v,v'}). \tag{1.32}$$

Cai et al. [25] claimed a negative answer to the question: Is it true that the inverse problem is solvable in polynomial time whenever the forward problem is solvable in polynomial time? They showed that

Theorem 1.6 ([25]) *The problem (Inv,V1C,Unbounded,l_1) is $\mathcal{NP}$-hard by a polynomial time transformation of the satisfiability problem to its decision problem.*

We can conclude that the inverse problem of a polynomially solvable problem is not necessarily polynomially solvable and may become a more challenging $\mathcal{NP}$-hard problem. Roughly speaking, only if the feasible region of the inverse problem is a *polytope*, can we ensure that the inverse problem under unit l_1 norm is solvable in polynomial time [25]. Unfortunately, the feasible region of the inverse location problem cannot be characterised by a set of linear equations/inequalities.

1.2.3 The Forward Problem Is $\mathcal{NP}$-Hard, So Is Its Inverse Problem

In most cases, if the forward problem is $\mathcal{NP}$-hard, then the its inverse problem (*Inv*) is also $\mathcal{NP}$-hard [30]. Indeed, the optimal value $\min_{\bar{w}} \|\bar{w} - w\|$ of the inverse problem is 0 if and only if the given feasible solution x^0 is already an optimal solution for the initial weight vector w. This leads to a *Turing-reduction* from the optimality test associated with the forward problem (that is, to decide if a given solution is optimal) to the inverse problem.

Theorem 1.7 ([30, Lemma 2]) *If the optimality test associated with the forward problem is co$\mathcal{NP}$-complete, then its inverse problem (Inv) is $\mathcal{NP}$-hard.*

1.2.3.1 (Inv, (metric) Traveling Salesman Problem,Unbounded, l_1)

Given an edge-weighted complete graph $G = (V, E, w)$, a Hamiltonian cycle is a cycle traversing each vertex once. Minimum *Traveling Salesman Problem (TSP)* is to find a minimum Hamiltonian cycle whose weight is minimum. The metric TSP requires that edge weights satisfy the triangular inequality.

Given a Hamiltonian cycle HC^0, the inverse (metric) Traveling Salesman Problem is to modify w to a new weight vector $\bar{w}$ under l_1 norm such that HC^0 is minimum Hamiltonian cycle with respect to w.

Lemma 1.3 ([31, Proposition 2]) *The optimality test associated with the (metric) TSP problem is co$\mathcal{NP}$-complete.*

It follows from Theorem 1.7 and Lemma 1.3 that

Theorem 1.8 ([31, Proposition 3]) *The problem (Inv, (metric) TSP, Unbounded, l_1) is $\mathcal{NP}$-hard.*

1.2.3.2 (Inv/IOV, Bin-Packing Number Problem, Unbounded, l_p)

In the usual Bin-Packing Number (BPN) problem, a given list $N = \{1, \ldots, n\}$ of items with an item size vector $w = (w_1, \ldots, w_n) \in \mathbb{Z}^{n+}$ must be packed into a minimum number of bins. It is well known to be $\mathcal{NP}$-hard [44]. In practice, certain items have to be packed separately (e.g., for security reasons), and the number of bins may be limited, insufficient to pack all the items. In such cases, item sizes may require be adjusted so that all the items can be packed in a prescribed way using the available bins. If a pre-selected packing way is required to be an optimal way, it arises the inverse (BPN) problem. It a target packing number is given, it renders the inverse optimal value problem on (BPN), denoted by (IOVBPN).

Chung and Park used a polynomial time reduction from the three-partition problem, known to be strongly $\mathcal{NP}$-hard [44], to show that

Theorem 1.9 ([32, Proposition 1]) *The problem (IOV, BPN, Unbounded, l_p) is strongly $\mathcal{NP}$-hard for any $p \in \{1, 2, \cdots, \infty\}$.*

Lemma 1.4 ([28]) *No approximation algorithm for (IOV, BPN, Unbounded, l_p) returns a solution of value z_{IOVBPN} such that $z_{IOVBPN} \leq OPT_{IOVBPN} + (\mu - 2)$ for any constant $\mu \geq 2$ for any $p \in \{1, 2, \cdots, \infty\}$.*

Lemma 1.5 ([32, Proposition 3]) *No approximation algorithm for the problem (IOV, BPN, Un-bounded, l_p) returns a solution of value z_{IOVBPN} such that $z_{IOVBPN} \leq OPT_{IOVBPN} + \log(\sum_{i=1}^{n} w_i)$ for any $p \in \{1, 2, \cdots, \infty\}$.*

The aforementioned lemma was proved by a polynomial time reduction from the partition problem, known to be $\mathcal{NP}$-hard [44]. These negative approximation results provided a motivation for considering an alternative framework for approximating (IOVBPN)—a differential approximation algorithm for (IOVBPN) under unit l_1-norm.

Lemma 1.6 ([32, Proposition 4]) *There exists a differential approximation scheme for the problem (IOV, BPN, Unbounded, l_1).*

Chung and Park [32] showed that the continuous version of the inverse (BPN) problem, each adjustment is a real number, does not admit an optimal solution unless the pre-selected solution is already optimal for the given bin-packing instance. Furthermore, they showed the discrete version with integer adjustments is $\mathcal{NP}$-hard by proving that the Optimization test for (BPN) is *co$\mathcal{NP}$*-complete.

Theorem 1.10 ([32, Lemma 6]) *The discrete version of (IOV, BPN,Integers,l_p) problem with integer adjustments is $\mathcal{NP}$-hard.*

1.2.3.3 (Inv, 0-1 Knapsack problem, Unbounded, l_1/l_∞)

The 0-1 *Knapsack problem (KP)* is a well-known classical linear 0-1 combinatorial optimization problem [93]. Let J denote a set of n items with profits $c_j \in \mathbb{N}_0 = \{0, 1, 2, \ldots\}$ and weights $w_j \in \mathbb{N}_0$, for each $j \in J$. Let $W \in \mathbb{N}_0$ denote the capacity of a knapsack. The 0-1 Knapsack problem consists of selecting a subset $S \subseteq J$, such that the sum of the profits of $j \in S$ is maximized, and the sum of their weights does not exceed the capacity W. The problem *(KP)* can be modeled as an integer LP problem as follows:

$$\begin{aligned} & \max \sum_{j\in J} c_j x_j \\ (KP)\ & \text{s.t.} \sum_{j\in J} w_j x_j \le W, \\ & x_j \in \{0, 1\},\ j \in J. \end{aligned}$$

Let $X = \left\{x \in \{0, 1\}^n : \sum_{j\in J} w_j x_j \le W\right\}$. Then an instance of (KP) is defined by (X, c). It is assumed that $\sum_{j\in J} w_j > W$ and $w_j \le W$, for all $j \in J$.

In an *Inverse 0-1 Knapsack Problem (IKP)*, a solution x^0 is given, we aim to modify c to $\bar{c} \in \mathbb{N}_0^n$ such that the x^0 is an optimal solution of $(X, \bar{c})$. Its mathematical model can be formulated as follows:

$$\begin{aligned} & \min \|\bar{c} - c\| \\ (IKP)\ & s.t.\ \bar{c}^T x^0 \ge \bar{c}^T x, \text{ for any } x \in X. \end{aligned}$$

Denote by (IKDP$_\infty$) the Inverse 0-1 Knapsack Decision Problem under unit l_∞ norm. Roland et al. [114] claimed the following theorem that the complement problem of (IKDP$_\infty$), denoted by ($\overline{\text{IKDP}_\infty}$), is $\mathcal{NP}$-Complete. The $\mathcal{NP}$-Hardness of ($\overline{\text{IKDP}_\infty}$)is provided by a polynomial time reduction from the Knapsack Decision Problem (KDP) to ($\overline{\text{IKDP}_\infty}$)

Theorem 1.11 ([114, Theorem 3.6]) *The inverse 0-1 knapsack decision problem (IKDP$_\infty$) under unit l_∞ norm is co$\mathcal{NP}$-Complete.*

Furthermore, they proposed a pseudo-polynomial time algorithm with time complexity $O(nW \log C)$, where $C = \max_{j\in J} (1 - x_j^0)c_j$, which implies that problem (IKDP$_\infty$) is fortunately a weakly *co$\mathcal{NP}$*-Complete problem. This algorithm performs a binary search method in $O(\log C)$ iterations to find the minimum value

$k^* \in [0, C]$ such that x^0 is an optimal solution of the problem KP$(X, \bar{c}^{k^*})$, where

$$\bar{c}_j^{k^*} = \begin{cases} \max\left\{0, c_j - k^*\right\}, & \text{if } x_j^0 = 0, \\ c_j + k^*, & \text{if } x_j^0 = 1. \end{cases} \tag{1.33}$$

In each iteration, a knapsack problem KP$(X, \bar{c}^k)$ needs to be solved in $O(nW)$ time by using a dynamic programming approach in [93].

The $co\mathcal{NP}$-Hardness of (IKDP$_1$) is provided by a polynomial time reduction from ($\overline{\text{KDP}}$) to (IKDP$_1$), where ($\overline{\text{KDP}}$) is the complement of (KDP). However, (IKDP$_1$) has not been proved to belong to $co\mathcal{NP}$ because the knapsack problem is required for building a certificate for the "No" instances.

Theorem 1.12 ([114, Theorem 4.3]) *The inverse 0-1 knapsack decision problem (IKDP$_1$) under unit l_1 norm is $co\mathcal{NP}$-hard.*

Moreover, they showed that the problem (IKDP$_1$) can be solved by an oracle Turing machine $\mathcal{NP}^{\text{KDP}}$. Hence, the problem (IKDP$_1$) should be harder to solve than (IKDP$_\infty$) and there is less hope of finding a pseudo-polynomial time algorithm to solve it.

By describing the optimality conditions in the dynamic programming algorithm for the 0-1 knapsack problem, they formulated the problem (IKP$_1$) as an integer linear program. Hence, if the profit vector is integer-valued, the inverse problem is reduced to solve an integer linear program. Otherwise, the inverse problem is solved by the linear relaxation of the integer linear program, which can be solved with a pseudo-polynomial time algorithm.

1.2.3.4 Inverse Integer Linear Programming Problems

The forward optimization problem is the *Mixed Integer Linear Programming (MILP) problem* below.

$$(MILP) \quad \max_{x \in \mathcal{S}} c^\top x$$

where $c \in \mathbb{Q}^n$ and $\mathcal{S} = \{x \in \mathbb{R}^n \mid Ax \leq b\} \cap \left(\mathbb{Z}^r \times \mathbb{R}^{n-r}\right)$ for $A \in \mathbb{Q}^{m \times n}$, $b \in \mathbb{Q}^m$ and $r \in \mathbb{N}_0$. When $r = 0$, (MILP) is just the problem (LP). We do not assume $x^0 \in \mathcal{S}$, and replace $\mathcal{S}$ with the augmented set $\mathcal{S}^+ = \mathcal{S} \cup \left\{x^0\right\}$. Let conv$(\mathcal{S})$ be the convex hull of $\mathcal{S}$.

Several Mathematical Formulations of the Inverse (MILP) Problem (IMILP)
A straightforward formulation of this problem is to compute the mathematical inverse of a function

$$\phi(d) = \text{argmax}_{x \in \mathcal{S}^+} d^\top x. \tag{1.34}$$

Then a formulation of (IMILP) in terms of the function ϕ was given by Bulut and Ralphs [17]

$$\text{(IMILP-}\phi\text{)} \qquad \min_{d \in \phi^{-1}(x^0)} \|c - d\|. \tag{1.35}$$

However, the formulation (1.35) did not suggest any direct connection to existing methodology for solving (IMILP). Bulut and Ralphs [17] then formulated the (IMILP) as a semi-infinite optimization problem as follows:

$$\begin{array}{rl} & \min_d \ \|c - d\| \\ \text{(IMILP-}\mathcal{S}\text{)} & \text{s.t.} \quad d^\top x \leq d^\top x^0 \quad \forall x \in \mathcal{S}. \end{array} \tag{1.36}$$

The problem (IMILP) can also be formulated as a conic optimization problem, which can best highlight the underlying mathematical structure. Let $\mathcal{K} := \{(y, d) \in \mathbb{R}^{n+1} \mid \|c - d\| \leq y\}$ be a norm cone, while

$$\mathcal{K}(\gamma) := \left\{d \in \mathbb{R}^n \mid (\gamma, d) \in \mathcal{K}\right\}$$

is a ball with center c and radius γ that contains vectors whose objective value in (IMILP) is at most γ. Let $\mathcal{K}^*(\gamma) := \left\{x \in \mathbb{R}^n \mid d^\top \left(x^0 - x\right) \leq 0, \forall d \in \mathcal{K}(\gamma)\right\}$, which can be seen as the radial cone obtained by translating the dual of cone($\mathcal{K}(\gamma)$) from the origin to x^0. We introduce $C\left(x^0\right) := \text{cone}\left(\mathcal{S}^+ - \left\{x^0\right\}\right) = \text{cone}\left(\left\{x - x^0 \mid x \in \mathcal{S}\right\}\right)$ as the radial cone with vertex at x^0 generated by rays $x - x^0$ for all $x \in \mathcal{S}$. Furthermore, the normal cone $C^\circ\left(x^0\right) := \left\{d \in \mathbb{R}^n \mid d^\top \left(x - x^0\right) \leq 0, \forall x \in \mathcal{S}\right\}$ is the set of feasible objectives of (IMILP) with target vector x^0 and is precisely the polar of $C\left(x^0\right)$.

Finally, in terms of the aforementioned cones and sets, (IMILP) can be reformulated as [17]

$$\text{(IMILP-}C\text{)} \qquad \min_{d \in \mathcal{K}(y) \cap C^\circ(x^0)} y. \tag{1.37}$$

Since the feasible region of (IMILP) can be regarded as the polar of $C\left(x^0\right)$, the so-called 1-polar of conv($\mathcal{S}$), defined as

$$\mathcal{P}^1 = \left\{d \in \mathbb{R}^n \mid d^\top x \leq 1, \forall x \in \text{conv}(\mathcal{S})\right\},$$

provides another formulation of (IMILP) below [17].

$$\begin{array}{rl} & \min_{d,\rho} \ \|c - \rho d\| \\ \text{(IMILP-}\mathcal{P}\text{)} & \text{s.t.} \quad d^\top x^0 \geq 1, \quad d \in \mathcal{P}^1, \\ & \qquad \rho \geq 0, \end{array} \tag{1.38}$$

where ρ is a multiplier that allows for scaling of the members of the 1-polar to improve the objective function value.

Now we clarify the close relationship between the inverse problem (IMILP) and the separation problem for $\text{conv}(\mathcal{S})$.

The Separation Problem

Given an $\hat{x} \in \mathbb{Q}^n$, *the separation problem* for $\text{conv}(\mathcal{S})$ is to determine whether $\hat{x} \in \text{conv}(\mathcal{S})$ and, if not, to generate a hyperplane, associated with *a valid inequality*, separating $\hat{x}$ from $\text{conv}(\mathcal{S})$.

Definition 1.3 ([17, Definition 1]) A *valid inequality* for a set Q is a pair $(a, b) \in \mathbb{Q}^{n+1}$ such that $Q \subseteq \left\{x \in \mathbb{R}^n \mid a^\top x \leq b\right\}$. The inequality is said to be *violated* by $\hat{x} \in \mathbb{Q}^n$ if $a^\top \hat{x} > b$.

We can see that $\hat{d}$ is feasible for (IMILP) if and only if $(\hat{d}^\top, \hat{d}^\top x^0)$ is an inequality valid for $\text{conv}(\mathcal{S})$. Furthermore, when $x^0 \notin \text{conv}(\mathcal{S})$, $\left(\hat{d}^\top, \max_{x \in \mathcal{S}} \hat{d}^\top x\right)$ is an inequality valid for $\text{conv}(\mathcal{S})$ that is violated by x^0, which also provides an informal argument for the equivalence of (IMILP) and the separation problem [17].

An in-depth investigation is conducted on the general inverse integer optimization problems in [115]. Employing the principle of super-additive duality, it successfully derives a polyhedral characterization of the ensemble of all feasible objective functions. This characterization, while exclusively comprising continuous variables, intriguingly entails an exponential multitude of constraints. Subsequently, a solution methodology that leverages this polyhedral framework is introduced, offering a novel approach to tackling the inverse optimization challenges.

In a related development, Wang [129] introduces a cutting-plane algorithm, analogous to the one delineated in the subsequent discourse, and furnishes a detailed exposition of computational outcomes on a spectrum of test problems. This is achieved through the practical implementation of the proposed algorithm, thereby substantiating its efficacy and providing a tangible demonstration of its problem-solving capabilities.

A Cutting-Plane Algorithm for Solving (IMILP) Under Unit l_1 and l_∞ Norms

We first formulate the unit l_1 and l_∞ norms explicitly as an LP using standard linearization techniques by introducing y_i and θ, respectively, [17]. Then replace the infinite constraints for $\mathcal{S}$ in (IMILP-$\mathcal{S}$) by the finite sets $\mathcal{E}$ and $\mathcal{R}$, which are the sets of the extreme points and rays of $\text{conv}(\mathcal{S})$, respectively.

$$
\begin{aligned}
& \min_{d,y} \sum_{i=1}^{n} y_i, \\
(\text{IMILP}_1) \quad \text{s.t.} \quad & c_i - d_i \leq y_i, \quad \forall i \in \{1, 2, \ldots, n\}, \\
& d_i - c_i \leq y_i, \quad \forall i \in \{1, 2, \ldots, n\}, \\
& d^\top x \leq d^\top x^0, \quad \forall x \in \mathcal{E}, \\
& d^\top r \leq 0, \quad \forall r \in \mathcal{R}.
\end{aligned}
$$

$$\begin{aligned}
&\min_{d,\theta}\ \theta \\
(\text{IMILP}_\infty)\quad &\text{s.t. } c_i - d_i \le \theta, \quad \forall i \in \{1,2,\ldots,n\}, \\
&\quad d_i - c_i \le \theta, \quad \forall i \in \{1,2,\ldots,n\}, \\
&\quad d^\top x \le d^\top x^0, \quad \forall x \in \mathcal{E}, \\
&\quad d^\top r \le 0, \quad \forall r \in \mathcal{R}.
\end{aligned} \tag{1.39}$$

It is still challenging to solve the problems (IMILP_1) and (IMILP_∞) since they may contain exponential constraints. The cutting-plane algorithm can avoid explicitly enumerating the inequalities by generating them dynamically in the standard way. We only describe Algorithm 1 for the problem (IMILP_∞), which can be easily extended to (IMILP_1).

Assume $\mathcal{S}$ is bounded. Then conv($\mathcal{S}$) has no extreme rays and $\mathcal{R} = \emptyset$. Grötschel et al. [48] showed that (IMILP_∞) can be solved efficiently using a cutting-plane algorithm, provided we can solve the problem of separating a given point from the feasible region efficiently. Therefore, (1.39) is equivalent to solving the separation problem for set $C^\circ\left(x^0\right)$, which is further equivalent to the forward problem. For an arbitrary $\hat{d} \in \mathbb{R}^n$, let $x^* \in \operatorname{argmax}_{x\in\mathcal{S}} \hat{d}^\top x$. If $\hat{d}^\top x^* > \hat{d}^\top x^0$, then x^* yields a new inequality valid for $C^\circ\left(x^0\right)$ that is violated by x^*. Otherwise, we show that $\hat{d} \in C^\circ\left(x^0\right)$.

Algorithm 1.1 A cutting-plane algorithm for (IMILP_∞) [17, Algorithm 1]

Input: a cost vector c and a vector x^0, and the feasible region $\mathcal{S}$ of (MILP).
Output: an optimal solution (θ^*, d^*).

Initialize $k := 0, \mathcal{E}_0 := \emptyset, loop := 1$.
while $loop = 1$ **do**
  Let $k := k + 1$.
  Solve the master problem with $\mathcal{E}_{k-1} = \left\{x^1, \ldots, x^{k-1}\right\}$ to determine d^k, θ^k.

$$\begin{aligned}
&\min_{d,\theta}\ \theta \\
(\text{Inv}\mathcal{P}_k)\quad &\text{s.t. } c_i - d_i \le \theta, \quad \forall i \in \{1,2,\ldots,n\}, \\
&\quad d_i - c_i \le \theta, \quad \forall i \in \{1,2,\ldots,n\}, \\
&\quad d^\top x \le d^\top x^0, \quad \forall x \in \mathcal{E}_{k-1}.
\end{aligned}$$

  Solve the separation problem by $(\mathcal{P}_k)$ $\max_{x\in\mathcal{S}} d^{k\top} x$.
  if $(\mathcal{P}_k)$ is unbounded **then return** $\theta^* := \|c\|_\infty, d^* := 0$.
  else let x^k be an optimal solution of $(\mathcal{P}_k)$.
  end if
  if $d^{k\top}\left(x^k - x^0\right) > 0$ **then** let $loop := 0$.
    return $\theta^* := \left\|c - d^k\right\|_\infty, d^* := d^k$.

else update $\mathcal{E}_{k+1} := \mathcal{E}_k \cup \{x^k\}$.
end if
end while

Complexity of Inverse MILP

The authors in [17] delved into the general complexity of the problem (IMILP). They establish that the decision versions of the IMILP are $co\mathcal{NP}$-complete for primal bound verification, $\mathcal{NP}$-complete for dual bound verification, and $\mathcal{DP}$-complete for optimal value verification. This classification is derived through a series of theorems and lemmas that construct polynomial-time reductions between the IMILP and its corresponding forward MILP problems.

The authors first demonstrate that the primal bound verification problem for the inverse problem (IMPVP) is in $co\mathcal{NP}$ by providing a short certificate for the "NO" output instances. They then prove that the IMPVP is $co\mathcal{NP}$-complete by showing a Karp reduction from the MILP Dual Bound Verification Problem (MDVP), a known $co\mathcal{NP}$-complete problem.

Subsequently, this chapter proves that the dual bound verification problem for the inverse problem (IMDVP) is in $\mathcal{NP}$, with a short certificate for the YES output instances. The completeness of the IMDVP for $\mathcal{NP}$ is established through a reduction from the MILP Primal Bound Verification Problem (MPVP), which is a known $\mathcal{NP}$-complete problem.

Finally, the authors illustrate that the optimal value verification problem for the inverse problem (IMOVP) is $\mathcal{DP}$-complete. This is accomplished by showing that the IMOVP is in $\mathcal{DP}$ and that it can be reduced from the MILP Optimal Value Verification Problem (MOVP), which is a $\mathcal{DP}$-complete problem.

The chapter's complexity analysis underscores the computational challenge of the IMILP, highlighting the polynomial-time hierarchy's relevance to both the inverse and forward MILP problems. The results contribute to a deeper understanding of the complexity landscape for inverse optimization problems and suggest potential directions for developing efficient algorithms to tackle these complex combinatorial problems.

1.2.4 An Open Problem: Is there an Inverse $\mathcal{NP}$-Hard Problem Polynomially Solvable?

Yes. It is still an interesting open problem: Is there an inverse $\mathcal{NP}$-hard problem polynomially solvable?

Although some papers [29, 31] considered some polynomially solvable cases of an inverse $\mathcal{NP}$-hard problem, one can see that their associated forward problems are actually polynomially solvable cases of an $\mathcal{NP}$-hard problem. Bulut and Ralphs [17] indicated that if a forward mixed integer LP problem is polynomially solvable, then the associated inverse problem is also polynomially solvable.

Chung [29] presented two polynomially solvable cases of the Inverse Optimal value problem on (BPN). Let B be the bin capacity. One case is the offline (BPN) problem with two item sizes $w_1 \leq B$ and $w_2 \leq B$, where the number of w_i-item is $n_i > 0$ satisfying $n_1 + n_2 = n$. This problem can be solved in polynomial time using the algorithm of McCormick et al. [94]. The other case is that item sizes are all greater than $B/3$. The algorithm First Fit Decreasing (FFD) can find its optimal solution in $O(n \log n)$ time [29].

Lemma 1.7 ([29, Proposition 1]) *The inverse optimal value problem on (BPN) with two item sizes can be solved in $O(n^2 \log B)$ time.*

Lemma 1.8 ([29, Proposition 1]) *the Inverse Optimal value problem on (BPN) when item sizes are all greater than $B/3$ can be solved in $O(n \log n + K)$ time, where K is the given target packing number of bins.*

Chung and Demange [31] considered a polynomially solvable case of the inverse traveling salesman problem on a line, in which one changes the positions of requests instead of the weights of edges between requests. Its associate forward problem, $\mathscr{L}$-TSP, is defined as follows. Given n items (requests) $1, \ldots, n$, dispersed along a line $\mathscr{L}$, it defines a sequence $\pi = [\pi(1), \ldots, \pi(n)]$ of item labels where $\pi(i), 1 \leq i \leq n$, denotes the label of the i-th item when reading from the left to the right. Considering the line $\mathscr{L}$ as the axis of real numbers, there is a one-to-one correspondence between each request i and a number x_i on thereal axis. We call $x = (x_1, \cdots, x_n)$ the position vector. Obviously, $\mathscr{L}$-TSP is trivially solved by a tour containing one two-way trip of length $2 \times max_{i,j} d(x_i, x_j)$.

Theorem 1.13 ([31, Proposition 10]) *The problem (Inv, $\mathscr{L}$-TSP,Unbounded,l_1) can be solved in $O(n^3 \log n)$ time.*

1.3 Applications of Inverse Optimization Problems

In the domain of operations research and management science, the burgeoning field of inverse optimization has captured considerable scholarly interest. As technology progresses and application fields expand, inverse optimization has risen to prominence, offering innovative challenges and solutions to traditional optimization methodologies. This approach, which reconstructs the original optimization parameters from known solutions, provides critical insights into underlying decision-making processes that enhance prescriptive analytics.

The utility of inverse optimization is particularly pronounced in its capacity to retrodict, enabling a novel pathway from observed outcomes to their causal antecedents. This capability is invaluable in disciplines such as economics, logistics, and biomedicine, where understanding the rationale behind decisions is paramount. By tapping into the "why" behind decisions, inverse optimization facilitates the extraction of richer information from existing data, thereby broadening the horizons for knowledge discovery.

The application spectrum of inverse optimization is remarkably expansive, ranging from supply chain management and financial market analysis to healthcare. Its interdisciplinary applications are underpinned by the method's versatility and its potential to address a multitude of practical issues across varied sectors. This broad applicability underscores the burgeoning importance of inverse optimization research and signals its significant prospects for future interdisciplinary inquiry and practical utility.

Looking ahead, with the continuous development of big data and artificial intelligence technologies, inverse optimization problems are poised to have an even broader range of applications and research opportunities. It is reasonable to believe that in the near future, inverse optimization will play a significant role in a greater variety of fields and scenarios, contributing to the advancement of technology and the development of society.

1.3.1 Geophysical Sciences

In the field of geophysical sciences, an important application of inverse combinatorial optimization problems is the prediction of earthquake paths [5]. A geological area can be discretized into a number of square grid cells, which can be regarded as a network structure. Each cell in these grids contains a node representing an earthquake observation station. The arcs connecting these nodes represent the seismic wave transmission time from one observation station to another. When considering the propagation of seismic waves along the shortest path, by observing the shortest time of seismic waves and considering the geological characteristics of the area, we can reconstruct the propagation time of the earthquake and thus predict its trajectory. It is evident that this is an inverse shortest path problem.

The chapter [26] by Chan, Mahmood, and Zhu delves into the theoretical underpinnings and a broad spectrum of applications of inverse optimization, offering a comprehensive review that spans various domains.

1.3.2 Healthcare Applications

One of the pivotal areas where inverse optimization demonstrates its utility is in healthcare. It is instrumental in understanding clinical decisions, such as treatment plans or patient triage, by estimating the criteria that healthcare providers consider. This understanding can lead to the development of more efficient protocols and improved patient outcomes.

In the field of medical science, we mainly discuss clinical pathway concordance measurement. Clinical pathway provides standardized processes for a specific group of patients. Patient pathways that follow these pathways means "concordance." Patient pathways are modeled as a walk on a graph representing the healthcare system. Clinical pathways are shortest paths on this graph. We want the clinical pathway to be the shortest path on the graph, and let θ be the corresponding arc costs [26]. So the model can be expressed as follows:

$$\begin{aligned}
\min_{\boldsymbol{\theta},\boldsymbol{\lambda},\boldsymbol{\epsilon}} \quad & \sum_{i=1}^{N} \left(\varepsilon_i^r\right)^2 \\
\text{s.t.} \quad & A^T\lambda \le \theta, \\
& \theta^T \hat{x}_i^r = b^T\lambda + \varepsilon_i^r, i = 1, 2, \ldots, N, \\
& \|\theta\|_\infty = 1, \\
& A\theta = 0.
\end{aligned}$$

Among them, A and b encode the flow balance constraints in an SP with one source and one sink. Suppose we have N kinds of clinical pathway, $\hat{x}_1^r, \hat{x}_2^r, \ldots, \hat{x}_N^r$ between the source and sink. The constraint $\|\theta\|_\infty = 1$ is to prevent $\theta = 0$ from being optimal. The constraint $A\theta = 0$ is to deal with the lower dimensionality of the forward feasible region and ensure that the cost vector does not point orthogonal to the entire feasible region.

In the realm of medical diagnostics, the efficacy of disease screening protocols is inherently contingent upon the discerning capabilities of the diagnostic tests employed, namely their sensitivity and specificity. Ayer [14] conducted an inquiry into the theoretical construct of an ideal diagnostic assay, deducing the spectrum of sensitivity and specificity indices that would render a specific screening strategy most efficacious. The progression of disease within the purview of a predefined screening regimen is encapsulated via a model of a partially observable Markov process, thereby imbuing the ensuing inverse optimization dilemma with nonlinear characteristics. Confronted with this complexity, a comprehensive solution algorithm has been articulated to address the optimization conundrum effectively.

1.3.3 Transportation and Logistics

In transportation systems, inverse optimization plays a crucial role in analyzing routing decisions made by drivers or in optimizing supply chain networks. By inferring the objectives of drivers from their chosen routes, companies can better align incentives with desired behaviors, potentially reducing costs and improving service quality.

In traffic networks, there are typically various routes available for users to choose from in order to connect different locations [5]. The cost incurred by users is directly proportional to the traffic flow. In most cases, lower costs are more attractive.

Therefore, to study this issue, we introduce two concepts of traffic flow: user equilibrium flow and system optimal flow. The former implies that no user within the flow can reduce their travel cost by changing their route, while the latter refers to the flow with the lowest total travel cost within the network. Generally, we aim to achieve a system optimal flow. If our goal is to transform a user equilibrium flow into a system optimal flow, we need to impose tolls on certain path segments. If the objective is to minimize the total tolls to achieve a balance between the two, then this involves an inverse optimization problem under the l_1 norm. On the other hand, if the objective is to minimize the maximum toll imposed on any road, then it pertains to the l_∞ norm.

1.3.4 Energy Systems

The energy sector has also embraced inverse optimization to analyze decisions made by energy operators in market-clearing bidding problems. It aids in comprehending the strategies employed by market participants, which is vital for the design of effective energy policies and the operation of sustainable energy systems.

1.3.5 Agent Demand Models

Forward Model [26] Let y be a variable representing a level of electricity consumption. Given a price p_t at time period t, we assume that an agent's electricity consumption behavior is the solution to an unobserved constrained utility maximization problem

$$\max_y \{u_t(y) - p_t y l_t \leq y \leq r_t\}, \tag{1.40}$$

where $u_t(y)$ is the agent's utility function, and (l_t, r_t) represent known lower and upper bounds on consumption. We assume that the utility function is a stepwise function

$$u_t(y) := \begin{cases} c_{1,t}y & \text{if } y < s \\ c_{1,t}s_1 + \cdots + c_{i,t}(y - (i-1)s) & \text{if } (i-1)s \leq y \leq is, i \geq 2, \end{cases}$$

where each "step" has equal width s for a total of I steps. Furthermore, we assume that $c_i^t \geq c_j^t$ for all $i < j$, which implies that the marginal utility is nonincreasing in the consumption amount. With this function, (1.40) can be rewritten as a linear program

$$\max_{\mathbf{x}\in[0,s]^I} \left\{ \left(\sum_{i=1}^{I} c_{i,t} x_i \right) - p_t \mathbf{1}^\top \mathbf{x} \mid l_t \le \mathbf{1}^\top \mathbf{x} \le r_t \right\}$$

with decision vector x representing the amount consumed in each step.

Inverse Model [26] The inverse optimization model estimates the components $c_{i,t}$ of the utility function for all i. Let the vector $\pi_t = (\pi_{t,1}, \ldots, \pi_{t,m})$ describe a set of m "features" at period t. Features can include recent consumption behavior or descriptions of weather conditions. Consider a dataset of N observations of consumption features-price tuples $\{(\hat{y}_t, p_t, \pi_t)\}_{t=1}^{N}$. We first transform each $\hat{y}_t$ into a vector $\hat{x}_t$ by setting the first $\lfloor y_t/s \rfloor$ elements of the vector as s, the $\lceil y_t/s \rceil$-th element as $\lfloor y_t/s \rfloor - s$, and the last $I - \lceil y_t/s \rceil$ elements as zero. We then define the utility function components as a linear regression of the features

$$c_{i,t} = \theta_{i,0} + \theta_{i,1}\pi_{t,1} + \cdots + \theta_{i,m}\pi_{t,m}, \quad \forall i \in \{1, \ldots, I\}.$$

Let $\kappa > 0$ be a regularization parameter. The regularized inverse optimization problem is

$$\min_{\boldsymbol{\theta}\in\Theta} \frac{1}{N} \sum_{t=1}^{N} \ell\left(\hat{\mathbf{x}}_t, \mathcal{X}_t^{\text{opt}}(\boldsymbol{\theta})\right) + \kappa \|\boldsymbol{\theta}\|_p$$

which estimates the regression parameters of the utility function subject to the utility taking the shape of a stepwise nonincreasing linear function. Both the Minimum Distance loss function $\ell_{\text{D}}\left(\hat{\mathbf{x}}_t, \mathcal{X}_t^{\text{opt}}(\boldsymbol{\theta})\right)$ and the Absolute Suboptimality loss function $\ell_{\text{ASO}}\left(\hat{\mathbf{x}}_t, \mathcal{X}_t^{\text{opt}}\right)$ have been examined in the literature. The regularization hyperparameter κ can be tuned using cross-validation. Because Model (1.40) is a linear optimization model, strong duality can be used to formulate and solve the inverse problem.

1.3.6 Design Problems

Inverse optimization plays a transformative role in design problems, transcending its utility in mere operational contexts. It serves as a powerful analytical tool that, by examining historical decisions culminating in successful outcomes, informs the configuration of systems or processes. This retrospective yet proactive approach is fundamentally about learning from the past to enhance future designs.

The method leverages data from prior decisions to discern patterns and parameters that contributed to their effectiveness. By identifying these factors, inverse optimization aids designers in replicating the conditions that yielded optimal results, thereby streamlining the design process. It is particularly instrumental in scenarios where the objective function or constraints that led to a successful design are not fully understood or are too complex to articulate explicitly.

In engineering and systems design, for instance, inverse optimization can deduce the performance criteria that made previous designs effective. This understanding can then be systematically integrated into new design protocols, ensuring that they are not only efficient but also robust across a range of conditions. It allows for the optimization of complex systems where traditional forward optimization might be intractable or insufficiently informative.

Moreover, in the context of product design, inverse optimization can analyze consumer choices or market trends reflected in the success of past products. This analysis can guide the development of new products that align with consumer preferences and market demands, thus enhancing market competitiveness.

The chapter [26] underscores the broad applicability of inverse optimization in design problems, highlighting its potential to revolutionize the way systems and processes are conceptualized and engineered. By infusing design processes with the insights gleaned from historical data, inverse optimization paves the way for more agile, informed, and successful design strategies.

1.3.7 Estimation Problems

In the realm of estimation problems, inverse optimization emerges as an invaluable analytical technique for discerning the underlying parameters of a model, especially when direct observation or measurement of these parameters is impeded by practical challenges or exorbitant costs. This inverse approach is particularly pertinent within the domain of econometric models, where the influence of latent variables on observable outcomes necessitates sophisticated inferential methods.

The application of inverse optimization in such contexts is grounded in its ability to retrospectively deduce the parameters that likely governed the generation of observed data. By analyzing the outcomes that have resulted from a given set of decisions or behaviors, this method provides a pathway to infer the unobservable factors that may have contributed to these phenomena. This is instrumental in econometric analyses, where the interplay between latent constructs and their observable manifestations is crucial for understanding economic behaviors and outcomes.

Moreover, inverse optimization offers a robust alternative to traditional estimation techniques that may be limited by data availability or the complexity of the model at hand. It is adept at handling scenarios characterized by model uncertainty or incomplete information, where the true parameters are not directly measurable but can be inferred from the decisions or outcomes that are assumed to be optimal or near-optimal given certain conditions.

The chapter [26] elucidates the profound utility of inverse optimization in estimation problems, highlighting its capacity to uncover the latent structures and parameters that drive complex systems. By harnessing the power of inverse optimization, researchers and practitioners can glean deeper insights into the mechanisms that shape economic and social phenomena, thereby enhancing the precision and reliability of their models and forecasts. This methodological advancement, as discussed in the chapter, marks a significant stride in the field of operations research and management science, offering a versatile framework for tackling a myriad of estimation challenges across diverse domains.

1.3.8 Machine Learning and Data Science

The integration of inverse optimization with the burgeoning fields of machine learning and data science presents a compelling convergence of methodologies that enhances our capacity to interpret and predict complex decision-making phenomena. The chapter [26] adeptly navigates this intersection, elucidating the synergistic potential of these domains.

In the context of machine learning, where algorithms iteratively improve their performance based on data, inverse optimization offers a mechanism to interpret the outcomes of these algorithms by retrodicting the decision-making criteria. It allows for the identification of the relative importance or "weights" assigned to various factors within the historical decisions that have led to successful outcomes. This capability is invaluable for refining predictive models by providing a deeper understanding of the underlying decision processes that are not always transparent in conventional machine learning approaches.

Data science, with its focus on extracting knowledge from large datasets, benefits significantly from the application of inverse optimization. It enables data scientists to move beyond mere correlation and toward a causal understanding of the factors influencing decisions. This is particularly pertinent in "big data" scenarios where the volume and complexity of data can obscure the decision-making logic. By employing inverse optimization, researchers can discern patterns and parameters that are pivotal in shaping decisions, thereby enhancing the explanatory power of data-driven insights.

Moreover, this chapter discusses the utility of inverse optimization in the presence of high-dimensional data, where traditional optimization techniques may be computationally prohibitive. In such instances, inverse optimization can efficiently deduce the most influential variables, thereby facilitating more focused and effective data analysis and model building.

In essence, the amalgamation of inverse optimization with machine learning and data science, as highlighted in the chapter, marks a significant advancement in our analytical arsenal. It provides a robust framework for enhancing model refinement, predictive analytics, and decision-making across a spectrum of applications, from finance to healthcare, logistics to public policy. This interdisciplinary approach exemplifies the evolving nature of scientific inquiry, where the cross-pollination of ideas and methods paves the way for innovative solutions to complex problems.

1.3.9 Structural Estimation and Bilevel Optimization

The scholarly work [26] meticulously explores the intricate relationship between inverse optimization and structural estimation, revealing a confluence of methodologies that enriches the analytical landscape of operations research. Structural estimation, traditionally concerned with the development of parametric models that encapsulate agent behavior, finds a complementary approach in inverse optimization. This latter technique, through its ability to infer optimal decision-making criteria from observed outcomes, provides a novel lens through which to examine and model agent behavior in economic and operational contexts.

This chapter further delves into the nexus between inverse optimization and bilevel optimization, elucidating how the former can be conceptualized as a special case of the latter under certain conditions. In bilevel optimization, a hierarchical structure is present, where the solution to an upper-level optimization problem informs the parameters of a lower-level model. Inverse optimization, with its retrospective deductive capabilities, can effectively estimate these parameters by analyzing the outcomes of decisions that are presumed to be optimal or near-optimal solutions to the lower-level problem.

This reduction of inverse optimization to a parameter estimation task within a bilevel framework is particularly pertinent in scenarios where the upper-level model's solution provides critical context for the lower-level decision-making process. It allows for a more nuanced understanding of the interplay between different levels of decision-making and the factors that influence them, thereby enhancing the predictive accuracy and practical applicability of the models.

Moreover, this chapter underscores the methodological commonalities between these fields, such as the use of reformulations based on duality, which are employed to transform and solve complex optimization problems. The insights drawn from this intersection of disciplines offer a rich vein for future research, encouraging the cross-pollination of ideas and techniques that can lead to more sophisticated models and solutions.

In summary, the integration of inverse optimization with structural estimation and bilevel optimization exemplifies the multidisciplinary nature of modern operations research. It highlights the potential for inverse optimization to not only inform the modeling of agent behavior but also to contribute to the solution of hierarchical optimization problems, thereby broadening its applicability and impact across diverse domains.

1.3.10 Future Research Directions

The scholarly article [26] by Chan, Mahmood, and Zhu not only provides a comprehensive review of the current theoretical and application landscape of inverse optimization but also astutely identifies several promising avenues for future research. It is replete with insights that chart the course for the evolution of this field, emphasizing the necessity for more sophisticated models capable of accommodating the imperfections inherent in model-data congruence.

One of the pivotal directions for future research highlighted is the development of advanced models that can adeptly manage scenarios where the model–data fit is not perfect. This involves creating algorithms and methodologies that can handle uncertainties and discrepancies between the theoretical models and the empirical data, thereby enhancing the robustness and applicability of inverse optimization in real-world contexts.

Furthermore, this chapter underscores the potential of integrating inverse optimization with cutting-edge technologies such as artificial intelligence (AI) and machine learning. This integration is envisioned to augment the predictive capabilities of inverse optimization, enabling it to process and analyze large volumes of data more efficiently. The synergy between AI and inverse optimization could lead to the creation of adaptive models that continuously improve their accuracy through learning from new data.

Additionally, the authors advocate for the exploration of new applications in emerging domains. The chapter points to finance and environmental management as areas ripe for the application of inverse optimization. In finance, for instance, it could be utilized to understand and predict market behaviors and investment decisions. In environmental management, it could play a crucial role in optimizing resource allocation and policymaking based on ecological outcomes.

This chapter also suggests that future research should consider the ethical implications and societal impacts of deploying inverse optimization models, especially in high-stakes decision-making scenarios. This includes examining issues related to data privacy, model transparency, and the potential for algorithmic bias.

In essence, the future research directions proposed in the chapter reflect a forward-looking approach that seeks to expand the theoretical underpinnings of inverse optimization, enhance its computational capabilities, and broaden its practical applications. These directions collectively aim to solidify the role of inverse optimization as a key instrument in the decision-making toolkit across various scientific and professional disciplines.

1.4 Conclusions

In this chapter, we have endeavored to construct a comprehensive and cohesive framework for Inverse Combinatorial Optimization Problems (ICOP), which is designed to accommodate a broad spectrum of problem types. This includes not only the traditional ICOPs but also inverse optimal values problems, partial inverse problems, network improvement problems, and network interdiction problems. The development of such an all-encompassing framework is pivotal as it provides clarity to the intricate and often convoluted interrelations that exist among these various problem types.

Our examination of classical combinatorial optimization problems, which are typically solvable in polynomial time, serves as a foundation upon which the complexities of ICOPs can be better understood. Problems such as the shortest path, minimum spanning tree, and center/median location have been meticulously analyzed to demonstrate their relevance and applicability within the ICOP paradigm. These well-established problems form the bedrock of combinatorial optimization, and their inverse counterparts present a fertile ground for exploring new methodologies and algorithms.

To effectively communicate the multifaceted nature of a specific ICOP, we have introduced a standardized format that enables a detailed representation of the diverse attributes inherent in such problems. This format is not only systematic but also flexible enough to accommodate the unique characteristics of different ICOPs. A table has been meticulously crafted to detail each attribute according to this standardized format, thereby providing a clear and structured means of comparison and analysis.

Through this unified approach, we have been able to identify a plethora of open questions that remain to be addressed in the field of inverse combinatorial optimization. These questions are not merely academic curiosities but represent significant challenges that, once overcome, could lead to substantial advancements in the field. Our exploration of the complexity interplay between forward and inverse optimization problems has been particularly enlightening. We have posited that if a forward problem is polynomially solvable ($\mathcal{P}$), its inverse is likely to reside in $\mathcal{P}$ as well, albeit with the potential to be $\mathcal{NP}$-hard. This duality underscores the nuanced relationship between forward and inverse problems and opens up new avenues for research.

Conversely, the guarantee that an $\mathcal{NP}$-hard forward problem will result in an $\mathcal{NP}$-hard inverse problem raises a fundamental question: Can an inverse problem of a deemed $\mathcal{NP}$-hard problem be solved in polynomial time? This question stands as an open challenge in the field and invites further investigation. It is a question that strikes at the heart of computational complexity and has implications that extend beyond the realm of combinatorial optimization.

The applications of inverse optimization are far-reaching, spanning across diverse domains such as healthcare, transportation, logistics, and energy systems. In each of these domains, inverse optimization offers critical insights into decision-making processes and enhances data-driven prescriptive analytics. The method's utility in understanding observed behaviors and guiding future tasks cannot be overstated, as it highlights the potential to refine models and predict outcomes in complex systems. This capability to retrodict and predict is a powerful tool in the hands of researchers and practitioners alike, enabling them to make more informed decisions and to anticipate and mitigate potential challenges.

In conclusion, this chapter has laid the groundwork for a deeper understanding of ICOPs. It has not only provided a structured framework for analyzing these problems but has also sparked new questions and opened up new frontiers for exploration. The journey from hereon will involve tackling the open challenges identified, developing new algorithms and methodologies, and applying the insights gained to real-world problems. The potential of inverse optimization to transform the way we approach complex systems is immense, and the future of this field looks promising as we stand on the cusp of new discoveries and innovations.

Chapter 2
Generalized Inverse Bottleneck Optimization Problems

Abstract This chapter provides an in-depth examination of Generalized Inverse Bottleneck Optimization Problems (GIBOPs), presenting a unified framework for understanding and solving these complex combinatorial optimization issues. We explore various problem formulations, develop efficient algorithms under different norm constraints, and discuss the computational complexities involved. Practical applications in network optimization and other fields are highlighted, along with a set of open problems that pave the way for future research. It is a significant contribution to the field of (GIBOP), offering both theoretical insights and practical solutions.

Keywords Inverse bottleneck optimization problems · Budget constrained inverse bottleneck optimization problems · Budget constrained improvement bottleneck capacity problem · Budget constrained interdiction bottleneck capacity problem · Inverse constrained bottleneck problems

2.1 Inverse Bottleneck Optimization Problems

A *Bottleneck Optimization Problem (BOP)* is to find a *CS* $S \in \mathcal{S}$ whose bottleneck weight $w(S) = \max_{e \in S} w(e)$ is minimum, which can be formulated as (2.1).

$$(BOP) \quad \min_{S \in \mathcal{S}} \max_{e \in S} w(e) \tag{2.1}$$

Definition 2.1 We define *a cut* of a family $\mathcal{S}$ as a subset of E, which has a nonempty intersection with any set $S \in \mathcal{S}$.

Next we introduce a system $(E, \mathcal{K})$, which is dual to the system $(E, \mathcal{S})$ in some sense. Let $\tilde{\mathcal{K}}$ be the class of all cuts of $\mathcal{S}$, which is nonempty, since E itself is a cut of $\mathcal{S}$. We denote the class of all minimal cuts (with respect to set inclusion) in $\tilde{\mathcal{K}}$ by

Xiucui Guan contributed to this chapter.

X. Guan et al., *Inverse Combinatorial Optimization Problems*, Springer Optimization and Its Applications 225, https://doi.org/10.1007/978-3-031-91175-0_2

$\mathcal{K}$. The class $\mathcal{K}$ is uniquely defined by $\mathcal{S}$ and is a clutter, which means that no two distinct sets in S are comparable. Moreover, for any $e \in E$, there is a cut K with $e \in K$. We can assume without loss of generality that the system $\mathcal{S}$ forms a clutter, and $E = \cup_{S \in \mathcal{S}} S = \cup_{K \in \mathcal{K}} K$. Since $\mathcal{S}$ is a system of cuts for $\mathcal{K}$ with exactly the same properties as $\mathcal{K}$ has with respect to $\mathcal{S}$, the systems $(E, \mathcal{S})$ and $(E, \mathcal{K})$ are dual to each other. We call them *a blocking pair* and give some examples further [20].

- If $(E, \mathcal{S})$ is the system of all paths between two vertices s and t in an undirected graph G, then the cut system $(E, \mathcal{K})$ consists of all minimal s, t-cuts.
- If $(E, \mathcal{S})$ is the system of all spanning trees in an undirected graph G, then the cut system $(E, \mathcal{K})$ consists of all minimal cuts in G.
- If $(E, \mathcal{S})$ is the system of all bases of a matroid, then the cut system $(E, \mathcal{K})$ consists of all circuits of the dual matroid.
- If $(E, \mathcal{S})$ is the system of all perfect matchings (assignments) in a complete bipartite graph $G = (X, Y; E)$ with vertex sets X and Y, $|X| = |Y| = n$, then a minimal cut K is the edge set of a subgraph of G induced by $V \cup W$ with $V \subseteq X$, $W \subseteq Y$, and $|V| + |W| = n + 1$.

Lemma 2.1 ([58, Lemm 1]) *A set S is an optimal bottleneck set under a weight vector w if and only if $E^{\geq} := \{e \in E | w(e) \geq \max_{e \in S} w(e)\}$ is a cut of the system $\mathcal{S}$.*

Given a *CS* S^0, an *Inverse Bottleneck Optimization Problem (IBOP)* is to find a new weight vector $\bar{w}$ satisfying the bounded constraint to make a given set S^0 is an optimal bottleneck set under $\bar{w}$. Its objective is to minimize the modification under some norm. Its mathematical model is given as follows:

$$\begin{aligned} & \min \|\bar{w} - w\| \\ (IBOP)\ & \text{s.t. } \max_{e \in S^0} \bar{w}(e) \leq \max_{e \in S} \bar{w}(e), \forall\, S \in \mathcal{S}, \\ & w - l \leq \bar{w} \leq w + u. \end{aligned}$$

We denote the problems *(IBOP)* under weighted l_1, l_∞ norms and bottleneck-type and sum Hamming distance by *($IBOP_1$)*, *($IBOP_\infty$)*, *($IBOP_{\text{bH}}$)*, *($IBOP_{\text{wsH}}$)*, respectively.

The inverse BOPs have a big application potential [133]. In some computer networks, the real capacity of each link in a network differs from its designed one because of wearing-down of the link after years of operation. As the maximum capacity path between a pair of nodes can be observed by sending big data through the network, if we reasonably assume that the real capacities of links are not far from their designed ones, then we may use the inverse maximum capacity path problem to estimate the real capacities of the links.

Let $S^w = \arg\min_{S \in \mathcal{S}} \max_{e \in S} w(e)$ and

$$\underline{w} = \max\{w(S^w), max\{w(e) - l(e) | e \in S^0\}\}. \tag{2.2}$$

Then we know that $\underline{w} \le \bar{w}(S^0) \le w(S^0)$ [56].

Sort the set $(\{w(e)|e \in E\} \cup \{w(e)+u(e)|e \in E\} \cup \{\underline{w}\}) \cap [\underline{w}, w(S^0)]$ of values in a strictly increasing order, say $\underline{w} = q_1 < q_2 < \cdots < q_t = w(S^0)$, where $t = O(m)$.

2.1.1 (Inv, BOP, Bounded, Weighted l_1 Norm/Sum Hamming)

Theorem 2.1 ([56, Theorem 1, 4]) *The problem (IBOP$_1$)/(IBOP$_{wsH}$) has an optimal solution $\bar{w}^1$ / $\bar{w}^{wsH}$ such that the bottleneck weight* $\max_{e \in S^0} \bar{w}(e) = q_h$ *for* $1 \le h \le t$*, where*

$$\bar{w}^1(e) = \begin{cases} q_h, & \text{if } \bar{w}^1(e) \neq w(e), \\ w(e), & \text{otherwise.} \end{cases} \tag{2.3}$$

$$\bar{w}^{wsH}(e) = \begin{cases} q_h, & \text{if } \bar{w}^{wsH}(e) \neq w(e), w(e) - l(e) \le q_h \le w(e) + u(e), \\ w(e) - l(e), & \text{if } \bar{w}^{wsH}(e) \neq w(e), q_h < c(e) - l(e), \\ w(e) + u(e), & \text{if } \bar{w}^{wsH}(e) \neq w(e), q_h > w(e) + u(e), \\ w(e), & \text{otherwise.} \end{cases} \tag{2.4}$$

Given a value p, let $E^{\ge}(p) = \{e \in E|w(e) \ge p\}$, $E(p) = \{e \in E|w(e) < p\}$ and $\mathcal{S}(p) := \{S \in \mathcal{S}|S \subseteq E(p)\}$, then $\mathcal{S}(p) \neq \emptyset$ when $E^{\ge}(p)$ is not a cut of $\mathcal{S}$. Let $S^0(p) = \{e \in S^0|w(e) > p\}$. Hence, we need to decrease weight $w(e)$ to p for edges in $S^0(p)$ and simultaneously increase weight $w(e)$ of at least one edge in each $S \in \mathcal{S}(p)$ to p. Then we solve the *Inverse Objective Value on Bottleneck Optimization Problem (IOVBOP(p))* defined on $E(p)$ with respect to p as follows:

$$\begin{aligned} &\min \|\bar{w}_p - w\| \\ (\mathit{IOVBOP}(p)) \quad &\text{s.t. } \max_{e \in S} \bar{w}_p(e) = p, \forall S \in \mathcal{S}(p), \\ &\bar{w}_p(e) \le w(e) + u(e), \ \forall e \in E(p). \end{aligned} \tag{2.5}$$

Let $\bar{E}(p) := \{e \in E(p)|p \le w(e) + u(e)\}$. Then the problem (IOVBOP(p)) can be solved by a *Minimum Cut (MC(p))* of family $\mathcal{S}(p)$ under the capacity vector $c_p : E(p) \to R$ defined as follows.

$$\begin{cases} c_p^1(e) = c(e)(p - w(e)), \ c_p^{\text{wsH}}(e) = c(e), \ \forall e \in \bar{E}(p), \\ c_p^1(e) = +\infty, \qquad\qquad\quad c_p^{\text{wsH}}(e) = +\infty, \ \forall e \in E(p) \backslash \bar{E}(p). \end{cases} \tag{2.6}$$

$$\begin{aligned} &\min_K \sum_{e \in K} c_p(e) \\ (MC(p)) \quad &\text{s.t. } K \subseteq E(p) \text{ is a cut of subclass } \mathcal{S}(p). \end{aligned} \tag{2.7}$$

Lemma 2.2 ([58, Lemm 7]) *If $K(p)$ is not a cut of $\mathcal{S}(p)$, then $K(q)$ is not a cut of $\mathcal{S}(q)$ for all $q \geq p$.*

We have the following Remark 2.1 by Lemma 2.2.

Remark 2.1 ([133, Remark 1]) In the real implementation, we can call a binary search algorithm to determine the largest value $q_{\bar{t}}$ such that $c_{q_{\bar{t}}}(K(q_{\bar{t}})) < +\infty$. Then we check the values q_k one by one within the interval $[q_1, q_{\bar{t}}]$ until the optimal value $\phi(q_h) = \sum_{e \in S^0(q_h)} c(e)(w(e) - q_h) + \sum_{e \in K(q_h)} c(e)(q_h - w(e))$ under weighted l_1 norm or $\phi(q_h) = \sum_{e \in S^0(q_h) \cup K(q_h)} c(e)$ under weighted sum Hamming distance is obtained.

Theorem 2.2 ([56, Theorem 2]) *The problem (IBOP$_1$)/(IBOP$_{wsH}$) can be solved by $O(m)$ minimum cut problems, and the objective value corresponding to the value q_h is $\phi(q_h)$.*

2.1.2 (Inv, BOP, Bounded, Weighted l_∞ Norm)

Similarly, the problem *(IOVBOP(p))* defined on $E(p)$ under weighted l_∞ norm can be solved by a *Bottleneck Cut (BC(p))* of family $\mathcal{S}(p)$ under the capacity vector c_p^1 defined by (2.6).

$$\begin{aligned} &\min_K \max_{e \in K} c_p^1(e) \\ (BC(p)) \quad &\text{s.t. } K \subseteq E(p) \text{ is a cut of subclass } \mathcal{S}(p). \end{aligned} \tag{2.8}$$

Let $K(p)$ be an optimal bottleneck cut and $c_p^1(K(p)) = \max_{e \in K(p)} c_p^1(e)$ be the corresponding objective value. If $c_p^1(K(p)) = +\infty$, then the instance is infeasible; otherwise we define $\bar{w}_p$ as follows.

$$\bar{w}_p(e) = \begin{cases} p, & \forall e \in K(p), \\ w(e), & \forall e \in E(p) \backslash K(p). \end{cases} \tag{2.9}$$

Lemma 2.3 ([133, Lemma 8]) *The IOV problems (IOVBST(p)) on bottleneck spanning trees and (IOVCP(p)) on capacity paths under weighted l_∞ norm can be solved by finding a max-sum spanning tree under the edge weight $c_p^1(e)$ on $G(p) = (V, E(p))$.*

Using similar arguments as in [133, Section 3], we can claim that the objective value $\psi(p)$ given by (2.10) of the problem (IOVBOP(p)) with respect to p is a piecewise linear function in each interval $(q_k, q_{k+1}]$, $k = 1, \cdots, t-1$.

$$\psi(p) = \max \left\{ \max_{e \in S^0(q_{k+1})} \{c(e)(w(e) - p)\}, \max_{e \in K(q_{k+1})} \{c(e)(p - w(e))\} \right\}. \tag{2.10}$$

Notice that not like $\phi(p)$, the minimizer of function $\psi(p)$ on each $[q_k, q_{k+1}]$ is not necessary an endpoint of the interval, because the function is in general not concave in each interval. However, its minimizer over $(q_k, q_{k+1}]$ must be located at either an endpoint of the interval or some intersection points of two linear functions in the group $\{c(e)(w(e)-p) \mid e \in S^0(q_{k+1})\} \cup \{c(e)(p-w(e)) \mid e \in K(q_{k+1})\}$. Therefore, based on Remark 2.1, we first determine the largest value $q_{\bar{t}}$ such that $c_{q_{\bar{t}}}(K(q_{\bar{t}})) = \max_{e\in K(q_{\bar{t}})}\{c(e)(p-w(e))\} < +\infty$. Then check the sequence $\{q_i\}_1^{\bar{t}}$ of endpoints and the intersection points of linear functions in the following group:

$$\left\{c(e)(w(e)-p) \mid e \in S^0\right\} \cup \{c(e)(p-w(e)) \mid e \in E\}$$

within the interval $[q_1, q_{\bar{t}}]$. Obviously, the number of such intersection points is at most $O(m^2)$. Hence, the total number of values p to be checked in order to determine $\min \psi(p)$ is $O(m^2)$.

Theorem 2.3 ([133]) *The problem (IBOP$_\infty$) can be solved by $O(m^2)$ bottleneck cut problems, and the objective value corresponding to the value p^* is $\psi(p^*)$.*

Remark 2.2 ([133, Remark 2]) The problem (*Inv, BOP, Bounded, l_∞ norm*) can be solved by $O(m)$ bottleneck cut problems.

In this case, the cost function $\psi(p)$ over the subinterval $(q_k, q_{k+1}]$ reduces to

$$\psi(p) = \max\left\{w(S^0) - p, \max_{e\in K(q_{k+1})}\{(p-w(e))\}\right\}.$$

Hence, the minimizer of $\psi(p)$ over $[q_k, q_{k+1}]$ must be located at either an endpoint of the interval or an intersect point of the linear functions $w(S^0) - p$ and $p - w(e)$ for some edge $e \in K(q_{k+1})$, which is $\frac{1}{2}\left(w(S^0) + w(e)\right)$. So, the overall minimizer of $\psi(p)$ must be contained in the following set, whose cardinality is $O(m)$.

$$\Gamma = \{q_i\}_1^{\bar{t}} \cup \left\{\frac{1}{2}\left(w(S^0) + w(e)\right) \mid e \in E\right\}.$$

Remark 2.3 ([133, Remark 3]) The problem (*Inv, BOP, Unbounded, l_∞ norm*) can be solved by following optimal solution with the objective value $\frac{1}{2}(w(S^0) - w(S^w))$, where $p^* = \frac{1}{2}(w(S^0) + w(S^w))$.

$$\bar{w}(e) = \begin{cases} \min\{w(e), p^*\}, & \forall e \in S^0, \\ p^*, & \forall e \in K(p^*), \\ w(e), & \text{otherwise} \end{cases}$$

In this case, $\psi(p^*) \geq \max\left\{w(S^0) - \bar{w}(S^0), \bar{w}(S^w) - w(S^w)\right\}$. It follows from $w(S^w) \leq \bar{w}(S^0) \leq w(S^0)$ and $\bar{w}(S^0) = p^* \leq \bar{w}(S^w)$ that $\psi(p^*) \geq \frac{1}{2}(w(S^0) - w(S^w))$. It is easy to verify that $\bar{w}(S^0) = p^*$ and S^0 is an optimal bottleneck set.

Furthermore, the optimal objective value $\psi(p^*)$ can be calculated by

$$\begin{aligned}\psi(p^*) &= \max\left\{w(S^0) - p^*,\ \max_{e\in K(p^*)}\{(p^* - w(e))\}\right\}\\ &= \max\left\{\frac{1}{2}(w(S^0) - w(S^w)),\ p^* - \min_{e\in K(p^*)} w(e)\right\}\\ &= \max\left\{\frac{1}{2}(w(S^0) - w(S^w)),\ p^* - w(S^w)\right\}\\ &= \frac{1}{2}(w(S^0) - w(S^w)).\end{aligned}$$

2.1.3 (Inv, BOP, Bounded, Bottleneck Hamming Distance)

Given a *CS* S^0, the problem ($IBOP_{\mathrm{bH}}$) can be formulated as follows:

$$\begin{aligned} &\min\ c_{bH}(\bar{w}) = \max_{e\in E} c(e)H(\bar{w}(e), w(e))\\ (IBOP_{bH})\quad &\text{s.t. } \max_{e\in S^0}\bar{w}(e) \le \max_{e\in S}\bar{w}(e),\ \forall\, S\in\mathcal{S},\\ &\quad\quad w - l \le \bar{w} \le w + u.\end{aligned}$$

Sort all the costs $c(e)$ of edges $e \in E$ in a strictly increasing order: $c(e_{i_1}) < c(e_{i_2}) < \cdots < c(e_{i_\tau})$. For a given cost $c(e_{i_k})$, let $E_k = \{e \in E | c(e) \le c(e_{i_k})\}$,

$$w^k(e) := \begin{cases} w(e) - l(e), & \text{if } e \in E_k \cap S^0,\\ w(e) + u(e), & \text{if } e \in E_k \backslash S^0,\\ w(e), & \text{if } e \notin E_k,\end{cases} \tag{2.11}$$

$\max_{e\in S^0} w^k(e) = p_k$, and $E_k^{\ge} := \{e \in E | w^k(e) \ge p_k\}$. Let

$$w^{p_k}(e) := \begin{cases} p_k, & \text{if } e \in E_k \cap S^0,\ p_k \le w(e) + u(e),\\ w(e) - l(e), & \text{if } e \in E_k \cap S^0,\ p_k > w(e) + u(e),\\ p_k, & \text{if } e \in E_k\backslash S^0,\ w(e) - l(e) \le p_k \le w(e) + u(e),\\ w(e) + u(e), & \text{if } e \in E_k\backslash S^0,\ p_k < w(e) - l(e) \text{ or } p_k > w(e) + u(e),\\ w(e), & \text{if } e \notin E_k,\end{cases} \tag{2.12}$$

and $E_{p_k}^{\ge} := \{e \in E | w^{p_k}(e) \ge p_k\}$. Then we have the following properties.

Lemma 2.4 ([56, Lemma 5]) *If $E_k^{\ge}$ is a cut of $\mathcal{S}$, then $E_{p_k}^{\ge}$ is also a cut of $\mathcal{S}$.*

Lemma 2.5 ([56, Lemma 6]) *If $E^{\geq}_{\bar{p}_k}$ is not a cut of $\mathcal{S}$, then for any $\bar{w}$ with $w - l \leq \bar{w} \leq w + u$ and $c_{bH}(\bar{w}) = c(e_{i_k})$, $\bar{E}^{\geq} = \{e \in E | \bar{w}(e) \geq \bar{p} = \max_{e \in S^0} \bar{w}(e)\}$ is not a cut of $\mathcal{S}$.*

It follows from Lemma 2.5 that if $E^{\geq}_{\bar{p}_k}$ is not a cut of $\mathcal{S}$, then there is no feasible solution $\bar{w}$ of problem (*IBOP*$_{bH}$) with $c_{bH}(\bar{w}) \leq c(e_{i_k})$. Next we present Algorithm 2.1 to solve the problem (*IBOP*$_{bH}$) by performing a binary search method. In each iteration, we mainly verify whether the set $E^{\geq}_{\bar{p}_k}$ is a cut of $\mathcal{S}$ or not, which is equivalent to checking if $\bar{G}_k = (V, E \backslash E^{\geq}_{\bar{p}_k})$ contains a feasible solution of $\mathcal{S}$ or not. Let $\lfloor x \rfloor$ be the maximal integer not greater than x.

Algorithm 2.1 A binary search algorithm to solve (*IBOP*$_{bH}$) [56, Algorithm 5]

Input: A system $(V, E, \mathcal{S})$ with four functions w, c, l and u defined on E.
Output: an optimal solution $\bar{w}$ and the optimal objective value $c_{bH}(\bar{w})$.
1: Sort all the weights $c(e)$ of edges $e \in E$ in a strictly increasing order, $c(e_{i_1}) < c(e_{i_2}) < \cdots < c(e_{i_\tau})$.
2: Put $a := 1$ and $b := \tau$.
3: **if** $E^{\geq}_{\bar{p}_\tau}$ is not a cut of $\mathcal{S}$ **then**
4: the instance is infeasible.
5: **Return** $\bar{w} := []$ and $c_{bH}(\bar{w}) := +\infty$.
6: **else**
7: **while** $b - a > 1$ **do**
8: Let $k := \lfloor (a + b)/2 \rfloor$, $E_k := \{e \in E | c(e) \leq c(e_{i_k})\}$, $p_k := \max_{e \in S^0} w^k(e)$ and $E^{\geq}_{\bar{p}_k} := \{e \in E | w^{p_k}(e) \geq p_k\}$, where w^k and w^{p_k} are defined as in (2.11) and (2.12), respectively.
9: **if** $E^{\geq}_{\bar{p}_k}$ is a cut of $\mathcal{S}$ **then** update $b := k$.
10: **else** update $a := k$.
11: **end if**
12: **end while**
13: **Return** an optimal solution $\bar{w} := w^{bH}$ and its value $c_{bH}(\bar{w}) := c(e_{i_b})$.
14: **end if**

Theorem 2.4 ([56, Theorem 6]) *The inverse problem (IBOP$_{bH}$) under bottleneck-Hamming distance can be solved in $O(T_{bH} \log m)$ operations, where T_{bH} is the time to verify if $E^{\geq}_{\bar{p}_k}$ is a cut of $\mathcal{S}$ or not.*

2.2 Budget-Constrained Inverse Bottleneck Optimization Problem

Given a feasible solution S^0, a cost vector $c > 0$, and two bound vectors $l, u \geq 0$ defined on the edge set E, let $w(S) := \max_{e \in S} w(e)$ and

$$\mathcal{W} = \left\{ \bar{w} | w - l \leq \bar{w} \leq w + u, \bar{w}(S^0) \leq \bar{w}(S), \forall\, S \in \mathcal{S} \right\}$$

be the set of all feasible weight vectors $\bar{w}$ of (IBOP). Then the Budget-Constrained IBOP (*BCIBOP*) under *bH-wsH distance* is to find a new weight vector $\bar{w} \in \mathcal{W}$ to minimize the *bH-distance* such that the *wsH-distance* is upper bounded by a given value B. It can be mathematically formulated as follows:

$$\begin{aligned} &\min_{\bar{w}\in\mathcal{W}} c_{bH}(\bar{w}) = \max_{e\in E} c(e)H(\bar{w}(e), w(e)) \\ \text{(BCIBOP)}\quad & s.t.\ \ c_{sH}(\bar{w}) = \sum_{e\in E} c(e)H(\bar{w}(e), w(e)) \le B, \end{aligned}$$

Now we investigate the lower and upper bounds $\underline{C}, \bar{C}$ of the optimal objective value C_{BC} of (BCIBOP) based on the relationship among problems (BCIBOP), (IBOP$_{bH}$) and (IBOP$_{sH}$), whose optimal solutions are denoted by w^{BC}, w^{bH}, and w^{sH}, respectively.

Firstly, the lower bound $\underline{C}$ is given by $c_{bH}(w^{bH})$ since (BCIBOP) is a sub-problem of (IBOP$_{bH}$) by adding one more budget constraint on the wsH-distance. Secondly, the upper bound $\bar{C}$ is given by $c_{bH}(w^{sH})$ if $c_{sH}(w^{sH}) \le B$. The problem (BCIBOP) is infeasible if $c_{sH}(w^{sH}) > B$. As a result, C_{BC} is one of the edge costs in the interval $[\underline{C}, \bar{C}] = [c_{bH}(w^{bH}), c_{bH}(w^{sH})]$.

We invoke a binary search method to determine the optimal value C_{BC} quickly from the interval $[\underline{C}, \bar{C}]$. For the interval $[c(e_{i_{\underline{t}}}), c(e_{i_{\bar{t}}})]$ in the k-th iteration, we mainly check if there is a feasible solution $\bar{w}_k$ of (BCIBOP) with $c_{bH}(\bar{w}_k) \le c(e_{i_k})$. If yes, the interval is reduced to $[c(e_{i_{\underline{t}}}), c(e_{i_k})]$, otherwise, to $(c(e_{i_k}), c(e_{i_{\bar{t}}})]$. We first check if the optimal solution $\bar{w}^{p_k}$ of the problem (IOVBOP(p_k)) under wsH-distance suitable or not, where $p_k = w^k(S^0)$. Otherwise, we check a feasibility of a solution w^{q_i} satisfying $w^{q_i}(S^0) = q_i > p_k$ and $c_{sH}(w^{q_i}) \le B$ by Algorithm 2.2.

Algorithm 2.2 Find a feasible solution of (BCIBOP) [49, Algorithm 1]

Input: A system $(V, E; \mathcal{S})$ with four vectors c, w, l and u, a feasible solution S^0, an edge cost $c(e_{i_k})$, $E^k := \{e \in E | c(e) \le c(e_{i_k})\}$, two values p_k and B.
Output: the objective value $c_{sH}(\hat{w})$ and the solution $\hat{w}$.

Sort the values in $(\{w(e)|e \in E\} \cup \{w(e) + u(e)|e \in E^k\}) \cap [w(S^w), w(S^0)] \cap (p_k, w(S^0)]$ in a strictly increasing order, say $q_1 < q_2 < \cdots < q_t$.
Initialize $\hat{w} := w$ and $c_{sH}(\hat{w}) := +\infty$.
for $i = 1$ to t **do**
  Let $S^0(q_i) := \{e \in S^0 | w(e) > q_i\}$ and $c_{q_i}(S^0) := \sum_{e\in S^0(q_i)} c(e)$.
  if $c_{q_i}(S^0) > B$ **then return** $c_{sH}(\hat{w}) := +\infty$ and $\hat{w}(e) := []$.
  else let $E^{\ge}_{q_i} := \{e \in E | w(e) \ge q_i\}$.
    if $E^{\ge}_{q_i}$ is a cut of $\mathcal{S}$ **then**
      Return $c_{sH}(\hat{w}) = c_{q_i}(S^0)$ and $\hat{w}(e) := \begin{cases} q_i, & \text{if } e \in S^0(q_i), \\ w(e), & otherwise. \end{cases}$
    else

$$c_{q_i}(e) := \begin{cases} c(e), & \text{if } e \in E^k, w(e) < q_i \le w(e) + u(e), \\ +\infty, & otherwise. \end{cases} \tag{2.13}$$

```
            Find a minimum cut K(q_i) defined as in (2.7).
            if c_{q_i}(K(q_i)) = +∞ then return c_{sH}(ŵ) := +∞ and ŵ(e) := [].
            else if c_{q_i}(K(q_i)) + c_{q_i}(S^0) ≤ B then
                Return c_{sH}(ŵ) := c_{q_i}(K(q_i)) + c_{q_i}(S^0) and
```

$$\hat{w}(e) := \begin{cases} q_i, & if\ e \in K(q_i) \cup S^0(q_i), \\ w(e), & otherwise. \end{cases}$$

```
            else update i := i + 1.
            end if
        end if
    end if
end for
```

Algorithm 2.3 An algorithm to solve problem (BCIBOP) [49, Algorithm 2]

Input: A system $(V, E; \mathcal{S})$ with four vectors c, w, l, u, a value B and a feasible solution S^0.

Output: an optimal solution $\bar{w}$ and its objective value $c_{bH}(\bar{w})$.

Sort all the costs $c(e)$ of edges $e \in E$ in a strictly increasing order, i.e., $c(e_{i_1}) < c(e_{i_2}) < \cdots < c(e_{i_\tau})$.

Call Algorithm 2.1 to solve problem (IBOP$_{bH}$) and obtain its solution w^{bH} and its objective value $c_{bH}(w^{bH})$.

if $c_{bH}(w^{bH}) = +\infty$ **then**

  Return $\bar{w} := []$ and $c_{bH}(\bar{w}) := +\infty$. The problem (BCIBOP) is infeasible.

else find the index $\underline{t}$ such that $c(e_{i_{\underline{t}}}) = c_{bH}(w^{bH})$.

  Solve (IBOP$_{sH}$) by Remark 2.1 and obtain an optimal solution w^{sH} and its objective value $c_{sH}(w^{sH})$.

  if $c_{sH}(w^{sH}) > B$ **then**

    Return $\bar{w} := []$, $c_{bH}(\bar{w}) := +\infty$. The problem (BCIBOP) is infeasible.

  else find the index $\bar{t}$ such that $c(e_{i_{\bar{t}}}) = c_{bH}(w^{sH})$, and let $\bar{w}_{\bar{t}} := w^{sH}$.

    while $\bar{t} - \underline{t} > 1$ **do**

      Let $k := \lceil (\underline{t} + \bar{t})/2 \rceil$, define $E^k := \{e \in E | c(e) \le c(e_{i_k})\}$, $p_k := \max_{e \in S^0} w^k(e)$ and $E_{\ge}^{p_k} := \{e \in E | w^{p_k}(e) \ge p_k\}$, where w^k, w^{p_k} are defined as in (2.11),(2.12), respectively.

      if $E_{\ge}^{p_k}$ is not a cut of $\mathcal{S}$ **then** put $\underline{t} := k$;

      else if $\sum_{e \in S^0(p_k)} c(e) < B$ **then**

        Find a minimum cut $K(p_k)$, where c_{p_k} is defined as in (2.13) and q_i is replaced with p_k.

        if $\sum_{e \in S^0(p_k) \cup K(p_k)} c(e) \le B$ **then**

put $\bar{t} := k$, and $\bar{w}_k(e) := \begin{cases} p_k, & if\ e \in K(p_k) \cup S^0(p_k), \\ w(e), & otherwise. \end{cases}$

else call Algorithm 2.2 and obtain the value $c_{sH}(\hat{w})$ and $\hat{w}$.

if $c_{sH}(\hat{w}) \leq B$ **then** put $\bar{t} := k$, and $\bar{w}_k := \hat{w}$;

else put $\underline{t} := k$.

end if

end if

else put $\underline{t} := k$.

end if

end while

Return $\bar{w} := \bar{w}_{\bar{t}}$ and the objective value $c_{bH}(\bar{w}) := c(e_{i_{\bar{t}}})$.

end if

end if

Theorem 2.5 ([49, Theorem 3.4]) *The problem (BCIBOP) under bH-wsH distance can be reduced to $O(m \log m)$ minimum cut problems.*

2.3 Budget-Constrained Improvement Bottleneck Capacity Problem

In some networks, we need to determine the weakest part of a combinatorial structure, which is called a bottleneck capacity. It may be the weakest part of a scheme in transportation networks, or the slowest job in an assembly line of manufacturing systems, or the narrowest link in a route of communication networks. We may choose from all possible schemes the one with the largest bottleneck capacity. Mathematically, the *Bottleneck Capacity Problem (BCP)* is a max–min problem[150], which can be formulated further similar to (2.1).

$$(BCP) \quad \max_{S \in \mathcal{S}} \min_{e \in S} w(e) \tag{2.14}$$

There are many *Budget-Constrained Improvement problems on BCP (BC-Imp-BCP)* in real life. For example, a transportation network may need to increase its flow capacity, and a production system may need to enhance its productive ability [150].

2.3.1 (BC-Imp, BCP, Bounded, Weighted l_1 Norm)

We first consider the unbounded problem ($BC\text{-}Imp\text{-}BCP_1$) under weighted l_1 norm, then we extend its solving method to the bounded case. They can be mathematically formulated as follows:

$$
\begin{aligned}
&\max_{\bar{w}} \max_{S \in \mathcal{S}} \min_{e \in S} \bar{w}(e) \\
(BC\text{-}Imp\text{-}BCP_1)\ \ s.t.\ &\sum_{e \in E} c(e)(\bar{w}(e) - w(e)) \leq B, \qquad (2.15)\\
\textit{Unbounded}: \quad &\bar{w}(e) \geq w(e), \quad e \in E. \\
\textit{Bounded}: \quad &w(e) \leq \bar{w}(e) \leq w(e) + u(e), \quad e \in E.
\end{aligned}
$$

Burkard et al. [20, Section 3] pointed out that the *Hamiltonian cycle* problem can be expressed as a bottleneck capacity problem without the budget constraint. Let $G = (V, E)$ with $|V| = n$. Construct a complete graph $K_n = (V, E_n, \mathbf{w})$ with $w(e) = 1$ if $e \in E$ and $w(e) = 0$ otherwise. Then the bottleneck capacity $\max_{S \in \mathcal{S}} \min_{e \in S} w(e)$ is either 1 or 0, depending whether G has a Hamiltonian cycle or not.

Theorem 2.6 ([20]) *The class of problems $(BC\text{-}Imp\text{-}BCP_1)$ includes some $\mathcal{NP}$-hard problems.*

The *Improvement problems on BCP $(Imp\text{-}BCP_1)$* under weighted l_1 norm are formulated as follows:

$$
\begin{aligned}
&\min_{\bar{w}} \sum_{e \in E} c(e)(\bar{w}(e) - w(e)) \\
(Imp\text{-}BCP_1)\ \ s.t.\ &\max_{S \in \mathcal{S}} \min_{e \in S} \bar{w}(e) \geq r, \qquad (2.16)\\
\textit{Unbounded}: \quad &\bar{w}(e) \geq w(e), \quad e \in E. \\
\textit{Bounded}: \quad &w(e) \leq \bar{w}(e) \leq w(e) + u(e), \quad e \in E.
\end{aligned}
$$

For a given value r, let

$$
c^r(e) = \begin{cases} c(e)(r - w(e)), & \text{if } w(e) < r \leq w(e) + u(e), \\ +\infty, & \text{if } r > w(e) + u(e), \\ 0, & \text{otherwise.} \end{cases} \qquad (2.17)
$$

Then the problem $(Imp\text{-}BCP_1)$ is equivalent to solving the *min-sum problem* (2.18) [145] whose optimal objective value and optimal set are denoted by $h(r)$ and S^r, respectively. Hence, an optimal solution of (*Imp-BCP*) when $h(r) < +\infty$ is given by $\bar{\mathbf{w}}^r$ as follows:

$$
h(r) = \min_{S \in \mathcal{S}} \sum_{e \in S} c^r(e). \qquad (2.18)
$$

$$
\bar{w}^r(e) = \begin{cases} r, & \text{if } e \in S^r \text{ and } r > w(e), \\ w(e), & \text{otherwise.} \end{cases} \qquad (2.19)
$$

Theorem 2.7 ([145, Theorem 1]) *If* $(S^*, \bar{w}^*)$ *is an optimal solution of (BC-Imp-BCP) with maximum objective value* r^**, then* $h(r^*) = B$ *and* S^* *is an optimal set of problem (2.18) when* $r = r^*$*.*

Theorem 2.8 ([145, Theorem 2]) *Suppose* $h(\tilde{r}) = B$ *and the optimal set of problem (2.18) is* $S^{\tilde{r}}$ *when* $r = \tilde{r}$*. Then* $(S^{\tilde{r}}, \bar{\mathbf{w}}^{\tilde{r}})$ *is an optimal solution of (BC-Imp-BCP), and the maximum value is* $\tilde{r}$*.*

It follows from the Theorems 2.7 and 2.8 that the unbounded problem $(BC\text{-}Imp\text{-}BCP_1)$ can be solved by a series of the equations like

$$h(r) = \min_{S \in \mathcal{S}} \sum_{e \in S} c^r(e) = B. \tag{2.20}$$

Zhang and Liu [145] presented a two-phase *Binary-Newton (BN)* algorithm for solving (2.20). The first phase aims to obtain a small interval where the optimum value r^* is located by a binary method; the second phase finds a solution of (2.20) by Radzik's Newton method applied to a combinatorial equation [112].

Now we extend the *BN-algorithm* to the bounded problem $(BC\text{-}Imp\text{-}BCP_1)$. Denote the set $\{w(e), w(e) + u(e) | e \in E\}$ of values by $\hat{W} = \{\hat{w}_i\}$ and let $\hat{w}_1 = \min\left\{\hat{w}_i \mid \hat{w}_i \in \hat{W}\right\}$, $\hat{w}_{k_0} = \max\left\{\hat{w}_i \mid \hat{w}_i \in \hat{W}\right\}$. Let $c_{\min} = \min\{c(e) \mid e \in E\}$, $d = \min\{|S| \mid S \in \mathcal{S}\}$ and $\hat{w}_{k_0+1} = \hat{w}_{k_0} + \frac{B}{d \cdot c_{\min}}$. The lower and upper bounds of the solution r^* of (2.20) is given by $\hat{w}_1 \le r^* \le \hat{w}_{k_0+1}$.

Algorithm 2.4 An algorithm to solve the bounded problem (BC-Imp-BCP$_1$) [145, BN-Algorithm]

Input: A system $(V, E; \mathcal{S})$ with two vectors c, w and a value B.
Output: an optimal solution $(S^{r^*}, \bar{\mathbf{w}}^{\mathbf{r}^*})$.

Phase I Binary Method

Arrange all different values in $\hat{W}$ increasingly by $\hat{w}_1 < \hat{w}_2 < \cdots < \hat{w}_{k_0} < \hat{w}_{k_0+1}$.
Put $\underline{t} := 1$ and $\bar{t} := k_0 + 1$.
while $\bar{t} - \underline{t} > 1$ **do**
  Put $k := \lceil(\underline{t} + \bar{t})/2\rceil$ and $r := \hat{w}_k$.
  Solve the min-sum problem (2.18) for the fixed r to obtain the optimal value $h(r)$ and an optimal solution S^r.
  if $h(r) > B$ **then** put $\bar{t} := k$
  else if $h(r) < B$ **then** put $\underline{t} := k$
  else let $r^* := r$ and $S^{r^*} := S^r$.
    Return an optimal solution $(S^{r^*}, \bar{\mathbf{w}}^{\mathbf{r}^*})$ by (2.19).
  end if
end while
Let $\begin{cases} a(e) := c(e)w(e), & b(e) := c(e), & \text{for } w(e) \le \hat{w}_{\underline{t}}, \\ a(e) := 0, & b(e) := 0, & \text{for } w(e) > \hat{w}_{\underline{t}}. \end{cases}$

Phase II Radzik's Newton Algorithm

Put $k := 1$, $r_k := \hat{w}_{\underline{t}}$, $S^k := S^{r_k}$ and $h(r_k) := h(\hat{w}_{\underline{t}})$.
while $h(r_k) \neq B$ **do**
    Put $a\left(S^k\right) := \sum\limits_{e \in S^k} a(e), b\left(S^k\right) := \sum\limits_{e \in S^k} b(e), r_{k+1} := \left(B + a\left(S^k\right)\right) / \left(b\left(S^k\right)\right)$ and $k := k + 1$.
    Solve the min-sum problem $h(r_k) = \min_{S \in \mathcal{S}} \sum_{e \in S} \left(-a(e) + b(e) \cdot r_k\right)$. Denote its optimal set and optimal value, respectively, by S^k and $h\left(r_k\right)$.
end while
Put $r^* := r_k$ and $S^{r^*} := S^k$.
Return an optimal solution $(S^{r^*}, \bar{\boldsymbol{w}}^{r^*})$ by (2.19).

It follows from the unbounded result in [145, Theorem 4] that we can obtain the same time complexity for the bounded problem $(BC\text{-}Imp\text{-}BCP_1)$.

Theorem 2.9 ([145]) *The bounded problem $(BC\text{-}Imp\text{-}BCP_1)$ can be solved by BN-algorithm in $(T_{ms} m^2 \log m)$ time, where T_{ms} is the time to solve the min-sum problem (2.18).*

2.3.2 (BC-Imp, BCP, Bounded, Weighted l_∞ Norm)

The mathematical model of the bounded problem $(BC\text{-}Imp\text{-}BCP_\infty)$ under weighted l_∞ norm is given as follows:

$$
\begin{aligned}
&\max_{\bar{w}} \max_{S \in \mathcal{S}} \min_{e \in S} \bar{w}(e) \\
&s.t. \ \max_{e \in E} c(e)(\bar{w}(e) - w(e)) \leq B, \\
&\qquad w(e) \leq \bar{w}(e) \leq w(e) + u(e), \quad e \in E.
\end{aligned}
\tag{2.21}
$$

Let $\bar{u}(e) = \min\left\{w(e) + u(e), w(e) + \frac{B}{c(e)}\right\}$. Then the problem (2.21) becomes the following model [145]

$$
\begin{aligned}
&\max_{\bar{w}} \max_{S \in \mathcal{S}} \min_{e \in S} \bar{w}(e) \\
&s.t. \ w(e) \leq \bar{w}(e) \leq \bar{u}(e), \quad e \in E.
\end{aligned}
$$

which is equivalent to a bottleneck capacity problem with a new capacity vector $\bar{u}$.

2.3.3 (BC-Imp, SCP, Bounded, Weighted l_1 Norm)

Now let us consider the *Sum Capacity Problem (SCP)*

$$(SCP)\ \max_{S \in \mathcal{S}} \sum_{e \in S} w(e). \tag{2.22}$$

The mathematical model of the (*BC-Imp*) problem on (*SCP*) under weighted l_1 norm is given as follows:

$$\begin{aligned}
& \max_{\bar{w}} \max_{S \in \mathcal{S}} \sum_{e \in S} \bar{w}(e) \\
(BC\text{-}Imp\text{-}SCP_1)\ \ & s.t. \sum_{e \in E} c(e)(\bar{w}(e) - w(e)) \le B, \\
Unbounded: \ \ & \bar{w}(e) \ge w(e),\ \ e \in E. \\
Bounded: \ \ & w(e) \le \bar{w}(e) \le w(e) + u(e),\ \ e \in E.
\end{aligned} \tag{2.23}$$

The problem (*BC-Imp-SCP*) is useful in various circumstances [145]. For example, suppose each $S \in \mathcal{S}$ is regarded as an instrument, which is composed of some components e_i. The reliability of instrument S is defined to be the product $\prod_{e_i \in S} e^{w_i}$ of reliabilities $e^{w_i} \in (0, 1]$ of all components in S. Let c_i be the cost of increasing reliability of component e_i by per unit of w_i. The problem is how to increase reliabilities of the components to maximum, the highest reliability of the system, such that the total increasing cost is within a given budget B. Notice that

$$\max_{S \in \mathcal{S}} \prod_{e_i \in S} e^{w_i} \Leftrightarrow \max_{S \in \mathcal{S}} \sum_{e_i \in S} w(e_i)$$

Then the problem can be formulated as a case of (*BC-Imp-SCP*).

Theorem 2.10 ([145, Corollary]) *The unbounded problem* $(BC\text{-}Imp\text{-}SCP_1)$ *contains an optimal solution* $(\bar{S}, \tilde{w})$ *with the following form.*

$$\tilde{w}(e_j) = \begin{cases} w(e_j) + \frac{B}{c_{\bar{i}}},\ if\ j = \bar{i},\ and\ c_{\bar{i}} = \min\{c(e_i) | e_i \in \bar{S}\}, \\ w(e_j), \qquad otherwise. \end{cases} \tag{2.24}$$

It follows from Theorem 2.10 that the problem $(BC\text{-}Imp\text{-}SCP_1)$ is equivalent to solving a problem (*SCP*) with a new weight vector $\tilde{w}$.

Algorithm 2.5 An algorithm to solve the unbounded problem (BC-Imp-SCP$_1$) [145]

Input: A system $(V, E; S)$ with two vectors c, w and a value B.
Output: an optimal solution $(\bar{S}, \bar{w})$ and the objective value θ.
1: Sort the values in the set $\{c(e)|e \in E\}$ in a non-decreasing order, say $c_{j_1} \le c_{j_2} \le \cdots \le c_{j_m}$.
2: Initialize $\theta := 0$, $\bar{S} := \emptyset$, and $\bar{w} := []$.
3: **for** $i = 1$ to m **do**
4: Let $\tilde{w}^i$ is defined as in (2.24), where $c_{\bar{i}}$ is replaced with c_{j_i}.
5: Solve the problem (SCP) with a weight vector $\tilde{w}^i$ to obtain its optimal set S^i and $\tilde{w}^i(S^i) := \sum_{e \in S^i} \tilde{w}^i(e)$.
6: **if** $\tilde{w}^i(S^i) > \theta$ **then** update $\theta := \tilde{w}^i(S^i)$, $\bar{S} := S^i$ and $\bar{w} := \tilde{w}^i$.
7: **end if**
8: **Return** an optimal solution $(\bar{S}, \bar{w})$ and the objective value θ.
9: **end for**

For the bounded case, we impose an upper bound $w(e) + u(e)$ on $\bar{w}(e)$. Then in Algorithm 2.5, we need to replace the weight vector $\tilde{w}^i$ in Line 4 by a subroutine in Algorithm 2.6.

Algorithm 2.6 A subroutine to determine weight vector $\tilde{w}^i$ in Line 4 for the bounded problem (BC-Imp-SCP) [145]

1: Let $k := i$ and $\tilde{w}^i := w$.
2: **while** $B > 0$ **do**
3: **if** $\frac{B}{c(e_{j_k})} \le u(e_{j_k})$ **then** let $\tilde{w}^i(e_{j_k}) := w(e_{j_k}) + \frac{B}{c(e_{j_k})}$ and $B := 0$.
4: **else** let $\tilde{w}^i(e_{j_k}) := w(e_{j_k}) + u(e_{j_k})$, $B := B - c(e_{j_k})u(e_{j_k})$, $k := k + 1$.
5: **end if**
6: **end while**

It follows from [145, Theorem 3] that the bounded problem (BC-Imp-SCP_1) has the same time complexity as that of the unbounded case.

Theorem 2.11 ([145, Theorem 3]) *The bounded problem (BC-Imp-SCP_1) under weighted l_1 norm can be solved by Algorithms 2.5 and 2.6 in $O(mT_{SCP})$ time, where T_{SCP} is the time to solve the problem (SCP).*

2.3.4 (BC-Imp, SCP, Bounded, Weighted l_∞ Norm)

The mathematical model of the bounded problem (*BC-Imp-SCP*) under weighted l_∞ norm is given as follows:

$$\begin{aligned} &\max_{\bar{w}} \max_{S \in \mathcal{S}} \sum_{e \in S} \bar{w}(e) \\ s.t.\ &\max_{e \in E} c(e)(\bar{w}(e) - w(e)) \le B, \\ &w(e) \le \bar{w}(e) \le w(e) + u(e), \quad e \in E. \end{aligned} \tag{2.25}$$

Similar to discussion in Sect. 2.3.2, the problem (2.25) is equivalent to solving a sum capacity problem with a new capacity vector $\bar{u}$.

2.3.5 (BC-Imp, BOP, Bounded, Weighted l_1/l_∞ Norm)

When the weights $w(e)$ represent consumption or expenses instead of efficiencies, we get the mathematical model of (*BC-Imp*) problem on (*BOP*), which can be similarly solved by the algorithms in the previous subsections.

$$\begin{aligned} &\min_{\bar{w}} \min_{S \in \mathcal{S}} \max_{e \in S} \bar{w}(e) \\ (\textit{BC-Imp-BOP})\ s.t.\ &\|\bar{w} - w\| \le B, \\ \textit{Unbounded}: \quad &\bar{w}(e) \le w(e), \quad e \in E. \\ \textit{Bounded}: \quad &w(e) - l(e) \le \bar{w}(e) \le w(e), \quad e \in E. \end{aligned} \tag{2.26}$$

Its *Improvement problems on BOP (Imp-BOP)* is formulated as follows:

$$\begin{aligned} &\min_{\bar{w}} \|\bar{w} - w\| \\ (\textit{Imp-BOP})\ s.t.\ &\min_{S \in \mathcal{S}} \max_{e \in S} \bar{w}(e) \le r, \\ \textit{Unbounded}: \quad &\bar{w}(e) \le w(e), \quad e \in E. \\ \textit{Bounded}: \quad &w(e) - l(e) \le \bar{w}(e) \le w(e), \quad e \in E. \end{aligned} \tag{2.27}$$

2.4 Budget-Constrained Interdiction Bottleneck Capacity Problem

In many production processes, there are the so-called critical tasks, which are particularly urgent. Assume that a critical task can be performed by any of the machines $e \in S$ in time $w(e)$ and the sets S are mutually disjoint. The fastest production is achieved, if every critical task is performed by that machine e, which yields the minimum value $w(e)$ in the corresponding set S. The other machines of S are used in the production of noncritical tasks. When the machines work in parallel, the completion time of all critical tasks is given by $\max_{S\in\mathcal{S}} \min_{e\in S} w(e)$. A budget B is allocated for improving the machines to speed up the production process of the critical tasks. The aim of *Budget-Constrained Interdiction problems on BCP (BC-Int-BCP)* is to minimize the completion time of all critical tasks subject to the given budget constraint [20]. In computer networks, improving the speed of a message from one headquarter station to all substations leads exactly to the problem (*BC-Int-BCP*) [20].

However, in some adversarial networks, an interdictor may wish to interdict a network property by reducing the capacity of a network. We take the maximum capacity path interdiction problem (MCPIP) as an example [125]. This problem consists of two players, user and interdictor, that compete in a directed and capacitated network. The user solves an MCP problem, whereas the interdictor decreases arc capacities to reduce the quality of the user's chosen path within a constrained budget. The interdictor may represent a hostile agent or a network failure event that reduces arc capacities. Under the assumption that both agents have complete knowledge of the network topology and parameters, then the agents interact in a Stackelberg game fashion [118], where the interdictor plays first by reducing arc capacities, and then the user determines its maximum capacity path after observing the interdictor's actions. Applications of the MCPIP include information transmission planning (e.g., telecommunication networks using multi-protocol label switching) in the presence of a proactive adversary (i.e., interdictor), whose intention is to reduce the bandwidth or reliability of a subset of links.

2.4.1 (Int, BCP, Bounded, Weighted l_1 Norm)

We first consider the problem $(Int\text{-}BCP_1)$ under weighted l_1 norm, which can be mathematically formulated as follows:

$$
\begin{aligned}
&\min_{\bar{w}} \sum_{e\in E} c(e)(w(e)-\bar{w}(e))\\
(Int\text{-}BCP_1)\ \ s.t.\ &\max_{S\in\mathcal{S}} \min_{e\in S} \bar{w}(e) \le r, \qquad (2.28)\\
&w(e)-l(e) \le \bar{w}(e) \le w(e), \ \ e \in E.
\end{aligned}
$$

For a given value r, let

$$c_r(e) = \begin{cases} c(e)(w(e) - r), & \text{if } w(e) - l(e) \leq r \leq w(e), \\ +\infty, & \text{if } r < w(e) - l(e), \\ 0, & \text{otherwise.} \end{cases} \tag{2.29}$$

Then the problem $(Int\text{-}BCP_1)$ is equivalent to solving the *min cut problem* in the system $(E, \mathcal{K})$ defined by (2.30) whose optimal objective value and optimal set are denoted by $\phi(r)$ and K^r, respectively. Hence, an optimal solution of $(Int\text{-}BCP_1)$ when $\phi(r) < +\infty$ is given by $\bar{\boldsymbol{w}}_r$ below.

$$\phi(r) = \min_{K \in \mathcal{K}} \sum_{e \in K} c_r(e). \tag{2.30}$$

$$\bar{w}_r(e) = \begin{cases} r, & \text{if } e \in K^r \text{ and } r < w(e), \\ w(e), & \text{otherwise.} \end{cases} \tag{2.31}$$

2.4.2 (BC-Int, BCP, Bounded, Weighted l_1 Norm)

We now consider the problem $(BC\text{-}Int\text{-}BCP_1)$ under weighted l_1 norm, which can be mathematically formulated as follows:

$$\begin{aligned} & \min_{\bar{\boldsymbol{w}}} \max_{S \in \mathcal{S}} \min_{e \in S} \bar{w}(e) \\ (BC\text{-}Int\text{-}BCP_1) \quad & s.t. \sum_{e \in E} c(e)(w(e) - \bar{w}(e)) \leq B, \\ & \quad w(e) - l(e) \leq \bar{w}(e) \leq w(e), \quad e \in E. \end{aligned} \tag{2.32}$$

Similar to the solution method for the problem $(BC\text{-}Imp\text{-}BCP_1)$, the problem $(BC\text{-}Int\text{-}BCP_1)$ can be solved by a series of problems $(Int\text{-}BCP_1)$. We first consider the lower bound $r_{\min}$ and the upper bound $r_{\max}$ for the parameter r.

It is obvious that $\max_{S \in \mathcal{S}} \min_{e \in S} \bar{w}_r(e) = r$. Then there exists a feasible set S with $w(e) > r$ for all $e \in S$ and a cut K with $l(e) < r$ for all $e \in K$. Define

$$E_r := \{e \in E | l(e) \leq r\}, \qquad E^r := \{e \in E | w(e) \geq r\};$$
$$r_{\min} := \min\{l(e) | \exists K \in \mathcal{K} : K \subseteq E_{l(e)}\},\ r_{\max} := \min\{w(e) | \exists S \in \mathcal{S} : S \subseteq E^{w(e)}\}.$$

Then a parameter r is feasible if there is set $S \in \mathcal{S}$ with $S \subseteq E^r$ and a cut $K \in \mathcal{K}$ with $K \subseteq E_r$. Furthermore, the feasible parameters are in the range $[r_{\min}, r_{\max}]$ [20].

First, we sort all values in the set $\{w(e) - l(e), w(e) | e \in E\}$, which lie in the feasible range $[r_{\min}, r_{\max}]$ increasingly: $r_{\min} = r_1 < r_2 < \cdots < r_q = r_{\max}$.

Lemma 2.6 ([20, Lemma 2.4]) *The function value $\phi(r)$ is a decreasing, piecewise linear function in $[r_{\min}, r_{\max}]$. In every subinterval $[r_i, r_{i+1}), i = 1, 2, \cdots, q-1$, it is concave and continuous. It may be discontinuous in some value r_i.*

Theorem 2.12 ([20, Key Theorem 2.5]) *The optimum value r^* of $(BC\text{-}Int\text{-}BCP_1)$ is attained by $r^* = \min\left\{r \in [r_{\min}, r_{\max}] \,\middle|\, \min_{K \in \mathcal{K}} \sum_{e \in K} c_r(e) \le B\right\}$. And its optimal solution is given by $\bar{w}_{r^*}$ defined as in (2.31).*

Note that the Key Theorem exhibits a striking similarity between $(BC\text{-}Int\text{-}BCP_1)$ and $(BC\text{-}Imp\text{-}BCP_1)$, whose subproblems are $\min_{S \in \mathcal{S}} \sum_{e \in S} c^r(e)$ and $\min_{K \in \mathcal{K}} \sum_{e \in K} c_r(e)$, respectively.

We can similarly call BN-Algorithm 2.4 to solve the problem $(BC\text{-}Int\text{-}BCP_1)$ by making the following five revisions.

1: In Line 1, arrange all values in the set $\hat{W} := \{w(e) - l(e), w(e) | e \in E\}$ within $[r_{\min}, r_{\max}]$ increasingly: $r_{\min} := \hat{w}_1 < \hat{w}_2 < \cdots < \hat{w}_{k_0} < r_{\max} := \hat{w}_{k_0+1}$.

5: Replace the min-sum problem (2.18) with (2.30).

15: Let $\begin{cases} a(e) := c(e)w(e), & b(e) := c(e), \quad \text{for } w(e) > \hat{w}_{\underline{t}}, \\ a(e) := 0, & b(e) := 0, \quad\quad \text{for } w(e) \le \hat{w}_{\underline{t}}. \end{cases}$

18: Put $a\left(S^k\right) := \sum_{e \in S^k} a(e)$, $b\left(S^k\right) := \sum_{e \in S^k} b(e)$, $r_{k+1} := \left(a\left(S^k\right) - B\right) / \left(b\left(S^k\right)\right)$ and $k := k + 1$.

19: Solve the min-cut problem $h(r_k) = \min_{K \in \mathcal{K}} \sum_{e \in K} (a(e) - b(e) \cdot r_k)$. Denote its optimal set and optimal value, respectively, by S^k and $h(r_k)$.

We point out that the bounded problem $(BC\text{-}Int\text{-}BOP_1)$ under weighted l_1 norm can be solved by a method similar to the earlier analysis.

$$
\begin{aligned}
& \max_{\bar{w}} \min_{S \in \mathcal{S}} \max_{e \in S} \bar{w}(e) \\
(BC\text{-}Int\text{-}BOP_1) \quad s.t. \ & \sum_{e \in E} c(e)(\bar{w}(e) - w(e)) \le B, \\
& w(e) \le \bar{w}(e) \le w(e) + u(e), \quad e \in E.
\end{aligned} \tag{2.33}
$$

2.4.3 *Complexities of the Problems $(BC\text{-}Int\text{-}BCP_1)$ and $(BC\text{-}Imp\text{-}BCP_1)$*

Burkard et al. [20] proposed a polynomial transformation of the *$\mathcal{NP}$-complete vertex cover problem* to the *decision version* of the problem $(BC\text{-}Int\text{-}BCP_1)$.

Theorem 2.13 ([20, Proposition 3.1]) *The class of problems $(BC\text{-}Int\text{-}BCP_1)$ includes some $\mathcal{NP}$-hard problems.*

Theorem 2.14 ([20, Theorem 3.2]) *The problem (BC-Imp-BCP_1) concerning class $\mathcal{S}$ can be solved in polynomial time, if the min-sum problem (2.18) is polynomially solvable. The problem (BC-Int-BCP_1) concerning class $\mathcal{S}$ can be solved in polynomial time, if the problem (2.30) is polynomially solvable.*

It follows from Theorem 2.14 and their dual relationship that concerning class $\mathcal{S}$ one of the problems (BC-Imp-BCP_1)and (BC-Int-BCP_1) is polynomially solvable, whereas the other is $\mathcal{NP}$-hard. For example, consider the system $(V, \mathcal{S})$, where V is the set of vertices in a graph G and the class $\mathcal{S}$ consists of all edges of G. A cut of this system is a vertex cover of G. In this case, the problem (BC-Imp-BCP_1) is polynomially solvable, whereas (BC-Int-BCP_1) is $\mathcal{NP}$-hard. On the other hand, the problem (BC-Imp-BCP_1) concerning class $\mathcal{K}$ of all vertex covers is $\mathcal{NP}$-hard, whereas the corresponding (BC-Int-BCP_1) is polynomially solvable.

When the system $(E, \mathcal{S})$ is the class of all spanning trees of G, the system $(E, \mathcal{K})$ is the class of all cuts $(R, \bar{R})$ of G, which contains all edges $e \in E$ with one endpoint in R and the other endpoint in $\bar{R}$, where $R, \bar{R} \subseteq V, R \cap \bar{R} = \emptyset, R \cup \bar{R} = V$. A minimum cut can be found in $O(mn + n^2 \log n)$ time by an algorithm given by Stoer and Wagner [119].

Theorem 2.15 ([20, Theorem 4.2]) *The BN- algorithm solves the (BC-Int) problem on bottleneck capacity spanning trees in $O(m^2 \log^2 n(mn + n^2 \log n))$ time.*

2.5 Budget-Constrained Interdiction Bottleneck Capacity Problem by Upgrading Nodes

Different from the edge upgrading model discussed in Sects. 2.3–2.4, the node upgrading model [75–77, 79] is to reduce the weights of all edges incident to the node. In communication networks, to improve the communication quality of the network, installing a device with better performance at a node can reduce the communication delay on its adjacent edge. In general, a cost $c(v)$ should be paid for improving each node v by a unit amount. For each edge $e \in E$, we are given three nonnegative values $w_0(e) \geq w_1(e) \geq w_2(e) \geq 0$, where $w_i(e)$ represents the weight of the edge e if exactly $i (i = 0, 1, 2)$ of its endpoints are upgraded. The edge weight vector w_Ω resulting from upgrading the node set Ω is calculated by (1.22). Therefore, the corresponding interdiction problem (*Int-Node-BCP*) and the Budget-Constrained interdiction problem (*BC-Int-Node-BCP*) by upgrading nodes can be formulated by (2.34) and (2.35), respectively.

$$\min_{\Omega \subseteq V} \sum_{v \in \Omega} c(v)$$

$$(\textit{Int-Node-BCP}) \quad s.t. \quad \max_{S \in \mathcal{S}} \min_{e \in S} \boldsymbol{w}_\Omega(S) \leq D. \tag{2.34}$$

$$\begin{aligned} &\min_{\Omega\subseteq V}\ \max_{S\in\mathcal{S}}\min_{e\in S} \boldsymbol{w}_{\Omega}(S) \\ (BC\text{-}Int\text{-}Node\text{-}BCP)\ \ &s.t.\ \sum_{v\in\Omega} c(v)\leq B. \end{aligned} \tag{2.35}$$

Recall that system $(V, E, \mathcal{S})$ includes a lot of combinatorial problems. A special case is that $\mathcal{S}$ is just the edge set E, which means each set $S = \{e\}, \forall e \in E$. We showed that it is $\mathcal{NP}$-hard by presenting a polynomial transformation from the weighted node cover problem to the decision version of the case.

Lemma 2.7 ([57, Proposition 2.1]) *The problem (BC-Int-Node-BCP) includes some strongly* ***NP****-hard cases, even if* $|S| = 1$ *for all* $S \in \mathcal{S}$.

Lemma 2.8 ([57, Proposition 2.2]) *The problem (Int-Node-BCP) includes some* ***MaxSNP****-hard cases, even if* $|S| = 1$ *for all* $S \in \mathcal{S}$.

Given a value D, we can partition the edge set E into four subsets according to how many endpoints of the edge must be upgraded in order to decrease its weight to be below the threshold D [76]. An edge e is *uncritical* if $w_0(e) \leq D$, *1-critical* if $w_0(e) > D \geq w_1(e)$, *2-critical* if $w_1(e) > D \geq w_2(e)$ and *useless* if $w_2(e) > D$. Denote by E^1, E^2, E^3 the set of 1-critical, 2-critical, and useless edges, respectively.

2.5.1 A Linear Time Algorithm for Problem (Int-Node-BCP) on a Tree-path System

A tree-path system $(V, E, \mathcal{S})$ is defined as follows: $T = (V, E)$ is a tree rooted at v_0, and $\mathcal{S}$ is the set of paths from root v_0 to all the leaves of T. Let Y be the set of all the leaf nodes and $A(v) := \{u|(v, u) \in E\}$ be the set of child nodes of v.

Next we proposed a dynamic programming method to solve the problem in linear time. Let T_v be a subtree rooted at v including v and all its descendants. Let $\Omega^+(v)$ and $\Omega^-(v)$ be the minimum cost upgrading set in T_v that includes and excludes v, respectively. For simplicity, let $f^+(v) := c(\Omega^+(v))$, $f^-(v) := c(\Omega^-(v))$. Then the minimum cost upgrading set $\Omega(v)$ for the subtree T_v can be determined by

$$\Omega(v) := \begin{cases} \Omega^+(v), & \text{if } f^+(v) \leq f^-(v), \\ \Omega^-(v), & \text{otherwise.} \end{cases} \tag{2.36}$$

In order to find the minimum cost upgrading set $\Omega(v_0)$ for a tree-path system, we need to calculate $\min\{f^-(v_0), f^+(v_0)\}$.

Algorithm 2.7 Solve (*Int-Node-BCP*) on a tree-path system [57, Algorithm 1]

Input: A tree T rooted at v_0 with three edge weight vectors w_0, w_1, and w_2, a node cost vector c, and a positive value D.

Output: minimum cost upgrading set $\Omega(v_0)$ and its cost $c(\Omega(v_0))$.

Let $E^1 := \{e|w_1(e) \leq D < w_0(e)\}$, $E^2 := \{e|w_2(e) \leq D < w_1(e)\}$ and $E^3 := \{e|w_2(e) > D\}$.

Run a Breadth-First-Search starting from the root v_0, and let the ordered node set be V_0 after deleting its leaves.

for $v \in Y$ **do**

  Initialize $f^+(v) := +\infty$, $S^+(v) := \emptyset$, $f^-(v) := +\infty$, $S^-(v) := \emptyset$.

end for

while $V_0 \neq \emptyset$ **do**

  Choose the last node v in V_0.

  if $\exists(v, u) \in E^3$ with $u \in Y$ **then** $f^+(v) := +\infty$ and $\Omega^+(v) := \emptyset$.

  else

$$f^+(v) := c(v) + \sum_{(v,u)\in E^2} \min\{c(u), f^-(u)\} + \sum_{u\notin Y:(v,u)\in E^3} \min\{f^+(u), f^-(u)\},$$

$$\Omega^+(v) := \{v\} \cup \bigcup_{(v,u)\in E^2} \widehat{\Omega}(u) \cup \bigcup_{u\notin Y:(v,u)\in E^3} \Omega(u);$$

  end if

  if $\exists(v, u) \in E^3 \cup E^2$ with $u \in Y$ **then** $f^-(v) := +\infty$, $\Omega^-(v) := \emptyset$,

  else

$$f^-(v) := \sum_{(v,u)\in E^1} \min\{c(u), f^-(u)\} + \sum_{u\notin Y:(v,u)\in E^3\cup E^2} \min\{f^+(u), f^-(u)\},$$

$$\Omega^-(v) := \bigcup_{(v,u)\in E^1} \widehat{\Omega}(u) \cup \bigcup_{u\notin Y:(v,u)\in E^3\cup E^2} \Omega(u);$$

$$\widehat{\Omega}(u) := \begin{cases} \{u\}, & if\ c(u) \leq f^-(u), \\ \Omega^-(u), & if\ c(u) > f^-(u). \end{cases}$$

  end if

  Update $V_0 := V_0 - \{v\}$.

end while

Let $c(\Omega(v_0)) := \min\{f^-(v_0), f^+(v_0)\}$.

if $c(\Omega(v_0)) = +\infty$ **then** the instance is infeasible and $\Omega(v_0) := \emptyset$.

end if

Theorem 2.16 ([57, Theorem 3.2]) *The problem (Int-Node-BCP) on a tree-path system can be solved in $O(n)$ operations by Algorithm 2.7.*

We further extend the problem (*Int-Node-BCP*) on a tree-path system to a tree with multiple sources and multiple terminals in [54, 59]. Let $T = (V, E)$ be an undirected tree satisfying $V_s \cup V_t \supseteq Y$, where V_s and V_t are the sets of sources and terminals, respectively. Assume that there are no edges connecting two sources or two terminals. We call $\overrightarrow{T} = (V, \overrightarrow{E})$ an orientated tree of T if it satisfies the conditions as follows: (i) each source $s_i \in V_s$ has only out-arcs and each terminal $t_j \in V_t$ has only in-arcs; (ii) $\overrightarrow{T}$ is a union of directed paths (not necessary disjoint) from each source to all its reachable terminals (not necessary from every source to every terminal) and each arc of $\overrightarrow{E}$ must be lied on one of such paths. We call $\overrightarrow{T} = (V, \overrightarrow{E})$ a directed tree with multiple sources and multiple terminals (*MSMT tree* for simplicity), and we omit the arrow over T and E in the sequel.

We showed that the problem (*Int-Node-BCP*) on an *MSMT tree* is $\mathcal{NP}$-hard and $\mathcal{MAXSNP}$-hard for $K \geq 2$ on general directed graphs, where K is the number of source- terminal pairs. We proposed a general algorithm with time complexity $O(K \cdot n + 2^K \cdot (K^2 + n))$ for the problem on MSMT trees in [59], which is a polynomial algorithm when the number K is upper bounded by a given value. Thus, we turned to consider a special case when there are no two-critical edges and proposed two polynomial algorithms with time complexities $O(n^2 \log n)$ and $O(n^2)$ when $c(v)$ is a general cost or not, respectively [54]. These complexities only concern the number n of nodes and are irrespective of the number K.

2.5.2 An $O(n \log n)$ Algorithm for Problem (BC-Int-Node-BCP) on a Tree-Path System

There is a close relationship between the algorithms for problems (2.34) and (2.35). We can solve problem (2.35) by using a binary search over an appropriate set of admissible values for a series of problems (2.34), which is a common technique for treating problems when the objectives are interchanged [4].

Now we consider the admissible values of the optimal objective value OPT of problem (2.35). Then OPT is an integer satisfying $w_2(S) \leq OPT \leq w_0(S)$ in the set $\mathcal{D} := \{w_0(e), w_1(e), w_2(e) | e \in E\}$.

Algorithm 2.8 Solve *(BC-Int-Node-BCP)* on a tree-path system [57, Algorithm 2]

Input: A tree T rooted at v_0 with three edge weight vectors w_0, w_1, and w_2, a node cost vector c, and a positive value B.

Output: the bottleneck weight D^b and the upgrading set Ω^b.

1: Let $\mathcal{D} := \{w_0(e), w_1(e), w_2(e) | e \in E\}$. Delete the values that are greater than $w_0(S)$ or smaller than $w_2(S)$ from $\mathcal{D}$. Sort $\mathcal{D}$ nondecreasingly, and denote the k-th element in $\mathcal{D}$ by D^k.

2: Put $a := 1, b := |\mathcal{D}|$.

while $b - a > 1$ **do**
    put $k := \lfloor (a+b)/2 \rfloor$.
    Call Algorithm 2.7 to solve problem (2.34) with respect to the value D^k and let Ω^k be the minimum cost upgrading set.
    if $c(\Omega^k) > B$ **then** put $a := k$;
    else put $b := k$.
    end if
end while

Theorem 2.17 ([57, Theorem 4.1]) *The problem (BC-Int-Node-BCP) on a tree-path system can be solved in $O(n \log n)$ operations by Algorithm 2.8.*

2.6 Inverse Constrained Bottleneck Problems on Networks

In a *Constrained Bottleneck Problem (CBP)* on a network $G = (V, E)$, two weights $w^1(e)$ and $w^2(e)$ are defined for each edge e. The objective is to find a feasible solution $S \in \mathcal{S}$ that minimizes the w^1-bottleneck weight subject to a bounded constraint on the w^2-sum weight, which is formulated below.

$$\begin{aligned} &\min_{S \in \mathcal{S}} \max_{e \in S} w^1(e) \\ (CBP)\ &\text{s.t.} \sum_{e \in S} w^2(e) \leq B. \end{aligned} \tag{2.37}$$

We focused on the inverse problem $(ICBP_\infty)$ under weighted l_∞ norm by finding two vectors $\bar{w}^1$ and $\bar{w}^2$ to make a given set S^0 an optimal set. Its mathematical model can be formulated as follows:

$$\begin{aligned} &\min \max\{\max_{e \in E}\{c^1(e)|\bar{w}^1(e) - w^1(e)|\}, \max_{e \in E}\{c^2(e)|\bar{w}^2(e) - w^2(e)|\}\} \\ &\text{s.t. } \max_{e \in S^0} \bar{w}^1(e) \leq \max_{e \in S} \bar{w}^1(e),\ \forall S \in \mathcal{S}, \\ (ICBP_\infty)\quad &\sum_{e \in S^0} \bar{w}^2(e) \leq B, \\ &w^1 - l^1 \leq \bar{w}^1 \leq w^1 + u^1, w^2 - l^2 \leq \bar{w}^2 \leq w^2 + u^2. \end{aligned} \tag{2.38}$$

The problem *(ICBP)* on *ST* denoted by *(ICBST)* has applications in communication networks and transportation networks [58]. In a communication network G, we denote by $r(e)$ $(0 < r(e) \leq 1)$ a link reliability and by $t(e)$ a transit time for each link e. Given a spanning tree T^0, we aim to modify the reliability and transit time vectors such that the maximum transit time of T^0 is minimum while the reliability $\prod_{e \in T^0} r(e)$ is at least R, while the modification is as small as possible under weighted l_∞ norm. This is an inverse problem (*ICBST*) when choosing $w^1(e) := t(e)$, $w^2(e) := -\log(r(e))$, and $B := -\log(R)$. In transportation networks, let $w^1(e)$ and $w^2(e)$ be a construction time and cost, respectively. Assume that all edges are constructed simultaneously. We aim to modify the vectors as small as possible such that T^0 can be constructed in the least amount of time satisfying a prescribed budget B on the construction cost under the new parameters. If we want to reduce the construction time, then we need more construction costs to employ more workers and use more machines. Hence, it is meaningful to consider the case that the modifications of weight vectors w^1 and w^2 are proportional as in

$$\bar{w}_p^2(e) - w^2(e) = \Delta c(e)(w^1(e) - \bar{w}_p^1(e)), \ \forall e \in E. \tag{2.39}$$

2.6.1 The Problem (ICBP$_\infty$) Under the Proportional Case

We use the notation $\bar{w}^1(S) = \max_{e \in S} \bar{w}^1(e)$. Taking the bounds and proportional conditions on the modifications of weights into consideration, let $\alpha(e) := \max\{w^1(e) - l^1(e), w^1(e) - \frac{u^2(e)}{\Delta c(e)}\}$ and $\beta(e) := \min\{w^1(e) + u^1(e), w^1(e) + \frac{l^2(e)}{\Delta c(e)}\}$, then the possible value of $\bar{w}^1(S^0)$ is in the interval $[\alpha(S^0), \beta(S^0)]$. Hence, we need to solve a sub-problem, Inverse Optimal Value Problem on (*CBP*), denoted by (*IOVCBP*(p)), with respect to a value $p \in [\alpha(S^0), \beta(S^0)]$ to find the weight vector $(\bar{w}_p^1, \bar{w}_p^2)$ satisfying the following problem, where $\rho^1(e) := \max\{c^1(e), \Delta c(e) c^2(e)\}$.

$$\begin{aligned}
&C(p) := \min \max_{e \in E} \rho^1(e)|\bar{w}_p^1(e) - w^1(e)| \\
&\text{s.t.} \quad \max_{e \in S^0} \bar{w}_p^1(e) = p \le \max_{e \in S} \bar{w}_p^1(e), \ \forall S \in \mathcal{S}, \\
(IOVCBP(p)) \quad &\sum_{e \in S^0} \bar{w}_p^2(e) \le B, \\
&\bar{w}_p^2(e) - w^2(e) = \Delta c(e)(w^1(e) - \bar{w}_p^1(e)), \ \forall e \in E, \\
&\alpha(e) \le \bar{w}_p^1(e) \le \beta(e), \ \forall e \in E.
\end{aligned}$$

Therefore, we will determine an optimal solution to the inverse problem (*ICBP*) by the properties of function $C(p)$ on the interval $[\alpha(S^0), \beta(S^0)]$.

2.6.2 Solve the Inverse Optimal Value Problem $(IOVCBP(p))$

We summarized the proposed method in [58] for the problem $(IOVCBP(p))$ in the following algorithm.

Algorithm 2.9 An algorithm to solve the problem $(IOVCBP(p))$ [58]

Input: A system $(V, E, \mathcal{S})$ with six edge weight vectors w^1, w^2, ρ^1, $\Delta c, \alpha$, and β, a given structure S^0 and two positive values B, p.
Output: the weight vectors $\bar{w}_p^1$, $\bar{w}_p^2$ and the cost $C(p)$.

1: Let $E^+(p) := \{e \in E | w^1(e) \geq p\}$.

2: **if** $E^+(p)$ is a cut of $\mathcal{F}$ **then** let $S^0(p) := \{e \in S^0 | w^1(e) \geq p\}$, $c_p^0(S^0(p)) := \max_{e\in S^0(p)} \rho^1(e)(w^1(e) - p)$ and

$$\begin{cases} w_p^1(e) := p, & w_p^2(e) := w^2(e) + \Delta c(e)(w^1(e) - p), e \in S^0(p), \\ w_p^1(e) := w^1(e), \; w_p^2(e) := w^2(e), & e \in E \backslash S^0(p). \end{cases}$$

3: **if** $\sum_{e\in S^0} w_p^2(e) \leq B$ **then return** $\bar{w}_p^1 := w_p^1, \bar{w}_p^2 := w_p^2$ and $C(p) := c_p^0(S^0(p))$.

4: **else** let $S_<^0(p) := \{e \in S^0 | w^1(e) < p\}$, $B_p := B - \sum_{e\in S^0(p)} w_p^2(e)$, $\beta_p(e) := \min\{p, \beta(e)\}$, $\mu_p(e) := w^2(e) - \Delta c(e)(\beta_p(e) - w^1(e))$ and $\rho^2(e) := \max\{c^2(e), \frac{c^1(e)}{\Delta c(e)}\}$.

5: **if** $\sum_{e\in S^0(p)} \mu_p(e) > B_p$ **then return** "the instance is infeasible".

6: **else** Sort the costs $\{\rho^2(e)(w^2(e) - \mu_p(e)) | e \in S^0(p)\}$ in a strictly increasing order and denote the ordered set by $\mathrm{R}(p) := \{r_1, \cdots, r_i, \cdots, r_\tau\}$, where $\tau \leq |S^0(p)|$. For any value x, let $S_p^0(x) := \{e \in S_<^0(p) | \rho^2(e)(w^2(e) - \mu_p(e)) \leq x\}$ and $h_p(x) = \sum_{e\in S_p^0(x)} \mu_p(e) + \sum_{e\in S_<^0(p)\backslash S_p^0(x)} (w^2(e) - x/\rho^2(e))$.

7: Solve the Inverse Constraint Problem ($\mathrm{ICP}_\infty(p)$) under weighted l_∞ norm with value p by a binary search in $\mathrm{R}(p)$ to find a subinterval $[r_i, r_{i+1})$ such that $h_p(r_i) > B_p$ and $h_p(r_{i+1}) \leq B_p$.

$$(ICP_\infty(p)) \quad \begin{aligned} \min \; & \max_{e\in S_<^0(p)} \rho^2(e)(w^2(e) - \widehat{w}_p^2(e)) \\ \text{s.t.} \; & \sum_{e\in S_<^0(p)} \widehat{w}_p^2(e) \leq B_p, \\ & \widehat{w}_p^2(e) \geq \mu_p(e), \; \forall e \in S_<^0(p), \end{aligned} \tag{2.40}$$

$$\hat{c}_p(S^0_<(p)) := \begin{cases} \dfrac{\sum\limits_{e\in S^0_<(p)} w^2(e)-B_p}{\sum\limits_{e\in S^0_<(p)} \frac{1}{\rho^2(e)}}, & \text{if } h_p(r_1) < B_p, \\ r_1, & \text{if } h_p(r_1) = B_p, \\ r_{i+1}, & \text{if } h_p(r_{i+1}) = B_p, \\ \dfrac{\sum\limits_{e\in S^0_p(r_i)} \mu_p(e) + \sum\limits_{e\in S^0_<(p)\setminus S^0_p(r_i)} w^2(e)-B_p}{\sum\limits_{e\in S^0_<(p)\setminus S^0_p(r_i)} \frac{1}{\rho^2(e)}}, & \text{if } h_p(r_i) > B_p > h_p(r_{i+1}), \end{cases}$$

$$\widehat{w}^2_p(e) := \max\{\mu_p(e), w^2(e) - \hat{c}_p(S^0_<(p))/\rho^2(e)\}, \forall e \in S^0_<(p). \tag{2.41}$$

Return $C(p) := \max\{\hat{c}_p(S^0_<(p)), c^0_p(S^0(p))\}$ and

$$\begin{cases} \bar{w}^2_p(e) := \widehat{w}^2_p(e), \ \bar{w}^1_p(e) := w^1(e) + \frac{1}{\Delta c(e)}(w^2(e) - \widehat{w}^2_p(e)), \ \forall e \in S^0_<(p) \\ \bar{w}^2_p(e) := w^2_p(e), \ \bar{w}^1_p(e) := w^1_p(e), \quad \textit{otherwise}, \end{cases}$$

end if

end if

else let $E(p) := \{e \in E | w^1(e) < p\}$, $\mathcal{S}(p) := \{S \in \mathcal{S} | S \subseteq E(p)\}$, $\overline{E}(p) := \{e \in E(p) | p \le \beta(e)\}$ and

$$c_p(e) := \begin{cases} \rho^1(e)(p - w^1(e)), & \forall e \in \overline{E}(p), \\ +\infty, & \forall e \in E(p)\setminus\overline{E}(p). \end{cases} \tag{2.42}$$

Solve the Bottleneck Cut $K(p)$ of family $\mathcal{S}(p)$ under the capacity vector c_p.

$$\min_K \max_{e\in K} c_p(e)$$

$$(BC(p)) \text{ s.t. } K \subseteq E(p) \text{ is a cut of family } \mathcal{S}(p).$$

if $c_p(K(p)) := \max_{e\in K(p)} c_p(e) = +\infty$ **then**

Return the instance is infeasible.

else let $\widetilde{C}(p) := \max\{c^0_p(S^0(p)), c_p(K(p))\}$,

$$\tilde{w}^2_p(e) := \begin{cases} w^2(e) - \Delta c(e)(p - w^1(e)), \ \forall e \in K(p), \\ \max\{\mu_p(e), w^2(e) - \frac{\widetilde{C}(p)}{\rho^2(e)}\}, \ \forall e \in S^0_<(p)\setminus K(p). \end{cases}$$

if $\sum_{e\in S^0_<(p)} \tilde{w}^2_p(e) \le B_p$ **then**

Return $\bar{w}_p^2(e) := \begin{cases} \tilde{w}_p^2(e), \forall e \in S_<^0(p) \cup K(p) \\ w_p^2(e), \text{ for other edges} \end{cases}$, $\bar{w}_p^1(e) := w^1(e) + \frac{w^2(e)-\bar{w}_p^2(e)}{\Delta c(e)}$ and $C(p) := \widetilde{C}(p)$.
else solve an inverse constraint problem (ICP$_\infty(p)$) defined as in (2.40) by replacing $S_<^0(p)$ with $S_<^0(p)\backslash K(p)$ and B_p with $B_p' := B_p - \sum_{e \in S_<^0(p) \cap K(p)} \tilde{w}_p^2(e)$. Let the optimal solution obtained by (2.41) be $\widehat{w}_p^2(e)$ for any $e \in S_<^0(p)\backslash K(p)$.
Return $C(p) := \max_{e \in S_<^0(p)\backslash K(p)} \rho^1(e)(\bar{w}_p^1(e) - w^1(e))$ and

$$\begin{cases} \bar{w}_p^1(e) := p, & \bar{w}_p^2(e) := w^2(e) - \Delta c(e)(p - w^1(e)), \forall e \in K(p), \\ \bar{w}_p^2(e) := \widehat{w}_p^2(e), & \bar{w}_p^1(e) := w^1(e) + \frac{w^2(e)-\widehat{w}_p^2(e)}{\Delta c(e)}, \forall e \in S_<^0(p)\backslash K(p), \\ \bar{w}_p^1(e) := p, & \bar{w}_p^2(e) := w^2(e) + \Delta c(e)(w^1(e) - p), \forall e \in S^0(p), \\ \bar{w}_p^1(e) := w^1(e), & \bar{w}_p^2(e) := w^2(e), \text{for other edges.} \end{cases}$$

end if
end if
end if

2.6.3 *Solve the Inverse Problem ($ICBP_\infty$)*

Algorithm 2.10 An algorithm for the problem (ICBP$_\infty$) [58]

Input: A system (V, E, S) with 9 edge weight vectors $w^1, w^2, l^1, l^2, u^1, u^2, c^1, c^2$, and Δc, a given structure S^0 and a positive value B.
Output: the weight vectors $\bar{w}^1$, $\bar{w}^2$ and the objective value C.
Let $\alpha(e) := \max\{w^1(e) - l^1(e), w^1(e) - \frac{u^2(e)}{\Delta c(e)}\}$, $\beta(e) := \min\{w^1(e) + u^1(e), w^1(e) + \frac{l^2(e)}{\Delta c(e)}\}$ and $\rho^1(e) := \max\{c^1(e), \Delta c(e)c^2(e)\}$.
Let IP be the set of all intersection points of these linear functions including $\{\rho^1(e)(w^1(e) - p)|e \in S^0\} \cup \{\rho^1(e)(p - w^1(e))|e \in E\}$.
Sort these values $(\{w^1(e)|e \in E\} \cup \{w^1(e)+u^1(e)|e \in E\} \cup \{w^1(e)+\frac{l^2(e)}{\Delta c(e)}|e \in E\} \cup \{\alpha(S^0)\} \cup IE) \cap [\alpha(S^0), \beta(S^0)]$ in a strictly increasing order, say $\alpha(S^0) = q_1 < q_2 < \cdots < q_t = \beta(S^0)$, where $t = O(|E|^2) = O(m^2)$.
Put $a := 1, b := t$.
while $b - a > 1$ **do**
put $k := \lfloor (a + b)/2 \rfloor$.
Call Algorithm 2.9 to solve the inverse optimal value problem (*IOVCBP*(q_k)) with respect to the value q_k and let $C(q_k)$ be the cost.
if $C(q_k) = +\infty$ **then** put $b := k$;
else put $a := k$.
end if

11: **end while**
12: Initialize $C := +\infty$.
13: **for** $i = 1$ to b **do**
14: Call Algorithm 2.9 to solve the inverse optimal value problem (*IOVCBP*(q_i)) and let $C(q_i)$ be the cost.
15: **if** $C(q_i) < C$ **then** update $C := C(q_i)$, $\bar{w}^1 := \bar{w}^1_{q_i}$, $\bar{w}^2 := \bar{w}^2_{q_i}$
16: **end if**
17: **end for**

Theorem 2.18 ([58, Theorem 6]) *The inverse constrained bottleneck problem under weighted l_∞ norm and the proportional condition (2.39) can be solved by Algorithm 2.10 in $O(m^2(|S|\log|S| + T_{bc}))$, where $O(|S|\log|S|)$ and T_{bc} are the complexities for the inverse constrained problem (ICP_∞) and the bottleneck cut problem, respectively.*

Corollary 2.1 ([58]) *The inverse constrained bottleneck spanning tree problem can be solved in $O(m^2(m + n\log n))$ time. The inverse constrained maximum capacity path problem can be solved in $O(m^3\log m)$ time. The inverse constrained bottleneck assignment problem can be solved in $O(m^2n^3\log m)$ time.*

We can also consider the problem under weighted l_1 norm, denoted by (ICBP_1), where the objective function is $\min\sum_{e\in E}[c^1(e)\,|\bar{w}^1(e) - w^1(e)| + c^2(e)|\bar{w}^2(e) - w^2(e)|]$. Similar to the case of l_∞ norm, the main problem to consider is the inverse optimal value problem ($\text{IOVCBP}_1(p)$) when $E^+(p)$ is not a cut of $\mathcal{S}$, which is equivalent to solving a minimum constrained cut problem defined as follows:

$$\begin{aligned}
\min\ & \sum\nolimits_{e\in E(p)}(c^2(e) + c^1(e)/\Delta c(e))(w^2(e) - \bar{w}^2_p(e))\\
\text{s.t.}\ & \widetilde{E}(p) := \{e \in \overline{E}(p)|\bar{w}^2_p(e) = w^2(e) - \Delta c(e)(p - w^1(e))\}\ \text{is a cut of}\ \mathcal{S}(p),\\
& \sum\nolimits_{e\in S^0_<(p)} \bar{w}^2_p(e) \le B_p, \qquad (2.43)\\
& \bar{w}^2_p(e) \ge \mu_p(e),\ \forall e \in E(p).
\end{aligned}$$

Let $K(p) \subseteq \overline{E}(p)$ be the minimum cut of $\mathcal{S}(p)$ under the capacity function c_p, which is defined by (2.42) by replacing $\rho^1(e)$ with $c^1(e) + \Delta c(e)c^2(e)$. Note that when we reduce the weights $w^2(e)$ on edges $e \in S^0_<(p)\backslash K(p)$, $(S^0_<(p) \cup K(p))\cap\widetilde{E}(p)$ may contain other cuts of $\mathcal{S}(p)$ except $K(p)$, thus we may obtain less modification costs to satisfy all the constraints in (2.43). This is the main dilemma to solve such a problem. Therefore, it is difficult to present a general method, and we hope to obtain some specific method for a concrete problem based on its combinatorial properties in future research.

Remark 2.4 (Open Problem) It is an open problem to devise a general algorithm to solve the inverse constrained bottleneck problem under weighted l_1 norm and the proportional condition (2.39).

2.7 Conclusions

This chapter delves into the intricate domain of Generalized Inverse Bottleneck Optimization Problems (GIBOPs). It is a testament to the complexity and richness of inverse bottleneck optimization within combinatorial optimization. We have meticulously explored various facets of GIBOPs, providing a thorough analysis of their structural properties and developing algorithms tailored to address the unique challenges posed by these problems (Table 2.1).

The chapter begins by defining the concept of a bottleneck optimization problem and extends this to the inverse setting, where the goal is to determine the parameters of the model that make a given solution optimal. The authors present a systematic approach to formulating GIBOPs, highlighting the importance of understanding the dual relationships between systems and cuts in network optimization.

Table 2.1 The outline of generalized inverse bottleneck optimization problems

Class of GIBOPs	Norm	Bound	*Transformed Problem*	References
IOV	*weighted* l_1	*Bounded*	*Minimum Cut*	[56, 133] 2.1.1
	weighted sum-Ham	*Bounded*	*Minimum Cut*	[56] 2.1.1
	weighted l_∞	*Bounded*	*Bottleneck Cut*	[58, 133] 2.1.2
	bottleneck-Ham	*Bounded*	*Minimum Cut*	[56] 2.1.3
Inv	*weighted* l_1	*Bounded*	$O(m)$ *Min Cuts*	[56, 133] 2.1.1
	weighted sum-Ham	*Bounded*	$O(m)$ *Min Cuts*	[56, 89] 2.1.1
	weighted l_∞	*Bounded*	$O(m^2)$ *Bot Cuts*	[58, 133] 2.1.2
	weighted l_∞	*Unbounded*	$O(m)$ *Bot Cuts*	[133] 2.1.2
	unit l_∞	*Unbounded*	*One Bot Cut*	[133] 2.1.2
	bottleneck-Ham	*Bounded*	$O(\log m)$ *Cut Ver.*	[56] 2.1.3
BC-Inv	*bottleneck-Ham*	Bounded	$O(m \log m)$ *Min Cuts*	[49] 2.2
	weighted sum-Ham			
Imp-BCP	weighted l_1	*Bounded*	*One Min Sum*	[145]
BC-Imp-BCP	*weighted* l_1	*Unbounded*	$O(m^2 \log m)$ *Min Sum*	[145]
		Bounded		2.3.1
	weighted l_∞	*Bounded*	*One BCP*	2.3.2
BC-Imp-SCP	*weighted* l_1	*Unbounded*	$O(m)$ *SCPs*	[145]
		Bounded		2.3.3
	weighted l_∞	*Bounded*	*One SCP*	2.3.4
Int-BCP	weighted l_1	*Bounded*	*One Min Cut*	[20] 2.4.1
BC-Int-BCP	weighted l_1	*Bounded*	$O(m^2 \log m)$ *Min Cut*	[20] 2.4.2
Int-N-BCP	weighted l_1	*Bounded*	$O(n)$ *DP*	[57] 2.5.1
BC-Int-N-BCP	weighted l_1	*Bounded*	$O(n \log n)$ *BSM*	[57] 2.5.2
IOV-CBP	weighted l_∞	*Bounded*	*1 ICP + 1 Bot Cut*	[58] 2.6.2
Inv-CBP	weighted l_∞	*Bounded*	$O(m^2)$ *IOV-CBPs*	[58] 2.6.3

A significant portion of the chapter is dedicated to the development of solution methodologies for GIBOPs under different norm constraints, such as the weighted ℓ_1 norm, weighted ℓ_∞ norm, and Hamming distance. The algorithms proposed are grounded in sound mathematical principles and are supported by rigorous complexity analyses. The authors demonstrate the applicability of these algorithms through various examples and case studies, showcasing their efficacy in solving real-world problems.

The chapter also addresses the computational complexity of GIBOPs, identifying conditions under which these problems are polynomially solvable and those that lead to $\mathcal{NP}$-hard scenarios. This discussion is pivotal for understanding the limitations and potential of current computational methods in tackling inverse bottleneck problems.

Throughout this chapter, we emphasize the practical implications of GIBOPs, illustrating their relevance in diverse fields such as network design, logistics, and supply chain management. The chapter concludes with a series of open problems and future research directions, setting the stage for further exploration and advancement in this area of study.

Chapter 3
Generalized Inverse Maximum Capacity Path Problems

Abstract In this chapter, we investigate the restricted inverse optimal value problem on the maximum capacity path under weighted l_1 and l_∞ norms and address the interdiction problem with bounded capacity adjustments under the l_1 norm. Without altering the capacity values of given paths, we modify the algorithm (**AIMCPM**) proposed in Deaconu and Tayyebi (IEEE Access 8:225957–225966;2020). For these problems, we first present the mathematical models, then design solution algorithms, prove the optimality of the algorithms, and analyze their time complexity. The interdiction problem is addressed with a binary search method, resulting in an $O(m^2 \log m)$ time complexity algorithm, which greatly improved the algorithm's complexity proposed in Deaconu and Tayyebi (IEEE Access 8:225957–225966;2020). For the restricted inverse optimal value problem, it is tackled with a minimum cost cut approach in $O(mn)$ time under l_1 norm, while by finding maximum capacity paths in $O(m)$ time under l_∞ norm.

Keywords Interdiction maximum capacity path problem · Restricted inverse optimal value problem · Binary search method · Minimum cost cut · Weighted l_1 norm · Weighted l_∞ norm

3.1 (Int, Maximum Capacity Path, Bounded, Weighted l_1 Norm)

3.1.1 Introduction

Let $G = (V, E, w)$ be a directed weighted network with a source s and a terminal t, and w is a capacity vector defined on the edge set E. We introduce a variable x_{ij} to denote the reduction of capacity of e_{ij}. Let l_{ij} be the lower bound of the updated capacity. Given a budget B, the Interdiction problem on Maximum Capacity Path (**IntMCP**) aims at minimizing the capacity of the network by reducing the capacities

Junhua Jia, Wenjing Du, and Xiucui Guan have equally contributed to this chapter.

X. Guan et al., *Inverse Combinatorial Optimization Problems*, Springer Optimization and Its Applications 225, https://doi.org/10.1007/978-3-031-91175-0_3

of edges as much as possible under weighted l_1 norm within a total cost B.

$$\begin{aligned}
&\min \max_{P \in \mathcal{P}} \min_{e_{ij} \in P} (w_{ij} - x_{ij}) \\
&s.t. \sum_{(i,j) \in E} c_{ij} x_{ij} \leq B, \\
&\quad\; 0 \leq x_{ij} \leq w_{ij} - l_{ij}, \quad \forall (i,j) \in E.
\end{aligned}$$

where $\mathcal{P}$ is the set of all $s - t$ paths and c_{ij} is the cost required to adjust one unit of w_{ij}. Based on the mathematical model given by Tayyebi et al. [125], we can formulate the problem **(IntMCP)** as follows.

$$\textbf{(IntMCP)} \quad \begin{aligned}
&\min z && (3.1)\\
&s.t. \sum_{(i,j) \in E} c_{ij} x_{ij} \leq B, && (3.2)\\
&\quad 0 \leq x_{ij} \leq w_{ij} - l_{ij}, \quad \forall (i,j) \in E, && (3.3)\\
&\quad z = \max\ w_{\min}, && (3.4)\\
&\quad s.t.\ \ w_{ij} - x_{ij} + M_{ij}(1 - y_{ij}) \geq w_{\min}, \quad \forall (i,j) \in E, && (3.5)\\
&\qquad \sum_{j:(i,j) \in E} y_{ij} - \sum_{j:(j,i) \in E} y_{ji} = \begin{cases} 1, & i = s, \\ 0, & i \in V \setminus \{s,t\}, \\ -1, & i = t, \end{cases} && (3.6)\\
&\qquad y_{ij} \in \{0,1\}, \quad \forall (i,j) \in E. && (3.7)
\end{aligned}$$

Notice that we imposed a lower bound l_{ij} for the updated capacity in (3.3); however, the lower bound is 0 in the model (3c) considered in [125]. Hence, our model is more general.

Definition 3.1 In a weighted directed connected graph $G = (N, E, c, w, s, t)$, an **s-t cut** $[X, \bar{X}]$ is the division of the vertex set N into two sets X and $\bar{X}$, where both X and $\bar{X}$ are non-empty, and $s \in X, t \in \bar{X}$. The set of forward arcs is $(X, \bar{X}) = \{e = (i,j) \in E | i \in X, j \in \bar{X}\}$, and the set of backward arcs is $(\bar{X}, X) = \{e = (i,j) \in E | i \in \bar{X}, j \in X\}$. An $s-t$ cut $[X, \bar{X}] = (X, \bar{X}) \cup (\bar{X}, X)$ is the union of the forward and backward arc sets. Define the **capacity** $c(X, \bar{X}) = \sum_{e \in (X,\bar{X})} c(e)$ of an **s-t cut** as the sum of the capacities of the forward arcs in the cut, where $c(e)$ is the capacity of arc e. Similarly, define the **adjustment cost** of an **s-t cut** as the sum of the costs of the forward arcs, that is $\alpha(X, \bar{X}) = \sum_{e \in (X,\bar{X})} \alpha(e)$, where $\alpha(e) = w(e)d(e)$ is the adjustment cost of arc e when the adjustment amount is $d(e)$ and the unit adjustment cost is $w(e)$. A minimum capacity or cost **s-t cut** refers to an **s-t cut** whose capacity or cost is the smallest among all **s-t cuts**, also known as the **minimum cut problem**.

The minimum cut problem can be solved in $O(mn)$ time using the maximum flow algorithm in [104].

3.1.2 Some Properties of the Problem (IntMCP)

Theorem 3.1 *Let* $\boldsymbol{x}^*$ *be an optimal solution of the problem* **(IntMCP)** *and* $E^* = \{e \in E | x^*(e) \neq 0\}$. *Then there is an* $s-t$ *cut* $\Bbbk^*$ *such that* $E^* \subseteq \Bbbk^*$.

Proof Assume that $\boldsymbol{x}^*$ is an optimal solution corresponding to the capacity value z^*. Consider the set $D^* = \{e \in E | w(e) - x^*(e) \leq z^*\}$. There is at least one $s-t$ cut $\Bbbk^* \subset D^*$. Otherwise, there exists an $s-t$ path P_0 with capacity $w(P_0) - x^*(P_0) > z^*$, which contradicts that the optimal objective value under $\boldsymbol{x}^*$ is z^*. Define $\bar{\boldsymbol{x}}$ as follows:

$$\bar{x}(e) = \begin{cases} x^*(e), & e \in \Bbbk^*, \\ 0, & e \notin \Bbbk^*. \end{cases} \tag{3.8}$$

For any $e \in E$ with $\bar{x}(e) \neq 0$, we have $x^*(e) \neq 0$, hence $\sum_{e \in E} c(e)\bar{x}(e) \leq \sum_{e \in E} c(e)x^*(e) \leq B$. Moreover, since any $s-t$ path intersects with $\Bbbk^*$, the maximum capacity value of a path under $\bar{x}$ is also z^*. Therefore, $\bar{\boldsymbol{x}}$ is also an optimal solution, and the corresponding adjusted edge set $E^* \subseteq \Bbbk^*$. □

Let P_j be the j-th path and $W_{\max} = \max_{P_j \in \mathcal{P}} w(P)$, $L_{\max} = \max_{P_j \in \mathcal{P}} l(P)$. For any $z \in [L_{\max}, W_{\max}]$, define some sets $E_1^z = \{e \in E | l(e) \leq z \leq w(e)\}$, $E_2^z = \{e \in E | l(e) > z\}$, $E_3^z = \{e \in E | z > w(e)\}$ and a cost vector

$$\bar{c}^z(e) = \begin{cases} c(e)(w(e) - z), & z \in E_1^z, \\ +\infty, & z \in E_2^z, \\ 0, & z \in E_3^z. \end{cases} \tag{3.9}$$

Then the cost of any $s-t$ cut $\Bbbk$ is defined by $\bar{\boldsymbol{c}}^z(\Bbbk) = \sum_{e \in \Bbbk} \bar{c}^z(e)$. Let $\mathcal{K}$ be the set of all $s-t$ cuts in $G = (V, E)$. Define a function

$$\begin{aligned} f(z) &= \min_{\Bbbk \in \mathcal{K}} \{\bar{\boldsymbol{c}}^z(\Bbbk)\} - B \\ &= \min_{\Bbbk \in \mathcal{K}} \left\{ \sum_{e \in \Bbbk} \max\{0, \min\{c(e)(w(e) - l(e)), c(e)(w(e) - z)\}\} - B \right\}. \end{aligned}$$

Arranging the values in $\mathcal{Z} = \{w(e) | e \in E\} \cap [L_{\max}, W_{\max}]$ non-decreasingly, we can obtain $z_0 = L_{\max} \leq z_1 \leq \cdots \leq z_\tau = W_{\max}$, where $\tau \geq 1$ is an integer. We can conclude that

Lemma 3.1 ([125, Lemma 1]) *The function* $f(z) : \mathcal{R} \to \mathcal{R}$ *is piecewise linear and non-increasing. Furthermore,* $f(z) : [z_{i-1}, z_i] \to \mathcal{R}$ *is a piecewise-linear decreasing concave function for any* $i = 1, 2, \cdots, \tau$.

Proof Since B is a constant, $f(z) + B = \min_{\Bbbk \in \mathcal{K}}\{\bar{c}^z(\Bbbk)\}$ is a minimum function about $\bar{c}^z(\Bbbk)$, and hence, we only need to consider the cuts $\Bbbk$ contained in E_1^z. For any $s-t$ cut $\Bbbk$ in G contained in E_1^z, the function

$$\begin{aligned}\bar{c}^z(\Bbbk) &= \sum_{e \in \Bbbk} \max\{0, \min\{c(e)(w(e) - l(e)), c(e)(w(e) - z)\}\} \\ &= \sum_{e \in \Bbbk} \max\{0, c(e)(w(e) - z)\}\end{aligned}$$

is a concave non-increasing piecewise linear function about z on $[L_{\max}, W_{\max}]$.

For any $z \in [z_{i-1}, z_i]$, where $i = 1, 2, \cdots, \tau$, let $b_{\Bbbk}^i = \sum_{e \in \Bbbk \cap E_1^z} c(e)w(e) - B$ and $a_{\Bbbk}^i = \sum_{e \in \Bbbk \cap E_1^z} c(e)$. Then the function $f(z)$ can be written in the following form:

$$f(z) = \min_{\Bbbk \in \mathcal{K}} \left\{ \sum_{e \in \Bbbk \cap E_1^z} c(e)(w(e) - z) - B \right\} = \min_{\Bbbk \in \mathcal{K}} \left\{ b_{\Bbbk}^i - a_{\Bbbk}^i z \right\} \tag{3.10}$$

Since $f(z)$ is defined as the minimum of a set of linear functions in (3.10), $a_{\Bbbk}^i > 0$ and $a_{\Bbbk}^i$ and $b_{\Bbbk}^i$ are constants on $z \in [z_{i-1}, z_i]$. Hence, $f(z)$ is concave and non-increasing on $[z_{i-1}, z_i]$. □

For any edge $e \in E$, to describe the relationship between the modified capacity z and the modification cost $\tilde{B} = c(e)(w(e) - z)$, we introduce a function $z_i^-(\tilde{B}) = w_i - \frac{1}{c_i}\tilde{B}$, $\tilde{B} \in [0, c_i(w_i - l_i)]$, where $w_i = w(e_i)$, $c_i = c(e_i)$, $l_i = l(e_i)$.

Consider a family of linear functions

$$\mathbb{Z} = \left\{ z_i^-(\tilde{B}) = w_i - \frac{1}{c_i}\tilde{B} \,\middle|\, \tilde{B} \in [0, c_i(w_i - l_i)],\ \ i = 1, 2, \cdots, m \right\}.$$

Let $(B_k, z_i^-(B_k))$ be an intersection point of any two lines in $\mathbb{Z}$. Denote the set of vertical coordinates of intersection points by

$$Z^- = \bigcup_{i=1,\cdots,m} \left\{ z_i^-(B_k), w_i, l_i \right\} \cap [L_{\max}, W_{\max}]. \tag{3.11}$$

Sort the values in Z^- non-decreasingly and denote the k-th smallest value in Z^- by z_k^-.

Lemma 3.2 ([67, Lemma 12]) *Suppose* $Y = \{y_i = a_i x + b_i | i = 1, 2, ..., N_1\}$ *is a family of different lines defined on* x, $X = [X_1, X_2, ..., X_{N_2}]$ *is the set of* $x-$*coordinates of intersection points of all lines in* Y *satisfying* $X_j < X_{j+1}, j =$

$1, 2, ..., N_2 - 1$. *For any* $x_j \in [X_j, X_{j+1}]$, $j = 1, 2, ..., N_2 - 1$, *let* $\hat{x}_j = \frac{X_j + X_{j+1}}{2}$, *if* $y_{i_1}(\hat{x}_j) < y_{i_2}(\hat{x}_j) < \cdots < y_{i_{N_1}}(\hat{x}_j)$, *then* $y_{i_1}(x_j) \leq y_{i_2}(x_j) \leq \cdots \leq y_{i_{N_1}}(x_j)$.

Theorem 3.2 *Let* z_k^- *and* z_{k+1}^- *be any two adjacent vertical coordinates in* Z^- *and* $\hat{z}_k = \frac{z_k^- + z_{k+1}^-}{2}$. *Suppose for any* $z \in [z_k^-, z_{k+1}^-]$, *there is an cut* $\Bbbk_z$ *satisfying* $0 \leq l(e) \leq z_k^- < \hat{z}_k < z_{k+1}^- \leq w(e)$ *for any* $e \in \Bbbk_z$. *Let* $\hat{\Bbbk}_k$ *be the minimum cut of the network* $\hat{G}_k = (V, E, \hat{z}_k, s, t)$ *with respect to the modified capacity* $\hat{z}_k$. *Then* $\hat{\Bbbk}_k$ *is also the minimum cut of the network* $\hat{G}_k = (V, E, z, s, t)$ *with respect to any capacity* $z \in [z_k^-, z_{k+1}^-]$.

Proof Suppose that when the capacity value is $z \in [z_k^-, z_{k+1}^-]$, $\hat{\Bbbk}_k$ is not the minimum cut, and $\Bbbk_z$ is the minimum cut. Suppose $e^1 \in \hat{\Bbbk}_k$ and $e^0 \in \Bbbk_z$, and only e^0 and e^1 are different between $\hat{\Bbbk}_k$ and $\Bbbk_z$, without loss of generality, we assume that $c(e^0) \neq 0$ and $c(e^1) \neq 0$, and $0 \leq l(e) \leq z_k^- < \hat{z}_k < z_{k+1}^- \leq w(e)$ is satisfied. Then we have

$$\begin{aligned} & K(\hat{\Bbbk}_k, z) - K(\Bbbk_z, z) \\ =& c(e^1)(w(e^1) - z) - c(e^0)(w(e^0) - z) \\ =& c(e^1)w(e^1) - c(e^0)w(e^0) + (c(e^0) - c(e^1))z > 0 \end{aligned} \tag{3.12}$$

Case 1 When $c(e^0) = c(e^1)$, since $\Bbbk_z$ is the minimum cut for the capacity value of z according to (3.12), we have $w(e^1) > w(e^0)$. When the capacity value is $\hat{z}_k$,

$$\begin{aligned} & K(\hat{\Bbbk}_k, \hat{z}_k) - K(\Bbbk_z, \hat{z}_k) \\ =& \sum_{e \in \hat{\Bbbk}_k, w(e) > \hat{z}_k} c(e)\left(w(e) - \hat{z}_k\right) - \sum_{e \in \Bbbk_z, w(e) > \hat{z}_k} c(e)\left(w(e) - \hat{z}_k\right) \\ =& c(e^1)w(e^1) - c(e^1)\hat{z}_k - c(e^0)w(e^0) + c(e^0)\hat{z}_k \\ =& c(e^1)w(e^1) - c(e^0)w(e^0) > 0 \end{aligned} \tag{3.13}$$

Therefore, $K(\hat{\Bbbk}_k, \hat{z}_k) > K(\Bbbk_z, \hat{z}_k)$ implies that $\hat{\Bbbk}_k$ is not the minimum cut with capacity $\hat{z}_k$.

Case 2 When $c(e^0) > c(e^1)$, since $\Bbbk_z$ is the minimum cut with capacity z, then from (3.12), we have

$$z > \frac{c(e^0)w(e^0) - c(e^1)w(e^1)}{c(e^0) - c(e^1)}$$

Since $z \in [z_k^-, z_{k+1}^-)$, we have $z_k^- > \frac{c(e^0)w(e^0) - c(e^1)w(e^1)}{c(e^0) - c(e^1)}$, which implies that $c(e^0)z_k^- - c(e^1)z_k^- > c(e^0)w(e^0) - c(e^1)w(e^1)$, or equivalently,

$c(e^0)\left(z_k^- - w(e^0)\right) > c(e^1)\left(z_k^- - w(e^1)\right)$. Therefore,

$$c(e^0) < \frac{c(e^1)\left(z_k^- - w(e^1)\right)}{z_k^- - w(e^0)} \tag{3.14}$$

since $c(e^1)\left(w(e^1) - z\right) - c(e^0)\left(w(e^0) - z\right) > 0$, $w(e^0) - z < w(e^1) - z$, $w(e^0) < w(e^1)$, then we have

$$\begin{aligned}
&K(\hat{\mathbb{k}}_k, \hat{z}_k) - K(\mathbb{k}_z, \hat{z}_k) \\
=& \sum_{e \in \hat{\mathbb{k}}_k, w(e) > \hat{z}_k} c(e)(w(e) - \hat{z}_k) - \sum_{e \in \mathbb{k}_z, w(e) > \hat{z}_k} c(e)(w(e) - \hat{z}_k) \\
=& c(e^1)(w(e^1) - \hat{z}_k) - c(e^0)(w(e^0) - \hat{z}_k) \\
>& c(e^1)(w(e^1) - \hat{z}_k) - \frac{c(e^1)(z_k^- - w(e^1))}{z_k^- - w(e^0)}(w(e^0) - \hat{z}_k) \quad \text{since (3.14)} \\
=& \frac{c(e^1)(w(e^1) - \hat{z}_k)(z_k^- - w(e^0)) - c(e^1)(z_k^- - w(e^1))(w(e^0) - \hat{z}_k)}{z_k^- - w(e^0)} \\
=& \frac{c(e^1)\left(z_k^- w(e^1) - z_k^- \hat{z}_k - w(e^0)w(e^1) + w(e^0)\hat{z}_k\right)}{z_k^- - w(e^0)} \\
+& \frac{c(e^1)\left(-z_k^- w(e^0) + z_k^- \hat{z}_k + w(e^0)w(e^1) - w(e^1)\hat{z}_k\right)}{z_k^- - w(e^0)} \\
=& \frac{c(e^1)\left(z_k^- w(e^1) + w(e^0)\hat{z}_k - z_k^- w(e^0) - w(e^1)\hat{z}_k\right)}{z_k^- - w(e^0)} \\
=& \frac{c(e^1)\left(\hat{z}_k - z_k^-\right)\left(w(e^0) - w(e^1)\right)}{z_k^- - w(e^0)}
\end{aligned}$$

Since $\hat{z}_k = \frac{z_k^- + z_{k+1}^-}{2}$, we have $z_k^- < \hat{z}_k$. Also, since $c(e^1) > 0$, $w(e^0) < w(e^1)$, $z_k^- - w(e^0) < 0$, we can conclude that $K(\hat{\mathbb{k}}_k, \hat{z}_k) > K(\mathbb{k}_z, \hat{z}_k)$, and therefore, $\hat{\mathbb{k}}_k$ is not the minimum cut with capacity $\hat{z}_k$.

Case 3 When $c(e^0) < c(e^1)$, since $\mathbb{k}_z$ is the minimum cut with capacity z, it follows from (3.12) that

$$z < \frac{c(e^0)w(e^0) - c(e^1)w(e^1)}{c(e^0) - c(e^1)}$$

Since $z \in [z_k^-, z_{k+1}^-)$, we have $z_{k+1}^- \le \frac{c(e^0)w(e^0)-c(e^1)w(e^1)}{c(e^0)-c(e^1)}$, which implies that $c(e^0)z_{k+1}^- - c(e^1)z_{k+1}^- \ge c(e^0)w(e^0) - c(e^1)w(e^1)$. Therefore, we have

$$w(e^0) \le \frac{c(e^0)z_{k+1}^- - c(e^1)z_{k+1}^- + c(e^1)w(e^1)}{c(e^0)} \tag{3.15}$$

$$\begin{aligned}
&K(\hat{\Bbbk}_k, \hat{z}_k) - K(\Bbbk_z, \hat{z}_k)\\
=& \sum_{e\in\hat{\Bbbk}_k, w(e)>\hat{z}_k} c(e)(w(e)-\hat{z}_k) - \sum_{e\in\Bbbk_z, w(e)>\hat{z}_k} c(e)(w(e)-\hat{z}_k)\\
\ge& c(e^1)(w(e^1)-\hat{z}_k) - c(e^0)\left(\frac{c(e^0)z_{k+1}^- - c(e^1)z_{k+1}^- + c(e^1)w(e^1)}{c(e^0)} - \hat{z}_k\right)\\
=& c(e^1)w(e^1) - c(e^1)\hat{z}_k - c(e^0)z_{k+1}^- + c(e^1)z_{k+1}^- - c(e^1)w(e^1) + c(e^0)\hat{z}_k\\
=& c(e^1)\left(z_{k+1}^- - \hat{z}_k\right) - c(e^0)\left(z_{k+1}^- - \hat{z}_k\right)\\
=& \left(c(e^1) - c(e^0)\right)\left(z_{k+1}^- - \hat{z}_k\right)
\end{aligned}$$

It follows that $z_{k+1}^- > \hat{z}_k == \frac{z_k^- + z_{k+1}^-}{2}$. Furthermore, since $c(e^1) > c(e^0)$, it follows that $K(\hat{\Bbbk}_k, \hat{z}_k) > K(\Bbbk_z, \hat{z}_k)$. Therefore, $\hat{\Bbbk}_k$ is not the minimum cut with $\hat{z}_k$. □

3.1.3 Solve the Problem (IntMCP)

We need to determine which minimum cut edges correspond to the vertical coordinate z_k^- at each turning point. At the turning point $(B_k, z_i^-(B_k))$, due to the existence of bounds, we define the arc sets $E_k^1 = \{e \in E | l(e) \le z_k^- \le w(e)\}$, $E_k^2 = \{e \in E | l(e) > z_k^-\}$. Let the network $G_k = (N, E, \rho_k, s, t)$ be an auxiliary network, where N is the set of vertices of G, E is the set of arcs of G, and the cost $\rho_k(e)$ when the capacity of arc e is reduced to z_k^- is defined as follows:

$$\rho_k(e) = \begin{cases} c(e)(w(e) - z_k^-), & \text{if } e \in E_k^1;\\ +\infty, & \text{if } e \in E_k^2;\\ 0, & \text{otherwise.} \end{cases} \tag{3.16}$$

Note that $L_{\max} = \max_{P_j \in \mathcal{P}} l(P_j)$ is the capacity value of the maximum capacity path in the directed connected graph G under the capacity l.

Similarly, when replacing z_k^- with $L_{\max}$ in (3.16), we obtain two edge sets: $\underline{E}^1 = \{e \in E | l(e) \leq L_{\max} \leq w(e)\}$, $\underline{E}^2 = \{e \in E | l(e) > L_{\max}\}$, and the cost of arc e when the capacity is reduced to $L_{\max}$ is

$$\underline{\rho}(e) = \begin{cases} c(e)(w(e) - L_{\max}), & \text{if } e \in \underline{E}^1; \\ +\infty, & \text{if } e \in \underline{E}^2; \\ 0, & \text{otherwise.} \end{cases} \tag{3.17}$$

The corresponding network $\underline{G} = (V, E, \underline{\rho}, s, t)$ is the cost network when the capacity of each arc in G becomes $L_{\max}$. Let the minimum cut of $\underline{G}$ be $\underline{\Bbbk}$, and the cost of the minimum cut $\underline{\Bbbk}$ is $\underline{f}$.

Lemma 3.3 *If the cost $B \geq \underline{f}$, then the optimal objective value $z^* = L_{\max}$.*

Proof Since $B \geq \underline{f}$, and $\underline{f}$ is the cost of the minimum cut in $\underline{G} = (V, E, \underline{\rho}, s, t)$, and $\underline{\rho}$ is the cost vector when the capacity is reduced to $L_{\max}$, this implies that $z^* = L_{\max}$.

Theorem 3.3 $\Upsilon(\widetilde{B}) = \frac{z_a^- - z_b^-}{B_a - B_b}\widetilde{B} + \frac{B_b z_a^- - B_a z_b^-}{B_b - B_a}$ *is a monotonically decreasing function defined on* $[B_a, B_b]$.

Proof Since $z_a^- > z_b^-$ and $B_a < B_b$, we have $\frac{z_a^- - z_b^-}{B_a - B_b} < 0$. Therefore, $\Upsilon(\widetilde{B})$ is a monotonically decreasing function defined on $[B_a, B_b]$. □

Let $G_b^- = (N, E, \rho_b^-, s, t)$ be an auxiliary network, where N, E are the node set and edge set of graph G, respectively. Let $\rho_b^-(e)$ be the cost of the capacity value on arc e to be reduced to z_b^-.

$$\rho_b^-(e) = \begin{cases} c(e)(w(e) - z_b^-), & l(e) < z_b^- \leq w(e); \\ +\infty, & l(e) \geq z_b^-; \\ 0, & otherwise. \end{cases} \tag{3.18}$$

Theorem 3.4 *Let the optimal value be $z^* \in [z_a^-, z_b^-]$, where z_a^- and z_b^- are the y-coordinates of two adjacent turning points with $z_a^- < z_b^-$. If $B_b < B < B_a$, let the minimum cut in network $G_b^- = (N, E, \rho_b^-, s, t)$ be $\Bbbk_b^-$, with the cost of the minimum cut being B_b^-. In network $G_b = (N, E, \rho_b, s, t)$, the minimum cut is $\Bbbk_b$, with the cost of the minimum cut being B_b. If $B_b = B_b^-$, then the optimal value is $z^* = \Upsilon(B)$; if $B_b \neq B_b^-$, then the optimal value is $z^* = z_b^-$.*

Proof Since z_a^- and z_b^- are the y-coordinates of two adjacent turning points, for any capacity value $z \in (z_a^-, z_b^-)$, by Theorem 3.2, the set of edges in the minimum cut is the same. Let the minimum cut be $\Bbbk_a$, with the corresponding cost of the minimum cut being B_a. Suppose when the capacity value drops to $z = z_b^-$, a certain minimum cut is $\Bbbk_b$, with the corresponding cost of the minimum cut being B_b. If there exists some $\Bbbk_b = \Bbbk_a$, it indicates that in some $\Bbbk_b$, there is no edge with a lower bound value of z_b^-. At this point, by Theorem 3.1, the optimal objective value z is a piecewise linear decreasing convex function with respect to B. Therefore, the optimal objective value z^* can be obtained by substituting B into the line $\Upsilon(\widetilde{B}) = \frac{z_a^- - z_b^-}{B_a - B_b}\widetilde{B} + \frac{B_b z_a^- - B_a z_b^-}{B_b - B_a}$ determined by points $A = (B_a, z_a^-)$ and $B = (B_b, z_b^-)$. If there is no such $\Bbbk_b = \Bbbk_a$, it indicates that in all minimum cuts $\Bbbk_b$, there is at least one edge with a lower bound value of z_b^-. At this time, since the given cost B satisfies $B_b < B < B_a$, if the edges in the cut B_a, when the capacity value drops to z_b^-, the cost must be greater than B_b; otherwise, it does not satisfy Theorem 3.1. Therefore, the capacity value can drop at most to z_b^-, that is, the optimal objective value is z_b^-. □

Algorithm 3.1 An algorithm to solve (IntMCP) under weighted l_1 norm

---

Input: A network $G(V, E)$ with a source s and a terminal t; three vectors w, c, l; and a value B.

Output: An optimal solution (x^*, y^*) and its objective value z^*.

Calculate $W_{\max} := \max_{P_j \in \mathcal{P}} w(P)$, $L_{\max} := \max_{P_j \in \mathcal{P}} l(P)$.

Calculate $\underline{\rho}$ by (3.17) and get the cost network $\underline{G}$.

Find the minimum cut $\underline{\Bbbk}$ of $\underline{G}$, whose cost is $\underline{f}$.

if $B \geq \underline{f}$ **then return** $z^* := L_{\max}$,

$$x^*(e) := \begin{cases} w(e) - z^*, & e \in \underline{\Bbbk},\ w(e) > z^*, \\ 0, & otherwise. \end{cases} \tag{3.19}$$

else calculate Z^- by (3.11).

    Sort the values in Z^- non-decreasingly and denote the k-th smallest value in Z^- by z_k^-.

    Initialize $a := 1, b := |Z^-|$.

    while $b - a > 1$ **do**

        Let $k := \lfloor \frac{a+b}{2} \rfloor$. Calculate ρ_k by (3.16).

        Find the minimum cut of the network G_k under ρ_k, whose cost is f_k.

        if $f_k \leq B$ **then** $b := k$,

        else $a := k$.

        end if

    end while

    Calculate ρ_b by (3.16) and the cost f_b of minimum cut of the network $G_b := (V, E, \rho_b, s, t)$ under ρ_b.

    Calculate ρ_b^- by (3.18) and the cost f_b^- of minimum cut $\Bbbk_b^-$ of the network $G_b^- := (V, E, \rho_b^-, s, t)$ under ρ_b^-.

    if $f_b = f_b^-$ **then**

        Calculate ρ_a by (3.16) and the cost f_a of minimum cut of the network $G_a := (V, E, \rho_a, s, t)$ under ρ_a.

        Return $z^* := \frac{z_a^- - z_b^-}{f_a - f_b} B + \frac{f_b z_a^- - f_a z_b}{f_b - f_a}$.

20: **else Return** $z^* := z_b^-$.
21: **end if**
22: **Return**

$$x^*(e) := \begin{cases} w(e) - z^*, \; e \in \mathbb{k}_b^-, \, w(e) > z^*, \\ 0, \qquad\qquad otherwise. \end{cases} \tag{3.20}$$

23: Choose an edge e^* satisfying $x^*(e^*) \neq 0$ and a path P^* traversing edge e^*. Let $y^*(e) := 1, e \in P^*$, $y^*(e) := 0, e \notin P^*$.
24: **end if**

Theorem 3.5 *The z^* output by Algorithm 3.1 is the optimal value for problem (IntMCP), and the x^*, y^* defined by Eqs. (3.19)–(3.20) and line 23 are the optimal solutions.*

Proof Let $\mathbb{k}_b^-$, f_b^-, f_b be obtained from lines 15–16. For arcs $e \in \mathbb{k}_b^- \cap P_j$ with $w(e) > z^*$, it holds that $0 \leq x^*(e) \leq w(e) - l(e)$, hence $w(e) - x^*(e) = w(e) - (w(e) - z^*) = z^*$. Since any $s-t$ path must intersect with $\mathbb{k}_b^-$, there exists a path P such that $w(P) = z^*$ and $w(P) \leq z^*$.

From the construction of x^*, it is clear that for the arcs in the cut $\mathbb{k}_b^-$ with $w(e) > z^*$, the capacity will become $w(e) - x^*(e) = z^*$. Correspondingly, the capacity of any path P with $w(P) > z^*$ will become z^*, thus z^* is the capacity of the maximum capacity path after updating x^*.

When $f_b = f_b^-$, it indicates that when the capacity of the maximum capacity path is z_b^-, there exists at least one $s-t$ cut $\mathbb{k}$ where all arc capacities satisfy $0 \leq l(e) \leq z_a^- < z_b^- \leq w(e)$. Let $\mathbb{k}_a^> = \{e \in \mathbb{k}_a | w(e) > z_a^-\}$, $\mathbb{k}_b^> = \{e \in \mathbb{k}_b | w(e) > z_b^-\}$. By Theorem 3.2, $\mathbb{k}_a = \mathbb{k}_b$, hence $\mathbb{k}_a^> = \mathbb{k}_b^>$. Since

$$f_a = \sum_{e \in \mathbb{k}_a^>} c(e)(w(e) - z_a^-) = \sum_{e \in \mathbb{k}_b^>} c(e)w(e) - \left(\sum_{e \in \mathbb{k}_b^>} c(e) \right) z_a^-, \tag{3.21}$$

$$f_b = \sum_{e \in \mathbb{k}_b^>} c(e)(w(e) - z_b^-) = \sum_{e \in \mathbb{k}_b^>} c(e)w(e) - \left(\sum_{e \in \mathbb{k}_b^>} c(e) \right) z_b^-, \tag{3.22}$$

we have

$$\frac{z_b^- - z_a^-}{f_a - f_b} = \frac{1}{\sum_{e \in \mathbb{k}_b^>} c(e)}. \tag{3.23}$$

Multiplying (3.21) by z_b^- and (3.22) by z_a^-, and then subtracting them, we can obtain

$$f_b z_a^- - f_a z_b^- = \left(\sum_{e \in \mathbb{k}_b^>} c(e)w(e) \right) (z_a^- - z_b^-). \tag{3.24}$$

It follows from (3.23) and (3.24) that

$$\frac{f_b z_a^- - f_a z_b^-}{f_b - f_a} = \frac{\sum_{e \in \mathbb{k}_b^>} c(e)w(e)}{\sum_{e \in \mathbb{k}_b^>} c(e)}. \tag{3.25}$$

Thus, we have

$$\begin{aligned}
&\sum_{e \in \mathbb{k}_b^>} c(e)x^*(e) \\
=&\sum_{e \in \mathbb{k}_b^>} c(e)(w(e) - z^*) \\
=&\sum_{e \in \mathbb{k}_b^>} c(e) \left(w(e) - \frac{z_a^- - z_b^-}{f_a - f_b} B - \frac{f_b z_a^- - f_a z_b^-}{f_b - f_a} \right) \\
=&\sum_{e \in \mathbb{k}_b^>} c(e)w(e) - \left(\sum_{e \in \mathbb{k}_b^>} c(e) \right) \left(\frac{z_a^- - z_b^-}{f_a - f_b} B + \frac{f_b z_a^- - f_a z_b^-}{f_b - f_a} \right) \\
=&\sum_{e \in \mathbb{k}_b^>} c(e)w(e) - \left(\sum_{e \in \mathbb{k}_b^>} c(e) \right) \left(-\frac{B}{\sum_{e \in \mathbb{k}_b^>} c(e)} + \frac{\sum_{e \in \mathbb{k}_b^>} c(e)w(e)}{\sum_{e \in \mathbb{k}_b^>} c(e)} \right) \\
=&\sum_{e \in \mathbb{k}_b^>} c(e)w(e) + B - \sum_{e \in \mathbb{k}_b^>} c(e)w(e) \\
=&B.
\end{aligned}$$

When $f_b \neq f_b^-$, it indicates that when $z^* = z_b^-$, all $s - t$ cuts have at least one arc with a lower bound $l(e) = z_b^-$, that is, it is saturated. At this point, since $f_a < B < f_b$, the remaining cost is $B - f_b$, which means the sum of the weighted x^* values for all arcs in E is less than or equal to B. If there exists a smaller target value $\hat{z} < z^*$, then $\hat{B} > B$, which contradicts feasibility. Therefore, the value z^* obtained by Algorithm 3.1 is the optimal value.

The x^*, y^* defined by (3.19)–(3.20) and line 23 are the optimal solutions. □

Theorem 3.6 *The time complexity of Algorithm 3.1 is* $O(m^2 \log m)$.

Proof The time complexity for calculating the capacity of the maximum capacity path in the first line is $O(m)$. The time complexity for calculating $\underline{\rho}$ in the second

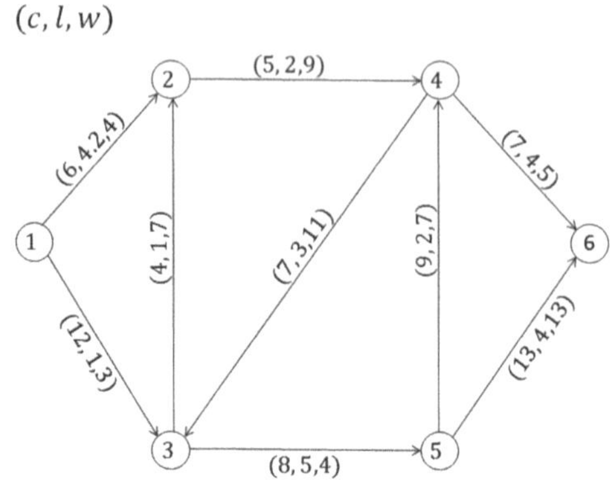

Fig. 3.1 An example network G

line is $O(m)$. The time complexity for calculating the minimum cut in the third line is $O(mn)$. The time complexity for calculating the set of all turning point y-coordinates Z^- in line 5 is $C_m^2 = O(m^2)$. The time complexity for sorting the set Z^- in line 6 is $O(m^2 \log(m^2)) = O(m^2 \log m)$. The number of iterations for the while-loop in lines 8-14 is $O(\log m^2) = O(\log m)$, thus the time complexity is $O(mn \log m^2) = O(mn \log m)$. Therefore, the overall time complexity of the algorithm is $O(m^2 \log m)$. □

It is worthy noting that this time complexity greatly reduces the time complexity $O(m^2 \log^2 m \Lambda(n, m))$ of the algorithm solving the unbounded problem proposed by Javad Tayyebi et al. [125], where $\Lambda(n, m)$ represents the complexity of finding the minimum cost cut. If the complexity $\Lambda(n, m) = O(mn)$ is commonly used, our complexity is reduced by $O(mn \log m)$ compared to theirs.

3.1.4 Numerical Experiments

In this subsection, we will first present a concrete example to demonstrate the operation of Algorithm 3.1 and then provide numerical experiments to verify the effectiveness of the algorithm.

Example 3.1 When all the lower bounds of the arcs are bounded, we provide a specific example to verify the effectiveness of Algorithm 3.1, as shown in Fig. 3.1.
$V := \{1, 2, \cdots, 6\}$,
$E := \{(1, 2), (1, 3), (2, 4), (3, 2), (3, 5), (4, 3), (4, 6), (5, 4), (5, 6)\}$,
$B := 31$,
$w := (6, 12, 5, 4, 8, 7, 7, 9, 13)$,
$l := (4.2, 1, 2, 1, 5, 3, 4, 2, 4)$,
$c := (4, 3, 9, 7, 4, 11, 5, 7, 13)$.
1. Calculate $W_{\max} := 8$, $L_{max} := 2$.
2. Calculate $\underline{\rho}$ by (3.17), then we get network $\underline{G}$.

3. In the cost network $\underline{G}$, the minimum cut $\underline{\Bbbk} := \{(2,4),(1,3)\}$ is obtained by using the minimum cut algorithm. Its cost $\underline{f} := 27 + 30 := 57$. Since $B < \underline{f}$, the algorithm proceeds to line 5.

5–6. Calculate the set $Z^- := \{2, 3, 3.5, 4, 4.2, 5, 5.125, 6, 6.4286, 6.75, 7, 8\}$ by (3.11).

7. $a := 1, b := 12$.

9. $k := 6$. Calculate ρ_6 by (3.16), then we get the network G_6.

10. The minimum cut of network G_6 under ρ_6 is $\tilde{C} := \{(2,4),(3,5)\}$. Its cost $f_6 := 0 + 12 := 12$. Since $f_6 < B$, the algorithm proceeds to line 11.

11. $a := 1, b := 6$ (After the first loop ends, it proceeds to line 8.).

After four iterations, we have that $a := 4, b := 5$. Since $b - a = 1$, the while loop ends, and the algorithm proceeds to line 15. At this point, the optimal value z^* is in the interval (4, 4.2), therefore, $z_a^- := 4$, $z_b^- := 4.2$.

15. Calculate ρ_5 by (3.16), and the cost of the minimum cut in $G_5 := (N, E, \rho_5, s, t)$ is $f_5 := 30.6$.

16. Calculate ρ_5^- by (3.18) and the cost $f_5^- := f_5 := 30.6$ of minimum cut $\Bbbk_5^-$. Since $f_5 := f_5^-$, the algorithm proceeds to line 18.

18. Calculate ρ_4 by (3.16) and the cost $f_4 := 24 + 9 := 33$ of the minimum cut $\Bbbk_4$.

19. Output the optimal value $z^* := \frac{z_4^- - z_5^-}{f_4 - f_5} B + \frac{f_5 z_4^- - f_4 z_5^-}{f_5 - f_4} := -\frac{1}{12} * 31 + \frac{27}{4} := 4.167$.

22. Output an optimal solution $x^* := (0, 7.833, 0.833, 0, 0, 0, 0, 0, 0)$.

Now we present the results of the numerical experiment of Algorithm 3.1. These programs were all written in Matlab R2016a.lnk and run on a computer with Windows 7, an Intel(R) Core(TM) i3-6100U CPU @2.30 GHz 2.30 GHz processor. The number of nodes n varies between 200 and 10,000, and the number of edges varies between 500 and 40,000, with 200 instances generated randomly. In each instance, four vectors c, l, w and a constant W are randomly generated, where c is evenly distributed over the interval [20,50], l over the interval [5,25], and w over the interval [30,60]. For each randomly generated graph, under the weighted l_1 norm, Atime, min_time, and max_time, respectively, represent the CPU time average, minimum, and maximum when solving the problem of the maximum capacity path's blocking under the weighted l_1 norm using Algorithm 3.1.

It can be seen from Table 3.1 that for each category of random graphs, the running time of Algorithm 3.1 is consistent with the time complexity $O(m^2 \log m)$, thereby verifying the effectiveness of the algorithm.

3.2 (RIOV, Maximum Capacity Path, Bounded, Weighted l_1 Norm)

This section will specifically introduce the inverse optimal value problem on maximum capacity path under weighted l_1 norm (RIOVMCP_1). We first present

Table 3.1 The experimental results of Algorithm 3.1

n	200	500	1500	2000	6000	10,000
m	500	800	4000	9000	16,000	40,000
Atime	0.0698	0.2931	4.1544	13.0528	68.1600	287.5219
min_time	0.0572	0.2682	4.0278	12.3019	65.1218	265.6891
max_time	0.0734	0.3156	4.5562	14.3417	70.4131	295.5241

its mathematical model, then design an algorithm with time complexity $O(mn)$ by analyzing some properties of the problem. Finally, numerical experiments are illustrated to demonstrate the specific process of the algorithm and check its effectiveness.

3.2.1 The Mathematical Model of the Problem (RIOVMCP$_1$)

Let $G = (N, E, c, w, s, t)$ be a weighted directed connected graph and $\bar{P}$ be a given $s - t$ path in graph G. The Restricted Inverse Optimal Value Problem on Maximum Capacity Path (RIOVMCP) is to make the given path $\bar{P}$ the maximum capacity path in graph G with its capacity being a fixed value L by adjusting the capacities of arcs as little as possible. Let $\bar{c}(e)$ be the adjusted capacity, $l(e)$ and $u(e)$ be the values to which the capacity $c(e)$ of arc e can be reduced and increased, respectively. Let $w(e)$ be the cost of adjusting the unit capacity. The problem (RIOVMCP$_1$) under weighted l_1 norm can be stated as follows.

$$\min \sum_{e\in E} w(e)|\bar{c}(e) - c(e)|$$

$$\textbf{(RIOVMCP}_1\textbf{)}\quad s.t.\ \bar{c}(\bar{P}) = L, \tag{3.26}$$

$$l(e) \le \bar{c}(e) \le u(e), \tag{3.27}$$

$$\max_P \bar{c}(P) = \bar{c}(\bar{P}). \tag{3.28}$$

3.2.2 Solve the Problem (RIOVMCP$_1$)

Under the weighted l_1 norm, in order to make $\bar{P}$ the maximum capacity path and to equalize its capacity value to L, we only need to alter the capacity of certain arcs within the path to achieve this goal.

In $G = (N, E, c, w, s, t)$, due to the existence of bounds, we define two sets $E^< = \{e \in E|c(e) < L\}$; $E^> = \{e \in E|l(e) > L\}$. Furthermore, since L is an arbitrary value, and due to the existence of bounds and the need to make the path

$\bar{P}$ the maximum capacity path, we need to consider two values $\underline{C} = \min_{e \in \bar{P}}\{l(e)\}$, $\bar{C} = \min_{e \in \bar{P}}\{u(e)\}$.

Given a path $\bar{P}$ in graph G, denote the set of arcs with capacity values less than L in path $\bar{P}$ as $\bar{P}^{<} = \{e \in \bar{P} | c(e) < L\}$. Construct an auxiliary network $G_L = (N, E, \alpha_L, s, t)$, where the cost $\alpha_L(e)$ on arc e is defined as:

$$\alpha_L(e) = \begin{cases} 0, & \text{if } e \in E^{<}; \\ +\infty, & \text{if } e \in E^{>}; \\ w(e)(c(e) - L), & \text{otherwise.} \end{cases} \tag{3.29}$$

Lemma 3.4 *For the given value L, if $L < \underline{C}$ or $L > \bar{C}$, the problem ($RIOVMCP_1$) is infeasible.*

Proof For the given value L, when $L < \underline{C}$, it is not possible to make $\bar{c}(\bar{P}) = L$; when $L > \bar{C}$, according to the definition of the maximum capacity path, to make path $\bar{P}$ the maximum capacity path, it is necessary to increase the capacity of all arcs in path $\bar{P}$ with capacities less than L to L, and since $L > \bar{C}$, it is also not possible to make $\bar{c}(\bar{P}) = L$. Therefore, when $L < \underline{C}$ or $L > \bar{C}$, the problem (RIOVMCP_1) is infeasible. □

Lemma 3.5 *For a given value L, if there exists an $s - t$ path P such that $P \subseteq E^{>}$, then the problem ($RIOVMCP_1$) is infeasible.*

Proof For a given value L, if there exists an $s - t$ path P satisfying $P \subseteq E^{>}$, then $l(P) > L$. Since in the optimal network $\bar{G} = (N, E, \bar{c}, s, t)$, it holds that $L = \bar{c}(\bar{P})$, therefore, $\bar{c}(P) \geq l(P) > L = \bar{c}(\bar{P})$, which contradicts the fact that path $\bar{P}$ is the maximum capacity path under capacity $\bar{c}$. Hence, the problem (RIOVMCP_1) is infeasible. □

Theorem 3.7 *For a given value L, the problem ($RIOVMCP_1$) is feasible if and only if the following two conditions are met simultaneously:*

1. *$L \in [\underline{C}, \bar{C}]$.*
2. *For any $s - t$ path P, $P \setminus E^{>} \neq \emptyset$.*

Proof Necessity: By the contrapositive of Lemmas 3.4 and 3.5, the necessity is proven.

Sufficiency is now proven. For a given value L, when conditions 1 and 2 are satisfied simultaneously, construct a solution

$$\bar{c}(e) = \begin{cases} L, & \text{if } l(e) \leq L \leq u(e); \\ c(e), & \text{otherwise.} \end{cases} \tag{3.30}$$

At this point, for any path P, it satisfies $\bar{c}(P) \leq L$, for path $\bar{P}$, it satisfies $\bar{c}(\bar{P}) = L$, and for any arc e, we have $l(e) \leq \bar{c}(e) \leq u(e)$. The $\bar{c}$ defined by (3.30) satisfies the constraint conditions (3.26)–(3.28); therefore, the problem (RIOVMCP_1) is feasible.

Further, we prove the properties of an optimal solution for problem (**RIOVMCP**$_1$). If the capacity of some arcs changes, then the changed capacity must become L.

Theorem 3.8 *If the problem (**RIOVMCP**$_1$) has an optimal solution $\bar{c}$, then its optimal solution satisfies:*

$$\bar{c}(e) = \begin{cases} L, & \textit{if } \bar{c}(e) \neq c(e); \\ c(e), & \textit{otherwise.} \end{cases}$$

Proof Assume $\bar{e} \in E$, and $\bar{c}(\bar{e}) \neq c(\bar{e})$, and consider two cases: $\bar{c}(\bar{e}) < c(\bar{e})$, $\bar{c}(\bar{e}) > c(\bar{e})$.

1. When $\bar{c}(\bar{e}) < c(\bar{e})$ and $\bar{c}(\bar{e}) \neq L$, there are two situations.

(a) $c(\bar{e}) > \bar{c}(\bar{e}) > L = \bar{c}(\bar{P})$. If there exists an $s-t$ path $\tilde{P}$ such that $\bar{e} \in \tilde{P} \setminus \bar{P}$ is the minimum arc of $\tilde{P}$, then $\bar{c}(\tilde{P}) = \bar{c}(\bar{e}) > L$, which contradicts the fact that $\bar{P}$ is the maximum capacity path. If $\bar{e} \in \bar{P}$ is the minimum arc of $\bar{P}$, then it contradicts $\bar{c}(\bar{P}) = L$. Otherwise, if $\bar{e}$ is not the minimum arc in any path, then the capacity of this arc does not need to be adjusted, and a solution better than $\bar{c}$ can be found.

(b) If $c(\bar{e}) > L = \bar{c}(\bar{P}) > \bar{c}(\bar{e})$, then construct another feasible solution

$$c^*(e) = \begin{cases} L, & \text{if } \bar{c}(e) < L < c(e); \\ \bar{c}(e), & \text{otherwise.} \end{cases}$$

At this point, $\sum_{e \in E} w(e)|c^*(e) - c(e)| < \sum_{e \in E} w(e)|\bar{c}(e) - c(e)|$. This shows that c^* is a better feasible solution than $\bar{c}$, which contradicts the optimality of $\bar{c}$.

2. When $\bar{c}(\bar{e}) > c(\bar{e})$ and $\bar{c}(\bar{e}) \neq L$, there are two cases.

(a) If $\bar{c}(\bar{e}) > L = \bar{c}(\bar{P}) > c(\bar{e})$, then construct another feasible solution

$$c^*(e) = \begin{cases} L, & \text{if } \bar{c}(e) < L < u(e); \\ \bar{c}(e), & \text{otherwise.} \end{cases}$$

At this point, $\sum_{e \in E} w(e)|c^*(e) - c(e)| < \sum_{e \in E} w(e)|\bar{c}(e) - c(e)|$. This shows that c^* is a better feasible solution than $\bar{c}$, which contradicts the assumption that $\bar{c}$ is the optimal solution.

(b) $L > \bar{c}(\bar{e}) > c(\bar{e})$. If $\bar{e} \in \bar{P}$, then $\bar{c}(\bar{P}) \leq \bar{c}(\bar{e}) < L$, which contradicts $\bar{c}(\bar{P}) = L$. Otherwise, if $\bar{e} \notin \bar{P}$, there is no need to increase the capacity value of $\bar{e}$.

This completes the proof. □

Algorithm 3.2 An algorithm for the problem (**RIOVMCP**$_1$)

Input: A directed connected graph $G := (V, E, c, w, s, t)$, a capacity vector c, a cost vector w, a path $\bar{P}$, and the capacity $c(\bar{P})$ of path $\bar{P}$.

Output: An optimal solution $\bar{c}$ and its optimal value $\bar{z}$.

1: Let $\underline{C} := \min_{e \in \bar{P}}\{l(e)\}$, $\bar{C} := \min_{e \in \bar{P}}\{u(e)\}$, $E^> := \{e \in E | l(e) > L\}$.

2: **if** $L < \underline{C}$ or $L > \bar{C}$, or there exists an $s - t$ path P, satisfying $P \subseteq E^{>}$ **then**
3: **Return** the instance is infeasible.
4: **else calculate** α_L from (3.29) and find the minimum cost cut $(X, \bar{X})$ of $G_L := (N, E, \alpha_L, s, t)$ under the cost vector α_L.
5: Let $A := \{e \in (X, \bar{X}) | \alpha_L(e) \neq 0\}$ and $\bar{P}^{<} := \{e \in \bar{P} | c(e) < L\}$.
6: **Compute**

$$\bar{c}(e) := \begin{cases} L, & e \in A \cup \bar{P}^{<}; \\ c(e), & \text{otherwise.} \end{cases} \tag{3.31}$$

7: **Return** $\bar{c}$ and $\bar{z} := \sum_{e \in (X, \bar{X})} \alpha_L(e) + \sum_{e \in \bar{P}^{<}} w(e)(L - c(e))$.
8: **end if**

Theorem 3.9 *If the problem (**RIOVMCP**$_1$) is feasible, then the vector $\bar{c}$ outputted by Algorithm 3.2 is an optimal solution.*

Proof Assume the problem (**RIOVMCP**$_1$) is feasible. To satisfy (3.26), by the definition of the maximum capacity path, we need to increase the capacity of arcs in $\bar{P}^{<} = \{e \in \bar{P} | c(e) < L\}$ to L. For other arcs $e \in E \setminus \bar{P}^{<}$, to satisfy constraints (3.27), (3.28), it is necessary to determine which arcs need to be changed. To this end, according to the definition of the auxiliary network G_L and the concept of the minimum cost cut, first obtain the minimum cost cut $(X, \bar{X})$ of the auxiliary network G_L under the cost vector α_L. Note that there may be arcs with a weight of 0 in the cut $(X, \bar{X})$; these arcs correspond to capacities $c(e) \leq L$ in G, that is, they do not need to be changed. Therefore, in the cut $(X, \bar{X})$, the arcs that need to be changed are those with a non-zero weight; these arcs correspond to arcs with $c(e) > L$ in network G. Denote the set of these arcs as $A = \{e \in (X, \bar{X}) | \alpha_L(e) \neq 0\}$. Thus, adjusting the capacities of arcs in sets $\bar{P}^{<}$ and A to L ensures both the feasibility of $\bar{c}$ defined by (3.31) and the optimality of the objective value under the weighted l_1 norm, with the minimum cost cut of G_L under the cost vector α_L. Additionally, the vector $\bar{c}$ defined by (3.31) also satisfies the properties of the optimal solution described in Theorem 3.8. □

Theorem 3.10 *The time complexity of Algorithm 3.2 is $O(mn)$.*

Proof The time complexity of computing the arc sets $\underline{C}, \bar{C}, E^{>}$ in line 1 is $O(m)$. Line 2 for finding an $s - t$ path can be solved using a depth-first search algorithm in $O(m)$ time. Line 4 for computing the minimum cut has a time complexity of $O(mn)$ [104], and the time complexity of computing arc sets A and $\bar{P}^{<}$ in line 5 is $O(m)$, and updating $\bar{c}$ in lines 6–7 has a time complexity of $O(m)$. Therefore, the time complexity of Algorithm 3.2 is $O(mn)$. □

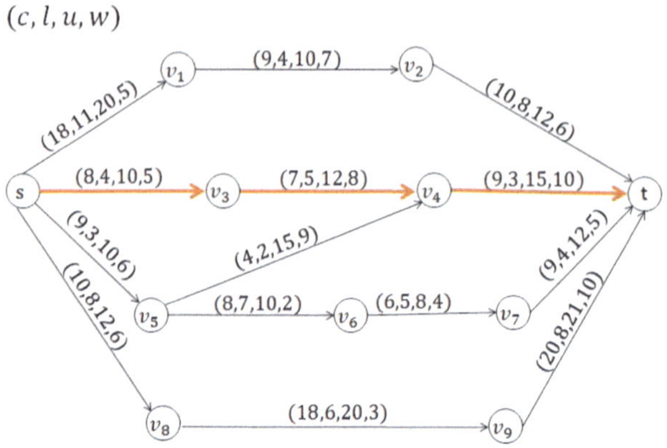

Fig. 3.2 An example of problem (RIOVMCP$_1$)

3.2.3 Numerical Experiments

In this subsection, we will first present a specific example to demonstrate the process of Algorithm 3.2 and then provide numerical experiments to verify the effectiveness of the algorithm.

Example 3.2 We now introduce a specific example to verify the correctness of Algorithm 3.2, as shown in Fig. 3.2:

$N := \{s, v_1, v_2, ..., v_9, t\}$,

$E := \{e_{s,1}, e_{1,2}, e_{2,t}, e_{s,3}, e_{3,4}, e_{4,t}, e_{s,5}, e_{5,4}, e_{5,6}, e_{6,7}, e_{7,t}, e_{s,8}, e_{8,9}, e_{9,t}\}$,

$c := \{18, 9, 10, 8, 7, 9, 9, 4, 8, 6, 9, 10, 18, 20\}$,

$l := \{11, 4, 8, 4, 5, 3, 3, 2, 7, 5, 4, 8, 6, 8\}$,

$u := \{20, 10, 12, 10, 12, 15, 10, 15, 10, 8, 12, 12, 20, 21\}$,

$w := \{5, 7, 6, 5, 8, 10, 6, 9, 2, 4, 5, 6, 3, 10\}$.

In Fig. 3.2, assume that the given path is $\bar{P} := \{e_{s,3}, e_{3,4}, e_{4,t}\}$, then $c(\bar{P}) := 7$.

(a) When $L := 1$, according to Algorithm 3.2, we have the following process.

1. Compute $\underline{C} := \min_{e \in \bar{P}}\{l(e)\} := 3$, $\bar{C} := \min_{e \in \bar{P}}\{u(e)\} := 10$, $E^{>} := \{e \in E | l(e) > L\} := \{e_{s,3}, e_{3,4}, e_{4,t}\}$.
2. Since $L < \underline{C} := 3$, at this point, no matter how the capacities of the arcs in path $\bar{P}$ are reduced, it is not possible to make $\bar{c}(\bar{P}) := L := 1$, and this instance is infeasible.

(b) When $L := 12$, according to Algorithm 3.2, we have the following process.

1. Compute $\underline{C} := \min_{e \in \bar{P}}\{l(e)\} := 3$, $\bar{C} := \min_{e \in \bar{P}}\{u(e)\} := 10$, $E^{>} := \{e \in E | l(e) > L\} := \emptyset$.

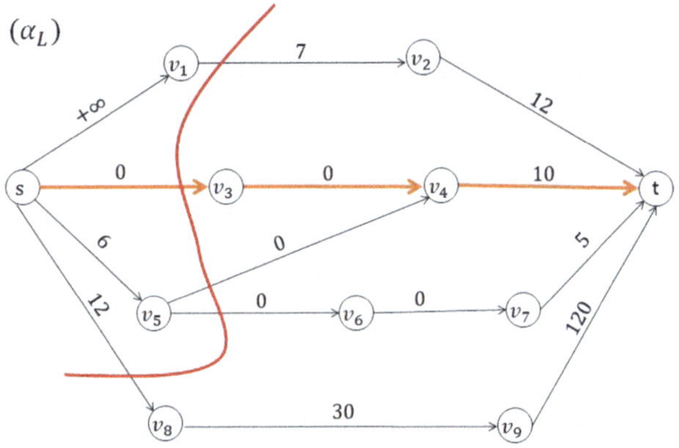

Fig. 3.3 The Network G_L and its minimum cost cut

2. Since $L > \bar{C} := 10$, arcs $e_{3,4}$ and $e_{4,t}$ can be increased to L, while the upper bound that arc $e_{s,3}$ can be increased to 10. At this point, the capacity value of path $\bar{P}$ cannot be equal to $L := 12$, and this instance is infeasible.

(c) When $L := 8$, according to Algorithm 3.2, we have the following process.

1. Compute $\underline{C} := \min_{e \in \bar{P}}\{l(e)\} := 3$, $\bar{C} := \min_{e \in \bar{P}}\{u(e)\} := 10$, $E^{>} := \{e \in E | l(e) > L\} := \{e_{s,1}\}$.
2. Since $\underline{C} \le L \le \bar{C}$, in network $G := (N, E, c, s, t)$, there is no $s - t$ path satisfying $P \subseteq E^{>}$; therefore, this problem is feasible, and the algorithm proceeds to line 4.
4. Compute α_L from (3.29) and obtain the minimum cost cut $(X, \bar{X}) = \{e_{1,2}, e_{s,3}, e_{5,4}, e_{5,6}, e_{s,8}\}$ of $G_L := (N, E, \alpha_L, s, t)$ under the cost vector α_L, as shown in Fig. 3.3.
5. Compute $A := \{e \in (X, \bar{X}) | \alpha_L(e) \neq 0\} := \{e_{1,2}, e_{s,8}\}$ and $\bar{P}^{<} := \{e \in \bar{P} | c(e) < L\} := \{e_{3,4}\}$.
6. Compute $\bar{c} := (18, 8, 9, 8, 8, 10, 8, 9, 4, 8, 6, 9, 18, 20)$, the optimal value $\bar{z} := \sum_{e \in (X, \bar{X})} \alpha_L(e) + \sum_{e \in \bar{P}^{<}} w(e)(L - c(e)) := 27$. It is known that path $\bar{P}$ is the maximum capacity path. The network $\bar{G}$ is shown in Fig. 3.4.

Now let us present the results of the numerical experiments for Algorithm 3.2, where the parameters are chosen similar to those for Algorithm 3.1.

In Table 3.2, it can be seen that for each category of random graphs, the running time of Algorithm 3.2 is consistent with the time complexity $O(mn)$ of the algorithm, thereby verifying the effectiveness of the algorithm.

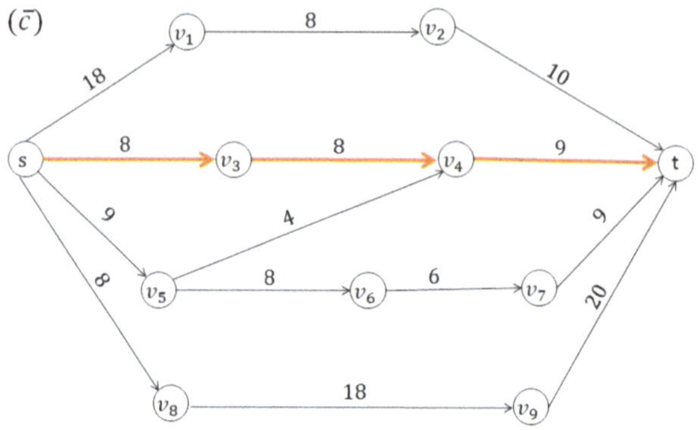

Fig. 3.4 The optimal network $\bar{G}$

Table 3.2 Experimental results of Algorithm 3.2

n	200	500	1000	2000	5000	10,000
m	500	800	5000	8000	18,000	40,000
Atime	0.0594	0.2498	3.9705	12.1372	64.7864	271.3476
min_time	0.0473	0.2138	3.7524	10.9868	61.3271	265.1438
max_time	0.0702	0.2583	4.3142	12.8953	67.0658	283.4628

3.3 (RIOV, Maximum Capacity Path, Bounded, Weighted l_∞ Norm)

This section will specifically introduce the restricted inverse optimal value problem on maximum capacity path under weighted l_∞ norm. While keeping the capacity value of the given path unchanged, an error in the algorithm (**AIMCPM**) proposed by Adrian Deaconu and Javad Tayyebi [35] is modified and a revised algorithm is proposed. When the capacity value of the given path can be changed, we first present the specific mathematical model, then design an algorithm with the same complexity of $O(m)$ as the former case. Finally, numerical experiments are conducted to illustrate the process and its effectiveness.

Let $G = (N, E, c, w, s, t)$ be a weighted directed connected graph and $\bar{P}$ be a given $s - t$ path in graph G. The problem (RIOVMCP$_\infty$) under the weighted l_∞ norm can be represented as

$$\min \max_{e\in E} w(e)|\bar{c}(e) - c(e)|$$

$$(\mathbf{RIOVMCP}_\infty)\ s.t.\ \bar{c}(\bar{P}) = L, \tag{3.32}$$

$$l(e) \le \bar{c}(e) \le u(e), \tag{3.33}$$

$$\max_P \bar{c}(P) = \bar{c}(\bar{P}). \tag{3.34}$$

3.3.1 The Capacity Value of the Given Path Remains Unchanged

Under the weighted l_∞ norm, with the capacity of the given path remaining at $c(\bar{P})$ and only allowing for the reduction of capacities, Deaconu and Tayyebi [35] proposed an algorithm with a time complexity of $O(m)$. They refer to this problem as the Inverse Maximum Capacity Path under Max-type distance, **(IMCPM)** and present its mathematical model as follows:

$$
\begin{aligned}
& \min \max_{e \in E} w(e)|\bar{c}(e) - c(e)| \\
\textbf{(IMCPM)}\quad & s.t.\ \bar{c}(\bar{P}) = c(\bar{P}), \\
& \bar{c}(e) \geq l(e), e \in E, \\
& \max_{P} \bar{c}(P) = \bar{c}(\bar{P}).
\end{aligned}
$$

By introducing the arc sets $\tilde{A}_1 = \{e \in E | l(e) > c(\bar{P})\}$ and $\tilde{A}_2 = \{e \in E | c(e) \leq c(\bar{P})\}$, and letting $\hat{A} = E - (\tilde{A}_1 \cup \tilde{A}_2)$, the aforementioned model can be transformed into the following model [35].

$$
\begin{aligned}
& \min \max_{e \in E} w(e)|\bar{c}(e) - c(e)| \\
& s.t.\ \bar{c}(\bar{P}) = c(\bar{P}), \\
& \quad \bar{c}(e) = c(e), e \in E - \hat{A}, \\
& \quad \bar{c}(e) \in \{c(e), c(\bar{P})\}, e \in \hat{A}, \\
& \quad \max_{P} \bar{c}(P) = \bar{c}(\bar{P}).
\end{aligned} \tag{3.35}
$$

Theorem 3.11 ([35]) *Problem (3.35) is feasible if and only if there is no s-t path in the directed graph $\tilde{G}_{\tilde{A}_1} = (N, \tilde{A}_1)$.*

Theorem 3.12 ([35]) *If problem (3.35) has an optimal solution, then there exists an optimal solution $\tilde{c}$ satisfying:*

1. *Arcs in sets $\tilde{A}_1$ and $\tilde{A}_2$ remain unchanged, that is, $\tilde{c}(e) = c(e),\ e \in \tilde{A}_1 \cup \tilde{A}_2$.*
2. *If the capacity of an arc is changed, then its value becomes $c(\bar{P})$, and*

$$
\tilde{c}(e) = \begin{cases} c(\bar{P}), \ \bar{c}(e) < c(e); \\ c(e), \ \textit{otherwise}. \end{cases}
$$

Theorem 3.13 ([35]) *The solution to problem (3.35) takes the following form:*

$$\bar{c}(e) = \begin{cases} c(\bar{P}), & e \in B; \\ c(e), & E - B. \end{cases}$$

Where B is the set of arcs from $\hat{A}$ that minimizes the objective function by eliminating some arcs.

In the original text [35], for any arc $e \in \hat{A}$, define $\hat{c}(e) = w(e)(c(e) - \bar{c}(P))$. Construct the auxiliary network $\hat{G} = (N, \hat{A}, \hat{c}, s, t)$. In their designed algorithm (**AIMCPM**), it is necessary to find an $s - t$ path or a maximum capacity path in network $\hat{G}$, which only contains arcs from $\hat{A}$ and does not include arcs from $\tilde{A}_1$ and $\tilde{A}_2$. *This causes an error* in the algorithm (**AIMCPM**) because there may not be an s-t path in $\hat{G}$, *leading to the solution obtained by the algorithm not necessarily being optimal.* Therefore, *we need to reconstruct the auxiliary network $\hat{G}_1 = (N, E, \hat{c}_1, s, t)$ and redefine $\hat{c}$ to ensure that there is an s-t path in $\hat{G}_1$.*

We construct the auxiliary network $\hat{G}_1 = (N, E, \hat{c}_1, s, t)$, where N is the set of vertices of G, E is the set of arcs of G, and the cost $\hat{c}_1(e)$ of arc e is defined as:

$$\hat{c}_1(e) = \begin{cases} +\infty, & e \in \tilde{A}_1; \\ 0, & e \in \tilde{A}_2; \\ w(e)(c(e) - \bar{c}(P)), & e \in \hat{A}. \end{cases} \tag{3.36}$$

The modified algorithm is as follows.

Algorithm 3.3 An algorithm to solve (**IMCPM**) under weighted l_∞ norm

---

Input: A weighted directed connected graph $G := (N, E, c, w, s, t)$, a capacity vector c, a cost vector w, a path $\bar{P}$, capacity of path $\bar{P}$, source s, sink t.

Output: An optimal solution $\bar{c}$ and its optimal value $\bar{z}$.

1: Let $\tilde{A}_1 := \{e \in E | l(e) > c(\bar{P})\}$.

2: **if** there is an $s - t$ path in $\tilde{G}_{\tilde{A}_1} := (N, \tilde{A}_1)$ **then** **return** "the instance is infeasible."

3: **else** let $\tilde{A}_2 := \{e \in E | c(e) \leq c(\bar{P})\}$, $\hat{A} := E - (\tilde{A}_1 \cup \tilde{A}_2)$.

4: Compute $\hat{c}_1$ from (3.36), find the maximum capacity path $\hat{P}$ in $\hat{G}_1 := (N, E, \hat{c}_1, s, t)$ under the cost vector $\hat{c}_1$, and denote its capacity as $\hat{c}_1(\hat{P})$.

5: Let $B := \{e \in E | 0 < \hat{c}_1(e) \leq \hat{c}_1(\hat{P})\}$.

6: Calculate

$$\bar{c}(e) := \begin{cases} c(\bar{P}), & e \in B; \\ c(e), & \text{otherwise}. \end{cases} \tag{3.37}$$

7: **Return** an optimal solution $\bar{c}$ and its value $\bar{z} := \max_{e \in B} w(e)(c(e) - \bar{c}(P))$.

8: **end if**

Theorem 3.14 *If problem* **(IMCPM)** *is feasible, then the solution $\bar{c}$ outputted by Algorithm 3.3 is an optimal solution to problem* **(IMCPM)**.

Proof Assume Problem **(IMCPM)** is feasible. By Theorem 3.12, it suffices to modify the capacities of some arcs to $c(\bar{P})$. Hence, we only need to change the capacities of the arcs corresponding to $\hat{c}(\hat{P})$.

Therefore, adjusting the capacities of the arcs in set B to $c(\bar{P})$ ensures both the feasibility conditions given by (3.37) and the optimality of the objective value under the weighted l_∞ norm, as ensured by the maximum capacity path in $\hat{G}_1$ under $\hat{c}_1(e)$. Additionally, the $\bar{c}$ defined by (3.37) also satisfies the properties of an optimal solution described in Theorem 3.12. □

Theorem 3.15 *The time complexity of Algorithm 3.3 is $O(m)$.*

Proof Line 2 for finding an $s-t$ path in $\tilde{G}_{\tilde{A}_1} = (N, \tilde{A}_1)$ can be solved using a depth-first search algorithm in $O(m)$. Line 4 for computing the maximum capacity path in $\hat{G}_1$ can be solved in $O(m)$ time [108]. This completes the proof. □

Next, we will first present a specific example to demonstrate the operation of Algorithm 3.3 and then provide numerical experiments to verify the effectiveness of the algorithm.

Example 3.3 We now introduce a specific example to illustrate the specific errors in [35] and to verify the correctness of Algorithm 3.3, as shown in Fig. 3.5.

$V := \{s, v_1, v_2, ..., v_9, t\}$,

$E := \{e_{s,1}, e_{1,2}, e_{2,t}, e_{s,3}, e_{3,4}, e_{4,t}, e_{s,5}, e_{5,4}, e_{5,6}, e_{6,7}, e_{7,t}, e_{s,8}, e_{8,9}, e_{9,t}\}$,

$c := (18, 9, 10, 8, 7, 9, 9, 4, 8, 6, 9, 10, 18, 20)$,

$l := (11, 4, 8, 4, 5, 3, 3, 2, 7, 5, 4, 8, 6, 8)$,

$w := (5, 7, 6, 5, 8, 10, 6, 9, 2, 4, 5, 6, 3, 10)$.

In Fig. 3.5, assume the given path is $\bar{P} := \{e_{s,3}, e_{3,4}, e_{4,t}\}$, then $c(\bar{P}) := 7$.

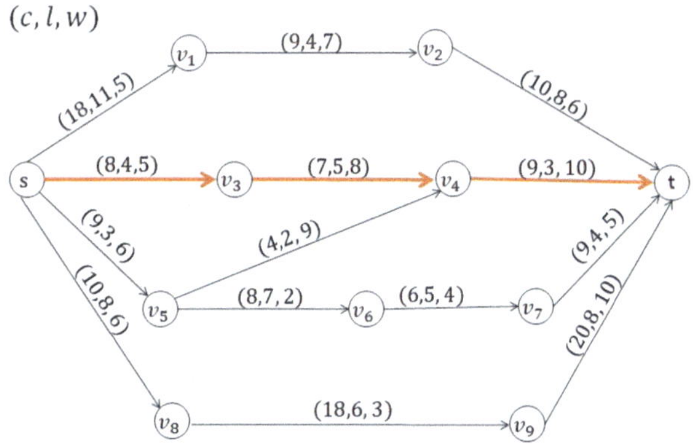

Fig. 3.5 An example of problem (IMCPM)

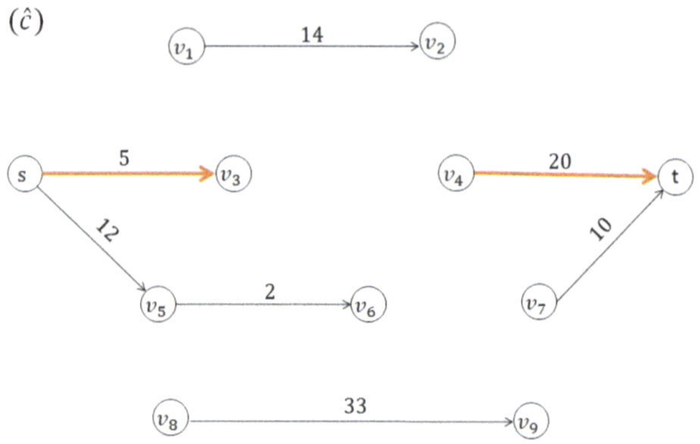

Fig. 3.6 The network $\hat{G}$

(a) The operation process of the original algorithm in [35] is as follows.

1. Compute $\tilde{A}_1 := \{e_{s,1}, e_{2,t}, e_{s,8}, e_{9,t}\}$.

2–4. Since there is no $s - t$ path in the network $\tilde{G}_{\tilde{A}_1} := (N, \tilde{A}_1)$, the problem is feasible.

5. Compute $\hat{A} := \{e_{1,2}, e_{s,3}, e_{4,t}, e_{s,5}, e_{5,6}, e_{7,t}, e_{8,9}\}$.

6–10. Draw the network $\hat{G} := (N, \hat{A})$ as shown in Fig. 3.6.

It follows from Fig. 3.6 that there is no $s - t$ path in network $\hat{G}$, and according to the original algorithm, $\bar{c} := c$ is the optimal solution. However, from Fig. 3.5, $c(\bar{P}) := 7$, and considering the path $s - v_1 - v_2 - t$ as P, then $c(P) := 9 > c(\bar{P}) := 7$, it is clear that $\bar{c} := c$ is not the optimal solution. Furthermore, in step 10 of the algorithm, the capacity of the maximum capacity path cannot be calculated because there is no $s - t$ path.

(b) According to Algorithm 3.3, we have the following process.

1. Compute $\tilde{A}_1 := \{e_{s,1}, e_{2,t}, e_{s,8}, e_{9,t}\}$.

2. There is no $s - t$ path in $\tilde{G}_{\tilde{A}_1}$, so the problem is feasible, and the algorithm proceeds to step 3.

3. Compute $\tilde{A}_2 := \{e_{3,4}, e_{5,4}, e_{6,7}\}$, $\hat{A} := \{e_{1,2}, e_{s,3}, e_{4,t}, e_{s,5}, e_{5,6}, e_{7,t}, e_{8,9}\}$.

4. Compute $\hat{c}_1$ from (3.36), and compute the maximum capacity path $\hat{P} := \{e_{s,8}, e_{8,9}, e_{9,t}\}$ in $\hat{G}_1$ under $\hat{c}_1$, with $\hat{c}_1(\hat{P}) := 33$, as shown in Fig. 3.7.

5. Compute $B := \{e_{1,2}, e_{s,3}, e_{4,t}, e_{s,5}, e_{5,6}, e_{7,t}, e_{8,9}\}$.

6–7. Return an optimal solution $\bar{c} := (18, 7, 7, 10, 7, 10, 7, 7, 4, 7, 6, 7, 7, 20)$ and the optimal value $\bar{z} = \max_{e \in B} w(e)(c(e) - \bar{c}(P)) = 33$. In this case, $\bar{P}$ is the maximum capacity path. The network $\bar{G}$ is as shown in Fig. 3.8.

Now let us present the results of the numerical experiments for Algorithm 3.3, where the parameters are chosen as those in Algorithm 3.1.

As can be seen in Table 3.3, for each category of random graphs, the running time of Algorithm 3.3 is consistent with the time complexity $O(m)$ of the algorithm.

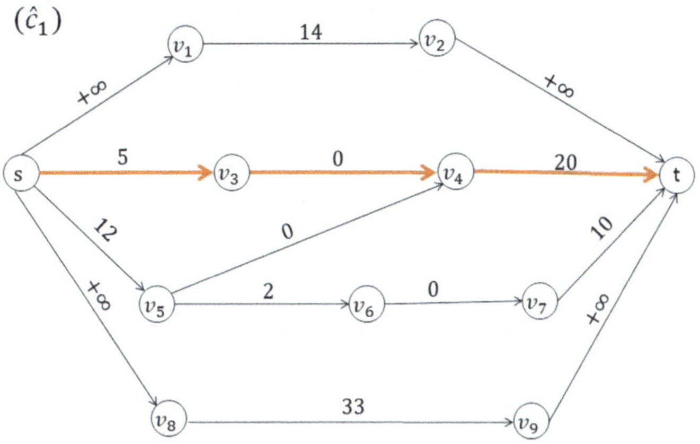

Fig. 3.7 The network $\hat{G}_1$

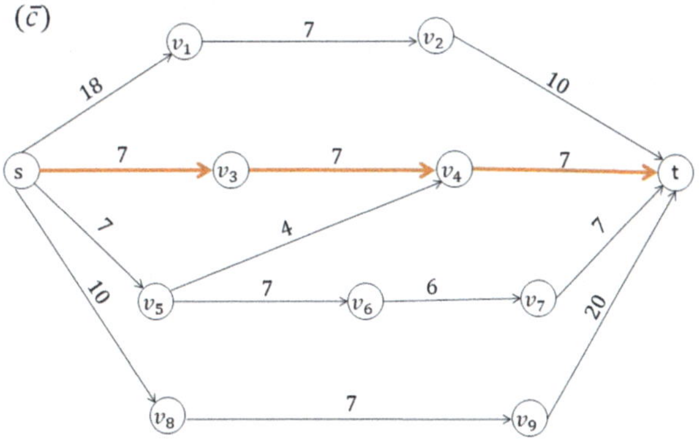

Fig. 3.8 The optimal network $\bar{G}$

Table 3.3 Experimental Results of Algorithm 3.3

n	200	500	1500	2000	6000	10,000
m	500	800	4000	9000	16,000	40,000
Atime	0.0592	0.1352	0.6032	1.3975	2.4217	6.0315
min_time	0.0378	0.1043	0.5431	1.1678	2.1386	5.7412
max_time	0.0684	0.1521	0.7421	1.7692	2.7912	6.4218

3.3.2 Solve the Problem (RIOVMCP$_\infty$)

This subsection investigates the problem (**RIOVMCP$_\infty$**) under the weighted l_∞ norm. We first present its mathematical model, then design an algorithm, followed by numerical experiments.

The mathematical model of the problem (**RIOVMCP**$_\infty$) is as follows.

$$
(\mathbf{RIOVMCP}_\infty)\quad
\begin{aligned}
&\min \max_{e\in E} w(e)|\bar{c}(e) - c(e)| \\
&s.t.\ \bar{c}(\bar{P}) = L, \\
&\qquad l(e) \le \bar{c}(e) \le u(e), \\
&\qquad \max_{P} \bar{c}(P) = \bar{c}(\bar{P}).
\end{aligned}
$$

The feasibility and optimality conditions for problem (**RIOVMCP**$_\infty$) can be ensured by Theorems 3.7 and 3.8. Hence, in the optimal solution $\bar{c}$, if the capacity of some arcs changes, the changed capacity will necessarily become L.

When the problem is feasible, under the weighted l_1 norm, in order to achieve $\bar{c}(\bar{P}) = L$, we need to increase the capacity of the arcs in $\bar{P}^<$ to L, and then determine the minimum cost cut $(X, \bar{X})$ in the auxiliary network $G_L = (N, E, \alpha_L, s, t)$, thereby determining which arcs in $(X, \bar{X})$ and $\bar{P}^<$ need to have their capacities changed. However, under the weighted l_∞ norm, we only need to find the maximum capacity path P_L in G_L, then $c(P_L)$ is the minimum value of the objective function, and define the set $B = \{e \in E|\alpha_L \le c(P_L)\}$. Due to the existence of upper and lower bounds for each arc, we define $A_1 = \{e \in E|l(e) \le L \le u(e)\}$. For the fixed value L, only the arcs in the set A_1 can change their capacity values, and changing all arcs in $A_1 \cap B$ will not affect the objective value while keeping all constraints. Therefore, we only need to adjust the capacities of arcs $e \in A_1 \cap B$ to L.

Algorithm 3.4 An algorithm for solving the problem (**RIOVMCP**$_\infty$)

Input: Weighted directed connected graph $G := (V, E, c, w, s, t)$, capacity vector c, cost vector w, path $\bar{P}$, capacity of path $\bar{P}$, fixed value L, source s, sink t.
Output: Optimal solution $\bar{c}$ and optimal value $\bar{z}$.

Compute $\underline{C} := \min_{e\in\bar{P}}\{l(e)\}$, $\bar{C} := \min_{e\in\bar{P}}\{u(e)\}$, $E^> := \{e \in E|l(e) > L\}$.
if $L < \underline{C}$ or $L > \bar{C}$, or there exists an $s-t$ path P, satisfying $P \subseteq E^>$ **then**
  return "the instance is infeasible."
else calculate α_L from (3.29), the maximum capacity path P_L in $G_L := (N, E, \alpha_L, s, t)$ under α_L, and denote its capacity as $c(P_L)$.
  Let $A_1 := \{e|l(e) \le L \le u(e)\}$ and $B := \{e \in E|\alpha_L(e) \le c(P_L)\}$.
  Calculate

$$
\bar{c}(e) := \begin{cases} L, & e \in A_1 \cap B; \\ c(e), & \text{otherwise}. \end{cases} \tag{3.38}
$$

  Return optimal solution $\bar{c}$ and its value $\bar{z} := \max_{e\in A_1\cap B} w(e)(L - c(e))$.
end if

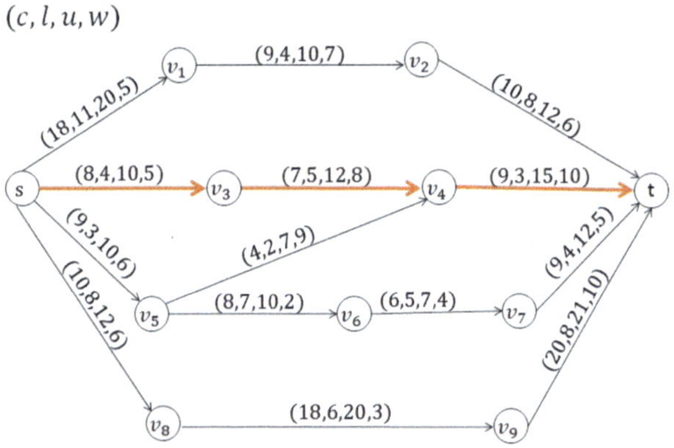

Fig. 3.9 An example of problem (RIOVMCP$_\infty$)

Based on the analysis before Algorithm 3.4, we can conclude that

Theorem 3.16 *If the problem (**RIOVMCP$_\infty$**) is feasible, then the vector $\bar{c}$ in* (3.38) *is an optimal solution to the problem (**RIOVMCP$_\infty$**).*

It follows from Theorem 3.15 that

Corollary 3.1 *The time complexity of Algorithm 3.4 is $O(m)$.*

3.3.3 Numerical Experiments

Next, we will first present a specific example to demonstrate the process of Algorithm 3.4 and then provide numerical experiments.

Example 3.4 We now introduce a specific example to verify the correctness of Algorithm 3.4, as shown in Fig. 3.9.

$V := \{s, v_1, v_2, ..., v_9, t\}$,

$E := \{e_{s,1}, e_{1,2}, e_{2,t}, e_{s,3}, e_{3,4}, e_{4,t}, e_{s,5}, e_{5,4}, e_{5,6}, e_{6,7}, e_{7,t}, e_{s,8}, e_{8,9}, e_{9,t}\}$,

$c := \{18, 9, 10, 8, 7, 9, 9, 4, 8, 6, 9, 10, 18, 20\}$,

$l := \{11, 4, 8, 4, 5, 3, 3, 2, 7, 5, 4, 8, 6, 8\}$,

$u := \{20, 10, 12, 10, 12, 15, 10, 7, 10, 7, 12, 12, 20, 21\}$,

$w := \{5, 7, 6, 5, 8, 10, 6, 9, 2, 4, 5, 6, 3, 10\}$.

In Fig. 3.9, assume that the given path is $\bar{P} := \{e_{s,3}, e_{3,4}, e_{4,t}\}$ and $c(\bar{P}) := 7$.

(a) When $L := 1$, according to Algorithm 3.4, we have the following process.

1. Compute $\underline{C} := \min_{e\in\bar{P}}\{l(e)\} := 3$, $\bar{C} := \min_{e\in\bar{P}}\{u(e)\} := 10$, $E^> := E$.
2. Since $L < \underline{C} := 3$, then the instance is infeasible since it is impossible to make $\bar{c}(\bar{P}) := L := 1$,

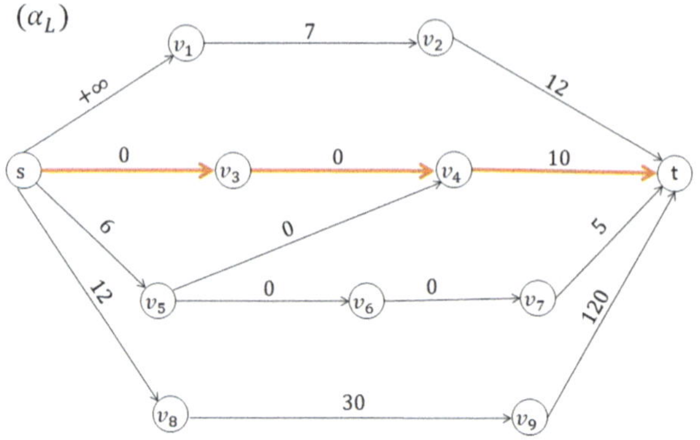

Fig. 3.10 The network G_L

(b) When $L := 12$, according to Algorithm 3.4, we have the following process.

1. Compute $\underline{C} := 3$, $\bar{C} := 10$, $E^{>} := \emptyset$.
2. Since $L > \bar{C}$, then arcs $e_{3,4}$ and $e_{4,t}$ can be increased to L, while the upper bound of $e_{s,3}$ can be increased to 10, which means that this instance is infeasible.

(c) When $L := 8$, according to Algorithm 3.4, we have the following process.

1. Compute $\underline{C} := 3$, $\bar{C} := 10$, $E^{>} := \{e_{s,1}\}$. Then $\underline{C} \le L \le \bar{C}$ and there is no $s-t$ path P satisfying $P \subseteq E^{>}$ in network $G := (N, E, c, s, t)$. Hence, the algorithm proceeds to line 4.
4. Compute α_L from (3.29), compute the maximum capacity path $P_L := \{e_{s,8}, e_{8,9}, e_{9,t}\}$ in $G_L := (N, E, \alpha_L, s, t)$ under α_L, and its capacity is $c(P_L) := 12$. The network G_L can be shown in Fig. 3.10.
5. Compute $A_1 := \{l(e) \le L \le u(e)\} := \{e_{1,2}, e_{2,t}, e_{s,3}, e_{3,4}, e_{4,t}, e_{s,5}, e_{5,6}, e_{7,t}, e_{s,8}, e_{8,9}, e_{9,t}\}$ and $B := \{e \in E|\alpha_L(e) \le c(P_L)\} := \{e_{1,2}, e_{2,t}, e_{s,3}, e_{3,4}, e_{4,t}, e_{s,5}, e_{5,4}, e_{5,6}, e_{6,7}, e_{7,t}, e_{s,8}\}$.
7. Compute $\bar{c} := (18, 8, 8, 8, 8, 8, 8, 8, 4, 8, 6, 8, 18, 20)$, and the optimal value $\bar{z} := \max_{e\in E} w(e)|\bar{c}(e) - c(e)| := 12$. The network $\bar{G}$ is shown in Fig. 3.11.

(d) When $L := 4$, according to Algorithm 3.4, we have the following process.

1. Compute $\underline{C} := 3$, $\bar{C} := 10$, $E^{>} := \{e_{s,1}, e_{2,t}, e_{3,4}, e_{5,6}, e_{6,7}, e_{s,8}, e_{8,9}, e_{9,t}\}$.
2. In network $G := (N, E, c, s, t)$, there exists an $s-t$ path $P := \{e_{s,8}, e_{8,9}, e_{9,t}\}$ satisfying $P \subseteq E^{>}$. Then this instance is infeasible.

Now let us present the results of the numerical experiments for Algorithm 3.4, where the parameters are chosen as those in Algorithm 3.1.

In Table 3.4, it can be seen that for each category of random graphs, the running time of Algorithm 3.4 is consistent with the time complexity $O(m)$ of the algorithm.

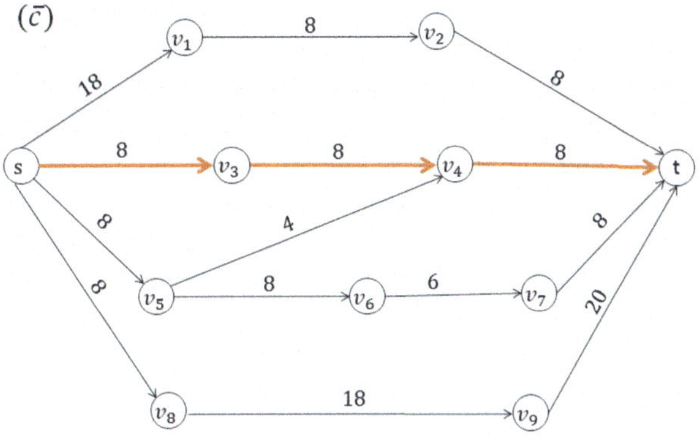

Fig. 3.11 The optimal network $\bar{G}$

Table 3.4 Experimental results of Algorithm 3.4

n	200	500	1500	2000	6000	10,000
m	500	800	4000	9000	16,000	40,000
Atime	0.0530	0.1136	0.5897	1.4281	2.3520	5.9813
min_time	0.0428	0.0987	0.5371	1.1638	2.1459	5.6210
max_time	0.0641	0.1306	0.6431	1.8673	2.5903	6.3642

3.4 Conclusions

This chapter primarily investigates the restricted inverse optimal value problem on the maximum capacity path under the weighted l_1 norm and l_∞ norm, as well as the interdiction problem on the maximum capacity path under the weighted l_1 norm. It is worthy noting that we modify the algorithm (**AIMCPM**) proposed by Deaconu and Tayyebi [35] by revising an error for the Problem (RIOVMCP$_\infty$ when the capacity value of the given path unchanged. For the aforementioned problems, we first present the mathematical models, then design solution algorithms, prove the optimality of the algorithms, and analyze their time complexities. Finally, numerical experiments are conducted to illustrate the specific process of the algorithms and validate their effectiveness.

For the interdiction problem under the weighted l_1 norm, we first prove a series of important properties. Then, based on the idea of binary search, in each iteration, we solve the minimum cost cut in the auxiliary network $G_k = (N, E, \rho_k, s, t)$ and determine whether the costs of the minimum cost cuts in networks $G_{\bar{b}} = (N, E, \rho_{\bar{b}}, s, t)$ and $G_k = (N, E, \rho_k, s, t)$ are the same when the capacity value is exactly $z_{\bar{b}}$. This leads to two methods for determining the optimal target value, resulting in an algorithm with a time complexity of $O(m^2 \log m)$. This time complexity greatly reduces the time complexity $O(m^2 \log^2 m\Lambda(n, m))$ of the algorithm solving the unbounded problem proposed by Javad Tayyebi et al. [35], where $\Lambda(n, m)$ represents the complexity of finding the minimum cost cut.

For the restricted inverse optimal value problem under the weighted l_1 norm, the main idea is to solve the minimum cost cut in the auxiliary network G_L, providing an algorithm with a time complexity of $O(mn)$. Under the weighted l_∞ norm with the capacity values of given paths remaining unchanged, the main idea is to find the maximum capacity path in the auxiliary network $\hat{G}_1$, offering an algorithm with a time complexity of $O(m)$. For the general case, under the weighted l_∞ norm, the main concept is to find the maximum capacity path in the auxiliary network G_L, with an algorithm having a time complexity of $O(m)$.

Table 3.5 shows the relationship between the research findings of this chapter and the existing research findings.

Table 3.5 Research results on generalized inverse maximum capacity path (MCP) problem

Problem type	Norm	Time complexity	Reference
Inverse MCP	Unit l_1 norm	$O(m\Lambda(n, m))$	[132]
	(Bounded, weighted l_∞ norm)	$O(m^3 \log n)$	[133]
	(Unbounded, weighted l_∞ norm)	$O(m^2 \log n)$	
	(Unbounded, l_∞ norm)	$O(m \log n)$	
	Weighted l_1 norm	$O(m\Lambda(n, m))$	[56]
	Bottleneck HD	$O(m \log^2 n)$	
	Sum-type HD	$O(m\Lambda(n, m))$	
	Any sum-type norm	$O(m^2 n)$	[35]
	Any maximum distance	$O(m^2)$	
RIOV on MCP	Weighted l_1 norm	$O(mn)$	[35] special case
		$O(mn)$	Section 3.2
	Weighted l_∞ norm	$O(m)$	[35] special case (**Incorrect**) (**algorithm**.)
		$O(m)$	Section 3.3
Interdiction MCP	Sum-type HD	$O(n^3 \sqrt{m} \log(n))$	[95]
	(Unbounded, l_1 norm)	$O(m^2 \log^2 m\Lambda(n, m))$	[125]
	(Bounded, l_1 norm)	$O(m^2 \log(m))$	Section 3.1

For future research, for the restricted inverse optimal value problem on the maximum capacity path, one can explore issues under the bottleneck Hamming distance and sum Hamming distance, as well as inverse optimal value problems under other norms.

Chapter 4
Some General Methods to Solve Inverse Linear Programming Problem Under Weighted l_1 Norm

Abstract In this chapter, we study the inverse (bounded) linear programming problem under weighted l_1 norm (**I(B)LP**). Given a feasible solution x^0 of the standard (**LP**) problem $\min\{c^T x | Ax = b, x \geq 0\}$, the objective is to modify the cost vector c to $\tilde{c}$ such that x^0 becomes an optimal solution of the modified (**LP**) problem, while minimizing $\|\tilde{c} - c\|$ under weighted l_1 norm. This chapter presents a comprehensive study on tackling inverse optimization problems through the design of general algorithms. For the inverse problem (**ILP**), a mathematical model of (**ILP**) in view of extreme vertices is examined, followed by two distinct solution approaches: the column generation method and the revised simplex method. Notably, the column generation method incorporates dual operator updates through complementary slackness conditions and a row generation technique. Subsequently, we delve into the resolution of the inverse bounded problem (**IBLP**), proposing two primal-dual algorithms grounded in duality theory. These algorithms consider the model (**IBLP**) from both primal and dual perspectives, with a focus on enhancing the dual-based algorithms' properties to yield an improved primal-dual algorithm. The generality of these algorithms allows for their application to various concrete combinatorial optimization issues, offering potential for the development of more efficient algorithms to inverse combinatorial optimization challenges.

Keywords Inverse linear programming problem · Inverse bounded linear programming problem · Column generation method · Row generation method · Primal-dual method

4.1 Introduction

In this chapter, we focus on some general methods to solve the inverse linear programming problem (**ILP**). Consider a standard linear programming problem (**LP**) of the form $\min\{c^T x | Ax = b, x \geq 0\}$ and define $F = \{x | Ax = b, x \geq 0\}$

Yunwei Zhang and Xiucui Guan have equally contributed to this chapter.

X. Guan et al., *Inverse Combinatorial Optimization Problems*, Springer Optimization and Its Applications 225, https://doi.org/10.1007/978-3-031-91175-0_4

as the domain of feasible solutions. We seek to modify the cost vector c to $\tilde{c}$ such that a given feasible solution $x^0 \in F$ becomes an optimal solution of the problem $\min_{x\in F}\{\tilde{c}^T x\}$. Our objective is to minimize the modification $\|c - \tilde{c}\|$ under some norm. The problem **(ILP)** can be formulated as follows:

$$\begin{aligned} &\min_{\tilde{c}} ||\tilde{c} - c|| \\ &\text{s.t.} \ \min_{x\in F} \tilde{c}^T x = \tilde{c}^T x^0. \end{aligned}$$

According to the given constraints, there is an equivalent mathematical model:

$$\begin{aligned} &\min_{\tilde{c}} ||\tilde{c} - c|| \\ &\text{s.t.} \ \ \tilde{c}^T x \geq \tilde{c}^T x^0, \ \forall x \in F \end{aligned} \tag{4.1}$$

There are many studies of inverse combinatorial optimization based on the model (4.1) and a main algorithm is a column generation method. Zhang et al. [149] studied the inverse shortest path problem under l_1 norm and designed a column generation method that transforms the problem of choosing the entering variable in each iteration into a shortest path problem with a new cost vector. Yang and Zhang [131] proposed two distinct general algorithms based on column generation and the ellipsoid method, respectively, to solve the inverse combinatorial optimization problem under l_1 norm. Zhang and Liu [146] proposed the Newton method to solve the inverse combinatorial optimization problem under l_∞ norm. Wang et al. [129] designed an algorithm called cutting plane method, which is based on column generation in principle, to solve the inverse integer linear programming problem under l_1 norm. Tayyebi and Aman [124] proposed an algorithm based on binary search technique to solve the inverse linear programming problem under the bottleneck-type weighted Hamming distance. Guan et al. [52] proposed a revised simplex method to solve the inverse max+sum spanning tree problem under weighted l_1 norm. Fathali [41] proposed a row generation method for the inverse continuous facility location problem under l_1 and l_∞ norm. The row generation method is also a revised variation of the column generation method.

There is another optimality condition according the duality theory. Let $J_0 = \{j | x_j^0 > 0\}$ be the index set of the non-zero component of x^0. The equivalent model of the problem **(ILP)** can be formulated as follows:

$$\begin{aligned} &\min \ \|c - \tilde{c}\| \\ &\text{s.t.} \ \ \pi A_j = \tilde{c}, \ j \in J_0, \\ &\qquad \pi A_j \leq \tilde{c}, \ j \notin J_0. \end{aligned} \tag{4.2}$$

where A_j is the j-th column vector of the matrix A.

Zhang and Liu [143] first studied some important properties of (4.2), then [144] extended the model to combinatorial structures including the shortest path problem and the assignment problem, and demonstrated that their corresponding inverse problems under unit l_1 norm can be transformed into a revised primal problem. Ahuja [5] systematically elaborated the connection between the primal problem and its inverse version and applied it to other combinatorial optimization problems. Karimi and Aman [71] considered the inverse minimum cost flow problem under Hamming distances and proposed a strong polynomial algorithm based on a binary search. Wang et al. [130] studied the inverse assignment problem under l_1 and l_∞ norms with restricted modification.

The structure of the chapter is as follows. In Sect. 4.2, we design some general methods to solve the inverse problem **(ILP)** under weighted l_1 norm. We first study a mathematical model **(ILP1)** and its properties. They we provide two methods to solve **(ILP1)**, employing the column generation method and the revised simplex method, respectively. In each iteration of the column generation method, we provide two ways to update a dual operator. One way is to use complementary slackness conditions, and the other way has been employed to derive a row generation. In Sect. 4.3, we propose two primal-dual algorithms to solve the inverse bounded linear programming **(IBLP)** problem based on duality theories. In such algorithms, we treat the model **(IBLP)** as a primal problem and a dual problem, respectively.

4.2 Some General Methods to Solve the Inverse Problem (ILP) Under Weighted l_1 Norm

4.2.1 A Mathematical Model of (ILP) Under Weighted l_1 Norm

Introduce a nonnegative weight vector of q in the weighted l_1 norm and let $\tilde{c} = c + \alpha - \beta$. Then we can transform the model (4.1) of (ILP) under weighted l_1 norm into the following form.

$$
\textbf{(ILP1)}\quad
\begin{aligned}
&\min\ q^T(\alpha+\beta)\\
&\text{s.t.}\ \ (c+\alpha-\beta)^T x^0 \le (c+\alpha-\beta)^T x,\ \forall x \in F,\\
&\qquad\ \alpha, \beta \ge 0.
\end{aligned}
$$

Firstly, we analyze the feasibility of problem **(ILP1)**.

Theorem 4.1 *There exists a feasible solution for the problem **(ILP1)**, and the objective function has a lower bound 0.*

Proof We define a solution below. Let $\beta_j = c_j, \alpha_j = 0$ for $\forall c_j \ge 0$ and $\alpha_j = -c_j, \beta_j = 0$ for $\forall c_j < 0$. It satisfies the constraints of problem **(ILP1)** since, for any $x \in F$, it holds

$$(c+\alpha-\beta)^T x^0 = (c+\alpha-\beta)^T x = 0.$$

Therefore, the feasible region of the problem is non-empty. Since $q \geq 0$ is a weight vector and $\alpha, \beta \geq 0$, the objective function $q^T(\alpha + \beta) \geq 0$. □

Let $J_0 = \{j | x_j^0 > 0\}$ be the index set of non-zero components of x^0. Next we introduce some properties of an optimal solution of problem **(ILP1)**.

Lemma 4.1 *There is an optimal solution* (α^*, β^*) *of problem* ***(ILP1)*** *satisfying*

$$\beta_j^* = 0, \textit{ for any } j \notin J_0.$$

Proof Suppose there exists an optimal solution (α^1, β^1) with a $\beta_{j_1}^1 > 0$ for some $j_1 \notin J_0$ and the rest values satisfy $\beta_j^1 = 0, j \notin J_0, j \neq j_1$. Then, we can construct a vector (α^2, β^2) such that

$$\begin{cases} \alpha^2 = \alpha^1, \\ \beta_{j_1}^2 = 0, \\ \beta_j^2 = \beta_j^1, \quad j \notin J_0, j \neq j_1. \end{cases}$$

and it implies

$$\begin{aligned} &\sum_{j=1}^{n}(c_j + \alpha_j^2 - \beta_j^2)(x_j^0 - x_j) \\ &= \sum_{j=1, j\neq j_1}^{n} (c_j + \alpha_j^1 - \beta_j^1)(x_j^0 - x_j) + (c_{j_1} + \alpha_{j_1}^1)(-x_{j_1}) \\ &\leq \sum_{j=1, j\neq j_1}^{n} (c_j + \alpha_j^1 - \beta_j^1)(x_j^0 - x_j) + (c_{j_1} + \alpha_{j_1}^1 - \beta_{j_1}^1)(-x_{j_1}) \\ &= \sum_{j=1}^{n}(c_j + \alpha_j^1 - \beta_j^1)(x_j^0 - x_j) \leq 0, \end{aligned}$$

and (α^2, β^2) satisfies feasibility. Additionally, the objective function is

$$\begin{aligned} \sum_{j=1}^{n} q_j(\alpha_j^2 + \beta_j^2) &= \sum_{j=1, j\neq j_1}^{n} q_j(\alpha_j^1 + \beta_j^1) + q_{j_1}\alpha_{j_1}^1 \\ &< \sum_{j=1, j\neq j_1}^{n} q_j(\alpha_j^1 + \beta_j^1) + q_{j_1}(\alpha_{j_1}^1 + \beta_{j_1}^1) = \sum_{j=1}^{n} q_j(\alpha_j^1 + \beta_j^1). \end{aligned}$$

It contradicts to the optimality of (α^1, β^1). □

Let $\underline{J}(x^0) = \{j | x_j^0 \leq x_j, \forall x \in F\}$ and $\bar{J}(x^0) = \{j | x_j^0 \geq x_j, \forall x \in F\}$. Then we can get the following property.

Lemma 4.2 *There is an optimal solution* (α^*, β^*) *of problem* ***(ILP1)*** *satisfying*

$$\beta_j^* = 0, \textit{ for any } j \in \underline{J}(x^0). \tag{4.3}$$

Proof Suppose there exists an optimal solution (α^1, β^1) with a $\beta_{j_1}^1 > 0$ for some $j_1 \in \underline{J}(x^0)$ and the rest values satisfy $\beta_j^1 = 0, j \in \underline{J}(x^0), j \neq j_1$. Then, we can construct a vector (α^2, β^2) such that

$$\begin{cases} \beta_{j_1}^2 = 0, \\ \beta_j^2 = \beta_j^1, \quad j \in \underline{J}(x^0), j \neq j_1, \\ \alpha^2 = \alpha^1. \end{cases}$$

and it implies

$$\begin{aligned}
&\sum_{j=1}^{n}(c_j + \alpha_j^2 - \beta_j^2)(x_j^0 - x_j) \\
&= \sum_{j=1, j \neq j_1}^{n} (c_j + \alpha_j^1 - \beta_j^1)(x_j^0 - x_j) + (c_{j_1} + \alpha_{j_1}^1)(x_{j_1}^0 - x_{j_1}) \\
&\leq \sum_{j=1, j \neq j_1}^{n} (c_j + \alpha_j^1 - \beta_j^1)(x_j^0 - x_j) + (c_{j_1} + \alpha_{j_1}^1 - \beta_{j_1}^1)(x_{j_1}^0 - x_{j_1}) \\
&= \sum_{j=1}^{n}(c_j + \alpha_j^1 - \beta_j^1)(x_j^0 - x_j) \leq 0, \forall x \in F,
\end{aligned}$$

and (α^2, β^2) satisfies feasibility. Similar to the relationship of their objective values in Lemma 4.1, we can obtain the contradiction with the optimality of (α^1, β^1). □

Lemma 4.3 *There is an optimal solution* (α^*, β^*) *of problem* ***(ILP1)*** *satisfying*

$$\alpha_j^* = 0, \textit{ for any } j \in \bar{J}(x^0). \tag{4.4}$$

In the model **(ILP1)**, x is considered as a parameter rather than a variable, so the number of constraint conditions depends on the number of feasible solutions in the feasible region F, which is infinite generally. Problem **(ILP1)** is a linear programming problem with infinite numbers of constraint conditions and $2n$ variables. Next we discuss that there exists an equivalent model of **(ILP1)** that has only a finite number of constraint conditions under certain conditions.

Lemma 4.4 *Assume that $c^T x$ of the problem **(LP)** has a lower bound for any $c \in \mathbb{R}^n$ and $x \in F$. Then problem **(ILP1)** is equivalent to the following model in the sense that they have the same optimal objective value.*

$$\begin{aligned} &\min\ q^T(\alpha+\beta) \\ &\text{s.t.}\ \ (c+\alpha-\beta)^T x^0 \le (c+\alpha-\beta)^T x,\ \forall x \in F_1, \\ &\qquad \alpha, \beta \ge 0, \end{aligned} \tag{4.5}$$

where $F_1 \subset F$ is a set containing all the extreme vertices of F and has finite elements.

Proof Let $F^I = \{(\alpha,\beta) | (c+\alpha-\beta)^T x^0 \le (c+\alpha-\beta)^T x, \forall x \in F, \alpha, \beta \ge 0\}$ and $F_1^I = \{(\alpha,\beta) | (c+\alpha-\beta)^T x^0 \le (c+\alpha-\beta)^T x, \forall x \in F_1, \alpha, \beta \ge 0\}$ be the feasible region of the problem **(ILP1)** and (4.5), respectively. Next we show that $F^I = F_1^I$. It is evident that $F^I \subseteq F_1^I$ since $F_1 \subset F$ and F^I has more constraints than F_1^I. Further, we proceed to demonstrate that $F_1^I \subseteq F^I$.

Recall that the optimal solution of the problem **(LP)** must be achieved at some extreme vertex of the feasible region F. Then we have $\min_{x \in F_1} \tilde{c}^T x = \min_{x \in F} \tilde{c}^T x$, which has a lower bound.

Let (α^1, β^1) be a feasible solution of (4.5), then we have

$$\begin{aligned} (c+\alpha^1-\beta^1)^T x^0 &\le \min_{x \in F_1}(c+\alpha^1-\beta^1)^T x = \min_{x \in F}(c+\alpha^1-\beta^1)^T x \\ &\le (c+\alpha^1-\beta^1)^T x, \forall x \in F, \end{aligned}$$

so (α^1, β^1) is the feasible solution **(ILP1)**, which means $F_1^I \subseteq F^I$.

Now we can conclude that $F^I = F_1^I$ and the problems **(ILP1)** and (4.5) are equivalent in the sense that their optimal values are the same. □

Let Γ be the index set F_1, that is, $F_1 = \{x^t | t \in \Gamma\}$. Then problem (4.5) can be transformed into the following form.

$$\begin{aligned} &\min \sum_{j=1}^n q_j(\alpha_j+\beta_j) \\ &\text{s.t.}\ \sum_{j=1}^n (x_j^0 - x_j^t)\alpha_j - \sum_{j=1}^n (x_j^0 - x_j^t)\beta_j \le \sum_{j=1}^n (x_j^t - x_j^0)c_j,\ t \in \Gamma, \\ &\qquad \alpha_j \ge 0, \qquad j = 1, \ldots, n, \\ &\qquad \beta_j \ge 0, \qquad j = 1, \ldots, n. \end{aligned} \tag{4.6}$$

This is a linear programming model, which has the following matrix form.

$$\begin{aligned} &\min\ q^T\alpha + q^T\beta \\ &\text{s.t.}\ \ H^T\alpha - H^T\beta \leq d, \\ &\qquad \alpha, \beta \geq 0. \end{aligned} \tag{4.7}$$

where the coefficient matrix $H_{n\times|\Gamma|}$ has elements $h_{jt} = x_j^0 - x_j^t$, $j = 1, \ldots, n$, $t \in \Gamma$, and the right-hand vector $d \in \mathbb{R}^{|\Gamma|}$ has elements $d_t = \sum_{j=1}^{n}(x_j^t - x_j^0)c_j$, $t \in \Gamma$.

Introducing a dual variable $-y^1$ and two slackness variables y^2 and y^3, the standard form of the dual problem of (4.7) is

$$(\mathbf{MP})\ \begin{aligned} &\min\ d^T y^1 + \mathbf{0}y^2 + \mathbf{0}y^3 \\ &\text{s.t.}\ \ -Hy^1 + y^2 \qquad = q, \\ &\qquad Hy^1 + \qquad y^3 = q, \\ &\qquad y^1, y^2, y^3 \geq 0. \end{aligned} \tag{4.8}$$

It is a linear programming problem with $|\Gamma| + 2n$ variables and $2n$ constraints.

To solve such linear programming problems with exponential variables, we employ the column generation method and the revised simplex method.

4.2.2 A Column Generation Method to Solve the Problem (ILP1)

The main idea of a column generation algorithm is to treat the primal problem as the Master Problem **(MP)** and only consider the Restricted Master Problem **(RMP)** with a subset of variables. Then, the subproblem is created to check whether the rest variables should be considered.

We treat the problem (4.8) as **(RMP)** and initialize the **(MP)** through only considering the variables y^2, y^3. Initialize the index set $\Gamma_0 = \emptyset$ of considered variables in y^1. The initial restricted master problem $(\mathbf{RMP}^{(0)})$ is formulated as follows.

$$(\mathbf{RMP}^{(0)})\ \begin{aligned} &\min\ \mathbf{0}y^2 + \mathbf{0}y^3 \\ &\text{s.t.}\ \ y^2 = q, \\ &\qquad y^3 = q, \\ &\qquad y^2, y^3 \geq 0. \end{aligned}$$

It is obvious that the optimal solution of $(\mathbf{RMP}^{(0)})$ is ${y^2}^0 = q, {y^3}^0 = q$. In addition, ${y^1}^0 = \mathbf{0}$ renders that $({y^1}^0, {y^2}^0, {y^3}^0)$ becomes a feasible solution of problem **(MP)**. According to the complementary slackness, the dual operator

(α^0, β^0) corresponding to $({y^1}^0, {y^2}^0, {y^3}^0)$ should satisfy $(-Hy^1 - q)\alpha^0 = 0$ and $(Hy^1 - q)\beta^0 = 0$. Therefore, $\alpha^0 = 0, \beta^0 = 0$.

The reduced cost for all unconsidered variables $y_t^1, t \in \Gamma$ is $z_t = d_t - {\alpha^0}^T H_t + {\beta^0}^T H_t = d_t$.

Therefore, if $d_t \geq 0$ for all $t \in \Gamma$, then (α^0, β^0) satisfies the constraints of problem (4.7), which means it is the optimal solution to **(ILP1)** actually. Otherwise, if there exists $t \in \Gamma$ such that $d_t < 0$, it means that the current optimal solution to the current restricted master problem ($\mathbf{RMP}^{(0)}$) is not the optimal solution to problem (4.7). We follow the principle of the minimum reduced cost and choose the corresponding variable $y_{t_0}^1$ to enter the variable set of ($\mathbf{RMP}^{(0)}$), so $\Gamma_1 = \Gamma_0 \cup \{t_0\}$ is updated to generate a new restricted master problem and update the corresponding dual operator.

In the $(k+1)$-th iteration, denote the variable set by $\{y_t^1 | t \in \Gamma_k\} \cup \{y_j^2 | j = 1, \ldots, n\} \cup \{y_j^3 | j = 1, \ldots, n\}$ and the corresponding restricted master problem ($\mathbf{RMP}^{(k)}$) is given below with its dual operator (α^k, β^k).

$$
(\mathbf{RMP}^{(k)}) \quad
\begin{array}{lll}
\min & \sum\limits_{t \in \Gamma_k} d_t y_t^1 & \\
\text{s.t.} & -\sum\limits_{t \in \Gamma_k} H_t y_t^1 + y^2 = q, & \\
& \sum\limits_{t \in \Gamma_k} H_t y_t^1 + y^3 = q, & \\
& y_t^1 \geq 0, & t \in \Gamma_k, \\
& y^2, y^3 \geq 0. &
\end{array}
$$

For all unconsidered variable $y_t^1, t \in \Gamma \backslash \Gamma_k$, the reduced cost is given by

$$
\begin{aligned}
z_t &= d_t - {\alpha^k}^T H_t + {\beta^k}^T H_t \\
&= \sum_{j=1}^{n} (x_j^t - x_j^0) c_j - \sum_{j=1}^{n} \alpha_j^k (x_j^0 - x_j^t) + \sum_{j=1}^{n} \beta_j^k (x_j^0 - x_j^t) \\
&= \sum_{j=1}^{n} (c_j + \alpha_j^k - \beta_j^k) x_j^t - \sum_{j=1}^{n} (c_j + \alpha_j^k - \beta_j^k) x_j^0,
\end{aligned}
\tag{4.9}
$$

where H_t is the t-th column of matrix H.

Notice that the latter term $\sum_{j=1}^{n} (c_j + \alpha_j^k - \beta_j^k) x_j^0$ in (4.9) is a constant and $x^t \in F_1 \subseteq F$, then the problem of calculating minimum reduced cost can be transformed into a modified primal linear programming problem ($\mathbf{P}^{(k)}$) with modified cost coefficients and the original feasible region.

$$
(\mathbf{P}^{(k)})\quad \begin{array}{ll} \min & \sum_{j=1}^{n}(c_j+\alpha_j^k-\beta_j^k)x_j \\ \text{s.t.} & Ax=b, \\ & x\geq 0. \end{array}
$$

Let $x^{t_k}(t_k \in \Gamma)$ be the optimal solution of the problem $(\mathbf{P}^{(k)})$ and update $\Gamma_{k+1} = \Gamma_k \cup \{t_k\}$. Then the corresponding column vector is

$$
\begin{pmatrix} -H_{t_k} \\ H_{t_k} \end{pmatrix} = \begin{pmatrix} -(x^0 - x^{t_k}) \\ x^0 - x^{t_k} \end{pmatrix}, \tag{4.10}
$$

and the corresponding cost coefficient is $d_{t_k} = \sum_{j=1}^{n}(x_j^{t_k} - x_j^0)c_j$. Then the updated restricted master problem is formulated as follows.

$$
(\mathbf{RMP}^{(k+1)})\quad \begin{array}{lll} \min & \sum\limits_{t\in\Gamma_{k+1}} d_t y_t^1 & \\ \text{s.t.} & -\sum\limits_{t\in\Gamma_{k+1}} H_t y_t^1 + y^2 = q, & \\ & \sum\limits_{t\in\Gamma_{k+1}} H_t y_t^1 + y^3 = q, & \\ & y_t^1 \geq 0, & t\in\Gamma_{k+1}, \\ & y^2, y^3 \geq 0. & \end{array}
$$

Next we continue to calculate the dual operator $(\alpha^{k+1}, \beta^{k+1})$ and update the corresponding minimum reduced cost problem $(\mathbf{P}^{(k+1)})$. Such a process continues until the minimum reduced cost $z_{t_{k^*}} = \sum_{j=1}^{n}(c_j + \alpha_j^{k^*} - \beta_j^{k^*})x_j^t - \sum_{j=1}^{n}(c_j + \alpha_j^{k^*} - \beta_j^{k^*})x_j^0 \geq 0$ is nonnegative for an iteration k^*. In this case, the corresponding dual operator $(\alpha^{k^*}, \beta^{k^*})$ is just the optimal solution of problem **(ILP1)**.

In summary, we propose an optimality condition of column generation method for the inverse problem of linear programming.

Lemma 4.5 (An Optimality Condition of Column Generation Method for (ILP1)) *Given a feasible solution (α^k, β^k) of the inverse problem (4.6), let x^{t_k} be the optimal solution of the linear programming problem* $\min\{(c+\alpha^k-\beta^k)^T x \mid Ax = b, x \geq 0\}$. *If the minimum reduced cost $z_k = (c+\alpha^k-\beta^k)^T(x^{t_k} - x^0) \geq 0$, then (α^k, β^k) is an optimal solution of the problem (4.6).*

A framework of the column generation method for **(ILP1)** is described as follows.

Algorithm 4.1 A framework of the column generation algorithm for (ILP1)

Input: A coefficient matrix A, a right-hand vector b, a coefficient vector c, a given feasible solution x^0, and a weight vector q.

Output: a modified cost vector $\tilde{c}$.

1: **[Initialization.]** Initialize $k := 0$, the set $\Gamma_0 := \emptyset$ of variables and the dual operators $\alpha^0 := 0$, $\beta^0 := 0$. Solve the problem $(\mathbf{P}^{(0)})$ with an optimal solution x^{t_0} and minimum reduced cost $z_{t_0} := c^T(x^{t_0} - x^0)$.

2: **while** $z_{t_k} < 0$ **do**

3: **[Update the dual operators.]**

Update the set $\Gamma_{k+1} := \Gamma_k \cup \{t_k\}$ of variables in the restricted master problem. Generate the column vector by (4.10) corresponding to x^{t_k} and calculate the cost coefficient

$$d_{t_k} := \sum_{j=1}^{n} (x_j^{t_k} - x_j^0) c_j.$$

Update the restricted master problem $(\mathbf{RMP}^{(k+1)})$. Calculate the dual operator $(\alpha^{k+1}, \beta^{k+1})$.

4: **[Solve the minimum reduced cost problem $(\mathbf{P}^{(k+1)})$.]**

Let $x^{t_{k+1}}$ be an optimal solution of $(\mathbf{P}^{(k+1)})$ and calculate the minimum reduced cost

$$z_{t_{k+1}} := (c + \alpha^{k+1} - \beta^{k+1})^T (x^{t_{k+1}} - x^0).$$

5: Update $k := k + 1$.

6: **end while**

7: **if** $z_{t_k} \geq 0$ **then** output $\tilde{c} := c + \alpha^k - \beta^k$ as the optimal solution of **(ILP1)**.

8: **end if**

4.2.2.1 Calculate the Dual Operator by Complementary Slackness Conditions

In the k-th iteration of column generation method, the restricted master problem $(\mathbf{RMP}^{(k)})$ is formulated as follows,

$$(\mathbf{RMP}^{(k)}) \quad \begin{array}{ll} \min \sum\limits_{t \in \Gamma_k} d_t y_t^1 & \\ \text{s.t.} \ -\sum\limits_{t \in \Gamma_k} H_t y_t^1 + y^2 = q, & \\ \sum\limits_{t \in \Gamma_k} H_t y_t^1 + y^3 = q, & \\ y_t^1 \geq 0, & t \in \Gamma_k, \\ y^2, y^3 \geq 0. & \end{array}$$

Actually the dual operator (α^k, β^k) is the optimal solution of the dual problem (**DRMP**$^{(k)}$) of (**RMP**$^{(k)}$).

$$
(\mathbf{DRMP}^{(k)})\quad \begin{array}{ll} \min & \sum_{j=1}^{n} q_j(\alpha_j+\beta_j) \\ \text{s.t} & \sum_{j=1}^{n}(x_j^0-x_j^t)\alpha_j-\sum_{j=1}^{n}(x_j^0-x_j^t)\beta_j \le d_t,\ t\in\Gamma_k. \\ & \alpha_j,\beta_j\ge 0, \qquad\qquad j=1,\ldots,n. \end{array} \tag{4.11}
$$

Next we can design two methods to calculate the dual operator. The first method involves solving problem (**RMP**$^{(k)}$) and constructing the linear equation groups based on complementary slackness conditions, which can be directly converted into a linear programming problem. The second method directly solves the dual problem (**DRMP**$^{(k)}$), which can be directly generated through the optimal solution of problem (**P**$^{(k)}$) without constructing the problem (**RMP**$^{(k)}$). We can also propose a new optimality condition and design an improved algorithm called row generation method.

Calculate the Dual Operator Based on Complementary Slackness Conditions Firstly, solve the problem (**RMP**$^{(k)}$) directly, and let $(y^{1^k}, y^{2^k}, y^{3^k})$ be an optimal solution. Let $\Gamma_k^0=\{t\in\Gamma|y_t^{1^k}=0\}$, $J_k^+=\{j|y_j^{2^k}=0\}$, $J_k^-=\{j|y_j^{3^k}=0\}$. We can obtain the following system of linear equation.

$$
\begin{cases} \alpha_j^k=0, & j\notin J_k^+, \\ \beta_j^k=0, & j\notin J_k^-, \\ \sum_{j=1}^{n}(x_j^0-x_j^t)\alpha_j-\sum_{j=1}^{n}(x_j^0-x_j^t)\beta_j=d_t, & t\notin\Gamma_k^0. \end{cases}
$$

The dual operator (α^k, β^k) can be obtained through solving the aforementioned system or solving the linear programming (4.12) formulated further. We introduce the artificial variables $d_t^+, d_t^-, t\notin\Gamma_k^0$. Notice that the right-hand vector $d_t, t\notin\Gamma_k^0$ is not nonnegative, and it may contain negative values.

$$
\begin{array}{ll} \min & \sum_{t\notin\Gamma_k^0}(d_t^+ + d_t^-) \\ \text{s.t.} & \sum_{j=1}^{n}(x_j^0-x_j^t)\alpha_j-\sum_{j=1}^{n}(x_j^0-x_j^t)\beta_j+d_t^+-d_t^-=d_t,\ t\notin\Gamma_k^0, \\ & \alpha^k,\beta^k,d_t^+,d_t^-\ge 0,\ t\notin\Gamma_k^0. \end{array} \tag{4.12}
$$

If the optimal value of the problem (4.12) is 0, then the dual operator (α^k, β^k) is just the optimal solution of problem (4.12). Otherwise, the system of linear equations has no feasible solution, and hence, the dual variables do not exist.

4.2.3 A Row Generation Method to Solve the Problem (ILP1)

In this subsection, we design a row generation method to solve the problem **(ILP1)**, which directly solves the dual problem (**DRMP**$^{(k)}$). We also provide some new optimality conditions.

We can also obtain the dual operator through solving the problem (**DRMP**$^{(k)}$), which is the relaxation version of problem (4.6) actually, since $\Gamma_k \subset \Gamma$. The model (4.11) of the problem (**DRMP**$^{(k)}$) is equivalent to the following form.

$$
(\mathbf{ILP}^{(k)}) \quad \begin{array}{ll} \min & \sum_{j=1}^{n} q_j(\alpha_j + \beta_j) \\ \text{s.t.} & \sum_{j=1}^{n} (c_j + \alpha_j - \beta_j)x_j^0 \le \sum_{j=1}^{n} (c_j + \alpha_j - \beta_j)x_j^t, \quad t \in \Gamma_k, \\ & \alpha_j, \beta_j \ge 0, \qquad\qquad j = 1, \ldots, n. \end{array} \tag{4.13}
$$

Thus, the optimal solution (α^k, β^k) can be obtained by solving problem (4.13). Let $c^k = c + \alpha^k - \beta^k$, then we can update the problem (**P**$^{(k)}$) below.

$$
(\mathbf{P}^{(k)}) \quad \begin{array}{ll} \min & {c^k}^T x \\ \text{s.t.} & Ax = b, \\ & x \ge 0. \end{array}
$$

Let x^{t_k} be an optimal solution of (**P**$^{(k)}$). Then, according to the column generation algorithm, we need to generate its corresponding column vector, update $\Gamma_{k+1} := \Gamma_k \cup \{t_k\}$, its corresponding problem (**RMP**$^{(k+1)}$), and its dual version (**DRMP**$^{(k+1)}$). Actually, the earlier operation is equivalent to adding the following constraint (4.14) to the linear programming problem (4.13).

$$
\sum_{j=1}^{n} (c_j + \alpha_j - \beta_j)x_j^0 \le \sum_{j=1}^{n} (c_j + \alpha_j - \beta_j)x_j^{t_k}, \tag{4.14}
$$

Thus, the problem (**DRMP**$^{(k+1)}$) is equivalent to (**ILP**$^{(k+1)}$) formulated as follows.

$$
(\mathbf{ILP}^{(k+1)}) \quad \begin{array}{ll} \min & \sum_{j=1}^{n} q_j(\alpha_j + \beta_j) \\ \text{s.t.} & \sum_{j=1}^{n} (c_j + \alpha_j - \beta_j)x_j^0 \le \sum_{j=1}^{n} (c_j + \alpha_j - \beta_j)x_j^t, \quad t \in \Gamma_k, \\ & \sum_{j=1}^{n} (c_j + \alpha_j - \beta_j)x_j^0 \le \sum_{j=1}^{n} (c_j + \alpha_j - \beta_j)x_j^{t_k}, \\ & \alpha_j, \beta_j \ge 0, \qquad\qquad j = 1, \ldots, n. \end{array}
$$

Notice that the two constraint conditions can be combined into one condition by choosing $t \in \Gamma_{k+1}$. Let $(\alpha^{k+1}, \beta^{k+1})$ be an optimal solution of $(\mathbf{ILP}^{(k+1)})$ and $c^{k+1} := c + \alpha^{k+1} - \beta^{k+1}$. Then we can update $k := k + 1$ to continue the aforementioned process.

According to the optimality condition in Lemma 4.5 of the column generation algorithm, it is evident that the following optimality conditions hold to terminate iterations.

Lemma 4.6 *If the optimal solution of* $(\mathbf{P}^{(k)})$ *satisfies* $x_j^{t_k} = x_j^0$, $j = 1, \ldots, n$, *then* $\tilde{c} = c + \alpha^k - \beta^k$ *is an optimal solution of the problem* **(ILP)**.

Proof For any non-basic variable $y_t^1, t \in \Gamma \backslash \Gamma_k$, the minimum reduced cost is

$$z_k = \sum_{j=1}^{n} (c_j + \alpha_j^k - \beta_j^k)(x_j^{t_k} - x_j^0) = 0,$$

which meets the optimality condition of the column generation method in Lemma 4.5, thus $\tilde{c} = c + \alpha^k - \beta^k$ is an optimal solution of **(ILP)**. □

Lemma 4.7 *If the optimal solution of* $(\mathbf{P}^{(k)})$ *satisfies* $\sum_{j=1}^n c_j^k x_j^{t_k} = \sum_{j=1}^n c_j^k x_j^0$, *then* $\tilde{c} = c^k = c + \alpha^k - \beta^k$ *is an optimal solution of the problem* **(ILP)**.

Proof For any non-basic variable $y_t^1, t \in \Gamma \backslash \Gamma_k$, the minimum reduced cost is

$$\begin{aligned} z_k &= \sum_{j=1}^{n} (c_j + \alpha_j^k - \beta_j^k)(x_j^{t_k} - x_j^0) \\ &= \sum_{j=1}^{n} (c_j + \alpha_j^k - \beta_j^k) x_j^{t_k} - \sum_{j=1}^{n} (c_j + \alpha_j^k - \beta_j^k) x_j^0 \\ &= \sum_{j=1}^{n} c_j^k x_j^{t_k} - \sum_{j=1}^{n} c_j^k x_j^0 = 0, \end{aligned}$$

which meets the optimality condition in Lemma 4.5, thus $\tilde{c} = c^k = c + \alpha^k - \beta^k$ is an optimal solution of **(ILP)**. □

Furthermore, we propose another specific optimality condition.

Theorem 4.2 (An Optimality Condition for Row Generation Method) *If* $c^k = c^{k+1}$ *holds for two consecutive iterations, then* $\tilde{c} = c^k$ *is the optimal solution to* ***(ILP1)***.

Proof Firstly, we show that $\tilde{c}$ is a feasible solution to **(ILP1)**. Let (α^k, β^k) and x^{t_k} be optimal solutions of problems $(\mathbf{ILP}^{(k)})$ and $(\mathbf{P}^{(k)})$, respectively. Then we have ${c^k}^T x^{t_k} = \min_{x \in F} {c^k}^T x$ and update $\Gamma_{k+1} = \Gamma_k \cup \{t_k\}$.

Moreover, let $(\alpha^{k+1}, \beta^{k+1})$ be an optimal solution of the problem $(\mathbf{ILP}^{(k+1)})$ and $c^{k+1} := c + \alpha^{k+1} - \beta^{k+1}$. Since $t_k \in \Gamma_{k+1}$, then c^{k+1} satisfies the feasibility constraint ${c^{k+1}}^T x^0 \leq {c^{k+1}}^T x^{t_k}$ of problem $(\mathbf{ILP}^{(k+1)})$. Given $c^{k+1} = c^k$, therefore, for any $x \in F$, we have

$$
{c^k}^T x^0 = {c^{k+1}}^T x^0 \leq {c^{k+1}}^T x^{t_k} = {c^k}^T x^{t_k} = \min_{x \in F} {c^k}^T x \leq {c^k}^T x.
$$

Hence, $\tilde{c} = c^k$ is the feasible solution of problem **(ILP1)**.

Then we show that $\tilde{c} = c^k$ is an optimal solution of problem **(ILP1)**. Let $(\hat{c}, \hat{\alpha}, \hat{\beta})$ be any feasible solution for **(ILP1)**. Then $\hat{c}^T x^0 \leq \hat{c}^T x$ holds for all $x \in F$. Since **(ILP1)** is a subproblem of $(\mathbf{ILP}^{(k)})$, $(\hat{c}, \hat{\alpha}, \hat{\beta})$ is also a feasible solution of $(\mathbf{ILP}^{(k)})$. Moreover, since (α^k, β^k) is an optimal solution for $(\mathbf{ILP}^{(k)})$, we have

$$
\sum_{j=1}^{n} q_j(\alpha_j^k + \beta_j^k) \leq \sum_{j=1}^{n} q_j(\hat{\alpha}_j + \hat{\beta}_j),
$$

Therefore, $\tilde{c} = c^k$ is an optimal solution of **(ILP1)**. □

In summary, we propose the general row generation method to solve the problem **(ILP1)** as follows.

Algorithm 4.2 A row generation method to solve the problem **(ILP1)**

Input: A linear programming problem **(P)** with a cost vector c, a right-hand vector b, a coefficient matrix A, a given feasible solution x^0, a weight vector q; an error $\epsilon > 0$.

Output: A modified cost vector $\tilde{c}$.

1: Initialize $k := 0$, the index set $\Gamma_k := \emptyset$ of basic variables, the dual operator $\alpha_j^k := 0, \beta_j^k := 0, j = 1, \ldots, n$ and $c^k := c + \alpha^k - \beta^k$. Take $\Delta_k := 1$ and initialize problem $(\mathbf{ILP}^{(k)})$ as

$$
(\mathbf{ILP}^{(k)}) \quad \begin{array}{ll} \min \sum_{j=1}^{n} q_j(\alpha_j + \beta_j) & \\ \text{s.t. } \ \alpha_j, \beta_j \geq 0, & j = 1, \ldots, n. \end{array}
$$

2: **while** $\Delta_k > \epsilon$ **do**

3: Solve problem $(\mathbf{P}^{(k)})$ $\min\{{c^k}^T x | Ax = b, x \geq 0\}$ and let the x^{t_k} be its optimal solution, update $\Gamma_{k+1} := \Gamma_k \cup \{t_k\}$.

4: **if** $\sum_{j=1}^{n}(x_j^{t_k} - x_j^0) < \epsilon$ **then** Output $\tilde{c} := c^k$ and **break**;

5: **else if** $|{c^k}^T x^{t_k} - {c^k}^T x^0| < \epsilon$ **then** Output $\tilde{c} := c^k$ and **break**;

6: **else** Add a constraint $\tilde{c}^T x^0 \leq \tilde{c}^T x^{t_k}$ to problem $(\mathbf{ILP}^{(k)})$ to obtain $(\mathbf{ILP}^{(k+1)})$. Let $(\alpha^{k+1}, \beta^{k+1})$ be an optimal solution of $(\mathbf{ILP}^{(k+1)})$,

7: Update $c^{k+1} := c + \alpha^{k+1} - \beta^{k+1}$.

8: Calculate $\Delta_{k+1} = \sum_{j=1}^{n}(c_j^{k+1} - c_j^k)$ and update $k := k + 1$.

```
        end if
    end while
```

4.2.4 A Revised Simplex Method to Solve the Problem (ILP1)

In this section, we employ the revised simplex method to solve the dual inverse problem **(MP)**, which can render an optimal solution to the problem **(ILP1)**.

Given the initial basic variables y^2 and y^3, the initial basic matrix and its inverse matrix are $B_0 = B_0^{-1} = I_{2n}$. Let $\bar{d}^0 = (0, \ldots, 0)_{2n}$ be the initial basic cost vector. Based on the rule of minimum reduced cost, the problem of choosing the entering variable is to calculate the minimum reduced cost z_t^1 for all non-basic variable $y_t^1, t \in \Gamma$.

$$z_t^1 = d_t - \bar{d}^0 B_0^{-1} H_t = d_t = \sum_{j=1}^{n} (x_j^t - x_j^0) c_j = c^T x^t - c^T x^0,$$

Therefore, the minimum reduced cost must be achieved at the optimal extreme vertex x^{t_0} of problem **(P)**, and the entering variable is $y_{t_0}^1$ for the first iteration.

In the k-th iteration, denote by $\bar{\Gamma}_{k-1}$, Λ_{k-1}^2 and Λ_{k-1}^3 the index sets of basic variables in y^{1^k}, y^{2^k} and y^{3^k}, respectively. Denote by (α^k, β^k) the dual operator. According to the principle of the simplex method, for any non-basic variables y_t^1, $t \in \Gamma \backslash \bar{\Gamma}_k$, the corresponding reduced cost is

$$\begin{aligned} z_t^1 &= d_t - \bar{d}^k B_k^{-1} H_t = c^T (x^t - x^0) + {\alpha^k}^T (x^t - x^0) - {\beta^k}^T (x^t - x^0) \\ &= (c + \alpha^k - \beta^k)^T (x^t - x^0), \end{aligned}$$

Then, minimum reduced cost for non-basic variables y_t^1, $t \in \Gamma \backslash \bar{\Gamma}_k$ can be obtained by solving the following linear programming problem:

$$(\mathbf{P}^{(k)}) \begin{array}{ll} & \min\ (c + \alpha^k - \beta^k)^T x \\ \text{s.t.} & Ax = b, \\ & x \geq 0. \end{array}$$

Let x^{t_k} be an optimal solution for problem $(\mathbf{P}^{(k)})$ and $\bar{z}_k^1 = (c + \alpha^k - \beta^k)^T (x^{t_k} - x^0)$ be the minimum reduced cost for y_t^1, $t \in \Gamma \backslash \bar{\Gamma}_k$.

For other non-basic variables y_j^2, $j \notin \Lambda_k^2$ and y_j^3, $j \notin \Lambda_k^3$, the corresponding reduced cost is :

$$z_j^2 = 0 - \bar{d}^k B_k^{-1} \begin{pmatrix} e_j \\ \mathbf{0} \end{pmatrix} - \alpha_j^k, \quad z_j^3 = 0 \quad \bar{d}^k B_k^{-1} \begin{pmatrix} \mathbf{0} \\ e_j \end{pmatrix} = \beta_j^k.$$

where e_j is a n-dimensional unit vector with the j-th component equal to 1. We denote the minimum reduced cost of y_j^2, $j \notin \Lambda_k^2$ and y_j^3, $j \notin \Lambda_k^3$ as

$$\bar{z}_k^2 = \min_{j \notin \Lambda_k^2} \{\alpha_j^k\} = \min_{j=1,\dots,n} \{\alpha_j^k\}, \quad \bar{z}_k^3 = \min_{j \notin \Lambda_k^3} \{\beta_j^k\} = \min_{j=1,\dots,n} \{\beta_j^k\}.$$

Thus, the minimum reduced cost for the k-th iteration can be obtained by $\bar{z}_k = \min\{\bar{z}_k^1, \bar{z}_k^2, \bar{z}_k^3\}$.

If $\bar{z}_k \geq 0$, then all reduced costs are non-negative, and (α^k, β^k) is an optimal solutions of problem **(ILP1)**. Hence, output $\tilde{c} = c + \alpha^k - \beta^k$.

If $\bar{z}_k < 0$, then we add the variable corresponding to the minimum reduced cost $\bar{z}_k$ to the basis and generate its column vector. Recall that there are three different cases for the value of $\bar{z}_k$, which implies different methods to generate the column vector.

If $\bar{z}_k = \bar{z}_k^1$, generate the corresponding entering column $\bar{H}_k$ as follows:

$$\bar{H}_k := B_k^{-1} \begin{pmatrix} -H_{t_k} \\ H_{t_k} \end{pmatrix} = B_k^{-1} \begin{pmatrix} -(x^0 - x^{t_k}) \\ (x^0 - x^{t_k}) \end{pmatrix}, \tag{4.15}$$

and the corresponding cost coefficient is denoted as $\tilde{d}_k = c^T (x^{t_k} - x^0)$.

Else if $\bar{z}_k = \bar{z}_k^2 = \alpha_{j_k}^k$, generate the corresponding entering column $\bar{H}_k$ as follows:

$$\bar{H}_k := B_k^{-1} \begin{pmatrix} e_{j_k} \\ \mathbf{0} \end{pmatrix}, \tag{4.16}$$

and the corresponding cost coefficient is denoted as $\tilde{d}_k = 0$.

Else if $\bar{z}_k = \bar{z}_k^3 = \beta_{j_k}^k$, generate the corresponding entering column $\bar{H}_k$ as follows.

$$\bar{H}_k := B_k^{-1} \begin{pmatrix} \mathbf{0} \\ e_{j_k} \end{pmatrix}, \tag{4.17}$$

and the corresponding cost coefficient is denoted as $\tilde{d}_k = 0$.

After generating the entering column and its corresponding cost coefficient, we update the right-hand vector by

$$\bar{b}^k = B_k^{-1} \begin{pmatrix} q \\ q \end{pmatrix}. \tag{4.18}$$

Based on the lexicographic rule, determine the pivot element $\bar{h}_{rk}$ by

$$\left(\frac{\bar{b}_r^k}{\bar{h}_{rk}}, \frac{(B_k^{-1})_r}{\bar{h}_{rk}}\right) := lex \min \left\{ \left(\frac{\bar{b}_i^k}{\bar{h}_{ik}}, \frac{(B_k^{-1})_i}{\bar{h}_{ik}}\right) \middle| \bar{h}_{ik} > 0 \right\}, \tag{4.19}$$

where $(B_k^{-1})_i$ is the i-th row vector of matrix B_k^{-1}.

According to the pivot element $\bar{h}_{rk}$, we can generate the rotation matrix

$$R_k := \begin{pmatrix} 1 \cdots & -\frac{\bar{h}_{1k}}{\bar{h}_{rk}} & \\ \ddots & \vdots & \\ & \frac{1}{\bar{h}_{rk}} & \\ & \vdots & \ddots \\ & \frac{-\bar{h}_{2n,k}}{\bar{h}_{rk}} & \cdots 1 \end{pmatrix}, \tag{4.20}$$

and update $B_{k+1}^{-1} = R_k B_k^{-1}$. Furtheremore, we update the cost coefficients of basic variables

$$\begin{cases} \bar{d}_j^{k+1} = \tilde{d}_k, & j = r, \\ \bar{d}_j^{k+1} = \bar{d}_j^k, & j \neq r. \end{cases}$$

Finally, we can update the dual operator

$$\begin{pmatrix} \alpha^{k+1} \\ \beta^{k+1} \end{pmatrix} := -(\bar{d}^k)^T B_{k+1}^{-1},$$

and $k := k + 1$ to continue the algorithms until there exists an iteration k^* in which $\bar{z}_{k^*} \geq 0$.

In summary, a revised simplex method to solve the problem **(ILP1)** is proposed.

Algorithm 4.3 A revised simplex algorithm to solve the problem **(ILP1)**

Input: A linear programming problem **(P)** with a cost vector c, a right-hand vector b, a coefficient matrix A, a given feasible solution x^0, a weight vector q; an error $\epsilon > 0$.

Output: A modified cost vector $\tilde{c}$.

1: Initialize $k := 0$, the initial basic matrix and its inverse matrix $B_k = B_k^{-1} := I_{2n}$, the initial basic cost vector $\bar{d}_j^k = 0, j = 1, \ldots, 2n$, and the initial dual operator $\alpha_j^k := \beta_j^k := 0, j = 1, \ldots, n$.

2: Solve the linear programming problem $(\mathbf{P}^{(0)}) \min\{c^T x | Ax = b, x \geq 0\}$. let x^{t_0} be its optimal solution.

Initialize $\bar{z}_0^1 := c^T(x^{t_0} - x^0)$, $\bar{z}_0^2 := 0$, $\bar{z}_0^3 := 0$, $\bar{z}_k := \min\{\bar{z}_0^1, \bar{z}_0^2, \bar{z}_0^3\} = \bar{z}_0^1$.
while $\bar{z}_k < 0$ **do**
  if $\bar{z}_k = \bar{z}_k^1$ **then**
    Generate the column vector of entering variable by (4.15) and its cost coefficient $\tilde{d}_k := c^T(x^{t_k} - x^0)$.
  else if $\bar{z}_k := \bar{z}_k^2$ **then**
    Let $\alpha_{j_k}^k = \bar{z}_k^2$. Generate the column vector of entering variable by (4.16) and its cost coefficient $\tilde{d}_k = 0$.
  else if $\bar{z}_k = \bar{z}_k^3$ **then**
    Let $\beta_{j_k}^k = \bar{z}_k^3$. Generate the column vector of entering variable by (4.17) and the cost coefficient $\tilde{d}_k = 0$.
  end if
  Update the right-hand side vector $\bar{b}^k := B_k^{-1}\begin{pmatrix} q \\ q \end{pmatrix}$.
  Select the pivot element $\bar{h}_{rk}$ in $\bar{H}_k$ based on the lexicographic rule by (4.19) and calculate the rotation matrix by (4.20).
  Update $B_{k+1}^{-1} := R_k B_k^{-1}$ and the cost coefficients $\bar{d}_r^{k+1} := \tilde{d}_k$ of basic variables while keeping other components unchanged.
  Update the dual operator $\begin{pmatrix} \alpha^{k+1} \\ \beta^{k+1} \end{pmatrix} := -(\bar{d}^{k+1})^T B_{k+1}^{-1}$.
  Update $c^{k+1} := c + \alpha^{k+1} - \beta^{k+1}$.
  Solve the problem ($\mathbf{P}^{(k+1)}$): $\min\{(c^{k+1})^T x | Ax = b, x \geq 0\}$ and let $x^{t_{k+1}}$ be its optimal solution.
  Calculate the minimum reduced cost $\bar{z}_{k+1} := \min\{\bar{z}_{k+1}^1, \bar{z}_{k+1}^2, \bar{z}_{k+1}^3\}$.

$$
\begin{aligned}
\bar{z}_{k+1}^1 &:= (c^{k+1})^T(x^{t_{k+1}} - x^0);\\
\bar{z}_{k+1}^2 &:= \min_{j=1,\ldots,n} \{\alpha_j^{k+1}\},\\
\bar{z}_{k+1}^3 &:= \min_{j=1,\ldots,n} \{\beta_j^{k+1}\};
\end{aligned}
$$

  if $\bar{z}_{k+1} \geq 0$ **then** output $\tilde{c} := c^{k+1}$.
  else update $k := k + 1$.
  end if
end while

4.3 Two Primal-Dual Algorithms to Solve the Inverse Bounded Linear Programming Problem Under Weighted l_1 Norm

In this section, we consider the Inverse Bounded Linear Programming **(IBLP)** problem and propose two primal-dual algorithms based on duality theories. In such algorithms, we treat the model **(IBLP)** as a primal problem and a dual problem, respectively.

Let the lower bound of variable x be $\mathbf{0}$ and the upper bound be u. Then the bounded linear programming **(BLP)** problem can be formulated as follows:

$$\textbf{(BLP)}\quad \min\{c^T x | Ax = b, \mathbf{0} \le x \le u\},$$

Based on the duality theory and Lemmas 4.2, 4.3, the inverse problem **(IBLP)** can be reformulated as follows:

$$\textbf{(IBLP)}\quad \begin{array}{lll} \min & \sum\limits_{j\in J\cup \bar{J}} q_j\alpha_j + \sum\limits_{j\in J\cup \underline{J}} q_j\beta_j & \\ \text{s.t.} & \pi A_j - \alpha_j \le c_j, & j \in \underline{J}, \\ & \pi A_j + \beta_j \ge c_j, & j \in \bar{J}, \\ & \pi A_j - \alpha_j + \beta_j = c_j, & j \in J, \\ & \alpha, \beta \ge 0, & \end{array}$$

where $\underline{J} = \{j | x_j^0 = 0\}$, $J = \{j | 0 < x_j^0 < u_j\}$, $\bar{J} = \{j | x_j^0 = u_j\}$.

Introducing the dual variable $-y_j$ for the j-th constraint of problem **(IBLP)**, we obtain its dual problem as follows:

$$\begin{array}{ll} \min & \sum\limits_{j=1}^{n} c_j y_j \\ \text{s.t.} & Ay = 0, \\ & -q_j \le y_j \le q_j, \ j \in J, \\ & 0 \le y_j \le q_j, \quad j \in \underline{J}, \\ & -q_j \le y_j \le 0, \ \ j \in \bar{J}. \end{array} \tag{4.21}$$

Taking $y = z - x^0$, we obtain

$$\begin{array}{lll} \min & \sum\limits_{j=1}^{n} c_j z_j & \\ \text{s.t.} & Az = b, & \\ & -q_j + x_j^0 \le z_j \le q_j + x_j^0, & j \in J, \\ & 0 \le z_j \le q_j, & j \in \underline{J}, \\ & -q_j + u_j \le z_j \le u_j & j \in \bar{J}. \end{array} \tag{4.22}$$

Thus, the following theorem can be directly obtained.

Theorem 4.3 *Given a bounded linear programming problem* **(BLP)***, if the given feasible solution* x^0*, its upper bound vector* u *and the weight vector* q *satisfy the following two conditions, then the dual inverse problem* (4.22) *can be transformed into the problem* **(IBLP)***.*

1. $x_j^0 \in \{0, u_j\}$, *that is, for any* $j \in J$, x_j^0 *can only take one of the values* 0 *and the upper bound* u_j;
2. $q_j = u_j$.

This theorem can be directly applied to a relaxed linear programming problem with 0–1 variables, for example, some combinatorial optimization problems including shortest path and maximum flow. In other words, the inverse problem of the aforementioned problems with unit weight vector q can be directly transformed into the original problem itself.

4.3.1 A Primal-Dual Algorithm by Considering (IBLP) as a Primal Problem

Considering the dual problem, we define the lower bound $\underline{l}$ and upper bound $\bar{u}$ of the variables as follows:

$$\underline{l}_j = \begin{cases} 0, & j \in \underline{J}, \\ -q_j, & j \in J \cup \bar{J}, \end{cases} \qquad \bar{u}_j = \begin{cases} 0, & j \in \bar{J}, \\ q_j, & j \in J \cup \underline{J}. \end{cases} \tag{4.23}$$

An equivalent model **(DIBLP)** of the dual problem (4.21) is obtained.

$$\textbf{(DIBLP)} \quad \begin{aligned} \min \ & \sum_{j=1}^{n} c_j y_j \\ \text{s.t.} \ & Ay = 0, \\ & \underline{l}_j \le y_j \le \bar{u}_j, \ j = 1, \ldots, n. \end{aligned}$$

Assuming that a feasible solution y^k of **(DIBLP)** is given, according to the complementary slackness property of the dual theory, if $\alpha^k, \beta^k \ge 0$ satisfy the following system (4.24) of equations, then $(\pi^k, \alpha^k, \beta^k)$ and y^k are, respectively, the optimal solutions of **(IBLP)** and **(DIBLP)**.

$$\begin{cases} \alpha_j^k(y_j^k - \bar{u}_j) = 0, & j \in J \cup \underline{J}, \\ \beta_j^k(y_j^k - \underline{l}_j) = 0, & j \in J \cup \bar{J}, \\ (\pi^k A_j - \alpha_j^k - c_j)y_j^k = 0, & j \in \underline{J}, \\ (\pi^k A_j + \beta_j^k - c_j)y_j^k = 0, & j \in \bar{J}, \\ \pi^k A_j - \alpha_j^k \le c_j, & j \in \underline{J}, \\ \pi^k A_j + \beta_j^k \ge c_j, & j \in \bar{J}, \\ \pi^k A_j - \alpha_j^k + \beta_j^k = c_j, & j \in J. \end{cases} \tag{4.24}$$

Let $J_+^k = \{j | y_j^k = \bar{u}_j\}$ and $J_-^k = \{j | y_j^k = \underline{l}_j\}$ be the index sets of the components of y^k. Without loss of generality, supposing that the weight coefficients $q_j > 0$, $j = 1, \ldots, n$. It is obvious that J_+^k and J_-^k are mutually disjoint. Let $J_=^k = \{j | y_j^k = 0\}$. Then the system (4.24) of equations is equivalent to

$$\alpha_j^k = 0, \qquad j \in J \cup \underline{J} \setminus J_+^k, \tag{4.25}$$

$$\beta_j^k = 0, \qquad j \in J \cup \bar{J} \setminus J_-^k, \tag{4.26}$$

$$\pi^k A_j - \alpha_j^k + \beta_j^k = c_j, \; j \in J, \tag{4.27}$$

$$\pi^k A_j - \alpha_j^k = c_j, \qquad j \in \underline{J} \setminus J_=^k, \tag{4.28}$$

$$\pi^k A_j - \alpha_j^k \le c_j, \qquad j \in J_=^k \cap \underline{J}, \tag{4.29}$$

$$\pi^k A_j + \beta_j^k = c_j, \qquad j \in \bar{J} \setminus J_=^k, \tag{4.30}$$

$$\pi^k A_j + \beta_j^k \ge c_j, \qquad j \in J_=^k \cap \bar{J}. \tag{4.31}$$

Considering different cases, we can substitute α, β into the system of equations.

Case 1 $j \in J, -q_j \le y_j^k \le q_j$.

1. If $y_j^k = q_j$, then (4.26) and (4.27) imply that $\beta_j^k = 0$ and $\pi^k A_j - \alpha_j^k = c_j$.
2. If $y_j^k = -q_j$, then (4.25) and (4.27) imply that $\alpha_j^k = 0$ and $\pi^k A_j + \beta_j^k = c_j$.
3. If $-q_j < y_j^k < q_j$, then the constraints (4.25), (4.26) and (4.27) imply that $\alpha_j^k = \beta_j^k = 0$ and $\pi^k A_j = c_j$.

Case 2 $j \in \underline{J}, 0 \le y_j^k \le q_j$

1. If $y_j^k = q_j$, then the constraint (4.28) implies that $\pi^k A_j - \alpha_j^k = c_j$.
2. If $y_j^k = 0$, then (4.25) and (4.29) imply that $\alpha_j^k = 0$ and $\pi^k A_j \le c_j$.
3. If $0 < y_j^k < q_j$, then the constraints (4.25) and (4.28) imply that $\alpha_j^k = 0$ and $\pi^k A_j = c_j$.

Case 3 $j \in \bar{J}, -q_j \leq y_j^k \leq 0$

1. If $y_j^k = -q_j$, then the constraint (4.30) implies that $\pi^k A_j + \beta_j^k = c_j$.
2. If $y_j^k = 0$, then (4.26) and (4.31) imply that $\beta_j^k = 0$ and $\pi^k A_j \geq c_j$.
3. If $0 < y_j^k < q_j$, then (4.26) and (4.30) imply that $\beta_j^k = 0$ and $\pi^k A_j = c_j$.

In summary, Eqs. (4.25)–(4.31) are equivalent to the following system of linear equations.

$$\begin{cases} \pi^k A_j - \alpha_j^k = c_j, & j \in J_+^k \cap (J \cup \underline{J}), \\ \pi^k A_j + \beta_j^k = c_j, & j \in J_-^k \cap (J \cup \bar{J}), \\ \pi^k A_j \leq c_j, & j \in J_-^k \cap \underline{J}, \\ \pi^k A_j = c_j, & j \notin (J_-^k \cup J_+^k), \\ \pi^k A_j \geq c_j, & j \in J_+^k \cap \bar{J}. \end{cases} \tag{4.32}$$

It is obvious that the system can be divided into

$$\alpha_j^k = \begin{cases} \pi^k A_j - c_j, & j \in J_+^k \cap (J \cup \underline{J}), \\ 0, & \text{otherwise}. \end{cases},$$

$$\beta_j^k = \begin{cases} c_j - \pi^k A_j, & j \in J_-^k \cap (J \cup \bar{J}), \\ 0, & \text{otherwise}. \end{cases} \tag{4.33}$$

and

$$\begin{cases} \pi^k A_j \leq c_j, & j \in J_-^k, \\ \pi^k A_j = c_j, & j \notin (J_-^k \cup J_+^k), \\ \pi^k A_j \geq c_j, & j \in J_+^k. \end{cases} \tag{4.34}$$

Thus, we can convert the problem of solving system (4.24) of equations into solving the system (4.34) of equations and then substitute the solution π^k into Eq. (4.33) to calculate α^k, β^k.

In order to solve the system of Eqs. (4.34), we introduce artificial variables γ^+, γ^-, and transform it into a linear programming problem as follows.

$$
(\mathbf{RP1}^{(k)}) \quad \begin{array}{ll} \min & \sum_{j=1}^{n} (\gamma_j^+ + \gamma_j^-) \\ \text{s.t.} & \pi^k A_j + \gamma_j^+ - \gamma_j^- \le c_j, \ j \in J_-^k, \\ & \pi^k A_j + \gamma_j^+ - \gamma_j^- = c_j, \ j \notin (J_-^k \cup J_+^k), \\ & \pi^k A_j + \gamma_j^+ - \gamma_j^- \ge c_j, \ j \in J_+^k, \\ & \gamma_j^+, \gamma_j^- \ge 0, \qquad j = 1, \ldots, n. \end{array}
$$

Taking the dual variable $-\bar{y}$, the dual problem of $(\mathbf{RP1}^{(k)})$ can be formulated as follows:

$$
(\mathbf{DRP1}^{(k)}) \quad \begin{array}{ll} \min & \sum_{j=1}^{n} c_j \bar{y}_j^k \\ \text{s.t.} & A\bar{y}^k = 0, \\ & -1 \le \bar{y}_j^k \le 1, \ j \notin (J_-^k \cup J_+^k), \\ & 0 \le \bar{y}_j^k \le 1, \quad j \in J_-^k, \\ & -1 \le \bar{y}_j^k \le 0, \ j \in J_+^k. \end{array} \tag{4.35}
$$

Let $\bar{y}^k$ be the optimal solution of problem $(\mathbf{DRP1}^{(k)})$. If the optimal value is negative, we need to determine the adjustment coefficient θ^k and update $y^{k+1} := y^k + \theta^k \bar{y}^k$, which is also a feasible solution of problem **(DIBLP)**. In order to improve feasible solution to attach to the optimal solution, the objective value is required to decrease by $c^T y^{k+1} = c^T (y^k + \theta^k \bar{y}^k) = c^T y^k + \theta^k c^T \bar{y}^k \le c^T y^k$. Hence, $\theta^k c^T \bar{y} \le 0$. There exists that $c^T \bar{y} = -(-c^T \bar{y}) = -\sum_{j=1}^{n} (\gamma_j^+ + \gamma_j^-) \le 0$, which implies $\theta^k \ge 0$.

To ensure that updated y^{k+1} remains feasibility of problem **(DIBLP)**, we need to calculate the adjustment coefficient $\theta^k = \min\{\theta_1^k, \theta_2^k\}$, where

$$
\theta_1^k = \min_{j \notin J_+^k, \bar{y}_j^k > 0} \left\{ \frac{\bar{u}_j - y_j^k}{\bar{y}_j^k} \right\}, \quad \theta_2^k = \min_{j \notin J_-^k, \bar{y}_j^k < 0} \left\{ \frac{\underline{l}_j - y_j^k}{\bar{y}_j^k} \right\}. \tag{4.36}
$$

Substituting $y^{k+1} := y^k + \theta^k \bar{y}^k$ and updating J_+^{k+1} and J_-^{k+1}, we can obtain the updated problem $(\mathbf{DRP1}^{(k)})$ by updating $k := k + 1$.

We continue the aforementioned process until the optimal objective value of the problem $(\mathbf{DRP1}^{(k^*)})$ is 0, whose optimal solution is $\bar{y}^{k^*}$. Then the corresponding solution $y^{k^*+1} := y^{k^*} + \theta^{k^*} \bar{y}^{k^*}$ is an optimal solution to problem **(DIBLP)**, and its dual variables, an optimal solution $(\pi^*, \alpha^*, \beta^*)$ to problem **(IBLP)** can be obtained.

Above all, the primal-dual algorithm to solve **(IBLP)** is proposed as follows:

Algorithm 4.4 A primal-dual algorithm by regarding **(IBLP)** as a primal problem

Input: A problem **(BLP)** with a cost vector c, a right-hand vector b, a coefficient matrix A, an upper bound vector u, and a given feasible solution x^0 and a weight vector q.

Output: A modified cost vector $\tilde{c}$.

1: Initialize $k := 0$, a dual feasible solution $y^k := (0, \ldots, 0)$, $z^k_{OPT} := -1$. Define the index sets $\underline{J} := \{j | x^0_j = 0\}$, $J := \{j | 0 < x^0_j < u_j\}$, $\bar{J} := \{j | x_j = u_j\}$, and calculate the vectors $\underline{l}$ and $\bar{u}$ by (4.23).

2: **while** $z^k_{OPT} < 0$ **do**

3: Let $J^k_+ := \{j | y^k_j = \bar{u}_j\}$ and $J^k_- := \{j | y^k_j = \underline{l}_j\}$. Let $\bar{y}^k$ be an optimal solution of problem (**DRP1**$^{(k)}$) by (4.35) and calculate $z^k_{OPT} := c^T \bar{y}^k$.

4: **if** $z^k_{OPT} \geq 0$ **then**

5: Calculate π^k based on the dual problem (**RP1**$^{(k)}$) of (**DRP1**$^{(k)}$). Calculate α^k and β^k by (4.33).

6: Output $\tilde{c} = c + \alpha^k - \beta^k$;

7: **else**

8: Calculate $\theta^k := \min\{\theta^k_1, \theta^k_2\}$ by (4.36).

9: Update $y^{k+1} := y^k + \theta^k \bar{y}^k$ and $k := k + 1$;

10: **end if**

11: **end while**

4.3.2 *A Primal-Dual Algorithm by Considering (IBLP) as a Dual Problem*

In this subsection, we treat the problem **(IBLP)** as a dual problem and propose some properties of this algorithm to design an improved algorithm.

Firstly, based on the model proposed by [144], an equivalent model of problem **(IBLP)** can be formulated as follows.

$$
\textbf{(IBLP2)} \quad
\begin{array}{lll}
\min & \sum\limits_{j \in J \cup \underline{J}} q_j \alpha_j + \sum\limits_{j \in J \cup \bar{J}} q_j \beta_j & \\
\text{s.t.} & \pi A_j - \alpha_j \leq c_j, & j \in J \cup \underline{J}, \\
& -\pi A_j - \beta_j \leq -c_j, & j \in J \cup \bar{J}, \\
& \alpha, \beta \geq 0. &
\end{array}
\tag{4.37}
$$

Considering (4.37) as a dual problem, we can transform it into the standard dual form.

$$
\textbf{(IBLP3)}\quad
\begin{array}{lll}
\max & \sum_{j\in J\cup \underline{J}} q_j(-\alpha_j) + \sum_{j\in J\cup \bar{J}} q_j(-\beta_j) & \\
\text{s.t.} & \pi A_j + (-\alpha_j) \le c_j, & j \in J\cup \underline{J}, \\
 & -\pi A_j + (-\beta_j) \le -c_j, & j \in J\cup \bar{J}, \\
 & -\alpha_j \le 0, & j \in J\cup \underline{J}, \\
 & -\beta_j \le 0, & j \in J\cup \bar{J}.
\end{array}
$$

Given a feasible solution $(\pi^k, \alpha^k, \beta^k)$ of **(IBLP3)**, let

$$
\begin{aligned}
J_1^k &:= \{j \in J\cup \underline{J} | \pi^k A_j - \alpha_j^k = c_j\}, \\
J_2^k &:= \{j \in J\cup \bar{J} | \pi^k A_j + \beta_j^k = c_j\}, \\
J_3^k &:= \{j \in J\cup \underline{J} | \alpha_j^k = 0, \}, \\
J_4^k &:= \{j \in J\cup \bar{J} | \beta_j^k = 0\}.
\end{aligned}
\tag{4.38}
$$

The dual problem of the restricted primal problem is formulated below.

$$
\textbf{(DRP2}^{(k)}\textbf{)}\quad
\begin{array}{lll}
\min & \sum_{j\in J\cup \underline{J}} q_j\bar{\alpha}_j + \sum_{j\in J\cup \underline{J}} q_j\bar{\beta}_j & \\
\text{s.t.} & \bar{\pi} A_j - \bar{\alpha}_j \le 0, & j \in J_1^k, \\
 & -\bar{\pi} A_j - \bar{\beta}_j \le 0, & j \in J_2^k, \\
 & \bar{\pi}_i \le 1, & i = 1, \ldots, m, \\
 & \bar{\alpha}_j \ge 0, & j \in J_3^k, \\
 & \bar{\alpha}_j \ge -1, & j \in J\cup \underline{J} \setminus J_3^k, \\
 & \bar{\beta}_j \ge 0, & j \in J_4^k, \\
 & \bar{\beta}_j \ge -1, & j \in J\cup \bar{J} \setminus J_4^k,
\end{array}
$$

Let $(\bar{\pi}^k, \bar{\alpha}^k, \bar{\beta}^k)$ be an optimal solution of $(\textbf{DRP2}^{(k)})$ and let

$$
\begin{aligned}
\theta_1^k &:= \min_{j\in J\cup \underline{J}\setminus J_1^k, \bar{\pi}^k A_j - \bar{\alpha}_j^k > 0} \left\{ \frac{c_j - \pi^k A_j + \alpha_j^k}{\bar{\pi}^k A_j - \bar{\alpha}_j^k} \right\}, \\
\theta_2^k &:= \min_{j\in J\cup \bar{J}\setminus J_2^k, \bar{\pi}^k A_j + \bar{\beta}^k < 0} \left\{ \frac{c_j - \pi^k A_j - \beta_j^k}{\bar{\pi}^k A_j + \bar{\beta}_j^k} \right\}, \\
\theta_3^k &:= \min_{j\in J\cup \underline{J}\setminus J_3^k, \bar{\alpha}_j^k < 0} \left\{ -\frac{\alpha_j^k}{\bar{\alpha}_j^k} \right\}, \\
\theta_4^k &:= \min_{j\in J\cup \bar{J}\setminus J_4^k, \bar{\beta}_j^k < 0} \left\{ -\frac{\beta_j^k}{\bar{\beta}_j^k} \right\}.
\end{aligned}
\tag{4.39}
$$

Then, we can obtain $\theta^k := \min\{\theta_1^k, \theta_2^k, \theta_3^k, \theta_4^k\}$, and update the feasible solution $(\pi^{k+1}, \alpha^{k+1}, \beta^{k+1})$ of problem **(IBLP3)**, where $\pi^{k+1} := \pi^k + \theta^k \bar{\pi}^k$, $\alpha^{k+1} := \alpha^k + \theta^k \bar{\alpha}^k$, $\beta^{k+1} := \beta^k + \theta^k \bar{\beta}^k$. Based on (4.38), $J_1^{k+1}, J_2^{k+1}, J_3^{k+1}, J_4^{k+1}$ are updated, and its corresponding problem (**DRP2**$^{(k+1)}$) can be also constructed.

We continue the aforementioned procedure until there exists an optimal solution $(\bar{\pi}^k, \bar{\alpha}^k, \bar{\beta}^k)$ whose optimal value equals 0 in problem (**DRP2**$^{(k)}$). Hence, we can obtain an optimal solution of problem **(IBLP)**.

4.3.2.1 An Improved Primal-Dual Algorithm

In this subsection, we introduce some simplifications and improvements in the actual execution of the aforementioned primal-dual algorithm.

Firstly, we propose a necessary condition for the problem **(IBLP3)**. Let $J_+(\pi) = \{j | \pi A_j > c_j\}$, $J_=(\pi) = \{j | \pi A_j = c_j\}$, $J_-(\pi) = \{j | \pi A_j < c_j\}$.

Lemma 4.8 *If* (π, α, β) *is an optimal solution of problem* ***(IBLP3)****, then the following condition holds.*

$$\alpha_j = \begin{cases} \pi A_j - c_j, & j \in J_+(\pi) \cap (J \cup \underline{J}), \\ 0, & \textit{otherwise}. \end{cases},$$

$$\beta_j = \begin{cases} c_j - \pi A_j, & j \in J_-(\pi) \cap (J \cup \bar{J}), \\ 0, & \textit{otherwise}. \end{cases} \tag{4.40}$$

Proof We firstly show that $(\pi^*, \alpha^*, \beta^*)$ satisfying (4.40) is a feasible solution of problem **(IBLP3)**, and then show that any optimal solution must satisfy this condition.

If $(\pi^*, \alpha^*, \beta^*)$ satisfies (4.40), then we have

$$\pi^* A_j - \alpha_j^* = \begin{cases} \pi^* A_j - (\pi^* A_j - c_j) = c_j \le c_j, & j \in J_+(\pi) \cap (J \cup \underline{J}), \\ \pi^* A_j \le c_j, & j \in J_-(\pi) \cup J_=(\pi) \cap (J \cup \underline{J}). \end{cases}$$

satisfying the constraints in problem **(IBLP3)**. Similarly, β^* also satisfies the constraints. Thus, $(\pi^*, \alpha^*, \beta^*)$ is a feasible solution of problem **(IBLP3)**.

Suppose that there is an optimal solution $(\pi^1, \alpha^1, \beta^1)$ that does not satisfy condition (4.40).

(1) Assume that there is only $j_1 \notin J_+(\pi^1) \cap (J \cup \underline{J})$ such that $\alpha_{j_1}^1 > 0$.

Next we construct a feasible solution $(\pi^2, \alpha^2, \beta^2)$ of the problem **(IBLP3)**, where $\pi^2 = \pi^1$, $\beta^2 = \beta^1$, $\alpha_j^2 = \alpha_j^1$ for $j \neq j_1$ and $\alpha_{j_1}^2 = 0 < \alpha_{j_1}^1$. It is obvious that, for j_1, we have $\pi^2 A_{j_1} - \alpha_{j_1}^2 = \pi^2 A_{j_1} \le c_{j_1}$, and for $j \neq j_1$,

$$\pi^2 A_j - \alpha_j^2 = \pi^1 A_j - \alpha_j^1 \le c_j,$$

$$\pi^2 A_j + \beta_j^2 = \pi^1 A_j + \beta_j^1 \ge c_j.$$

Its corresponding objective value is

$$\begin{aligned}\sum_{j\in J\cup \underline{J}} q_j\alpha_j^2 + \sum_{j\in J\cup \bar{J}} q_j\beta_j^2 &= \sum_{j\in J\cup \underline{J}, j\neq j_1} q_j\alpha_j^2 + \sum_{j\in J\cup \bar{J}} q_j\beta_j^2 \\ &< \sum_{j\in J\cup \underline{J}, j\neq j_1} q_j\alpha_j^1 + q_{j_1}\alpha_{j_1}^1 + \sum_{j\in J\cup \bar{J}} q_j\beta_j^1 \\ &= \sum_{j\in J\cup \underline{J}} q_j\alpha_j^1 + \sum_{j\in J\cup \bar{J}} q_j\beta_j^1,\end{aligned}$$

which contradicts the optimality of $(\pi^1, \alpha^1, \beta^1)$. □

(2) Suppose that there is only $j_2 \in J_+(\pi^1)\cap(J\cup\underline{J})$ such that $\alpha_{j_2}^1 \neq \pi^1 A_{j_2} - c_{j_2}$, and all other $j \neq j_2$ satisfy the given conditions. Obviously, $\alpha_{j_2}^1 > \pi^1 A_{j_2} - c_{j_2}$; otherwise, $\pi^1 A_{j_2} - \alpha_{j_2}^1 > c_{j_2}$ would violate the constraints of problem **(IBLP3)**.

Next we construct a feasible solution $(\pi^3, \alpha^3, \beta^3)$ where $\pi^3 = \pi^1$, $\beta^3 = \beta^1$, $\alpha_j^3 = \alpha_j^1$ for $j \neq j_2$ and $\alpha_{j_2}^3 = \pi^3 A_{j_2} - c_{j_2} < \alpha_{j_2}^1$. Then we have $\pi^3 A_{j_2} - \alpha_{j_2}^3 = c_{j_2}$ and, for $j \neq j_2$,

$$\begin{aligned}\pi^3 A_j - \alpha_j^3 &= \pi^1 A_j - \alpha_j^1 \le c_j, \\ \pi^3 A_j + \beta_j^3 &= \pi^1 A_j + \beta_j^1 \ge c_j.\end{aligned}$$

which means $(\pi^3, \alpha^3, \beta^3)$ satisfies the constraints of problem **(IBLP3)**. Furthermore, its corresponding objective value is

$$\begin{aligned}\sum_{j\in J\cup \underline{J}} q_j\alpha_j^3 + \sum_{j\in J\cup \bar{J}} q_j\beta_j^3 &= \sum_{j\in J\cup \underline{J}, j\neq j_2} q_j\alpha_j^3 + q_{j_2}\alpha_{j_2}^3 + \sum_{j\in J\cup \bar{J}} q_j\beta_j^3 \\ &< \sum_{j\in J\cup \underline{J}, j\neq j_2} q_j\alpha_j^1 + q_{j_2}\alpha_{j_2}^1 + \sum_{j\in J\cup \bar{J}} q_j\beta_j^1 \\ &= \sum_{j\in J\cup \underline{J}} q_j\alpha_j^1 + \sum_{j\in J\cup \bar{J}} q_j\beta_j^1,\end{aligned}$$

which contradicts the optimality of $(\pi^1, \alpha^1, \beta^1)$. Similarly, we can show that β^1 satisfies (4.40). □

We choose the initial feasible solution of problem **(IBLP3)** by $\pi^0 = \{0, \ldots, 0\}$ and α^0, β^0 defined as in (4.41). It is obvious that $(\pi^0, \alpha^0, \beta^0)$ satisfies the condition (4.40).

$$\alpha_j^0 = \begin{cases} -c_j, & j \in J\cup\underline{J}, c_j \le 0, \\ 0, & \text{otherwise.}\end{cases} \quad \beta_j^0 = \begin{cases} c_j, & j \in J\cup\underline{J}, c_j \ge 0, \\ 0, & \text{otherwise.}\end{cases} \tag{4.41}$$

In the k-th iteration, let $(\bar{\pi}^k, \bar{\alpha}^k, \bar{\beta}^k)$ be an optimal solution of (**DRP2**$^{(k)}$). Then, we show that if the feasible solution $(\pi^k, \alpha^k, \beta^k)$ satisfies condition (4.40), then the calculation of θ^k can be simplified. We can also adjust the updated feasible solution $(\pi^{k+1}, \alpha^{k+1}, \beta^{k+1})$ to satisfy (4.40).

The following lemma describes the relationship among the admissible column sets $J_1^k, J_2^k, J_3^k, J_4^k$.

Lemma 4.9 *Suppose the feasible solution $(\pi^k, \alpha^k, \beta^k)$ of problem* ***(IBLP3)*** *satisfies (4.40). Then, the admissible column sets $J_1^k, J_2^k, J_3^k, J_4^k$ constructed by (4.38) satisfy the following statements.*

1. $(J \cup \underline{J}) \setminus J_3^k \subseteq J_1^k$;
2. $(J \cup \bar{J}) \setminus J_4^k \subseteq J_2^k$;
3. $J \setminus J_4^k = J \setminus J_1^k$;
4. $J \setminus J_3^k = J \setminus J_2^k$.

Proof It follows from (4.40) that $\alpha_j^k = \pi^k A_j - c_j$ for all $j \in (J \cup \underline{J}) \setminus J_3^k$, which implies $j \in J_1^k$. Therefore, we have $(J \cup \underline{J}) \setminus J_3^k \subseteq J_1^k$. Similarly, we can show that $(J \cup \bar{J}) \setminus J_4^k \subseteq J_2^k$.

For all $j \in J \setminus J_4^k$, we have $\beta_j^k = c_j - \pi^k A_j > 0$, $\alpha_j^k = 0$, which implies $j \in J_3^k \cup J_2^k$. Therefore, $\pi^k A_j - \alpha_j = \pi^k A_j < c_j$ and $j \notin J_1^k$. Then $J \setminus J_4^k \subseteq J \setminus J_1^k$. Conversely, for all $j \in J \setminus J_1^k$, we have $\pi^k P_j - \alpha_j^k < c_j$, which implies $\alpha_j^k = 0$ and $\beta_j^k = c_j - \pi^k A_j$. Therefore, we have $j \in J \setminus J_4^k$. Similarly, we can show $J \setminus J_3^k = J \setminus J_2^k$. □

Lemma 4.10 *Suppose the feasible solution $(\pi^k, \alpha^k, \beta^k)$ of problem* ***(IBLP3)*** *satisfies the condition (4.40). Let $(\bar{\pi}^k, \bar{\alpha}^k, \bar{\beta}^k)$ be an optimal solution of problem* ***(DRP2***$^{(k)}$***)****. Then it satisfies the following statements.*

1. *For all $j \in J_3^k \setminus J_1^k$, $\bar{\alpha}_j^k = 0$;*
2. *For all $j \in J_4^k \setminus J_2^k$, $\bar{\beta}_j^k = 0$;*
3. *For all $j \in J_1^k \cap J_3^k$, if $\bar{\pi}^k A_j \geq 0$, then $\bar{\alpha}_j^k = \bar{\pi}^k A_j$; otherwise, $\bar{\alpha}_j^k = 0$;*
4. *For all $j \in J_2^k \cap J_4^k$, if $\bar{\pi}^k A_j \leq 0$, then $\bar{\beta}_j^k = -\bar{\pi}^k A_j$; otherwise, $\bar{\beta}_j^k = 0$;*
5. *For all $j \in J \cup \underline{J} \setminus J_3^k$, if $\bar{\pi}^k A_j \leq -1$, then $\bar{\alpha}_j^k = -1$; otherwise. $\bar{\alpha}_j^k = \bar{\pi}^k A_j$;*
6. *For all $j \in J \cup \bar{J} \setminus J_4^k$, if $\bar{\pi}^k A_j \geq 1$, then $\bar{\beta}_j^k = -1$; otherwise, $\bar{\beta}_j^k = -\bar{\pi}^k A_j$.*

Proof

1. For all $j \in J_3^k \setminus J_1^k$, the corresponding constraint in (**DRP2**$^{(k)}$) is $\bar{\alpha}_j^k \geq 0$, and the variable $\bar{\alpha}_j^k (j \in J_3^k)$ does not appear in any other constraints of (**DRP2**$^{(k)}$). To minimize the objective function, we have $\bar{\alpha}_j^k = 0$ in the optimal solution. Similarly, for all $j \in J_4^k \setminus J_2^k$, we have $\bar{\beta}_j^k = 0$.
2. For all $j \in J_1^k \cap J_3^k$, the corresponding constraints are $\bar{\pi}^k A_j - \bar{\alpha}_j^k \leq 0$ and $\bar{\alpha}_j^k \geq 0$. Obviously, if $\bar{\pi}^k A_j \geq 0$, we have $\bar{\alpha}_j^k = \bar{\pi}^k A_j$ in the optimal solution;

Otherwise, $\bar{\alpha}_j^k = 0$. Similarly, for all $j \in J_2^k \cap J_4^k$, if $\bar{\pi}^k A_j \leq 0$, we have $\bar{\beta}_j^k = \bar{\pi}^k A_j$; otherwise, $\bar{\beta}_j^k = 0$.

3. For all $j \in J \cup \underline{J} \setminus J_3^k$, we have $j \in J_1^k$ by Lemma 4.9. Therefore, the corresponding constraints are $\bar{\pi}^k A_j - \bar{\alpha}_j^k \leq 0$ and $\bar{\alpha}_j^k \geq -1$. According to the objective function, if $\bar{\pi}^k A_j \leq -1$, then $\bar{\alpha}_j^k = -1$; otherwise, $\bar{\alpha}_j^k = \bar{\pi}^k A_j$. Similarly, for any $j \in J \cup \bar{J} \setminus J_4^k$, if $\bar{\pi}^k A_j \geq 1$, then $\bar{\beta}_j^k = -1$; otherwise, $\bar{\beta}_j^k = -\bar{\pi}^k A_j$. □

According to Lemmas 4.9 and 4.10, we can resubstitute the optimal solution $(\bar{\pi}^k, \bar{\alpha}^k, \bar{\beta}^k)$ of problem (**DRP2**$^{(k)}$) to calculate the adjustment coefficient θ^k.

$$\theta_1^k = \min_{j \in J \cup \underline{J} \setminus J_1^k, \bar{\pi}^k A_j > 0} \left\{ \frac{c_j - \pi^k A_j + \alpha_j^k}{\bar{\pi}^k A_j - \bar{\alpha}_j^k} \right\} = \min_{j \in J \cup \underline{J} \setminus J_1^k, \bar{\pi}^k A_j > 0} \left\{ \frac{c_j - \pi^k A_j}{\bar{\pi}^k A_j} \right\},$$

$$\theta_2^k = \min_{j \in J \cup \bar{J} \setminus J_2^k, \bar{\pi}^k A_j + \bar{\beta}^k < 0} \left\{ \frac{c_j - \pi^k A_j - \beta_j^k}{\bar{\pi}^k A_j + \bar{\beta}_j^k} \right\} = \min_{j \in J \cup \bar{J} \setminus J_2^k, \bar{\pi}^k A_j < 0} \left\{ \frac{c_j - \pi^k A_j}{\bar{\pi}^k A_j} \right\}.$$

and $\theta_3^k = \min\{\theta_{3,1}^k, \theta_{3,2}^k, \theta_{3,3}^k\}$, $\theta_4^k = \min\{\theta_{4,1}^k, \theta_{4,2}^k, \theta_{4,3}^k\}$, where

$$\begin{aligned}\theta_{3,1}^k &= \min_{j \in J \setminus J_3^k, \bar{\pi}^k A_j \leq -1} \left\{ -\frac{\alpha_j^k}{\bar{\alpha}_j^k} \right\} = \min_{j \in J \setminus J_3^k, \bar{\pi}^k A_j \leq -1} \left\{ -\frac{\pi^k A_j - c_j}{-1} \right\} \\ &= \min_{j \in J \setminus J_3^k, \bar{\pi}^k A_j \leq -1} \left\{ \pi^k A_j - c_j \right\},\end{aligned}$$

$$\begin{aligned}\theta_{3,2}^k &= \min_{j \in \underline{J} \setminus J_3^k, \bar{\pi}^k A_j \leq -1} \left\{ -\frac{\alpha_j^k}{\bar{\alpha}_j^k} \right\} = \min_{j \in \underline{J} \setminus J_3^k, \bar{\pi}^k A_j \leq -1} \left\{ -\frac{\pi^k A_j - c_j}{-1} \right\} \\ &= \min_{j \in \underline{J} \setminus J_3^k, \bar{\pi}^k A_j \leq -1} \left\{ \pi^k A_j - c_j \right\},\end{aligned}$$

$$\theta_{3,3}^k = \min_{j \in J \cup \underline{J} \setminus J_3^k, -1 < \bar{\pi}^k A_j < 0} \left\{ -\frac{\alpha_j^k}{\bar{\alpha}_j^k} \right\} = \min_{j \in J \cup \underline{J} \setminus J_3^k, -1 < \bar{\pi}^k A_j < 0} \left\{ \frac{c_j - \pi^k A_j}{\bar{\pi}^k A_j} \right\},$$

$$\begin{aligned}\theta_{4,1}^k &= \min_{j \in J \setminus J_4^k, \bar{\pi}^k A_j \geq 1} \left\{ -\frac{\beta_j^k}{\bar{\beta}_j^k} \right\} = \min_{j \in J \setminus J_4^k, \bar{\pi}^k A_j \geq 1} \left\{ -\frac{c_j - \pi^k A_j}{-1} \right\} \\ &= \min_{j \in J \setminus J_4^k, \bar{\pi}^k A_j \geq 1} \left\{ c_j - \pi^k A_j \right\},\end{aligned}$$

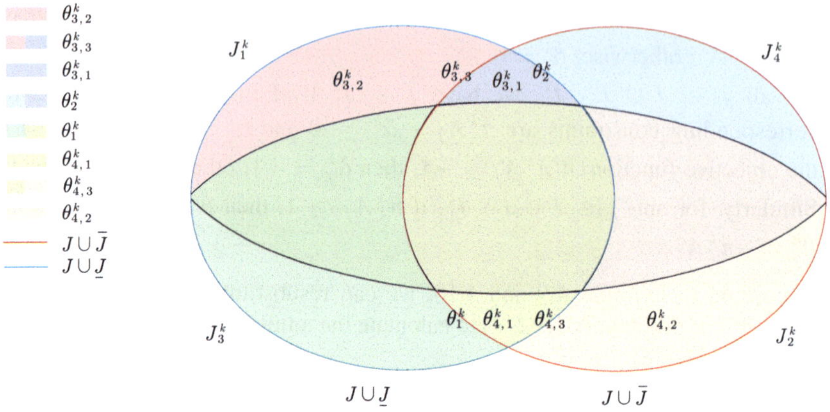

Fig. 4.1 Relationship among the sets and values

$$\theta^k_{4,2} = = \min_{j \in \bar{J} \setminus J_4^k, \bar{\pi}^k A_j \geq 1} \left\{ -\frac{\beta_j^k}{\bar{\beta}_j^k} \right\} = \min_{j \in \bar{J} \setminus J_4^k, \bar{\pi}^k A_j \geq 1} \left\{ -\frac{c_j - \pi^k A_j}{-1} \right\}$$

$$= \min_{j \in \bar{J} \setminus J_4^k, \bar{\pi}^k A_j \geq 1} \left\{ c_j - \pi^k A_j \right\},$$

$$\theta^k_{4,3} = \min_{j \in J \cup \underline{J} \setminus J_3^k, -1 < \bar{\pi}^k A_j < 0} \left\{ -\frac{\beta_j^k}{\bar{\beta}_j^k} \right\} = \min_{j \in J \cup \bar{J} \setminus J_4^k, 0 < \bar{\pi}^k A_j < 1} \left\{ \frac{c_j - \pi^k A_j}{\bar{\pi}^k A_j} \right\}.$$

We illustrate the relationship among the values and their correspongding sets of indexes in Fig. 4.1.

Furthermore, we can simplify the calculation. We first consider $\theta^k_{3,1}$, for $j \in J \setminus J_3^k$, we have $j \notin J_2^k$ by Lemma 4.8. Thus, $\{j \in J \setminus J_3^k | \bar{\pi}^k A_j \leq -1\} \subseteq \{j \in J \setminus J_2^k | \bar{\pi}^k A_j \leq -1\}$, and for $j \in J \setminus J_3^k, \bar{\pi}^k A_j \leq -1$, we have

$$\frac{c_j - \pi^k A_j}{\bar{\pi}^k A_j} \leq \pi^k A_j - c_j.$$

Thus, for $\forall j \in J \setminus J_3^k, \bar{\pi}^k A_j \leq -1$, there is a corresponding $j \in J \setminus J_2^k, \bar{\pi}^k A_j \leq -1$, with a smaller coefficient. Then we have $\theta_2^k \leq \theta^k_{3,1}$.

Then, considering $\theta^k_{3,2}$, we appropriately narrow it by taking

$$\hat{\theta}^k_{3,2} = \min_{j \in \underline{J} \setminus J_3^k, \bar{\pi}^k A_j \leq -1} \left\{ \frac{c_j - \pi^k A_j}{\bar{\pi}^k A_j} \right\},$$

which implies $\hat{\theta}^k_{3,2} \le \theta^k_{3,2}$. Similarly, we have $\theta^k_1 \le \theta^k_{4,1}$ and take

$$\hat{\theta}^k_{4,2} = \min_{j \in \bar{J} \setminus J^k_4, \bar{\pi}^k A_j \ge 1} \left\{ \frac{c_j - \pi^k A_j}{\bar{\pi}^k A_j} \right\},$$

which implies $\hat{\theta}^k_{4,2} \le \theta^k_{4,2}$.

Above all, we demonstrate that $\theta^k_2 \le \theta^k_{3,1}, \theta^k_1 \le \theta^k_{4,1}$ and take $\hat{\theta}^k_{3,2}, \hat{\theta}^k_{4,2}$ satisfying $\hat{\theta}^k_{4,2} \le \theta^k_{3,2}, \hat{\theta}^k_{4,2} \le \theta^k_{4,2}$. It implies that $\min\{\theta^k_1, \theta^k_2, \hat{\theta}^k_{3,2}, \theta^k_{3,3}, \hat{\theta}^k_{4,2}, \theta^k_{4,3}\} \le \theta^k$. Therefore, after obtaining the optimal solution $(\bar{\pi}^k, \bar{\alpha}^k, \bar{\beta}^k)$ of problem (**DRP2**$^{(k)}$), we can directly take

$$\begin{aligned}\hat{J}^k = &\{j \in J \cup \underline{J} \setminus J^k_1 | \bar{\pi}^k A_j > 0\} \cup \{j \in J \cup \bar{J} \setminus J^k_2 | \bar{\pi}^k A_j < 0\} \\ &\cup \{j \in \underline{J} \setminus J^k_3 | \bar{\pi}^k A_j < 0\} \cup \{j \in \bar{J} \setminus J^k_4 | \bar{\pi}^k A_j > 0\},\end{aligned}$$

and the improved adjustment coefficient

$$\hat{\theta}^k = \min_{j \in \hat{J}^k} \left\{ \frac{c_j - \pi^k A_j}{\bar{\pi}^k A_j} \right\}. \tag{4.42}$$

After improving the calculation of adjustment coefficient, if the feasible solution $(\pi^k, \alpha^k, \beta^k)$ satisfies (4.40), we need to ensure that the updating feasible solution $(\pi^{k+1}, \alpha^{k+1}, \beta^{k+1})$ satisfies (4.40) so that the simplification is promoted.

To do so, we construct $(\hat{\pi}^k, \hat{\alpha}^k, \hat{\beta}^k)$ by appropriately modifying $(\bar{\pi}^k, \bar{\alpha}^k, \bar{\beta}^k)$ and update $\pi^{k+1} := \pi^k + \hat{\theta}^k \hat{\pi}^k, \alpha^{k+1} := \alpha^{k+1} + \hat{\theta}^k \hat{\alpha}^{k+1}$ and $\beta^{k+1} := \beta^k + \hat{\theta}^k \hat{\beta}^k$, which is not only a feasible solution of problem (4.37), but also satisfies (4.40).

Denote the set of index $\tilde{J}^k = \{j \in J \cup \underline{J} \setminus J^k_3 | \bar{\pi} A_j \le -1\} \cap \{j \in J \cup \bar{J} \setminus J^k_4 | \bar{\pi} A_j \ge 1\}$, and we will show that $(\alpha^{k+1}, \beta^{k+1})$ satisfies (4.40) for any $j \notin \tilde{J}^k$.

Theorem 4.4 *Suppose that a feasible solution $(\pi^k, \alpha^k, \beta^k)$ of problem **(IBLP3)** satisfies (4.40). Let $(\bar{\pi}^k, \bar{\alpha}^k, \bar{\beta}^k)$ be an optimal solution of (**DRP2**$^{(k)}$) and $(\pi^{k+1}, \alpha^{k+1}, \beta^{k+1})$ be the updated feasible solution, where $\pi^{k+1} := \pi^k + \hat{\theta}^k \bar{\pi}^k, \alpha^{k+1} := \alpha^k + \hat{\theta}^k \bar{\alpha}^k, \beta^{k+1} := \beta^k + \hat{\theta}^k \bar{\beta}^k$. Then $(\alpha^{k+1}, \beta^{k+1})$ satisfies (4.40) for any $j \notin \tilde{J}^k$, where $\hat{\theta}^k$ is the improved adjustment coefficient obtained by Eq. (4.42).*

Proof **Case 1** $j \in J_+(\pi^k)$

1. Consider α^k_{j+1} corresponding to the index $j \in J \cup \underline{J}$. Note that $\alpha^k_j = \pi^k A_j - c_j > 0$ implies $j \notin J^k_3$. When $\bar{\pi}^k A_j \ge -1$, we have $\bar{\alpha}^k_j = \bar{\pi}^k A_j$ based on Lemma 4.10. Therefore,

$$\pi^{k+1} A_j - c_j - (\pi^k + \hat{\theta}^k \bar{\pi}^k) A_j - c_j = \pi^k A_j + \hat{\theta}^k \bar{\pi}^k A_j - c_j = \alpha^k_j + \hat{\theta}^k \bar{\alpha}^k_j = \alpha^{k+1}_j,$$

which implies that α_j^{k+1} satisfies (4.40). When $\bar{\pi}^k A_j \geq -1$, $j \in \tilde{J}^k$ is not included in our discussion.

2. Consider β_{j+1}^k corresponding to the index $j \in J \cup \bar{J}$. Note that $\beta_j^k = 0$ implies $j \in J_4^k \backslash J_2^k$. Based on Lemma 4.10, $\bar{\beta}_j^k = 0$ so that $\beta_j^{k+1} = 0$. Then, we show that $j \in J_+(\pi^{k+1})$ remains unchanged. When $\bar{\pi}^k A_j \geq 0$, we have $\pi^{k+1} A_j = \pi^k A_j + \hat{\theta}^k \bar{\pi}^k A_j \geq \pi^k A_j = c_j$; When $\bar{\pi}^k A_j < 0$, we have

$$\pi^{k+1} A_j = \pi^k A_j + \hat{\theta}^k \bar{\pi}^k A_j \geq \pi^k A_j + \theta_2^k \bar{\pi}^k A_j = \pi^k A_j + c_j - \pi^k A_j = c_j,$$

Thus, $j \in J_+(\pi^{k+1})$ keeps unchanged and $\beta_j^{k+1} = 0$ satisfies (4.40).

Case 2 $j \in J_=(\pi^k)$

1. Consider α_j^{k+1} corresponding to the index $j \in J \cup \underline{J}$. Then $j \in J_1^k \cap J_3^k$. If $\bar{\pi}^k A_j \geq 0$, based on Lemma 4.10, we have $\bar{\alpha}_j^k = \bar{\pi}^k A_j$ so that $c_j = \pi^k A_j \leq \pi^k A_j + \hat{\theta}^k \bar{\pi}^k A_j$, which implies $j \in J_+(\pi^{k+1})$. Thus,

$$\pi^{k+1} A_j - c_j = \pi^k A_j + \hat{\theta}^k \bar{\pi}^k A_j = c_j + \hat{\theta}^k \bar{\alpha}_j^k - c_j = \hat{\theta}^k \bar{\alpha}_j^k = 0 + \hat{\theta}^k \bar{\alpha}_j^k = \alpha_j^{k+1},$$

which satisfies (4.40). If $\bar{\pi}^k A_j \leq 0$, based on Lemma (4.10), we have $\bar{\alpha}_j^k = 0$ so that $c_j = \pi^k A_j \geq \pi^{k+1} A_j$, which implies $j \notin J_+(\pi^{k+1})$. Therefore, $\alpha_j^{k+1} = \alpha_j + \hat{\theta}^k \bar{\alpha}_j^k = 0$ satisfies (4.40).

2. Similarly, we can show that β_j^{k+1} for $j \in J \cup \bar{J}$ satisfies (4.40).

Case 3 $j \in J_-(\pi^k)$

1. Consider α_j^{k+1} corresponding to the index $j \in J \cup \underline{J}$. Note that $j \notin J_1^k$. Based on Lemma 4.10, $\bar{\alpha}_j^k = 0$ so that $\alpha_j^{k+1} = \alpha_j^k + \hat{\theta}^k \bar{\alpha}_j^k = 0$. Then, we show that $j \in J_-(\pi^{k+1})$ remains unchanged. When $\bar{\pi}^k A_j \leq 0$, there exists $\bar{\pi}^{k+1} A_j = \hat{\theta}^k \bar{\pi}^k A_j \leq \bar{\pi}^k A_j \leq c_j$. When $\bar{\pi}^k A_j \geq 0$, there exists

$$\pi^{k+1} A_j = \pi^k A_j + \hat{\theta}^k \bar{\pi}^k A_j \leq \pi^k A_j + \theta_1^k \bar{\pi}^k A_j = \pi^k A_j + c_j - \pi^k A_j = c_j.$$

Thus, $j \in J_-(\pi^{k+1})$ keeps unchanged and $\alpha_j^{k+1} = 0$ satisfies (4.40).

2. Consider β_j^{k+1} corresponding to the index $j \in J \cup \bar{J}$. When $\bar{\pi}^k A_j \geq 1$, based on Lemma 4.10, there exists $\bar{\beta}_j^k = -\bar{\pi}^k A_j$. Therefore,

$$c_j - \pi^{k+1} A_j = c_j - (\pi^k A_j + \hat{\theta}^k \bar{\pi}^k A_j) = \beta_j^k + \hat{\theta}^k \bar{\beta}_j^k = \beta_j^{k+1}.$$

which implies that β_j^{k+1} satisfies (4.40). When $\bar{\pi}^k A_j \leq 1$, $j \in \tilde{J}^k$ is not included in our discussion. □

Finally, we demonstrate that we can modify the corresponding $\bar{\alpha}_j^k$, $\bar{\beta}_j^k$ for $j \in \tilde{J}^k$ such that the following updated feasible solution not only maintains the feasibility but also satisfies (4.40).

For $j \in \tilde{J}^k$, let $\pi^{k+1} := \pi^k + \hat{\theta}^k \bar{\pi}^k$. For $j \in \{j \in J \cup \underline{J} \backslash J_3^k | \bar{\pi} A_j \leq -1\}$ and take

$$\tilde{\alpha}_j^k = \begin{cases} \bar{\pi}^k A_j, & \pi^{k+1} A_j - c_j > 0, \\ -\frac{\pi^k A_j - c_j}{\hat{\theta}^k}, & \text{otherwise.} \end{cases}$$

For $j \in \{j \in J \cup \bar{J} \backslash J_4^k | \bar{\pi} A_j \geq 1\}$, we take

$$\tilde{\beta}_j^k = \begin{cases} -\bar{\pi}^k A_j, & \pi^{k+1} A_j - c_j < 0, \\ -\frac{c_j - \pi^k A_j}{\hat{\theta}^k}, & \text{otherwise.} \end{cases}$$

For $j \in \tilde{J}^k$, we demonstrate that $(\alpha_j^{k+1} := \alpha_j^k + \hat{\theta}^k \tilde{\alpha}_j^k, \beta_j^{k+1} := \beta_j^k + \hat{\theta}^k \tilde{\beta}_j^k)$ satisfies the constraints of problem (**IBLP3**) and (4.40).

For $j \in \{j \in J \cup \underline{J} \backslash | \bar{\pi} A_j \leq -1\}$, if $\pi^{k+1} A_j - c_j > 0$, which implies that $j \in J_+(\pi^{k+1})$, then we have

$$\begin{aligned} \pi^{k+1} A_j - \alpha_j^{k+1} &= \pi^k A_j + \hat{\theta}^k \bar{\pi}^k A_j - \alpha_j^k - \hat{\theta}^k \tilde{\alpha}_j^k = \pi^k A_j - \alpha_j^k \\ &\quad + \hat{\theta}^k \bar{\pi}^k A_j - \hat{\theta}^k \bar{\pi}^k A_j = c_j, \end{aligned}$$

and it shows that α_j^{k+1} satisfies the constraints of problem (**IBLP3**) and (4.40). Otherwise, if $j \notin J_+(\pi^{k+1})$, then we have $\alpha_j^{k+1} = \alpha_j^k + \hat{\theta}^k \tilde{\alpha}_j^k = \pi^k A_j - c_j - (\pi^k A_j - c_j) = 0$, which satisfies (4.40). Additionally, considering the constraints of problem (**IBLP3**), we have $\pi^{k+1} A_j - \alpha_j^{k+1} = \pi^k A_j + \hat{\theta}^k \bar{\pi}^k A_j \leq c_j$. Similarly, for $j \in \{j \in J \cup \bar{J} \backslash J_4^k | \bar{\pi} A_j \geq 1\}$, the corresponding $\beta_j^{k+1} = \beta_j^k + \hat{\theta}^k \tilde{\beta}_j^k$ satisfies the constraints of problem (**IBLP3**) and (4.40).

Let $\tilde{\pi} := \bar{\pi}, \tilde{\alpha}_j^k := \bar{\alpha}_j^k, j \in J \cup \underline{J} \backslash \tilde{J}^k$ and $\tilde{\beta}_j^k = \bar{\beta}_j^k, j \in J \cup \bar{J} \backslash \tilde{J}^k$. Based on the above analysis, $(\pi^k + \hat{\theta}^k \tilde{\pi}^k, \alpha^k + \hat{\theta}^k \tilde{\alpha}^k, \beta^k + \hat{\theta}^k \tilde{\beta}^k)$ is a feasible solution for problem (**IBLP3**), and it satisfies (4.40).

Therefore, in the actual execution of the algorithm, the feasible solution of problem (**IBLP3**) in every iteration always satisfies (4.40), since simplification of the adjustment coefficient and modification of the updated feasible solution. We only need to calculate $\bar{\pi}^k A_j$, and update π^{k+1} to generate the index sets $J_-(\pi^{k+1})$, $J_=(\pi^{k+1})$, and $J_+(\pi^{k+1})$.

Algorithm 4.5 A primal-dual algorithm by considering **(IBLP)** as a dual problem

Input: A problem **(BLP)** with a cost vector c, a right-hand vector b, a coefficient matrix A, an upper bound vector u, and a solution x^0 and a weight vector q.
Output: A modified cost vector $\tilde{c}$.

1: Initialize $k := 0$, a dual feasible solution $\pi^k := (0, \ldots, 0)$ and $z^k_{OPT} := -1$. Denote index sets $\underline{J} =: \{j|x^0_j = 0\}$, $J := \{j|0 < x^0_j < u_j\}$, $\bar{J} := \{j|x^0_j = u_j\}$.
2: **while** $z^k_{OPT} < 0$ **do**
3: Let index sets $J_+(\pi^k) := \{j|\pi^k A_j > c_j\}$, $J_=(\pi^k) := \{j|\pi^k A_j = c_j\}$, $J_-(\pi^k) := \{j|\pi^k A_j < c_j\}$ and

$$J^k_1 := (J_=(\pi^k) \cup J_+(\pi^k)) \cap (J \cup \underline{J}),$$
$$J^k_2 := (J_=(\pi^k) \cup J_-(\pi^k)) \cap (J \cup \bar{J}),$$
$$J^k_3 := J_+(\pi^k) \cap (J \cup \underline{J}),$$
$$J^k_4 := J_-(\pi^k) \cap (J \cup \bar{J}).$$

4: Construct problem $(\mathbf{DRP2}^{(k)})$

$$\begin{aligned}
\min \ & \sum_{j \in J \cup \underline{J}} q_j \bar{\alpha}_j + \sum_{j \in J \cup \bar{J}} q_j \bar{\beta}_j \\
\text{s.t. } & \bar{\pi} A_j - \bar{\alpha}_j \le 0, && j \in J^k_1, \\
& -\bar{\pi} A_j - \bar{\beta}_j \le 0, && j \in J^k_2 \\
& \bar{\pi}_i \le 1, && i = 1, \ldots, m, \\
& \bar{\alpha}_j \ge 0, && j \in J^k_3, \\
& \bar{\alpha}_j \ge -1, && j \in (J \cup \underline{J}) \setminus J^k_3, \\
& \bar{\beta}_j \ge 0, && j \in J^k_4, \\
& \bar{\beta}_j \ge -1, && j \in (J \cup \bar{J}) \setminus J^k_4.
\end{aligned}$$

Let $(\bar{\pi}^k, \bar{\alpha}^k, \bar{\beta}^k)$ be an optimal solution of $(\mathbf{DRP2}^{(k)})$ and calculate

$$z^k_{OPT} := \sum_{j=1}^{n} q_j (\bar{\alpha}^k_j + \bar{\beta}^k_j);$$

5: **if** $z^k_{OPT} \ge 0$ **then** calculate

$$\alpha^k_j := \begin{cases} \pi^k A_j - c_j, & j \in J_+(\pi^k) \cap (J \cup \underline{J}), \\ 0, & \text{otherwise.} \end{cases}$$

$$\beta^k_j := \begin{cases} c_j - \pi^k A_j, & j \in J_-(\pi^k) \cap (J \cup \bar{J}), \\ 0, & \text{otherwise.} \end{cases}$$

```
        Output c̃ := c + α^k − β^k and break;
    else calculate π̄^k A_j and let
```

$$\hat{J}^k := \{j \in J \cup \underline{J} \backslash J_1^k | \bar{\pi}^k A_j > 0\} \cup \{j \in \bar{J} \cup J \backslash J_2^k | \bar{\pi}^k A_j < 0\}$$
$$\cup \{j \in \underline{J} \backslash J_3^k | \bar{\pi}^k A_j < 0\} \cup \{j \in \bar{J} \backslash J_4^k | \bar{\pi}^k A_j > 0\},$$

```
    Calculate θ̂^k := min_{j∈Ĵ^k} { (c_j − π^k A_j) / (π̄^k A_j) }.
        Let π^{k+1} := π^k + θ̂^k π̄^k and update k := k + 1.
    end if
end while
```

4.4 Conclusions

This chapter presents a comprehensive study on tackling inverse linear programming **(ILP)** problems within the framework of the weighted l_1 norm. We have systematically addressed the **(ILP)** by introducing a mathematical model and developing two distinct solution methodologies: the column generation method and the revised simplex method. These methods are innovative in their approach to handling the inverse problems by transforming them into a series of solvable linear programming problems.

The column generation method is particularly noteworthy for its ability to handle the infinite number of constraints typically associated with ILPs. By considering only a subset of variables in the master problem and solving a subproblem to identify potentially entering variables, we effectively reduce the computational complexity. The incorporation of complementary slackness conditions and the introduction of a row generation technique further enhance the algorithm's efficiency.

On the other hand, the revised simplex method provides an alternative approach by directly manipulating the dual space. This method ensures that the optimality conditions are met in a more direct manner, which is particularly useful for problems where the primal space is high-dimensional or complex.

We further extend to solve the inverse bounded linear programming (IBLP) problem by proposing a pair of primal-dual algorithms. These algorithms are not merely technical innovations but represent a significant advancement in the field, providing a dual perspective that enriches the understanding of IBLP and enhances the efficacy of the proposed solutions.

This chapter has provided a thorough examination of the inverse linear programming problem under the weighted l_1 norm, offering a suite of general methods to tackle the complexities of this optimization challenge. The mathematical models and algorithms presented herein represent a significant stride forward in the field of inverse combinatorial optimization, with the potential to inform and enhance future research endeavors. We have successfully demystified the intricacies of ILP and IBLP, arming the reader with the theoretical knowledge and practical tools necessary

Table 4.1 Research results on the problems (ILP) and (IBLP)

Problem type	Norm	Time complexity	Reference
The problem (ILP)	(Unbounded, Weighted l_1 norm)	Column generation alg.	Section 4.2.2
		Row generation alg.	Section 4.2.3
		Revised simplex method	Section 4.2.4
The problem (IBLP)	(Unbounded, Weighted l_1 norm)	Primal-dual alg.	Section 4.3.1
			Section 4.3.2

to navigate this nuanced domain. As the chapter draws to a close, it leaves behind a legacy of scholarly excellence and a springboard for further exploration into the depths of inverse optimization.

Table 4.1 shows the research results on the problems (ILP) and (IBLP).

Part II Generalized Inverse Shortest Path Problems

This part delves into the complexities of the generalized inverse shortest path problems, presenting a comprehensive analysis of the improvement/interdiction problems on shortest path (Imp-SP)/(Int-SP), interdiction problems on sum of root-leaf distance (Int-SRD) on trees, and restricted inverse optimal value problems on shortest paths on trees (RIOVSPT) under various normative conditions. The study encompasses a thorough investigation of the computational challenges and the development of efficient algorithms to address these challenges.

Chapter 5 on (Imp-SP) introduces the concept of enhancing existing shortest paths by incrementally adjusting edge weights, a problem with significant implications in transportation and communication networks. The authors present a detailed analysis under the l_1 norm and Hamming distance, highlighting the computational complexities and proposing efficient algorithms to tackle these. Notably, the complexity analysis for the problem (**Imp, SP, Bounded,** l_1) reveals its $\mathcal{NP}$-complete nature and offers a polynomial-time solution when path constraints are predefined.

In Chap. 6, the shortest path interdiction problems are explored, focusing on the budget-constrained (Int-SP) by upgrading edges on trees. The authors propose primal-dual algorithms to solve the problems under l_1 norm efficiently. Furthermore, the chapter addresses the $\mathcal{NP}$-hardness of the problems under weighted sum-type Hamming distance and provides dynamic programming algorithms for their unit cases.

In Chap. 7, the (Int-SRD) and its budget-constrained problem (BC-Int-SRD) are addressed under the l_∞ norm and bottleneck Hamming distance. The authors propose binary search algorithms to solve these problems, operating within pseudo-polynomial time. The inclusion of shortest path constraints leads to the formulation of the double interdiction problem ($\text{DIT}_{H\infty}$), which is also proven to be $\mathcal{NP}$-hard but solvable using a combination of dynamic programming and binary search techniques.

Lastly, Chap. 8 introduces the problem (RIOVSPT_1) under the weighted l_1 norm on trees. It meticulously examines the strategic adjustment of edge weights to achieve a minimal total cost while ensuring that the shortest root-leaf path length adheres to a predefined lower bound. The authors explore the relationship between the optimal solutions of (Int-SPT_1) and ($RIOVSPT_1$) and propose an algorithmic solution that incrementally refines an initial infeasible solution, leveraging a series of subproblems to converge upon an optimal configuration. The algorithm's efficiency is underscored by its $O(n^2)$ time complexity, highlighting the practicality of the approach.

We summarized the research results on generalized inverse shortest path problems on Table II, in which the abbreviations are listed below.

SRD, sum of root-leaf distance; **Card.,** cardinality constraint; **Db.,** double.

In summary, this part provides a systematic exposition of the generalized inverse shortest path problems, encapsulating a wealth of theoretical and algorithmic advancements. The text is replete with rigorous analyses, innovative methodologies, and a clear articulation of the implications for combinatorial optimization in network design and analysis. The contributions of this work are poised to significantly impact the field of inverse combinatorial optimization, offering a fertile ground for future research and application.

Table II Research results on generalized inverse shortest path problems

Problem type	Norm	Time complexity	Reference
Imp-SP	(Bounded, weighted l_1 norm)	General $\mathcal{NP}$-hard	Sect. 5.2.1 [142]
		Polynomial solvable: (s_i, t_i) are specified	Sect. 5.2.1 [142]
		Polynomial solvable: one pair of nodes	Sect. 5.2.2 [142]
		$O(n^2)$ on trees with a single source	Sect. 5.2.3 [142]
	(Bounded, weighted sum HD)	Strongly $\mathcal{NP}$-hard	Sect.5.3.1 [140]
	(Bounded, unit sum HD)	Exponential time on rooted trees **open problem** : **$\mathcal{NP}$-hard or not**?	Sect. 5.3.2 [135]

(continued)

Table II (continued)

Problem type	Norm	Time complexity	Reference
(BC)Int-SP on trees	(BC-Int, bounded, weighted l_1 norm)	$O(n^2)$	Sect. 6.2.2 [153]
	(BC-Int, bounded, unit l_1 norm)	$O(n)$	Sect. 6.2.3 [153]
	(Int, bounded, weighted l_1 norm)	$O(n^2)$	Sect. 6.3 [153]
	(Int, bounded, unit l_1 norm)	$O(n)$	Sect. 6.2.3 [153]
	(BC-Int, bounded, weighted sum HD)	$\mathcal{NP}$-hard	Sect. 6.4 [154]
	(BC-Int, bounded, unit sum HD)	$O(n^4)$	Sect. 6.4.1 [154]
	(Int, bounded, weighted sum HD)	$\mathcal{NP}$-hard	Sect. 6.4 [154]
	(Int, bounded, unit sum HD)	$O(n^4 \log n)$	Sect. 6.4.2 [154]
BC-Int-SRD on trees	l_1 norm	$O(n)$	Sect. 7.2.4 [151]
	l_∞ norm	$O(n)$	Sect. 7.3.1 [83]
	Unit sum HD	$O(n)$	Sect. 7.2.2 [151]
	Weighted sum HD	$\mathcal{NP}$-hard	Sect. 7.2.1 [151]
	Weighted node cost	$\mathcal{NP}$-hard	Sect. 7.2.7 [151]
	Unit node cost	$O(n)$	Sect. 7.2.8 [151]
	Bottleneck HD	$O(n)$	Sect. 7.3.3 [83]
Int-SRD on trees	l_1 norm	$O(n)$	Sect. 7.2.4 [151]
	l_∞ norm	$O(n \log n)$	Sect. 7.2.5 [83]
	Unit sum HD	$O(n \log n)$	Sect. 7.2.3 [151]
	Weighted sum HD	$\mathcal{NP}$-hard	Sect. 7.2.1 [151]
	Weighted node cost	$\mathcal{NP}$-hard	Sect. 7.2.7 [151]
	Unit node cost	$O(n \log n)$	Sect. 7.2.9 [151]
	Bottleneck HD	$O(n \log n)$	Sect. 7.2.6 [83]
BC-Int-SRD-Card. on trees	l_∞ norm	$O(n)$	Sect. 7.3.1 [83]
	Bottleneck HD	$O(n)$	Sect. 7.3.3 [83]
Int-SRD-Card. on trees	l_∞ norm	$O(Nn^2)$	Sect. 7.3.2 [83]
	Bottleneck HD	$O(n \log n)$	Sect. 7.3.4 [83]
BC-Db-Int-Card. on trees	l_∞ norm and bottleneck HD	$O(nN^2 \log_2 U)$	Sect. 7.4.1.4 [82]
Db-Int-Card. on trees		$O(nN^2 \log_2 U \log_2 K)$	Sect. 7.4.2.2 [82]
RIOV-SP on trees	(Bounded, weighted l_1 norm)	$O(n^2)$	Sect. 8.5 [152]

Chapter 5
Shortest Path Improvement Problems

Abstract We investigate the shortest path improvement problems (Imp-SPs) under different norms. We explore the intricacies of enhancing existing shortest paths in a graph by strategically modifying edge weights to improve the length of the shortest path. We introduce the problem (Imp, SP, Bounded, l_1) and its complexity, presenting a polynomial-time solution for specific cases. Furthermore, we extend the analysis to (Imp-SPs) under weighted sum Hamming distance, establishing its strong $\mathcal{NP}$-hardness and proposing effective heuristic algorithms. This chapter culminates in an exploration of (Imp-SPs) in tree networks, offering a combinatorial algorithm tailored to exploit the tree's structure. This work provides a significant contribution to the field, presenting both theoretical insights and practical algorithms for solving (Imp-SPs) efficiently.

Keywords Shortest path improvement problems · Time complexity · Weighted l_1 norm · Sum hamming distance · Trees

5.1 A Mathematical Model of Shortest Path Improvement Problems

Let $G = (V, E)$ be a graph with a set $V = \{v_1, v_2, \cdots, v_n\}$ of vertices and a set $E = \{e_1, e_2, \cdots, e_m\}$ of edges. Let $\boldsymbol{w} = (w_1, \cdots, w_m)$ and $\boldsymbol{c} = (c_1, \cdots, c_m)$ be a weight and cost vector defined on E, respectively. Let $\boldsymbol{l} = (l_1, \cdots, l_m)$ and $\boldsymbol{u} = (u_1, \cdots, u_m)$ be a lower and an upper bound vector for a modified weight vector $\bar{\boldsymbol{w}}$, respectively. For each pair (s_i, t_i) of vertices, s_i is called a source and t_i a terminal. We denote by $d_w(s_i, t_i)$ the weight of the shortest path from s_i to t_i under $\boldsymbol{w}$. Moreover, k pairs (s_i, t_i) of vertices and upper bounds $d_i (i = 1, 2, \cdots k)$ are specified with the request that the weight of the shortest path from s_i to t_i is not more than d_i. The shortest path improvement problem aims to shorten the lengths

Binwu Zhang and Xiucui Guan have equally contributed to this chapter.

X. Guan et al., *Inverse Combinatorial Optimization Problems*, Springer Optimization and Its Applications 225, https://doi.org/10.1007/978-3-031-91175-0_5

of edges with as less cost as possible under certain norm, which can be formulated as

$$\text{(Imp-SP)} \quad \begin{aligned} &\min_{\bar{\boldsymbol{w}}} \|\bar{\boldsymbol{w}} - \boldsymbol{w}\| \\ &s.t.\ d_{\bar{w}}(s_i, t_i) \le d_i, i = 1, 2, \cdots, k, \\ &\boldsymbol{l} \le \bar{\boldsymbol{w}} \le \boldsymbol{u}. \end{aligned} \tag{5.1}$$

This problem has a significant background [142]. In a transportation or communication network, the weight of a link may represent the transit time from one end to the other. So, the weight of a path is the traveling time from the source to the terminal. In order to improve the transmission efficiency of the network, one may want to reduce traveling times between some pairs of sources and terminals by decreasing weights of links. It is natural to produce costs when reducing traveling time in a link. We hope that the total cost is as less as possible to meet the requirement of speedup.

5.2 (Imp, SP, Bounded, Weighted l_1 Norm)

The problem (**Imp, SP, Bounded, Weighted l_1 Norm**), abbreviated by (**Imp-SP**$_{w1}$) can be formulated as follows:

$$\text{(Imp-SP}_{w1}) \quad \begin{aligned} &\min c(\bar{w}) = \sum_{j=1}^{m} c_j(w_j - \bar{w}_j) \\ &s.t.\ d_{\bar{w}}(s_i, t_i) \le d_i, i = 1, 2, \cdots, k, \\ &l_j \le \bar{w}_j \le w_j \quad j = 1, 2, \cdots, m. \end{aligned} \tag{5.2}$$

5.2.1 *Complexity of the Problem (Imp-SP$_{w1}$)*

The problem (**Imp-SP**$_{w1}$) was considered by Zhang and Lin [142]; they showed the $\mathcal{NP}$-completeness of the general model by transforming from the three-dimensional matching problem.

Definition 5.1 (Three-Dimensional Matching (3DM)) Given three disjoint sets U, V, W with $|U| = |V| = |W| = K$ and a subset Q of $U \times V \times W$, is there a subset M of Q with $|M| = K$ such that whenever (u, v, w) and (u', v', w') are distinct triples in M, $u \neq u'$, $v \neq v'$, and $w \neq w'$?

Theorem 5.1 ([142]) *The decision version of (**Imp-SP**$_{w1}$) is strongly $\mathcal{NP}$-complete even if l_j, u_j are 0 or 1, $c_j = 1$ for $j = 1, 2, \cdots, m$, and $d_i = 0$ for $i = 1, 2, \cdots, k$.*

Fig. 5.1 An example of graph G

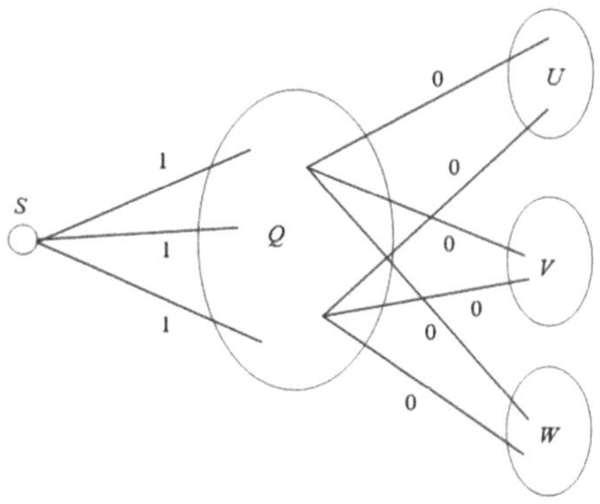

For proving this theorem, Zhang and Lin [142] constructed an instance of (**Imp-SP**$_{w1}$) as follows (Fig. 5.1).

(a) The vertex set of N is $U \cup V \cup W \cup Q \cup s$, where the new vertex s is a source, and all vertices in U, V, W are terminals.
(b) For every triple $q = (u, v, w) \in Q$, join edges $(q, u), (q, v), (q, w)$ where $u \in U, v \in V, w \in W$; also join (s, q) for all $q \in Q$.
(c) For edges $e = (q, u), (q, v)$ or (q, w), let $l(e) = u(e) = 0, c(e) = 1$; for edges $e = (s, q)$, let $u(e) = c(e) = 1, l(e) = 0$. Moreover, for all pairs of sources and terminals (s, u), (s, v) and (s, w), let the requested upper bound for these distances be $d_i = 0 (i = 1, 2, \cdots, k)$, and let the threshold $b = K$.

By proving that the instance of 3DM has a perfect matching M if and only if the instance of (**Imp-SP**$_{w1}$) (with respect to network N) has a feasible solution $\bar{w}$ such that $c(\bar{w}) \leq K$.

Zhang and Lin also provided a polynomial solvable case when the paths from s_i to $t_i (i = 1, \cdots, k)$ are specified.

Theorem 5.2 ([142]) *When the paths from s_i to $t_i (i = 1, \cdots, k)$ are specified, the problem (**Imp-SP**$_{w1}$) can be solved in polynomial time.*

Proof Let the path from s_i to t_i be denoted by $P_i (i = 1, 2, \cdots, k)$, and let $E = \{e_1, e_2, \cdots, e_m\}$. We define a $k \times m$ matrix $A = (a_{ij})$, as the path-edge incident matrix, by

$$a_{ij} = \begin{cases} 1, \text{ if edge } e_j \text{ is in path } P_i, \\ 0, \text{ otherwise}. \end{cases}$$

Let $d = (d_1, d_2, \cdots, d_k)^T$. Then the problem (**Imp-SP**$_{w1}$) can be formulated as a linear programming problem as follows:

$$\begin{aligned} &\min c^T(w - \bar{w}) \\ &s.t.\ A\bar{w} \le d, \\ &\qquad l \le \bar{w} \le w. \end{aligned} \tag{5.3}$$

It is well-known that there are polynomial-time algorithms for solving linear programming problems [106]. □

5.2.2 The Problem (Imp-SP$_{w1}$) with a Pair of Nodes

In this subsection, we consider the case when $k = 1$ in the following model.

$$\begin{aligned} &\min c(\bar{w}) = \sum_{j=1}^{m} c_j(w_j - \bar{w}_j) \\ &s.t.\ d_{\bar{w}}(s_1, t_1) \le d_1, \\ &\qquad l_j \le \bar{w}_j \le w_j \quad j = 1, 2, \cdots, m. \end{aligned} \tag{5.4}$$

For the vertex set $V = \{1, 2, \cdots, n\}$, assume that the source $s_1 = 1$, the terminal $t_1 = n$ and $d_1 = d$. For any non-negative scalar x, let

$$\begin{aligned} a_{ij}(x) &= \text{the cost for shortening edge } (i, j) \text{from } w_{ij} \text{ down to } x \\ &= \begin{cases} 0, & \text{if } x \ge w_{ij} \\ c_{ij}(w_{ij} - x), & \text{if } l_{ij} \le x \le w_{ij}. \end{cases} \end{aligned}$$

Define the optimality function as

$$f_j(y) = \text{the minimum cost such that the distance from 1 to } j \text{ is at most } y.$$

Then, $f_n(d)$ is the optimal value of (5.4). By the Optimality condition, Zhang and Lin [142] obtained the Dynamic Programming (DP) recursive equation.

$$f_j(y) = \min_{r \ne j} \min_{0 \le x \le y} \{f_r(y - x) + a_{rj}(x)\} \tag{5.5}$$

with the initial condition $f_1(y) = 0$ for all $y \ge 0$.

To avoid complicated notations and to concentrate attention on the structure of solutions, Zhang and Lin [142] start with the case that network N is acyclic. Recall that a directed graph is acyclic if and only if there exists a numbering of its vertices such that $(i, j) \in E \Rightarrow i < j$ [106]. So, they assume that the vertex set V has been arranged according to this numbering. For a vertex j, $j \geq 2$, denote its precedence set by $N^-(j) = \{r \in V | (r, j) \in E\}$. Then, (5.5) can be written as

$$f_j(y) = \min_{r \in N^-(j)} \min_{0 \leq x \leq y} \{f_r(y - x) + a_{rj}(x)\} \tag{5.6}$$

Zhang and Lin [142] obtained the next algorithm.

Algorithm 5.1 A DP Algorithm for (**Imp-SP**$_{w1}$) with a pair of nodes [142]

1: Let $f_1(y) = 0$ for all y.
2: **for** $j = 2, 3, \cdots, n$ **do**
3: For each $r \in N^-(j)$, compute $s_r(y) = \min_{0 \leq x \leq y} \{f_r(y - x) + a_{rj}(x)\}$ and record the optimal solution x as $x_{rj}(y)$.
4: Compute $f_j(y) = \min_{r \in N^-(j)} s_r(y)$ and save the optimal index r as $p_j(y)$.
5: **end for**
6: Suppose that $u = pn(d)$. [*Then u is the previous vertex of the terminal n (in a shortest path with length d).*]
7: The required length is $x^*_{un} = x_{un}(d)$ for edge (u, n).
8: Suppose that $v = p_u(d - x^*_{un})$. [*Then v is the previous vertex of u with $x^*_{vu} = x_{vu}(d - x^*_{un})$ as the adjusted length of edge (v, u).*]
9: Continuing in this way, we will eventually get $1 = p_i(d - x^*_{un} - \cdots - x^*_{ij})$ and $x^*_{1i} = x_{1i}(d - x^*_{un} - \cdots - x^*_{ij})$.
10: The shortest path from 1 to n is $(1, i, j, \cdots, v, u, n)$ with length $x^*_{1i} + x^*_{ij} + \cdots + x^*_{un} = d$ and the minimum cost $f_n(d)$.

Theorem 5.3 ([142]) *The **DP Algorithm** correctly solves the recursive equation (5.6), as well as the problem (5.4), in polynomial time.*

5.2.3 *The Problem (Imp-SP$_{w1}$) on Trees with a Single Source*

By (5.3), the problem (**Imp-SP**$_{w1}$) on trees network can be rewritten as

$$\begin{aligned} &\min c^T y \\ &s.t. \ Ay \geq b, \\ &\quad 0 \leq y \leq u. \end{aligned} \tag{5.7}$$

where $y = w - \bar{w} = (y_1, y_2, \cdots, y_m)^T$, $b = Aw - d = (b_1, b_2, \cdots, b_k)^T$ with $b_i = d_l(s_i, t_i) - d_i > 0$ representing the required reduction of the length of P_i; and $u = w - l = (u_1, u_2, \cdots, u_m)^T$ with $u_j \geq 0$ as the upper bound of the adjustment of edge e_j. The matrix $A = (a_{ij})$ is the $k \times m$ path-edge incident matrix and a_i^T is the i-th row of A. Zhang and Lin [142] obtained the next algorithm.

Algorithm 5.2 An algorithm for (**Imp-SP**$_{w1}$) on trees with a single source [142]

1: Initialize $y^0 := 0$ and $r := 0$.
2: **while** y^r is not feasible for (5.7) **do**
3: Let $B^r := \{i | b_i - a_i^T y^r > 0\}$ and update $c_j := +\infty$, if $y_j^r = u_j$.
4: Find a moving direction $z \in R^m$ by solving the subproblem

$$\begin{aligned} &\min c^T z \\ &s.t.\ A_i^T z \geq 1, \ \text{for } i \in B^r \\ &\quad 0 \leq z_j \leq 1 \text{ for } j = 1, 2, \cdots, m \end{aligned} \tag{5.8}$$

5: Let $y^{r+1} := y^r + \theta z$, where $\theta := \min\{\theta_1, \theta_2\}\}$, $\theta_1 := \min\{u_j - y_j^r | z_j = 1\} > 0, \theta_2 := \min\{\frac{b_i - a_i^T y^r}{a_i^T z} | i \in B^r\} > 0$.
6: Update $r := r + 1$.
7: **end while**
8: Output an optimal solution y^r.

For the subprobem (5.8), Zhang and Lin [142] proved the next two lemmas.

Lemma 5.1 ([142]) *Problem (5.8) has integer optimal solutions.*

A subset $W \subseteq E$ is called a cut with respect to B^r if $W \cap P_i \neq \emptyset (\forall i \in B^r)$. The cost of cut W is defined by $C(W) = \sum_{e_i \in W} c_i$. A minimum cut is a cut with the minimum cost.

Lemma 5.2 ([142]) *The incident vector z of a minimum cut is an optimal solution of the subproblem (5.8).*

For proving the problem (5.7) is polynomial time solvable, Zhang and Lin [142] obtained the next two results.

Lemma 5.3 ([142]) *There is an optimal solution y^* of problem (5.7) such that $y^* \geq y^1$, where y^1 is the first stage solution of Algorithm alg-CO-ImpSP when $r = 1$.*

Lemma 5.4 ([142]) *Algorithm 5.2 correctly solves the problem (**Imp-SP**$_{w1}$) on trees with a single source in $O(n^2)$ time.*

5.3 (Imp, SP, Bounded, Weighted Sum Hamming Distance)

The problem (**Imp, SP, Bounded, Weighted Sum Hamming Distance**), abbreviated by (**Imp-SP**$_{wsH}$), can be formulated as follows:

$$\begin{aligned} &\min \sum_{j=1}^{m} c_j H(\bar{w}_j, w_j) \\ (\textbf{Imp-SP}_{wsH})\ &s.t.\ d_{\bar{w}}(s_i, t_i) \le d_i, i = 1, 2, \cdots, k, \\ &l_j \le \bar{w}_j \le w_j \quad j = 1, 2, \cdots, m. \end{aligned} \tag{5.9}$$

The problem (**Imp-SP**$_{wsH}$) has practical background [140]. Consider a transportation network $G = (V, E, w)$, where V represents nodes corresponding to villages or towns, and E denotes edges representing roads with weights w indicating travel times. Given facilities at vertices s_i, $i = 1, 2, \ldots, k$, the goal is to minimize the travel time to community vertices t_i within a specified limit. This is achieved by strategically reducing travel times on select roads at a minimum cost, often involving road widening at specific narrow sections. The associated cost for road e_j may be a fixed c_j, rather than proportional to the reduction in travel time.

5.3.1 *Complexity of the Problem (Imp-SP$_{wsH}$)*

Zhang et al. [140] proved that the problem (**Imp-SP**$_{wsH}$) is strongly $\mathcal{NP}$-hard by reducing from the 3-Satisfiability (3-SAT) problem.

Suppose that x is a Boolean variable that can take only two values *true* and *false*. We call $\bar{x}$ the negation of Boolean variable x, where $\bar{x}$ =*true* if and only if x =*false*. Given a set of Boolean variables $X = \{x_1, x_2, \cdots, x_n\}$, a *truth assignment* for X is a function $t : X \longrightarrow$ {*true, false*}. If x is a variable in X, then Boolean variable x and its negation $\bar{x}$ are both called *literals* over X. A *clause* over X consists of some literals over X, and it is *true* under a truth assignment if and only if at least one of its members is true under that assignment.

The 3-SAT problem can be stated as follows. Given m clauses $C = \{C_1, C_2, \cdots, C_m\}$ on a set X of variables such that $|C_i| = 3$ for $1 \le i \le m$, is there a truth assignment for X such that all the clauses in C are *true*?

If such a truth assignment exists, then the 3-SAT problem is *satisfiable*.

Lemma 5.5 ([44]) *3-SAT problem is strongly $\mathcal{NP}$-hard.*

Theorem 5.4 ([140]) *Even if $c_i = w_i = 1$ and $l_i = 0$ for each edge of the network, the problem (**Imp-SP**$_{wsH}$) is strongly $\mathcal{NP}$-hard.*

To prove Theorem 5.4, Zhang et al. [140] first construct a reduction from the 3-SAT problem to a decision problem of (**Imp-SP**$_{wsH}$) as follows.

Let $X = \{x_1, x_2, \cdots, x_n\}$ be the set of variables and $C = \{C_1, C_2, \cdots, C_m\}$ be the set of clauses in an arbitrary instance of 3-SAT. Construct an undirected network $G = (V, E, w)$ in the following way.

(1) For each element x_i $(i = 1, 2, \cdots, n)$ in X, construct two *element vertices* of V, which are denoted by v_i and v_{n+i}. For each clause C_j $(j = 1, 2, \cdots, m)$,

construct a *clause vertex* in $q_j \in V$. If $x_i \in C_j$, then construct $2n$ vertices $q^1_{i,j}, q^2_{i,j}, \cdots, q^{2n}_{i,j}$. If $\bar{x}_i \in C_j$, then construct $2n$ vertices $q^1_{n+i,j}, q^2_{n+i,j}, \cdots, q^{2n}_{n+i,j}$. In addition, construct some vertices t_i $(i = 1, 2, \cdots, m + 1)$ and z_k $(k = 0, 1, 2, \cdots, n)$. Thus,

$$V = (\bigcup_{i=1}^{n}\{v_i, v_{n+i}\}) \cup (\bigcup_{i=0}^{n}\{z_i\}) \cup (\bigcup_{j=1}^{m}\{q_j\}) \cup (\bigcup_{j=1}^{m+1}\{t_j\}) \cup$$
$$(\bigcup_{k=1}^{2n}\bigcup_{j=1}^{m}\bigcup_{i=1}^{n}\{q^k_{i,j}|x_i \in C_j\}) \cup (\bigcup_{k=1}^{2n}\bigcup_{j=1}^{m}\bigcup_{i=1}^{n}\{q^k_{n+i,j}|\bar{x}_i \in C_j),$$

and there are $6mn + 3n + 2m + 2$ vertices in V.

(2) For each j, $j = 1, 2, \cdots, n$, construct edges $[z_{j-1}, v_j]$ and $[v_j, z_j]$. For each j, $j = n+1, n+2, \cdots, 2n$, construct edges $[z_{j-n-1}, v_j]$ and $[v_j, z_{j-n}]$. Thus

$$E_1 = \bigcup_{i=1}^{n}\{[z_{i-1}, v_i], [z_{i-1}, v_{i+n}], [z_i, v_i], [z_i, v_{i+n}]\},$$

and there are $4n$ edges in E_1.

If $x_i \in C_j$, then construct $2n + 1$ edges $[v_i, q^1_{i,j}], [q^1_{i,j}, q^2_{i,j}], \ldots, [q^{2n}_{i,j}, q_j]$. If $\bar{x}_i \in C_j$, then construct $2n+1$ edges $[v_{i+n}, q^1_{i+n,j}], [q^1_{i+n,j}, q^2_{i+n,j}], \ldots, [q^{2n}_{i+n,j}, q_j]$. Thus, there are $6mn + 3m$ edges in E_2.

$$E_2 = (\bigcup_{i=1}^{n}\bigcup_{j=1}^{m}\{\{[v_i, q^1_{i,j}], [q^{2n}_{i,j}, q_j], \bigcup_{k=1}^{2n-1}[q^k_{i,j}, q^{k+1}_{i,j}]\}|x_i \in C_j\}) \cup$$
$$(\bigcup_{i=1}^{n}\bigcup_{j=1}^{m}\{\{[v_{i+n}, q^1_{i+n,j}], [q^{2n}_{i+n,j}, q_j], \bigcup_{k=1}^{2n-1}[q^k_{i+n,j}, q^{k+1}_{i+n,j}]\}|\bar{x}_i \in C_j\}),$$

For each j, $j = 1, 2, \cdots, m$, construct edge $[q_j, t_j]$. In addition, we construct edge $[z_n, t_{m+1}]$. Thus, there are $m + 1$ edges in E_3

$$E_3 = \bigcup_{j=1}^{m}\{[q_j, t_j]\}) \cup \{[z_n, t_{m+1}]\}.$$

Hence, there are $6mn + 4n + 4m + 1$ edges in $E = E_1 \cup E_2 \cup E_3 = \{e_1, e_2, \cdots, e_f\}$.

(3) For each edge $e_i \in E$, $i = 1, 2, \cdots, f$, let $w(e_i) = 1$, $l(e_i) = 0$, and $c(e_i) = 1$. For $m + 1$ pairs of vertices (s_i, t_i), let s_i be z_0 and define the bound as $d_i = 1$.

Let $U = 2mn + 2n + m$. Since there are $6mn + 3n + 2m + 2$ vertices and $6mn+4n+4m+1$ edges in the graph G, which we constructed by the given 3-SAT instance, it is easy to see that the graph G can be constructed in a polynomial time.

Theorem 5.4 is proven by proving the next lemma.

Lemma 5.6 ([140]) *The 3-SAT problem is satisfiable if and only if the instance of the problem (**Imp-SP**$_{wsH}$) with respect to G has a feasible solution $\bar{w}$ satisfying*

$$\sum_{e_i \in E} H(\bar{w}(e_i), w(e_i)) \leq 2mn + 2n + m.$$

Zhang et al. [137] proved the inapproximability of **(Imp, SP, Bounded, wsH-distance)** by constructing an L-reduction from an instance of the Set-Cover problem to an instance of the problem (**Imp-SP**$_{wsH}$).

Definition 5.2 Set-Cover problem [105]: Let $T = \{t_1, t_2, \ldots, t_n\}$ be a finite set, and $S_1, S_2, \ldots, S_m$ be a collection of its subsets satisfying $\bigcup_{j=1}^{m} S_j = T$, find the minimum cardinality subset $\{S_{j_1}, S_{j_2}, \ldots, S_{j_q}\}$ in $\{S_1, S_2, \ldots, S_m\}$ that covers T, that is, $\bigcup_{\tau=1}^{q} S_{j_\tau} = T$ and q is minimized.

Lemma 5.7 ([113]) *It is $\mathcal{NP}$-hard to approximate the Set-Cover problem within a worst-case ratio of $O(\log n)$ if $m = O(p(n))$, where $p(n)$ is a polynomial function of n.*

Theorem 5.5 ([137]) *Even if $c_i = 1$ and $l_i = 0$ for each edge e_i, approximately solving the problem (**Imp-SP**$_{wsH}$) within a worst-case ratio of $O(\log |V|)$ is $\mathcal{NP}$-hard.*

For proving Theorem 5.5, [137] first introduce an L-reduction from an instance of the Set-Cover problem to an instance of the problem (**Imp-SP**$_{wsH}$) as follows.

Let I be an instance of the Set-Cover problem, that is, a collection of m subsets $\{S_1, S_2, \ldots, S_m\}$ of $T = \{t_1, t_2, \ldots, t_n\}$,, which satisfies $\bigcup_{j=1}^{m} S_j = T$. Zhang et al. [137] construct an undirected network $G = (V, E, w, l, c)$ and an instance $\tau(I)$ of the problem (**Imp-SP**$_{wsH}$) in the following three steps.

(1) For each element $t_i \in T$, introduce an *element vertex* t_i in V. For each subset S_j of instance I, introduce a *subset vertex* S_j in V. In addition, introduce $n + 1$ vertices $S, f_1, f_2, \ldots, f_n$ in V. Thus $V = F \cup \{S\} \cup S^* \cup T$, where $F = \{f_1, f_2, \ldots, f_n\}$ and $S^* = \{S_1, S_2, \ldots, S_m\}$.
(2) For each vertex $f_i \in F$, introduce an edge $(f_i, S) \in E$. For each subset vertex $S_j \in S^*$, $j = 1, 2, \ldots, m$, introduce an edge $(S, S_j) \in E$, and an edge $(S_j, t_i) \in E$ if and only if $t_i \in S_j$. Thus, $E = E_1 \cup E_2 \cup E_3$, where $E_1 = \{(f_i, S)| : f_i \in F\}$, $E_2 = \{(S, S_j)| : S_j \in S^*\}$, and $E_3 = \bigcup_{j=1}^{n} \bigcup_{i=1}^{m} \{(S_j, t_i)| : t_i \in S_j\}$.
(3) For each edge $e_i \in E$, let its length be $w_i = 1$, the lower bound be $l_i = 0$, and the cost be $c_i = 1$. For each source–terminal pair (f_k, t_k), $k = 1, 2, \cdots, n$, let the upper bound of its shortest distance be $d_k = 2$.

For a feasible solution w' of instance $\tau(I)$, define the cost (objective) function as $\beta_{w'}(\tau(I)) = \sum_{e_i \in E} H(w'_i, w_i)$, and let $\beta(I)$ and $\beta(\tau(I))$ be the optimal values of instance I and instance $\tau(I)$, respectively. [137] proved the next two lemmas.

Lemma 5.8 ([137]) *For each feasible solution w' of $\tau(I)$, we can construct a feasible solution SC of instance I in polynomial time such that $|SC| \leq \beta_{w'}(\tau(I))$, and therefore, $\beta(I) \leq \beta(\tau(I))$.*

Lemma 5.9 ([137]) *For any set-cover SC of T, we can define a feasible solution w' of $\tau(I)$ in polynomial time such that $|SC| = \beta_{w'}(\tau(I))$, and therefore, $\beta(\tau(I)) \leq \beta(I)$.*

It follows from Lemmas 5.8 and 5.9 that the problem (**Imp-SP**$_{wsH}$) is equivalent to the Set-Cover problem under the L-reduction, and hence, Theorem 5.5 is proven.

Zhang et al. [137] designed an $O(k^2n^3m)$ heuristic algorithm of the problem (**Imp-SP**$_{wsH}$), where k, n, m are the numbers of source–terminal pairs, vertices, and edges, respectively.

5.3.2 Algorithms of (Imp, SP, Bounded, sH-distance) on Rooted Trees

Zhang et al. [135] discussed a special problem **(Imp, SP, Bounded, sH-distance)** called the shortest path improvement problem on rooted trees under unit Hamming distance (denoted by (**SPIUH-RT**)). A rooted tree $T = (V, E)$ is rooted at source $s \in V$, where $V = \{s, v_1, v_2, \cdots, v_{n-1}\}$, and $E = \{e_1, e_2, \cdots, e_{n-1}\}$, $c_i = 1$ for all $i = 1, 2, \ldots, n-1$. Let d_k be the upper bound of the shortest distance connecting the source–terminal pair (s, t_k). The problem (**SPIUH-RT**) can be formulated as follows:

$$\begin{aligned} \min \ & \sum_{i=1}^{n-1} H(w_i^*, w_i) \\ s.t \ & d_{w^*}(s, t_k) \leq d_k, \ k = 1, 2, \ldots, r, \\ & l_i \leq w_i^* \leq w_i, \quad i = 1, 2, \ldots, n-1. \end{aligned} \tag{5.10}$$

Definition 5.3 ([135]) Let $\alpha(v)$ be the number of vertices whose degree greater than 2 on the only path from s to v (excluding v). The **Tab** of any node $v \in V$ is defined as follows:

$$Tab(v) := \begin{cases} \alpha(v) + 1, \ if \ degree(v) \leq 2, \\ \alpha(v) + 2, \ if \ degree(v) > 2. \end{cases} \tag{5.11}$$

Definition 5.4 ([135]) For each $e = (u, v) \in V$, if $Tab(u) \leq Tab(v)$, we define $L(e) := Tab(u)$ as the **layer number** of the edge e.

Definition 5.5 ([135]) For the source node s and node $u \in V$ that satisfy $degree(u) > 2$, we define a set of **critical children** (denoted by $CC(u)$) of the node u (including $u = s$) as follows. Let u be in the path from s to v. If

$degree(v) > 2$ and $Tab(v) = Tab(u) + 1$, then $v \in CC(u)$; if v is a leaf node with $Tab(v) = Tab(u)$, then $v \in CC(u)$.

Zhang et al. [135] proposed a dynamic programming algorithm **DP** to solve the problem (**SPIUH-RT**).

F_s : the optimal value of (**SPIUH-RT**).

$F_v(i_1, i_2, \ldots, i_p)$: the minimum number of modified edges in T_v when the edges in $S_v(i_1, \ldots, i_p)$ are modified, for any $v \in V^*$ ($i_1 \in \{0, 1, \ldots, s_1\}$, $i_2 \in \{0, 1, \ldots, s_2\}$, $\ldots$, $i_p \in \{0, 1, \ldots, s_p\}$), where T_v is the subtree of T rooted at v. $F_v(i_1, i_2, \ldots, i_p)$ is the optimal value of the following problem (**SPIUH-RT**) for T_v.

$$\begin{aligned} \min & \sum_{e_i \in T_v} H(w_i^*, w_i) \\ s.t \ & d_{w^*}(v, t_k) \le d_k', \quad \text{for each terminal } t_k \text{ of subtree } T_v, \\ & l_i \le w_i^* \le w_i, \quad e_i \in T_v, \end{aligned} \tag{5.12}$$

where

$$d_k' := d_k - \sum_{e_i \in S_v(i_1, \ldots, i_p)} l_i - \sum_{e_i \in P_v \setminus S_v(i_1, \ldots, i_p)} w_i. \tag{5.13}$$

If problem (5.12) is infeasible, then $F_v(i_1, i_2, \ldots, i_p) = +\infty$.

$G_{v_h}(i_1, i_2, \ldots, i_p)$: the minimum number of modified edges in T_{v,v_h} when the edges in $S_v(i_1, \ldots, i_p)$ are modified, for any $v \in V^*$ and $v_h \in CC(v)$ ($i_1 \in \{0, 1, \ldots, s_1\}$, $i_2 \in \{0, 1, \ldots, s_2\}$, $\ldots$, $i_p \in \{0, 1, \ldots, s_p\}$). Then the value $G_{v_h}(i_1, i_2, \ldots, i_p)$ is the optimal value of the following problem (**SPIUH-RT**) for the subtree T_{v,v_h} rooted at v.

$$\begin{aligned} \min & \sum_{e_i \in T_{v,v_h}} H(w_i^*, w_i) \\ s.t \ & d_{w^*}(v, t_k) \le d_k', \quad \text{for each terminal } t_k \text{ of subtree } T_{v_h}, \\ & l_i \le w_i^* \le w_i, \quad e_i \in T_{v,v_h}, \end{aligned} \tag{5.14}$$

where d_k' is defined as shown in (5.13).

The values for F and G can be computed by traversing the rooted tree in postorder (from leaves to the root s) using the following relations.

For each non-leaf node $v_h \in CC(v)$,

$$\begin{aligned} G_{v_h}(i_1, i_2, \ldots, i_p) = \min\{ & F_{v_h}(i_1, \ldots, i_p, i_{p+1}) + i_{p+1} : \\ & i_{p+1} \in \{0, 1, \ldots, s_{p+1}\},\ s_{p+1} = |P_{v_h} \cap E_{p+1}|\}. \end{aligned} \tag{5.15}$$

For each leaf node $v_h \in CC(v)$, T_{v,v_h} is just a chain from v to v_h. In this case, $G_{v_h}(i_1, i_2, \ldots, i_p)$ can be easily obtained by using a greedy algorithm 5.3 in [154].

For each node $v \in V^*$, we have

$$F_v(i_1, i_2, \ldots, i_p) = \sum_{v_h \in CC(v)} G_{v_h}(i_1, i_2, \ldots, i_p). \tag{5.16}$$

For $v = s$, there is only one non-leaf critical child v_s, $p = Tab(s) - 1 = 0$. Then,

$$G_{v_s} = \min\{F_{v_s}(i_1) + i_1 : i_1 \in \{0, 1, \ldots, s_1\}\}. \tag{5.17}$$

Therefore, the optimal value of the problem (**SPIUH-RT**) can be calculated as

$$F_s = G_{v_s}. \tag{5.18}$$

By computing the values of F and G alternately, Zhang et al. [135] designed Algorithm 5.4. It is an efficient algorithm for solving most instances of problems (**SPIUH-RT**), although its worst-case complexity is exponential time.

Algorithm 5.3 A greedy algorithm for Problem (**SPIUH**) on Chains [135]

Input: A chain P_1 with a source–terminal pair (s, t_1), two edge weight vectors l and w, and a value d'_1.
Output: An optimal solution $\tilde{w}$ or the infeasibility of the instance.
1: If $\Sigma_{e \in P_1} l(e) > d'_1$, then output that the instance is infeasible and stop.
2: Let $w^0 := w - l$. Rearrange the vector w^0 in a non-increasing order, that is, $w^0_{j_1} \geq w^0_{j_2} \geq \cdots \geq w^0_{j_m}$, where $m := |P_1|$. Let $i := 0$, and $W := \Sigma_{e \in P_1} w(e)$.
3: **while** $W > d'_1$ **do**
4: Put $i := i + 1$, and $W := W - w^0_{j_i}$.
5: **end while**
6: Output an optimal solution $\tilde{w}$: $\tilde{w}_{j_q} = l_{j_q}$ if $1 \leq q \leq i$ and w_{j_q} otherwise.

Algorithm 5.4 A DP algorithm for problem (**SPIUH-RT**) [135]

Input: A tree T rooted at s with source–terminal pairs (s, t_k), $k = 1, 2, \cdots, r$, two edge weight vectors l and w, and $d_1, d_2, \cdots, d_r$.
Output: The optimal value and an optimal solution of the problem (**SPIUH-RT**).
1: **for** $k = 1 \rightarrow r$ **do**
2: Call Algorithm GA for calculating all the values $G_{t_k}(i_1, \ldots, i_{p_k})$, where $i_1 \in \{0, 1, \ldots, s^k_1\}$, $i_2 \in \{0, 1, \ldots, s^k_2\}, \ldots, i_{p_k} \in \{0, 1, \ldots, s^k_{p_k}\}$, $s^k_j := |P_{t_k} \cap E_j|$, $j = 1, \cdots, p_k$ and

$$d'_k := d_k - \Sigma_{e_i \in S_{t_k}(i_1,\ldots,i_{p_k})} l_i - \Sigma_{e_i \in P_{t_k} \setminus S_{t_k}(i_1,\ldots,i_{p_k})} w_i.$$

3: **end for**
4: Let $L := \max\{L(e) : e \in E\}$ be the maximum layer number.
5: **for** $a = L - 1 \rightarrow 1$ **do**

Let $CC_a := \{v : Tab(v) = a + 1, v \notin Z$ is a critical child of some node$\}$.
while there is a node $v_\tau \in CC_a$ **do**
for $i_1 \in \{0, 1, \ldots, s_1^\tau\}$, $i_2 \in \{0, 1, \ldots, s_2^\tau\}$, $\ldots$, $i_a \in \{0, 1, \ldots, s_a^\tau\}$, where $s_j^\tau := |P_{v_\tau} \cap E_j|$, $j = 1, \cdots, a$ **do**
Calculate all the values

$$F_{v_\tau}(i_1, i_2, \ldots, i_a) := \Sigma_{v_h \in CC(v_\tau)} G_{v_h}(i_1, i_2, \ldots, i_a),$$

$$G_{v_\tau}(i_1, i_2, \ldots, i_{a-1}) := \min_{i_a \in \{0,1,\ldots,s_a^\tau\}} \{F_{v_\tau}(i_1, \ldots, i_a) + i_a, s_a^\tau = |P_{v_\tau} \cap E_a|\},$$

where the sets of modified edges are denoted by $S_{F_{v_\tau}}(i_1, i_2, \ldots, i_a)$ and $S_{G_{v_\tau}}(i_1, i_2, \ldots, i_{a-1})$, respectively.
Update $CC_a := CC_a \backslash v_\tau$.
end for
end while
end for
Let v_s be the only non-leaf critical child of the source node s.
Output the optimal value $F_s = G_{v_s}$ of the problem (**SPIUH-RT**) and an optimal solution is w^*, where $w_i^* := l_i$ if $e_i \in S_{G_{v_s}}$ and w_i otherwise.

5.4 Conclusions

This chapter delves into the nuances of the shortest path improvement problems (SPIPs), presenting a comprehensive analysis under various normative conditions. We introduce the concept of improving existing shortest paths by incrementally adjusting edge weights, a problem that finds relevance in diverse applications such as transportation and communication networks. We adeptly illustrate the complexities and algorithmic solutions for optimizing path lengths within a graph structure.

The chapter begins with a foundational exposition of SPIPs and progressively delves into more complex scenarios under different norms. Notably, we present a detailed investigation under the l_1 norm and Hamming distance, highlighting the computational challenges and proposing efficient algorithms to tackle these. The complexity analysis revealed the $\mathcal{NP}$-complete nature of the problem (**Imp, SP, Bounded**, l_1), while offering some polynomial-time algorithms when path constraints are predefined.

We further extend to the problem (**Imp, SP, Bounded, wsH-distance**), demonstrating the strong $\mathcal{NP}$-hardness of the problem, as well as its approximability. This chapter concludes with a detailed discussion on a special problem (**SPIUH-RT**) under a unit sum Hamming distance for rooted trees, presenting a combinatorial algorithm that leverages the tree's structure to optimize path improvements. The clarity in the presentation of algorithms, accompanied by a thorough analysis of their computational complexities, significantly contributes to the understanding and solvability of SPIPs. However, it is still *an open problem* that if the problem (**SPIUH-RT**) $\mathcal{NP}$-hard or not.

In summary, this chapter offers a rich tapestry of theoretical insights and practical solutions. It provides a solid foundation for future research in the field of inverse combinatorial optimization problems, especially on network improvement problems.

Chapter 6
Shortest Path Interdiction Problems on Trees

Abstract In this chapter, we introduce the shortest path interdiction problems on trees under different norms. We put forward two primal-dual algorithms both in $O(n^2)$ time to solve the (budget constrained) shortest path interdiction problem by upgrading edges under l_1 norm on trees. Additionally, we show that the problems under weighted sum Hamming distance are $\mathcal{NP}$-hard and propose two dynamic programming algorithms within $O(n^4)$ and $O(n^4 \log n)$ time for their unit norm cases, respectively.

Keywords Shortest path interdiction problems · Budget-constrained shortest path interdiction problem · l_1 Norm; Sum hamming distance · Trees

6.1 Introduction

Network interdiction problems by deleting critical edges (denoted by **(NIP-DE)**) have wide applications in communication networks, transportation networks, network war, and terrorist networks [39, 73]. The K-most-critical-edge problem [15, 78, 141] aims at making some network performance as poor as possible by deleting at most K edges. Most interdiction problems focus on shortest paths. For any K in the K-most-critical-edge shortest path problem, Khachiyan et al. [73] demonstrated that no approximation algorithm exists with an approximation ratio of 2, which remains the state-of-the-art result. For the scenario where $K = 1$, Nardelli et al. [78] introduced an $O(m\alpha(m, n))$ algorithm designed to address the problem on undirected networks. Here, α denotes the inverse Ackermann function, while m and n represent the counts of edges and vertices within the network, respectively. Zhang et al. [141] delved into the optimal shortest path set problem within an undirected graph, simulating a scenario where several vehicles traverse from a source vertex s to a destination vertex t. The objective is to identify the minimum ensemble of paths for these vehicles prior to their departure, ensuring that at least one vehicle arrives

Qiao Zhang and Xiucui Guan have equally contributed to this chapter.

X. Guan et al., *Inverse Combinatorial Optimization Problems*, Springer Optimization and Its Applications 225, https://doi.org/10.1007/978-3-031-91175-0_6

fastest by bypassing at least K edges. For the case when $K = 1$, they presented an $O(n^2)$ algorithm, and for $K > 1$, they devised a robust polynomial time algorithm. Focusing on the parameterized perspective, Bazgan et al. [15] provided a complexity and approximation analysis for the shortest path interdiction problem. This analysis hinges on the interplay among three pivotal parameters: the shortest path of length at least l, the path length increment b, and the diameter or a graph. When $b = 1$, they proposed an algorithm with a time complexity of $O(mn)$. However, the problem's complexity escalates significantly when $b \geq 2$.

Network interdiction typically involves edge or node deletion; however, in practical scenarios, modifying edge weights to disrupt network flow is often more feasible. For instance, in military contexts, delaying enemy support rather than outright blocking it is a common strategy. Terrorists may aim to impede fire truck routes by maximizing path lengths within a limited budget, effectively interdicting certain arcs [39]. Conversely, defenders must preemptively identify vulnerable arcs to establish safer alternative paths. This approach, which we term "upgrading critical edges," forms the basis of our study on shortest path interdiction in tree networks.

6.2 The Budget-Constrained Shortest Path Interdiction Problem by Upgrading Edges on Trees

Let $T = (V, E, w)$ be an edge-weighted tree rooted at v_1, where $V := \{v_1, v_2, \cdots, v_n\}$ and $E := \{e_2, e_3, \cdots, e_n\}$ are the sets of nodes and edges, respectively. Let $Y := \{t_0, t_1, \cdots, t_{\ell-1}\}$ be the set of leaves. Let $w(e)$ be an original length and $u(e)$ be an upper-bound of $w(e)$ after upgrade for each edge $e \in E$, where $w(e) \leq u(e)$. Denote by $\Delta u(e) = u(e) - w(e)$ the deviation between $u(e)$ and $w(e)$. Let $c(e)$ be a cost to upgrade the edge $e \in E$. Denoted by P_{v_i,v_j} the unique path from v_i to v_j on the tree T and record $P_k := P_{v_1,t_k}$ as the "root-leaf" path of the leaf node $t_k (k = 0, 1, 2, \cdots, \ell - 1)$ in brief.

The budget-constrained shortest path interdiction problem by upgrading edges on trees (denoted by **(BC-Int-SPT)**) aims to find an upgrade scheme $\bar{w}$ to maximize the length of the shortest root-leaf path from the root v_1 on the premise that the total upgrade cost under some norm is upper-bounded by a given value B. The mathematical model can be stated as follows.

$$\begin{aligned} &\max_{\bar{w}} \min_{t_k \in Y} \bar{w}(P_k) \\ \textbf{(BC-Int-SPT)} \quad &s.t. \quad \|\bar{w} - w\| \leq B, \\ &\qquad\quad w(e) \leq \bar{w}(e) \leq u(e), e \in E. \end{aligned}$$

We also considered the shortest path interdiction problem by upgrading edges on rooted trees **(Int-SPT)**, which aims to upgrade some edges to minimize the total cost under some norm on the premise that the length of the shortest root-leaf path is lower-bounded by a given value D.

$$\begin{aligned} & \min_{\bar{w}} \ \|\bar{w} - w\| \\ \textbf{(Int-SPT)} \quad & s.t. \ \min_{t_k \in Y} \bar{w}(P_k) \geq D, \\ & \qquad w(e) \leq \bar{w}(e) \leq u(e), e \in E. \end{aligned}$$

6.2.1 *Some Important Definitions*

We first present some important definitions to identify a storage structure of the rooted tree. Let $V^* = \{v \in V | deg(v) > 2\}$ be the set of vertices whose degree is more than 2.

Definition 6.1 ([153, Definition 2]) For $e_j = (v_i, v_j)$, we call v_i the **father** of v_j, denoted by $father(v_j) = v_i$. Define $Layer(v_1) := 1$ and the **Layer** of any other nodes $v \in V \setminus \{v_1\}$ as

$$Layer(v) := \begin{cases} Layer(father(v)), & \text{if } deg(v) \leq 2, \\ Layer(father(v)) + 1, & \text{if } deg(v) > 2. \end{cases}$$

Definition 6.2 ([152, Definition 2]) For each edge $e_j = (v_i, v_j) \in E$ with $Layer(v_i) \leq Layer(v_j)$, we define $LN(e_j) := Layer(v_i)$ as the **layer number** of the edge e_j.

Definition 6.3 ([153, Definition 3]) For a node $\bar{v} \in V^* \cup \{v_1\}$, we define a set $CD(\bar{v})$ of **critical descendant**. Let $\bar{v}$ be on the path from v_1 to $v \in V^* \setminus \{v_1\} \cup Y$. If $v \in V^* \setminus \{v_1\}$ and $Layer(v) = Layer(\bar{v}) + 1$, then $v \in CD(\bar{v})$; if $v \in Y$ and $Layer(v) = Layer(\bar{v})$, then $v \in CD(\bar{v})$. Correspondingly, if $v \in CD(\bar{v})$, we call $\bar{v}$ the **critical ancestor** of v, denoted by $CA(v) := \bar{v}$.

Definition 6.4 ([153, Definition 4]) For any node $v \in V^* \backslash \{v_1\} \cup Y$, define $\chi_v := P_{CA(v),v}$ as the chain from $CA(v)$ to v. The minimum cost of the chain is $\underline{C}(\chi_v) := \min_{e \in \chi_v^u} c(e) := c(\hat{e})$, where $\chi_v^u := \{e \in \chi_v | \Delta u(e) > 0\}$ and $\hat{e}$ is the edge closest to the root v_1 if there are multiple choices.

6.2.2 (BC-Int, SP-Trees, Bounded, Weighted l_1 Norm)

The problem **(BC-Int-SPT)** under weighted l_1 norm, denoted by **(BC-Int-SPT**$_1$**)**, can be formulated in the following form, where $r(e) = \bar{w}(e) - w(e)$, $e \in E$.

$$\begin{aligned} & \max_r \min_{t_k \in Y}(w(P_k) + r(P_k)) \\ \textbf{(BC-Int-SPT}_1\textbf{)} \quad s.t. \; & \sum_{e \in E} c(e)r(e) \le B, \\ & 0 \le r(e) \le d(e), e \in E. \end{aligned} \tag{6.1}$$

We solve the problem **(BC-Int-SPT**$_1$**)** through a primal-dual algorithm [106, Chapter 5-6] based on two sub-algorithms, the preprocessing algorithm[153, Algorithm 1] and the minimum cost cut algorithm[153, Algorithm 4].

Definition 6.5 ([153, Definition 18].) A set $E^u(T_v)$ of edges is called **a feasible cut** of a rooted tree T_v with respect to Δu if there is one and only one edge $e \in E^u(T_v)$ on every path from v to every leaf $t_j \in Y \cap T_v$ satisfying $\Delta u(e) > 0$. A **minimum cost cut** $\underline{E}^u(T_v)$ is a feasible cut whose cost $\underline{C}(E^u(T_v)) := \sum_{e \in E^u(T_v)} c(e)$ is minimum. For simplicity, let $\underline{C}(T_v) := \underline{C}(\underline{E}^u(T_v)) = \min_{E^u(T_v) \in \Gamma(T_v)} \underline{C}(E^u(T_v))$, where $\Gamma(T_v)$ is the set of feasible cuts of T_v.

For convenience, let $S_T = (Y, CD, CA, \chi, Layer)$ be the set of parameters for $T = T_{v_1}$. We proposed Algorithm $[\underline{E}^u(T_{v_1}), \underline{C}(T)] :=$ **MinCut**$(T, S_T, c, w, \Delta u, d)$ in [153, Algorithm 4] to obtain a minimum cost cut of T. The main idea is as follows. Firstly, for a given value d, we call the preprocessing algorithm [153, Algorithm 1] to delete some chains of the paths in T whose lengths are no less than d and denote the preprocessed tree still by T and for convenience, we denote this preprocessing algorithm by $[T, Y, E_{del}] :=$ **Preprocess**(T, S_T, w, D), which outputs the processed tree T with the relevant set Y of leaf nodes and the set E_{del} of deleted edges.

Secondly, for $v \in V^* \cup \{v_1\}$ in a decreasing order of $Layer(v)$, we determine a minimum cost cut for each subtree T_v as follows. Divide $T_v = \bigcup_{v_h \in CD(v)} \bar{T}_{v_h}$ into the union of all the branches $\bar{T}_{v_h} := \chi_{v_h} \cup T_{v_h}$. Calculate $\underline{C}(\bar{T}_{v_h}) := \min\{\underline{C}(\chi_{v_h}), \underline{C}(T_{v_h})\}$. Then $\underline{C}(T_v) := \sum_{v_h \in CD(v)} \underline{C}(\bar{T}_{v_h})$ is the value of the minimum cost cuts in T_v. Hence, we can calculate the value $\underline{C}(T_{v_1})$ of the minimum cost cut $\underline{E}^u(T_{v_1})$ for the tree T_{v_1}.

Then we present the main idea of the primal-dual algorithm to solve the problem **(BC-Int-SPT**$_1$**)** followed by the detailed Algorithm 6.1.

The model (6.1) of the problem **(BC-Int-SPT**$_1$**)** is equivalent to model below.

$$\begin{aligned} & \max z \\ & s.t. \; w(P_i) + r(P_i) \ge z, t_i \in Y \end{aligned}$$

$$(D) \qquad \sum_{e\in E} c(e)r(e) \leq B, \qquad (6.2)$$
$$0 \leq r(e) \leq d(e), e \in E,$$
$$z \geq 0.$$

Obviously, the problem (6.2) is a linear programming problem with $\ell + n$ constraints and n variables. Now we present the main idea of the primal-dual algorithm [106, Chapter 5-6]. Firstly, we consider the problem (6.2) as a dual problem **(D)**. Given a dual feasible solution $\pi^k = (r^k, z^k)$ of the problem **(D)**, we can determine a set J^k of admissible rows where equality constraints hold in the constraint conditions of **(D)** for π^k. Based on the set J^k, we can get the dual of the restricted primal problem, denoted by **(DRP^k)**. Secondly, we solve the problem **(DRP^k)** by finding a minimum cost cut and obtain its optimal solution $\bar{\pi}^k = (\bar{r}^k, \bar{z}^k)$. If the optimal objective value $\bar{z}^k$ of **(DRP^k)** is positive, then we determine the adjustment amount θ^k to obtain a better dual feasible solution $\pi^{k+1} = \pi^k + \theta^k\bar{\pi}^k$ whose objective value $z^{k+1} = z^k + \theta\bar{z}^k$ is larger than z^k and continue to find a set J^{k+1} for π^{k+1}. The aforementioned iteration terminates until the optimal objective value $\bar{z}^k$ of **(DRP)** is zero, we obtain an optimal dual solution π^k of **(D)**.

Algorithm 6.1 A primal dual algorithm to (**BC-Int-SPT**$_1$) [153, Algorithm 5].
$[r^k, z^k] :=$ (**BC-Int-SPT**$_1$)$(T, Y, CD, CA, \chi, LN, w, u, c, B)$

Input: A tree $T = (V, E)$, the set Y of leaves; the set CD of critical descendant, the set CA of critical ancestor, and the set χ of chains; the layer of vertices, three edge vectors w, u, c, and the maximum cost B.

Output: An optimal upgrade scheme r^k and the length z^k of the shortest path.

Calculate $\Delta u := u - w$ and $D_{max} := \min_{t_k\in Y}(w(P_k) + \Delta u(P_k))$.
Call $(T, Y, E_{del}) :=$ **Preprocess**(T, S_T, w, D_{max}).
Initialization: $k := 0$, $w(P_{i^*}) := \min_{t_i\in Y} w(P_i)$, $w(P_{i1^*}) := \min\{w(P_i)|w(P_i) > w(P_{i^*}), t_i \in Y\}$, $D^0 := \min\{w(P_{i1^*}), D_{max}\}$, $\Delta u^k := \Delta u$, $r^k := 0$, $g^k := 0$, $z^k := w(P_{i^*})$.
Call $(\underline{E}^{u^k}(T), \underline{C}(T)) :=$ **MinCut**$(T, S_T, c, w, \Delta u^k, z^k)$.
while $g^k < B$ **and** $z^k < D^0$ **do**
  Let $\Delta u^k(\tilde{e}) := \min_{e\in \underline{E}^{u^k}(T)} \Delta u^k(e)$, $\theta^k := \min\{\frac{B-g^k}{\underline{C}(T)}, \Delta u^k(\tilde{e}), D^0 - z^k\}$.
  Update $g^{k+1} := g^k + \theta^k\underline{C}(T)$, $z^{k+1} := z^k + \theta^k$, $r^{k+1}(e) := r^k(e) + \theta^k$ and $\Delta u^{k+1}(e) := \Delta u^k(e) - \theta^k$ for $e \in \underline{E}^{u^k}(T)$, $r^{k+1}(e) := r^k(e)$ and $\Delta u^{k+1}(e) := \Delta u^k(e)$ for $e \notin \underline{E}^{u^k}(T)$; $w^{k+1} := w + r^{k+1}$ and $k := k + 1$,
  if $z^k < D^0$ **then**
    call $(\underline{E}^{u^k}(T), \underline{C}(T)) :=$ **MinCut**$(T, S_T, c, w^k, d^k, z^k)$.
  end if
end while
if $g^k = B$ **then** **return** an optimal solution r^k and the length z^k.
else if $z^k = D^0$ **then**

let $J_1^k := \{t_i \in Y | -r^k(P_i)+z^k = w(P_i)\}$, $J_3^k := \{e_i \in E | r^k(e_i) = \Delta u(e_i)\}$.
end if
while $g^k < B$ **do**
 if $z^k = D_{max}$ **then** **return** an optimal solution r^k and z^k.
 else let $P_1^k := \{P_i | t_i \in J_1^k\}$ be a subset of T including all the paths for $t_i \in J_1^k$. Call $(\underline{E}^{u^k}(P_1^k), \underline{C}(P_1^k))$:= **MinCut**$(P_1^k, J_1^k, CD, CA, \chi, Layer, c, w^k, \Delta u^k, z^k)$.
  if $\underline{C}(P_1^k) > 0$ **then**
   The optimal solution of the problem **(DRP)** is $\bar{z}^k := 1, \bar{r}^k(e) := 1, e \in \underline{E}^{u^k}(P_1^k), \bar{r}^k(e) := 0$, otherwise.
   Calculate $\theta^k := \min\{\theta_1^k, \theta_2^k, \theta_3^k\}$, where

$$\theta_1^k := \min_{t_i \notin J_1^k, P_i \cap \underline{E}^{u^k}(P_1^k) = \emptyset} \{\theta_1(t_i) := w(P_i) + r^k(P_i) - z^k\}; \tag{6.3}$$

$$\theta_2^k := \frac{B - g^k}{\underline{C}(P_1^k)}; \tag{6.4}$$

$$\theta_3^k := \min_{e_j \notin J_3^k, e_j \in \underline{E}^{u^k}(P_1^k)} \{\Delta u(e_j) - r^k(e_j)\}. \tag{6.5}$$

   Let $g^{k+1} := g^k + \theta^k \underline{C}(P_1^k)$, $z^{k+1} := z^k + \theta^k$, $r^{k+1} := r^k + \theta^k \bar{r}^k$, $\Delta u^{k+1} := \Delta u^k - \theta^k \bar{r}^k$, $w^{k+1} := w + r^{k+1}$, $J_3^{k+1} := J_3^k \cup \{e_j \in \underline{E}^{u^k}(P_1^k) | \Delta u^{k+1}(e_j) = 0\}$.
   if $\theta^k = \theta_2^k$ **then** calculate $g^{k+1} := B$ and **return** (r^{k+1}, z^{k+1}).
   else if $\theta^k = \theta_1^k$ **then** let $J_1^{k+1} := J_1^k \cup \{t_j | \theta_1(t_j) = \theta_1\}$.
   end if
   Update $k := k + 1$.
  end if
 end if
end while

Theorem 6.1 ([153, Theorem 22]) *The problem* ***(BC-Int-SPT***$_1$***)*** *can be solved in* $O(n^2)$ *time by Algorithm 6.1.*

6.2.3 ((BC-)Int, SP-Trees, Bounded, Unit l_1 Norm)

Furthermore, Zhang et al. proposed [153, Algorithm 3] to solve the problem **(BC-Int-SPT**$_{u1}$**)** when $c(e) = 1$ for all $e \in E$ in (6.1). The main idea is in two steps. The first step is to determine a range $[L_{k-1}, L_k)$ of the optimal value $R(T_r^*)$ by a binary search method in $O(n)$ time without sorting the entries in the list $\mathcal{L} = \{L_j | v_j \in V\}$, where $L_j = w(v_1, v_j) + R_0(v_j, v_0) + d(v_1, v_j)$ [60]. The second step is to obtain the exact value $R(T_r^*)$ and an optimal solution r^* according to the two upgrade schemes $r_{k-1} = OPT(L_{k-1})$ and $r_k = OPT(L_k)$ by Algorithm 6.2.

It is worthy noting that we revised an error in [60]. Hambrusch and Tu considered a similar problem (EWRT) to (**BC-Int-SPT**$_{u1}$), in which some edge weights are reduced to minimize the length of the longest paths from the root to the leaves and the total cost under unit l_1 norm does not exceed a given value. We found a mistake in the formula to calculate r^* in the second step of the algorithm proposed in [60]. We revised the algorithm, whose time complexity is still $O(n)$ for (EWRT) as well as the problem (**BC-Int-SPT**$_{u1}$), in which we call a sub-routine Algorithm 6.2 for solving the problem (**Int-SPT**$_{u1}$).

Algorithm 6.2 An algorithm to solve the problem (**Int-SPT**$_{u1}$) $r^* = OPT(L)$. [153, Algorithm 2]

Input: An auxiliary tree T_{v_0}, the set E_{del} of deleted edges in the preprocessing algorithm; two edge weight vectors w, d, and two values L and L_{max}.

Output: An optimal canonical upgrade scheme r^*.

1: **if** $L > L_{max}$ **then**
2: Output "The upgraded length of the shortest path is impossible to be L, and its maximum length is L_{max}." **Return.**
3: **end if**
4: **if** $R_0(v_1, v_0) \geq L$ **then**
5: output $r^*(v_i, v_j) := 0$ for all $(v_i, v_j) \in E$ and **Return.**
6: **else**
7: Let $r^*(v_i, v_j) := 0$ for $(v_i, v_j) \in E_{del}$.
8: **for** $(v_i, v_j) \in E \setminus E_{del}$ **do**
9: Let $\Delta R_{ij} := L - w(v_1, v_i) - R_0(v_i, v_0)$ and $d(v_1, v_i) := \sum_{e \in P_{v_1, v_i}} d(e)$.

$$r^*(v_i, v_j) := \begin{cases} 0, & if\ d(v_1, v_i) \geq \Delta R_{ij}; \\ \Delta R_{ij} - d(v_1, v_i), & if\ d(v_1, v_i) < \Delta R_{ij} < d(v_1, v_i) \\ & \quad +d(v_i, v_j); \\ d(v_i, v_j), & if\ \Delta R_{ij} \geq d(v_1, v_i) + d(v_i, v_j). \end{cases} \tag{6.6}$$

10: **end for**
11: **end if**

Algorithm 6.3 An algorithm to solve (**BC-Int-SPT**$_{u1}$). [153, Algorithm 3]

Input: A tree $T = (V, E)$, the set Y of leaves; the set CC of critical children, the set CF of critical fathers and the chains; the layer of vertices, two edge weight vectors w, u, and the maximum cost M.

Output: The optimal canonical upgrade scheme r^*.

1: Calculate $d := u - w$ and $L_{max} := R(T_d)$.
2: Call $(T, Y, E_{del}) :=$ **Preprocess**$(T, Y, CC, CF, chain, Layer, w, L_{max})$.
3: Determine $r_{max} := OPT(L_{max})$ and the cost $M_{max} := \sum_{e \in E \setminus E_{del}} r_{max}(e)$.

if $M \geq M_{max}$ **then**
    output an optimal solution $r^* := r_{max}$ and the optimal value L_{max}, **return.**
else
    Construct an auxiliary network T_{v_0} by adding an artificial terminal v_0 and some edges (t, v_0) with $w(t, v_0) := 0$ and $d(t, v_0) := 0$ for every leaf $t \in Y$.
    Calculate $R_0(v_j, v_0) = \min_{P \in P_{v_j, v_0}} w(P)$ and $L_j = w(v_1, v_j) + R_0(v_j, v_0) + d(v_1, v_j)$ for each vertex $v_j \in V$. Determine the list $\angle$ of the efficient entries by L_j which is no more than L_{max}.
    Set $L_a := L_1$, $L_b := L_{max}$, $\angle_{ab} := \angle$, $n_{ab} := |\angle_{ab}|$, $r_a := OPT(L_a)$ and $r_b := OPT(L_b)$.
    For tree T, determine the sets $E_z := \{e_i \in E | r_a(e_i) = r_b(e_i) = 0\}$, $E_f = \{e_i \in E | r_a(e_i) = r_b(e_i) = d(e_i)\}$, $E_p = \{e_i \in E | 0 < r_a(e_i) < d(e_i), 0 < r_b(e_i) \leq d(e_i), r_b(e_j) = 0, \forall e_j \in P_{v_i,t}\}$ and $E_u = E \setminus (E_z \cup E_f \cup E_p)$.
    while $n_{ab} \neq 0$ **do**
        Let L_q be the $(\lfloor \frac{n_{ab}}{2} \rfloor)$th smallest element in list $\angle_{ab}$. Obtain $r_q(v_i, v_j)$ by (6.6).
        Let $\delta = L_b - L_q$. Let $E_{u,z}$, $E_{u,p}$ and $E_{u,f}$ be the sets of edges in E_u have zero, partial and full upgrade in r_q, respectively. Calculate $M(T_{r_q}) := M_f + (M_{p,b} - \delta \times |E_p|) + M_{u,f} + M_{u,p}$ with $M_f = \sum_{(v_i,v_j) \in E_f} d(v_i, v_j)$, $M_{p,b} = \sum_{e \in E_p} r_b(e)$, $M_{u,f} = \sum_{(v_i,v_j) \in E_{u,f}} d(v_i, v_j)$, $M_{u,p} = \sum_{(v_i,v_j) \in E_{u,p}} r_q(v_i, v_j)$.
        if $M(T_{r_q}) < M$ **then**
            let $L_a := L_q$, $E_f := E_f \cup E_{u,f}$, $E_u := E_u \setminus E_{u,f}$.
            Edges from $E_{u,p}$ that qualify for E_p are moved from set E_u to E_p.
            Update $\angle_{ab}$ by deleting the entries smaller than L_q. Let $n_{ab} := |\angle_{ab}|$.
        else if $M(T_{r_q}) > M$ **then**
            let $L_b := L_q$, $E_z := E_z \cup E_{u,z}$, $E_u := E_u \setminus E_{u,z}$.
            Edges from $E_{u,p}$ and $E_{u,f}$ that qualify for E_p are moved from set E_u to E_p.
            Update $\angle_{ab}$ by deleting the entries larger than L_q. Let $n_{ab} := |\angle_{ab}|$.
        else
            let $L_k := L_q$, **Return** $r^* := OPT(L_q)$.
        end if
    end while
    Let $L_{k-1} := L_a$, $L_k := L_b$ and $E_s := E \setminus (E_z \cup E_f)$.
    Determine $r_k := OPT(L_k)$ and $r_{k-1} := OPT(L_{k-1})$.
    Let $R(T_{r^0}) := \frac{M - M(T_{r_{k-1}}) + \sum_{(v_i,v_j) \in E_s} \hat{R}_{r_{k-1}}(P^*_{ij})}{|E_s|}$.
    Let $E_s^0 := E_s$, $E_s^1 := \{(v_i, v_j) \in E_s^0 | \hat{R}_{r_{k-1}}(P^*_{ij}) < R(T_{r^0})\}$, $R(T_{r^1}) := R(T_{r^0})$, $\kappa := 1$.
    while $E_s^{\kappa-1} \backslash E_s^{\kappa} \neq \emptyset$ **do**
        $R(T_{r^\kappa}) := \frac{M - M(T_{r_{k-1}}) + \sum_{(v_i,v_j) \in E_s^\kappa} \hat{R}_{r_{k-1}}(P^*_{ij})}{|E_s^\kappa|}$,
        Update $E_s^{\kappa+1} = \{(v_i, v_j) \in E_s^\kappa | \hat{R}_{r_{k-1}}(P^*_{ij}) < R(T_{r^\kappa})\}$ and $\kappa := \kappa + 1$.

end while
Let $E_s^* := E_s^\kappa$ and $R(T_{r^*}) := R(T_{r^\kappa})$.
Calculate $r^*(v_i, v_j) := 0$ for $(v_i, v_j) \in E_{del}$ and for $(v_i, v_j) \in E \setminus E_{del}$,

$$r^*(v_i, v_j) = \begin{cases} d(v_i, v_j), & if\ (v_i, v_j) \in E_f, \\ 0, & if\ (v_i, v_j) \in E_z, \\ r_{k-1}(v_i, v_j), & if\ (v_i, v_j) \in E_s \setminus E_s^*, \\ r_{k-1}(v_i, v_j) + R(T_{r^*}) - \hat{R}_{r_{k-1}}(P_{ij}^*), & if\ (v_i, v_j) \in E_s^*. \end{cases}$$

end if

Theorem 6.2 ([153, Theorem 16]) *The problem* **(BC-Int-SPT$_{u1}$)** *can be solved in $O(n)$ time by Algorithm 6.3.*

6.3 (Int, SP-Trees, Bounded, Weighted l_1 Norm)

The problem **(Int-SPT)** under weighted l_1 norm, denoted by **(Int-SPT$_1$)**, can be formulated as follows, where $r(e) = \bar{w}(e) - w(e), e \in E$.

$$\begin{aligned} & \min_{\bar{w}} \sum_{e \in E} c(e) r(e), \\ \textbf{(Int-SPT}_1\textbf{)} \quad s.t. \quad & \min_{t_k \in Y} (w(P_k) + r(P_k)) \geq D, \\ & 0 \leq r(e) \leq d(e), e \in E. \end{aligned} \tag{6.7}$$

Similar to Algorithm 6.1, replace (6.4) by $\theta_2^k = D - \max_{t_i \in J_1^k} w^k(P_i)$, which is the minimum length of the paths to be upgraded. Then we can obtain the relevant primal-dual algorithm to the problem **(Int-SPT$_1$)** in Algorithm 6.4.

Algorithm 6.4 The primal-dual algorithm to **(Int-SPT$_1$)**.[153, Algorithm 6]
$[r^k, g^k] :=$ **Int-SPT$_1$**$(T, Y, CD, CA, \chi, LN, w, u, c, D)$

Input: A tree $T = (V, E)$, the set Y of leaves; the set CD of critical descendant, the set CA of critical ancestor and the set χ of chains; the layer of vertices, three edge vectors w, u, c and the given length D.
Output: An optimal upgrade scheme r^k and the relevant upgrade cost g^k.
Calculate $\Delta u := u - w$ and $D_{max} := \min_{t_k \in Y}(w(P_k) + \Delta u(P_k))$.
Call $(T, Y, E_{del}) :=$ **Preprocess**(T, S_T, w, D_{max}).
Initialization: $k := 0$, $w(P_{i^*}) := \min_{t_i \in Y} w(P_i)$, $w(P_{i1^*}) := \min\{w(P_i) | w(P_i) > w(P_{i^*}), t_i \in Y\}$, $\Delta u^k := \Delta u$, $r^k := 0$, $w^k := w$, $g^k := 0$, $z^k := w(P_{i^*})$.
if $D > D_{max}$ **then**
return "The upgraded length of the shortest path is impossible to be D, and its maximum length is D_{max}."

end if
Initialize $J_1^k := \{t_i \in Y | w(P_i) = w(P_{i^*})\}$, $J_3^k := \emptyset$.
while $z^k < D$ **do**
  Let $P_1^k := \{P_i | t_i \in J_1^k\}$ be a subset of T including all the paths for $t_i \in J_1^k$.
  $(\underline{E}^{u^k}(P_1^k), \underline{C}(P_1^k)) :=$ **MinCut**$(P_1^k, J_1^k, CD, CA, \chi, \text{Layer}, c, w^k, \Delta u^k, z^k)$.
  if $\underline{C}(P_1^k) > 0$ **then**
    The optimal solution of the problem **(DRP)** is $\bar{z}^k := 1$, $\bar{r}^k(e) := 1$, $e \in \underline{E}^{u^k}(P_1^k)$, $\bar{r}^k(e) := 0, otherwise$.
    Calculate $\theta^k := \min\{\theta_1^k, \theta_2^k, \theta_3^k\}$, where θ_1^k, θ_3^k are defined as in (6.3), (6.5), $\theta_2^k := D - \max_{t_i \in J_1^k} w^k(P_i)$.
    Update $g^{k+1} := g^k + \theta^k \underline{C}(P_1^k)$, $z^{k+1} := z^k + \theta^k$, $r^{k+1} := r^k + \theta^k \bar{r}^k$, $\Delta u^{k+1} := \Delta u^k - \theta^k \bar{r}^k$, $w^{k+1} := w + r^{k+1}$, $J_3^{k+1} := J_3^k \cup \{e_j \in \underline{E}^{u^k}(P_1^k) | \Delta u^{k+1}(e_j) = 0\}$.
    if $\theta^k = \theta_1^k$ **then** let $J_1^{k+1} := J_1^k \cup \{t_j | \theta_1^k(t_j) = \theta_1^k\}$.
    end if
    Update $k := k + 1$.
  end if
end while
Return (r^k, g^k).

Corollary 6.1 ([153, Corollary 23]) *Algorithm 6.4 can solve the problem (MCSPIT$_1$) in $O(n^2)$ time.*

6.4 (BC-Int, SP-Trees, Bounded, Weighted sum Hamming Distance)

The problem **(BC-Int-SPT)** under weighted sum Hamming distance, denoted by **(BC-Int-SPT$_{wsH}$)**, can be formulated as the following form.

$$\textbf{(BC-Int-SPT}_{wsH}\textbf{)} \quad \begin{aligned} &\max_{\bar{w}} \min_{t \in Y} \bar{w}(P_t) \\ &s.t. \sum_{e \in E} c(e) H(\bar{w}(e), w(e)) \le B, \\ &\quad w(e) \le \bar{w}(e) \le u(e), e \in E. \end{aligned} \tag{6.8}$$

We first prove that the problem **(BC-Int-SPT$_{wsH}$)** defined on a chain is $\mathcal{NP}$-hard [154, Theorem 1] by transforming it into a 0-1 knapsack problem. Then we extend the $\mathcal{NP}$–hardness to the problem **(BC-Int-SPT$_{wsH}$)** [154, Corollary 2] defined on a chain tree [154, Definition 1], whose degrees (except the root and the leaves) are all

2. Finally, we can conclude that the problems **(BC-Int-SPT$_{wsH}$)** and **(Int-SPT$_{wsH}$)** defined on general rooted trees is still $\mathcal{NP}$-hard [154, Corollary 3].

Hence we focus on the problem **(BC-Int-SPT)** under unit sum Hamming distance, denoted by **(BC-Int-SPT$_{sH}$)**. We can only upgrade at most B critical edges in a rooted tree to maximize the shortest root-leaf distance of the tree.

$$\textbf{(BC-Int-SPT}_{sH}\textbf{)} \quad \begin{aligned} &\max_{\bar{w}} \min_{t \in Y} \bar{w}(P_t) \\ &s.t. \sum_{e \in E} H(\bar{w}(e), w(e)) \leq B, \\ &\qquad w(e) \leq \bar{w}(e) \leq u(e), e \in E. \end{aligned} \tag{6.9}$$

6.4.1 *Solve the Problem (BC-Int-SPT$_{sH}$)*

Notice that the problem **(BC-Int-SPT$_{sH}$)** only concerns whether an edge is upgraded or not. The larger the upgrade weight is, the better the objective value of the problem (6.8) is. Hence, the following theorem holds.

Theorem 6.3 ([154, Theorem 4]) *If $\bar{w}$ is an optimal solution to the problem **(BC-Int-SPT$_{wsH}$)**, so is w^* defined below.*

$$w^*(e) = \begin{cases} w(e), & \text{if } \bar{w}(e) = w(e), \\ u(e), & \text{if } \bar{w}(e) \neq w(e). \end{cases}$$

Lemma 6.1 ([154, Lemma 7]) *Suppose w' is an optimal solution of the problem (6.9). If $\Delta u(e_i) < \Delta u(e_j)$, $w'(e_i) > w(e_i)$, $w'(e_j) = w(e_j)$, $LN(e_i) = LN(e_j)$ and $L(e_i) = L(e_j)$, then w^* is an optimal solution of the problem (6.9), where*

$$w^*(e) = \begin{cases} w(e), & e = e_i, \\ u(e), & e = e_j, \\ w'(e), & otherwise. \end{cases}$$

Based on the aforementioned lemma, we can rearrange the edges with the same layer number on the same path such that their values of $\Delta u(e)$ are sorted non-increasingly. Notice that if the B upgraded edges have the same layer number on the same path, we can update the first B edges in this layer on this path.

For any $v \in V^* \cup \{s\}$, let $CD(v) = \{v_{h_1}, v_{h_2}, \cdots, v_{h_p}\}$ be the set of critical descendant of the node v, where

$$p = \begin{cases} deg(v), & v = v_1 \\ deg(v) - 1, & v \in V^* \end{cases}. \tag{6.10}$$

We define the left $q-$subtree as follows.

Definition 6.6 ([154, Definition 6]) Define the left $q-$subtree rooted at v as $T_v^{1:q} := P_{v,v_{h_1}} \cup T_{v_{h_1}} \cup P_{v,v_{h_2}} \cup T_{v_{h_2}} \cup \cdots \cup P_{v,v_{h_q}} \cup T_{v_{h_q}}$, $q = 1, 2, \cdots, p$, where $T_{v_{h_i}}$ is the subtree rooted at v_{h_i} and $T_{v_{h_i}} := \emptyset$ if v_{h_i} is a leaf. Specially, $T_v^{1:p}$ can be denoted by T_v and $T_v^{1:0} := \emptyset$.

Now we define the following two functions $g(P_{v,v_h}, k)$ and $f(T_v^{1:q}, k)$ to obtain a dynamic programming algorithm for the problem **(BC-Int-SPT$_{sH}$)**.

1. The function $g(P_{v,v_h}, k)$ defined on the chain P_{v,v_h}.

For any $v \in V^* \cup \{v_1\}$ and any $v_h \in CD(v)$, let $P_{v,v_h} := \{e_{i_1}, e_{i_2}, \cdots, e_{i_\beta}\}$ with $\Delta u(e_{i_1}) \geq \Delta u(e_{i_2}) \geq \cdots \geq \Delta u(e_{i_\beta})$, where $\beta = |E(P_{v,v_h})|$.

Define $g(P_{v,v_h}, k)$ as the sum of the edge lengths when the first k edges are upgraded on the chain P_{v,v_h}, where $0 \leq k \leq \beta$. Let $E(P_{v,v_h}, k)$ be the set of upgraded edges on the chain P_{v,v_h}.

We can calculate all the values $g(P_{v,v_h}, k)$ for all the chains in the tree T according to the formula below.

$$g(P_{v,v_h}, k) = \begin{cases} \sum_{j=1}^{\beta} w(e_{i_j}), & \text{if } k = 0, \\ \sum_{j=1}^{\beta} w(e_{i_j}) + \sum_{j=1}^{k} \Delta w(e_{i_j}), & \text{if } 1 \leq k \leq \beta. \end{cases} \tag{6.11}$$

$$E(P_{v,v_h}, k) = \bigcup_{j=1}^{k} \{e_{i_j}\}. \tag{6.12}$$

2. The function $f(T_v^{1:q}, k)$ defined on the non-chain $T_v^{1:q}$.

For all $v \in V^* \cup \{v_1\}$, define $f(T_v^{1:q}, k) := \max \min_{t \in T_v^{1:q} \cap Y} \bar{w}(P_t)$ as the optimal value of $T_v^{1:q}$ when k edges are upgraded, where $0 \leq k \leq |E(T_v^{1:q})|$ and let $E_v^{1:q}(k)$ be the corresponding set of k upgraded edges. Specially, denote $E_v^{1:p}(k)$ by $E_v(k)$ for simplicity.

We consider two cases to calculate the optimal value $f(T_v^{1:q}, k)$.

Case 1. When $q = 1$

We have $T_v^{1:1} = P_{v,v_{h_1}} \cup T_{v_{h_1}}$ as shown in Fig. 6.1.

$$\begin{aligned} f(T_v^{1:1}, k) = &\max\{g(P_{v,v_{h_1}}, k_1) + f(T_{v_{h_1}}, k_2)\} \\ s.t. \quad & k_1 + k_2 = k, \\ & k = 0, 1, 2, \cdots, |E(T_v^{1:1})|, \\ & k_1 = 0, 1, 2, \cdots, \min\{k, |E(P_{v,v_{h_1}})|\}, \\ & k_2 = 0, 1, 2, \cdots, \min\{k, |E(T_{v_{h_1}})|\}. \end{aligned} \tag{6.13}$$

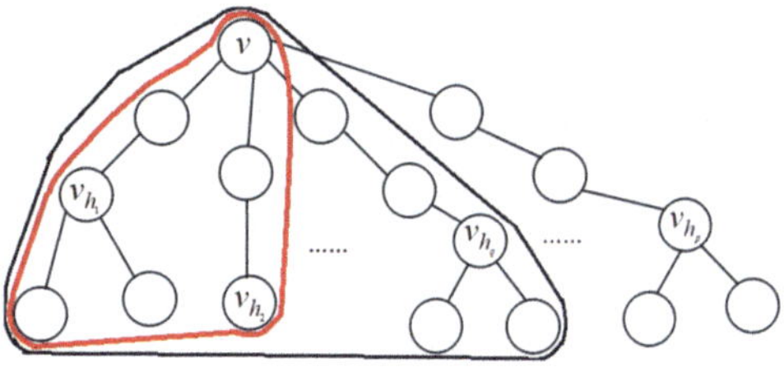

Fig. 6.1 The illustration of Case 1

If the optimal value $f(T_v^{1:1}, k) = g(P_{v,v_{h_1}}, k_1^*) + f(T_{v_{h_1}}, k - k_1^*)$ is obtained when $k_1 = k_1^*$, then the set of upgraded edges is

$$E_v^{1:1}(k) = E(P_{v,v_{h_1}}, k_1^*) \cup E_{v_{h_1}}(k - k_1^*). \tag{6.14}$$

Specially, when v_{h_1} is a leaf, $E(T_{v_{h_1}}) = \emptyset$ and $|E(T_{v_{h_1}})| = 0$, thus, $k_2 = 0$, $f(T_{v_{h_1}}, k_2) = f(T_{v_{h_1}}, 0) = 0$,

$$f(T_v^{1:1}, k) = g(P_{v,v_{h_1}}, k). \tag{6.15}$$

$$E_v^{1:1}(k) = E(P_{v,v_{h_1}}, k). \tag{6.16}$$

Case 2: when $q \geq 2$

For any $v \in V^* \cup \{v_1\}$, $CC(v) = \{v_{h_1}, v_{h_2}, \cdots, v_{h_p}\}$ is the set of critical children. Then $T_v^{1:q} = T_v^{1:(q-1)} \cup P_{v,v_{h_q}} \cup T_{v_{h_q}}, q = 2, \cdots, p$.

$$\begin{aligned} f(T_v^{1:q}, k) &= \max\min\{g(P_{v,v_{h_q}}, k_1) + f(T_{v_{h_q}}, k_2), f(T_v^{1:(q-1)}, k_3)\}, \\ s.t. \quad & k_1 + k_2 + k_3 = k, \\ & k = 0, 1, \cdots, |E(T_v^{1:q})|, \\ & k_1 = 0, 1, \cdots, \min\{k, |E(P_{v,v_{h_q}})|\}, \\ & k_2 = 0, 1, \cdots, \min\{k, |E(T_{v_{h_q}})|\}, \\ & k_3 = 0, 1, \cdots, \min\{k, |E(T_v^{1:(q-1)})|\}. \end{aligned} \tag{6.17}$$

If the optimal value

$$f(T_v^{1:q}, k) = \max\{g(P_{v,v_{h_q}}, k_1^*) + f(T_{v_{h_q}}, k_2^*), f(T_v^{1:(q-1)}, k - k_1^* - k_2^*)\}$$

is obtained when $k_1 = k_1^*$, $k_2 = k_2^*$, then its set of upgraded edges is

$$E_v^{1:q}(k) = E(P_{v,v_{h_q}}, k_1^*) \cup E_{v_{h_q}}(k_2^*) \cup E_v^{1:(q-1)}(k - k_1^* - k_2^*). \tag{6.18}$$

Specially, when v_{h_q} is a leaf, $E(T_{v_{h_q}}) = \emptyset$ and $|E(T_{v_{h_q}})| = 0$, thus, $k_2 = 0$, $f(T_{v_{h_q}}, k_2) = f(T_{v_{h_q}}, 0) = 0$, $T_v^{1:q} = T_v^{1:(q-1)} \cup P_{v,v_{h_q}}$,

$$f(T_v^{1:q}, k) = \max\min\{g(P_{v,v_{h_q}}, k_1), f(T_v^{1:(q-1)}, k - k_1)\}, \tag{6.19}$$

If the optimal value $f(T_v^{1:q}, k) = \max\{g(P_{v,v_{h_q}}, k_1^*), f(T_v^{1:(q-1)}, k - k_1^*)\}$ is obtained when $k_1 = k_1^*$, then its set of upgraded edges is

$$E_v^{1:q}(k) = E(P_{v,v_{h_q}}, k_1^*) \cup E_v^{1:(q-1)}(k - k_1^*). \tag{6.20}$$

To sum up, traverse the rooted tree T from leaves to the root s to calculate all the function values $g(P_{v,v_h}, k)$ and $f(T_v^{1:q}, k)$. Then $f(T_{v_1}, B)$ is the optimal value of the problem **(BC-Int-SPT$_{sH}$)**, $E_s(B)$ is the set of upgraded edges and an optimal solution is

$$\bar{w}(e) = \begin{cases} u(e), & e \in E_{v_1}(B), \\ w(e), & e \notin E_{v_1}(B). \end{cases} \tag{6.21}$$

According to the earlier analysis, we have the following dynamic programming algorithm for the problem **(BC-Int-SPT$_{sH}$)**.

Algorithm 6.5 A dynamic programming algorithm [154, Algorithm 2]
$[E_{v_1}(B), \bar{w}, f(T_{v_1}, B)] :=$ **BC-Int-SPT**$_{sH}(T, Y, w, u, B)$

Input: A tree T rooted at v_1, the set Y of leaves, two edge weight vectors w and u, and the number B of upgraded edges.

Output: The set $E_{v_1}(B)$ of upgraded edges, the upgraded weight vector $\bar{w}$, and the optimal value $f(T_{v_1}, B)$.

1: **(Breath-First Search (BFS).)** Let $V^* = \{v \in V\backslash\{v_1\} | deg(v) > 2\}$. Start from the root v_1 and label each node v by $Layer(v)$ when using the breath-first search strategy on T, where $Layer(v_1) = 1$. While executing the BFS, calculate $Layer(v)$ for each node v and $LN(e)$ for each edge e. Find the sets $CD(v)$ of critical descendant for each $v \in V^* \cup \{v_1\}$.

2: For each $v \in V^* \cup \{v_1\}$ and each $v_{h_j} \in CD(v)$, find all the chains $P_{v,v_{h_j}}$, and rearrange the edges in each chain $P_{v,v_{h_j}}$ such that the values $\Delta u(e) := u(e) - w(e)$ are in non-increasing order, where $j = 1, \cdots, p$, and p is as defined in (6.10).

3: **for** any $v \in V^* \cup \{v_1\}$ **do**

4:     **for** $v_h \in CD(v)$ **do**

5:         Calculate the value $g(P_{v,v_h}, k)$ and the set $E(P_{v,v_h}, k)$ of upgraded edges by (6.11) and (6.12), for each $k = 0, 1, \cdots, \min\{B, |E(P_{v,v_h})|\}$.

6:     **end for**

7: **end for**

8: **for** any $v \in V^* \cup \{v_1\}$ in descending order of the labels $Layer(v)$ of nodes **do**

  for $q = 1 : p$ **do**
    for $k = 0 : \min\{B, |E(T_v^{1:q})|\}$ **do**
      Calculate the value $f(T_v^{1:q}, k)$ by (6.13), (6.15), (6.17) and (6.19), and obtain the set $E_v^{1:q}(k)$ of upgraded edges by (6.14), (6.16), (6.18) and (6.20).
    end for
  end for
end for
For the root v_1, calculate the optimal value $f(T_{v_1}, B)$, the set $E_{v_1}(B)$ of upgraded edges and the optimal solution $\bar{w}$ given by (6.21).

Theorem 6.4 ([154, Theorem 8]) *The dynamic programming Algorithm 6.5 can solve the problem **(BC-Int-SPT$_{sH}$)** in $O(n^4)$ time.*

6.4.2 (Int, SP-Trees, Bounded, Sum Hamming Distance)

The shortest path interdiction problem **(Int-SPT$_{sH}$)** by upgrading edges on trees under unit sum Hamming distance aims to minimize the total number of upgrade edges on the premise that the shortest root-leaf distance of the tree is lower bounded by a given value D.

$$
\textbf{(Int-SPT}_{sH}\textbf{)} \quad
\begin{aligned}
&\min_{\bar{w}} \sum_{e \in E} H(\bar{w}(e), w(e)) \\
s.t.\ &\min_{t \in Y} \bar{w}(P_t) \geq D, \\
&w(e) \leq \bar{w}(e) \leq u(e), e \in E.
\end{aligned}
\tag{6.22}
$$

To solve the problem **(Int-SPT$_{sH}$)**, we aim to find the optimal value B^* among the values $\{1, 2, \cdots, n\}$ by a binary search method, and in each iteration, we solve a problem **(BC-Int-SPT$_{sH}(B_k)$)** by Algorithm 6.5, in which B_k is the median of the current interval $[k_1, k_2] \subseteq [1, n]$. Hence, the problem **(Int-SPT$_{sH}$)** can be solved in $O(n^4 \log n)$, as shown in Algorithm 6.6.

Algorithm 6.6 $[B^*, E^*, w^*] :=$ **Int-SPT**$_{sH}(T, Y, w, u, D)$ [154, Algorithm 3]

Input: A tree T rooted at v_1, the set Y of leaves, two edge weight vectors w and u and the lower bound D of the length of the shortest path.
Output: The optimal objective value B^*, the set E^* of upgrade edges and an optimal solution w^*.
Initialize $k_1 := 1, k_2 := n, i := 1$.
while $k_2 \neq k_1 + 1$ **do**
  Let $B_k := \lceil \frac{k_1+k_2}{2} \rceil$,
  Call $[E^*, w^*, D^*(i)] =$ **BC-Int-SPT**$_{sH}(T, Y, w, u, B_k)$.
  if $D^*(i) < D$ **then** let $k_1 := B_k$,
  else if $D^*(i) > D$ **then** let $k_2 := B_k$,

    else let $B^* := k$,
        Return (B^*, E^*, w^*).
    end if
    Update $i := i + 1$.
end while
Update $B^* := k_2$.
Call $[E^*, w^*, D^*(i)] =$ **BC-Int-SPT**$_{sH}(T, Y, w, u, B^*)$.
Return (B^*, E^*, w^*).

6.5 Conclusions

In this chapter, we introduce the (budget constrained) shortest path interdiction problems by upgrading edges. It resolved the difficulties encountered when deleting critical edges in general network interdiction problems. We measure the upgrade cost of edges under different norms. Under weighted l_1 norm, we put forward two primal-dual algorithms both in $O(n^2)$ time to solve the problems (**BC-Int-SPT**$_1$) and (**Int-SPT**$_1$). Furthermore, under unit l_1 norm, we proposed a linear time algorithm for the aforementioned two problems and as a by-product, we revised a mistake in the algorithm proposed in [60]. Under weighted sum Hamming distance, We show that the problems (**BC-Int-SPT**$_s H$) and (**Int-SPT**$_s H$) are $\mathcal{NP}$-hard. Under unit sum Hamming distance, we propose two dynamic programming algorithms within $O(n^4)$ and $O(n^4 \log n)$ time, respectively.

Chapter 7
Sum of Root-Leaf Distance Interdiction Problems on Trees

Abstract This chapter explores the interdiction problems on sum of root-leaf distance on trees (Int-SRD). First, the problems (Int-SRD) under weighted l_1 norm, bottleneck Hamming, and unit sum Hamming distance are solved in polynomial time and are shown to be $\mathcal{NP}$-hard under weighted sum Hamming distance and weighted node cost. Subsequently, recognizing that the original solution does not account for the shortest root-leaf distance, a shortest path constraint is added, leading to the formulation of the double interdiction problem ($\text{DIT}_{H\infty}$) and its minimal cost version. These problems are shown to be $\mathcal{NP}$-hard and are addressed through a combination of dynamic programming and binary search algorithms in pseudo-polynomial time.

Keywords Sum of root-leaf distance interdiction problems · Upgrading edges/nodes · Cardinality constraint · Double interdiction problem · Trees

7.1 Introduction

Network interdiction problems, particularly those involving the strategic upgrading of critical edges or nodes, have garnered significant attention due to their potential applications in mitigating the spread of infectious diseases. In the context of epidemiological modeling, the transmission dynamics of a disease can be represented as a network, where each node signifies an infected individual, and the edges represent potential transmission pathways with associated weights indicative of the infectious intensity. By implementing targeted interventions, such as disinfection, treatment, isolation, or vaccination, the infectious intensity along these edges can be effectively reduced, akin to upgrading the network's structure to impede the pathogen's progress.

Xiao Li, Qiao Zhang, and Xiucui Guan have equally contributed to this chapter.

X. Guan et al., *Inverse Combinatorial Optimization Problems*, Springer Optimization and Its Applications 225, https://doi.org/10.1007/978-3-031-91175-0_7

In this framework, the total infectious intensity of the network is often quantified by the sum of the distances from the root (source of infection) to all leaf nodes (individuals) in a rooted tree. This leads to the consideration of the sum of root-leaf distance (SRD) interdiction problem by upgrading edges or nodes on trees (Int, SRD-Trees). The edge-variant of this problem seeks to minimize the SRD by strategically reducing the weights of certain critical edges, subject to an upper bound on the upgrade cost, which is measured using various norms. Conversely, the node-variant focuses on upgrading a set of critical nodes to diminish the infectious intensity of adjacent edges, thereby affecting the overall network's transmission potential. As a related work, Magnouche and Martin [92] studied how to delete the least number of critical nodes so that the length of s-t path in the remained graph was at least d. They proved the $\mathcal{NP}$-hardness of the problem, presented an integer linear programming model with multiple exponential constraints and designed a branch-and-bound algorithm to solve it.

The minimum cost problems associated with these interdiction strategies aim to minimize the upgrade cost while ensuring that the sum of root-leaf distances does not exceed a predefined threshold. This approach not only provides a mathematical foundation for understanding disease transmission dynamics but also offers a strategic tool for public health officials to allocate resources effectively in the face of outbreaks.

The network interdiction problem is a strategic game between two entities: a leader and a follower, each with opposing goals. The follower seeks to optimize network usage for resource movement, such as supply chains or material transport, while the leader aims to thwart these efforts by disrupting network connectivity through targeted attacks on arcs or nodes. In scenarios like emergency response, transportation networks can be modeled as rooted trees with weighted edges. The root symbolizes critical infrastructure, internal nodes represent transportation hubs, and leaf nodes denote emergency facilities like fire stations and hospitals. Edge weights reflect road congestion levels, and strategic interdiction of these edges can significantly degrade network performance by impeding rescue operations. Decision-makers, constrained by limited resources, must identify key edges for interdiction to maximize the network's overall disruption.

The development of algorithms and mathematical models for these problems is an active area of research, with implications extending beyond epidemiology to other fields where network optimization is critical, such as cybersecurity, transportation, and logistics. As the world continues to grapple with the challenges posed by emerging infectious diseases, the study of network interdiction through edge and node upgrading remains a pertinent and pressing area of investigation.

7.2 The Sum of Root-Leaf Distance Interdiction Problem by Upgrading Edges/Nodes on Trees

Let $T = (V, E, w, u, c)$ be an edge-weighted tree rooted at vertex s, where $V = \{s, v_1, v_2, \ldots, v_n\}$ represents the set of vertices and $E = \{e_1, e_2, \ldots, e_n\}$ represents the set of edges. The tree also includes a set of leaf nodes $Y = \{t_1, t_2, \ldots, t_r\}$. Let $w(e)$ and $u(e)$ denote the original weight and the upper bound of the weight for edge $e \in E$, respectively, satisfying $u(e) \geq w(e) \geq 0$. Additionally, $c(e) > 0$ represents the unit modification cost for changing the weight of edge e.

Define $P_k = P_{t_k} = P_{s,t_k}$ as the path from the root s to the leaf t_k. The weight of path P_k is given by $w(P_k) = \sum_{e \in P_{s,t_k}} w(e)$, and the weight of T, denoted by $w(T) = \sum_{t_k \in Y} w(P_k)$, is sum of weights of all root-leaf paths.

To represent the SRD increment by upgrading one edge, we introduce the following definition.

Definition 7.1 ([151]) For any $e \in E$, define $L(e) := \left\{t_k \in Y | e \in P_{s,t_k}\right\}$ as the set of leaves t_k to which P_{s,t_k} passes through e. If $t_k \in L(e)$, then t_k is controlled by the edge e. Similarly, for any $v \in V$, define $L(v) := \left\{t_k \in Y | v \in P_{s,t_k}\right\}$ as the set of leaves controlled by the node v.

The sum of root-leaf distance interdiction problem by **upgrading edges** on trees, denoted by **(SRDIT)**, aims to find an upgrade scheme $\bar{w}$ to minimize the distance $d_{\bar{w}}(T)$ under $\bar{w}$ on the premise that the total upgrade cost under some norm is upper bounded by a given value K. Its mathematical model can be stated as follows.

$$\begin{aligned}
&\min_{\bar{w}} \sum_{t \in Y} d_{\bar{w}}(s,t) \\
\textbf{(SRDIT)} \quad s.t. \quad & \|\bar{w} - w\| \leq K, \\
& l(e) \leq \bar{w}(e) \leq w(e), e \in E.
\end{aligned}$$

The relevant minimum cost problem **(SRDIT)**, denoted by **(MCSRDIT)**, aims to find an upgrade scheme $\bar{w}$ to minimize the total upgrade cost such that the distance $d_{\bar{w}}(T)$ under $\bar{w}$ is upper bounded by a given value D. Its mathematical model can be stated as follows:

$$\begin{aligned}
&\min_{\bar{w}} \|\bar{w} - w\| \\
\textbf{(MCSRDIT)} \quad s.t. \quad & \sum_{t \in Y} d_{\bar{w}}(s,t) \leq D, \\
& l(e) \leq \bar{w}(e) \leq w(e), e \in E.
\end{aligned}$$

Note that the edge $e_j = (v_i, v_j)$ is labeled by the subscript of the endpoint v_j, which is further to the root s than v_i. Let $A(v_i) = \{e_j = (v_i, v_j) | e_j \in E\}$ be the set of edges adjacent to v_i. Let $\beta_1(e_j)$ and $\beta_0(e_j)$ be the weight when the node v_i is upgraded or not, respectively. Let $c(v_i)$ be the upgrade cost of the node v_i. The sum of root-leaf distance interdiction problem by **upgrading nodes** on trees, denoted by **(SRDIT**$_N$), aims to upgrade a subset $S \subseteq V$ of nodes to minimize the sum of root-leaf distance such that the total upgrade cost under some norm is upper bounded by a given value K. Its mathematical model can be stated as follows:

$$
\begin{aligned}
& \min_{S \subseteq V} \sum_{t \in Y} d_{\bar{w}}(s, t) \\
(\textbf{SRDIT}_N) \quad & s.t. \quad \sum_{v \in S} c(v) \leq K, \qquad (7.1) \\
& \bar{w}(e) = \begin{cases} \beta_1(e), & v \in S, e \in A(v), \\ \beta_0(e), & otherwise. \end{cases}
\end{aligned}
$$

The relevant minimum cost problem **(SRDIT**$_N$), denoted by **(MCSRDIT**$_N$), aims to upgrade a subset $S \subseteq V$ of nodes to minimize the total upgrade cost such that the sum of root-leaf distance is upper bounded by a given value D. Its mathematical model can be stated as follows:

$$
\begin{aligned}
& \min_{S \subseteq V} \sum_{v \in S} c(v) \\
(\textbf{MCSRDIT}_N) \quad & s.t. \quad \sum_{t \in Y} d_{\bar{w}}(s, t) \leq D. \qquad (7.2) \\
& \bar{w}(e) = \begin{cases} \beta_1(e), & v \in S, e \in A(v), \\ \beta_0(e), & otherwise. \end{cases}
\end{aligned}
$$

This section is organized as follows. First, we prove that the problems **(SRDIT)** and **(MCSRDIT)** under the weighted Hamming distance are $\mathcal{NP}$-hard. Then, we present two algorithms for solving the problems **(SRDIT)** and **(MCSRDIT)** under the unit Hamming distance, with time complexities of $O(n)$ and $O(n \log n)$, respectively. Next, we demonstrate that the problems **(SRDIT)** and **(MCSRDIT)** under the l_1 norm are equivalent to continuous knapsack problems, allowing them to be solved in $O(n)$ time. Additionally, we establish the equivalence between the problems **(SRDIT**$_N$), **(MCSRDIT**$_N$), and the 0-1 knapsack problem. For these problems with unit node cost, we develop two algorithms with time complexities of $O(n)$ and $O(n \log n)$, respectively.

7.2.1 ((BC-)Int, SRD-Trees, Bounded, Weighted Sum Hamming)

The problems **(SRDIT)** and **(MCSRDIT)** under weighted sum Hamming distance, denoted by **(SRDIT**$_{wsH}$) and **(MCSRDIT**$_{wsH}$), can be formulated as the following models (7.3) and (7.4), respectively.

$$\begin{aligned} &\min_{\bar{w}} \sum_{t\in Y} d_{\bar{w}}(s,t) \\ (\textbf{SRDIT}_{wsH}) \quad s.t. \quad &\sum_{e\in E} c(e)H\big(\bar{w}(e), w(e)\big) \leq K, \\ &l(e) \leq \bar{w}(e) \leq w(e), e \in E. \end{aligned} \tag{7.3}$$

$$\begin{aligned} &\min_{\bar{w}} \sum_{e\in E} c(e)H\big(\bar{w}(e), w(e)\big) \\ (\textbf{MCSRDIT}_{wsH}) \quad s.t. \quad &\sum_{t\in Y} d_{\bar{w}}(s,t) \leq D, \\ &l(e) \leq \bar{w}(e) \leq w(e), e \in E. \end{aligned} \tag{7.4}$$

In this subsection, we first prove a property of the optimal solution of the problems **(SRDIT**$_{wsH}$) and **(MCSRDIT**$_{wsH}$). Then show their $\mathcal{NP}$-hardness.

Theorem 7.1 ([151]) *If $\bar{w}$ is an optimal solution of the problem* **(*MCSRDIT***$_{wsH}$) *or* (***SRDIT***$_{wsH}$), *so is w^* defined as follows:*

$$w^*(e) = \begin{cases} w(e), \ if \ \bar{w}(e) = w(e), \\ l(e), \ \ if \ \bar{w}(e) \neq w(e). \end{cases} \tag{7.5}$$

Based on Theorem 7.1, we pursue an optimal solution w^* defined as in (7.5). If an edge e is upgraded, then the length of each path $P_{s,t_k}(t_k \in L(e))$ will decrease by $\Delta w(e)$.

Definition 7.2 ([151]) For any $e \in E$, let $Q(e) = |L(e)| \cdot \Delta w(e)$ be the **total reduction amount** of the edge e, which is the product of the upgrade amount of the edge e and the number of leaf nodes controlled by the edge e.

Next, we prove $\mathcal{NP}$-hardness of the problems **(SRDIT**$_{wsH}$) and **(MCSRDIT**$_{wsH}$) by showing the equivalence of the problems and the 0-1 knapsack problem. For convenience, we substitute $H\big(\bar{w}(e), w(e)\big)$ by $x(e)$, where

$$x(e) = \begin{cases} 0, \ if \ \bar{w}(e) = w(e), \\ 1, \ if \ \bar{w}(e) \neq w(e). \end{cases} \tag{7.6}$$

Theorem 7.2 ([151]) *The problems* **(SRDIT$_{wsH}$)** *and* **(MCSRDIT$_{wsH}$)** *are $\mathcal{NP}$-hard.*

7.2.2 (BC-Int, SRD-Trees, Bounded, Sum Hamming Distance)

The **(SRDIT$_{wsH}$)** problem under unit Hamming distance, denoted by **(SRDIT$_{sH}$)**, can be formulated as the following form.

$$\textbf{(SRDIT}_{sH}\textbf{)} \quad \begin{aligned} &\max_{\bar{w}} \sum_{e\in E} Q(e)H\big(\bar{w}(e), w(e)\big) \\ &s.t. \sum_{e\in E} H\big(\bar{w}(e), w(e)\big) \leq K, \\ &\quad l(e) \leq \bar{w}(e) \leq w(e), e \in E. \end{aligned} \tag{7.7}$$

The problem **(SRDIT$_{sH}$)** aims to upgrade K edges to be upgraded so that the sum of the total reduction amount is maximized. Thus, we can sort the edges by the values of $Q(e)$ in non-increasing order and upgrade the first K largest $Q(e)$-value edges.

Sort the edges by the values of $Q(e)$ in non-increasing order as follows.

$$Q(e_{i_1}) \geq Q(e_{i_2}) \geq \cdots \geq Q(e_{i_K}) \geq \cdots \geq Q(e_{i_n}).$$

Theorem 7.3 ([151]) *Let $\bar{E}_K = \{e_{i_\tau} | \tau = 1, 2, \cdots, K\}$ be the set of the first K largest $Q(e)$-value edges in E. Then $\bar{w}(e) = \begin{cases} l(e), & e \in \bar{E}_K \\ w(e), & e \notin \bar{E}_K \end{cases}$ is an optimal solution of the problem* **(SRDIT$_{uH}$)**.

Next, we present a linear-time algorithm, as described in Algorithm 7.1, for finding the first K largest values in an array with duplicate elements.

We begin by identifying the K-th largest element, denoted as q, in an array Q using the selection algorithm $q :=$ Selection(Q, K), as detailed in [33, pp. 220–222]. Unlike the partition algorithm outlined in [33, pp. 170–173], which assumes distinct elements, our method accommodates arrays Q that may contain duplicate values. To ensure the accurate identification of the top K elements, we define two sets: E_1, consisting of edges with values greater than q, and E_2, consisting of edges with values equal to q.

The final set of the first K largest edges, $\bar{E}_K$, is computed as $\bar{E}_K := \bar{E}_K := E_1 \cup E_2(1 : K - |E_1|)$. This set includes the top K largest $Q(e)$-value edges in the array Q. Both steps of this procedure can be executed in $O(n)$ time.

Algorithm 7.1 A greedy algorithm to solve the problem (**SRDIT**$_{sH}$) [151]

Input: A tree $T(V, E)$ rooted at s, the number n of edges, the set Y of leaf nodes, two edge weight vectors w and l, and the number K of upgrade edges.

Output: The set $\bar{E}_K$ of upgrade edges, the optimal solution $\bar{w}$, and the relevant value $d_{\bar{w}}(T)$.

1: **if** $K > n$ **then return** "The problem has no feasible solution!"
2: **end if**
3: For any $e \in E$, calculate the set $L(e)$ of leaf nodes controlled by the edge e, compute $\Delta w(e) := w(e) - l(e)$ and $Q(e) := |L(e)| \cdot \Delta w(e)$.
4: Call $q := Selection(Q, K)$.
5: Let $E_1 := \{e_i | Q(e_i) > q\}$, $E_2 := \{e_i | Q(e_i) = q\}$ and $\bar{E}_K := E_1 \cup E_2(1 : K - |E_1|)$.
6: The optimal solution is $\bar{w}(e) := \begin{cases} l(e), & e \in \bar{E}_K, \\ w(e), & e \notin \bar{E}_K. \end{cases}$ and the objective value is
7: $d_{\bar{w}}(T) := d_w(T) - \sum_{e \in \bar{E}_K} Q(e)$.

Theorem 7.4 ([151]) *Algorithm 7.1 can solve the problem (**SRDIT**$_{sH}$) in $O(n)$ time.*

7.2.3 (Int, SRD-Trees, Bounded, Sum Hamming Distance)

We consider the problem (**MCSRDIT**$_{sH}$) under unit Hamming distance, which can be formulated as follows.

$$\begin{aligned} &\min_{\bar{w}} \sum_{e \in E} H\big(\bar{w}(e), w(e)\big) \\ (\textbf{MCSRDIT}_{sH}) \quad &s.t. \sum_{e \in E} Q(e) H\big(\bar{w}(e), w(e)\big) \geq D', \\ &l(e) \leq \bar{w}(e) \leq w(e), e \in E. \end{aligned} \tag{7.8}$$

The problem (**MCSRDIT**$_{sH}$) aims to upgrade the least number of edges such that the sum of the total reduction amount is no less than D'. Obviously, we can first sort the edges by the $Q(e)-$values in non-increasing order and find the minimum number of edges to be upgraded by a binary search such that the sum of the total reduction amount achieves the lower bound D'.

Theorem 7.5 ([151]) *Let $\bar{E}_k = \{e_{i_\tau} | \tau = 1, \cdots, k\}$ be the set with minimum number of edges satisfying $\sum_{\tau=1}^{k} Q(e_{i_\tau}) \geq D'$. Then $\tilde{w}(e) = \begin{cases} l(e), & e \in \tilde{E}_k, \\ w(e), & e \notin \tilde{E}_k. \end{cases}$ is* *an optimal solution of* **(MCSRDIT$_{sH}$)** *for any* $\tilde{E}_k = \{e_{\alpha_\tau} | \tau = 1, \cdots, k\}$ *with* $\sum_{\tau=1}^{k} Q(e_{\alpha_\tau}) \geq D'$.

To find the set $\bar{E}_k = \{e_{i_\tau} | \tau = 1, 2, \cdots, k\}$ with the minimum number of edges satisfying $\sum_{\tau=1}^{k} Q(e_{i_\tau}) \geq D'$, we perform a method in two steps. In the first step, we sort the edges $e_{i_1}, \cdots, e_{i_n}$ by the values of $Q(e)$ non-increasingly. In the second step, we run a binary search algorithm to determine the minimum number k^* satisfying $\sum_{\tau=1}^{k^*} Q(e_{i_\tau}) \geq D'$ and $\sum_{\tau=1}^{k^*-1} Q(e_{i_\tau}) < D'$. Finally, we upgrade the first k^* largest $Q(e)$-value edges, which is just the set $\bar{E}_{k^*}$.

Algorithm 7.2 A greedy algorithm to solve the problem (**MCSRDIT$_{sH}$**) [151]

Input: A tree $T(V, E)$ rooted at s, two edge weight vectors w and l, and the value D.

Output: The set $\bar{E}_{k^*}$ of upgrade edges, the optimal solution $\bar{w}$, and the objective value k^*.

1: Calculate the set $L(e)$ of leaf nodes controlled by the edge $e \in E$. Calculate $\Delta w(e) := w(e) - l(e)$ and $Q(e) := |L(e)| \cdot \Delta w(e)$ for any $e \in E$. Calculate $D' := d_w(T) - D$.
2: **if** $D' > \sum_{e \in E} Q(e)$ **then return** "The problem has no feasible solution."
3: **end if**
4: Sort and relabel the edges in non-increasing order by their values of $Q(e)$ as follows.

$$Q(e_{i_1}) \geq Q(e_{i_2}) \geq \cdots \geq Q(e_{i_n}).$$

5: Initialization: $a := 1, b := n, k := \lfloor \frac{a+b}{2} \rfloor$, OPT='NO'.
6: **while** OPT='NO' **do**
7:     **if** $\sum_{j=1}^{k} Q(e_{i_j}) < D'$ **then** update $a := k$.
8:     **else if** $\sum_{j=1}^{k-1} Q(e_{i_j}) < D'$ **then** OPT='YES', **return** $k^* := k$.
9:     **else** update $b := k$.
10:     **end if**
11:     Update $k := \lfloor \frac{a+b}{2} \rfloor$.
12: **end while**
13: The optimal solution is $\bar{w}(e) := \begin{cases} l(e), & e \in \bar{E}_{k^*}, \\ w(e), & e \notin \bar{E}_{k^*}. \end{cases}$ where $\bar{E}_{k^*} := \{e_{i_\tau} | \tau = 1, 2, \cdots, k^*\}$.

Theorem 7.6 ([151]) *The problem* **(MCSRDIT$_{sH}$)** *can be solved in* $O(n \log n)$ *time by Algorithm 7.2.*

7.2.4 ((BC-)Int, SRD-Trees, Bounded, Weighted l_1 Norm)

When the weighted l_1 norm is applied to the upgrade cost, the problems **(SRDIT)** and **(MCSRDIT)**, denoted by **(SRDIT$_1$)** and **(MCSRDIT$_1$)**, can be formulated as the following models (7.9) and (7.10), respectively.

$$
\begin{aligned}
& \min_{\bar{w}} \sum_{t \in Y} d_{\bar{w}}(s,t) \\
\textbf{(SRDIT}_1\textbf{)} \quad s.t. & \sum_{e \in E} c(e)|\bar{w}(e) - w(e)| \leq K, \\
& l(e) \leq \bar{w}(e) \leq w(e), e \in E.
\end{aligned}
\tag{7.9}
$$

$$
\begin{aligned}
& \min_{\bar{w}} \sum_{e \in E} c(e)|\bar{w}(e) - w(e)| \\
\textbf{(MCSRDIT}_1\textbf{))} \quad s.t. & \sum_{t \in Y} d_{\bar{w}}(s,t) \leq D, \\
& l(e) \leq \bar{w}(e) \leq w(e), e \in E.
\end{aligned}
\tag{7.10}
$$

Next, the models (7.9) and (7.10) can each be transformed into a continuous knapsack problem. This allows the problems **(SRDIT$_1$)** and **(MCSRDIT$_1$)** to be solved in $O(n)$ time. For convenience, let $|\bar{w}(e) - w(e)| = w(e) - \bar{w}(e) = \Delta w(e)x(e)$, where $0 \leq x(e) \leq 1$. The following theorem holds:

Theorem 7.7 ([151]) *The problems* **(SRDIT$_1$)** *and* **(MCSRDIT$_1$)** *can be transformed into continuous knapsack problems and be solved in* $O(n)$ *time.*

7.2.5 (BC-Int, SRD-Trees, Bounded, Weighted l_∞ Norm)

We consider the problem (**MCSRDIT$_\infty$**) in view of increasing edge weights with minimum cost under weighted l_∞ norm to make the SRD at least the value D.

$$
\begin{aligned}
& \min_{\bar{w}} C(\bar{w}) := \max_{e \in E} c(e)(\bar{w}(e) - w(e)) \\
\textbf{(MCSRDIT}_\infty\textbf{)} \quad s.t. & \sum_{t_k \in Y} \bar{w}(P_k) \geq D, \\
& w(e) \leq \bar{w}(e) \leq u(e), \quad e \in E.
\end{aligned}
\tag{7.11}
$$

We can first solve two special cases based on the relationship between $u(T)$ and D, as well as $w(T)$ and D.

Lemma 7.1 ([83]) *If $D \leq w(T)$, then w is an optimal solution to (**MCSRDIT**$_\infty$).*

Lemma 7.2 ([83]) *If $u(T) < D$, then the problem (**MCSRDIT**$_\infty$) is infeasible.*

Then we consider the situation that $w(T) < D \leq u(T)$. First, we will prove an interesting property of the optimal solution.

Lemma 7.3 ([83]) *Let $\bar{w}$ be an optimal solution of (**MCSRDIT**$_\infty$), then $\bar{w}(T) = D$.*

Here, we construct an optimal solution for the problem (**MCSRDIT**$_\infty$) where $w(T) < D \leq u(T)$. Leveraging the properties of the l_∞ norm, we recognize that, with fixed update costs, updating each edge can increase the SRD by the maximum possible amount. For each edge $e \in E$, let $F(e) = c(e)(u(e) - w(e))$ represent the total cost required to achieve the maximum possible update. We then sort the $F(e)$ values for all edges in non-decreasing order and renumber them to obtain the sequence $\left\{F(e_{m_k})\right\}$ as follows:

$$F(e_{m_1}) \leq F(e_{m_2}) \leq \cdots \leq F(e_{m_n}). \tag{7.12}$$

For an index k, define a function

$$f(k) = \sum_{e \in E} |L(e)| \cdot \min\left\{u(e) - w(e), \frac{F(e_{m_k})}{c(e)}\right\} \tag{7.13}$$

to represent the maximum increment of SRD when the update cost is $F(e_{m_k})$.

Next, using the binary search method on the sequence $\left\{F(e_{m_k})\right\}$, we can determine a critical index k^* such that $f(k^*) \leq D - w(T) \leq f(k^* + 1)$. This allows us to obtain a unique optimal solution to the problem (**MCSRDIT**$_\infty$).

Theorem 7.8 ([83]) *When $w(T) < D \leq u(T)$, let $E_{\leq} := \{e \in E | F(e) \leq F(e_{m_{k^*}})\}$ and $\Delta D := D - w(T)$, then the weight vector*

$$\bar{w}(e) = \begin{cases} u(e), & e \in E_{\leq} \\ w(e) + \frac{F(e_{m_{k^*}})}{c(e)} + \frac{\Delta D - f(k^*)}{c(e) \sum_{j=k^*+1}^{n} \frac{|L(e_{m_j})|}{c(e_{m_j})}}, & e \in E \backslash E_{\leq} \end{cases} \tag{7.14}$$

*is the unique optimal solution to the problem (**MCSRDIT**$_\infty$).*

Algorithm 7.3 $[\bar{w}, C(\bar{w})] := \text{MCSRDIT}_{\infty}(T, n, w, u, c, D)$ [83]

Input: A rooted tree $T(V, E)$, the number n of edges, the weight vector w, an upper bound vector u, a cost vector c, and given value D.
Output: An optimal vector $\bar{w}$ and the cost $C(\bar{w})$.

For each edge $e \in E$, determine the set $L(e)$ and calculate $F(e) := c(e)(u(e) - w(e))$, $u(T) := \sum_{t_k \in Y} u(P_k)$, and let $\Delta D := D - w(T)$
if $u(T) < D$ **then**
    return "The problem is infeasible" and $\bar{w} = [\ ]$, $C(\bar{w}) = +\infty$
else if $w(T) > D$ **then** **return** $[w, 0]$.
else
    Sort $F(e)$ in a non-decreasing order by (7.12).
    Initialize $a := 1, b := n, k^* := 0$.
    while $b - a > 1$ and $k^* = 0$ **do**
        Let $k := \lfloor \frac{a+b}{2} \rfloor$.
        Calculate the values of $f(k)$ and $f(k+1)$ using (7.13).
        if $f(k) \le \Delta D \le f(k+1)$ **then** $k^* := k$.
        else if $f(k+1) < \Delta D$ **then** let $a := k$.
        else let $b := k$.
        end if
    end while
    Calculate $\bar{w}$ using (7.14), $C(\bar{w}) := F(e_{m_{k^*}}) + \frac{\Delta D - f(k^*)}{\sum_{j=k^*+1}^{n} \frac{|L(e_{m_j})|}{c(e_{m_j})}}$.
    return $[\bar{w}, C(\bar{w})]$.
end if

Theorem 7.9 ([83]) *The problem (**MCSRDIT**$_{\infty}$) can be solved by Algorithm 7.3 in $O(n \log n)$ time.*

7.2.6 (Int, SRD-Trees, Bounded, Bottleneck Hamming Distance)

We first consider the problem (**MCSRDIT**$_{bH}$), which can be formulated as follows.

$$
\begin{aligned}
& \min_{\bar{w}} C(\bar{w}) := \max_{e \in E} c(e) H(\bar{w}(e), w(e)) \\
(\textbf{MCSRDIT}_{bH})\ \ & s.t. \sum_{t_k \in Y} \bar{w}(P_k) \ge D, \\
& w(e) \le \bar{w}(e) \le u(e), \quad e \in E.
\end{aligned}
\tag{7.15}
$$

Lemma 7.4 ([83]) *When $u(T) < D$, the problem (**MCSRDIT**$_{bH}$) is infeasible.*

Lemma 7.5 ([83]) *If $w(T) \ge D$, then w is an optimal solution to (**MCSRDIT**$_{bH}$).*

When $u(T) \geq D > w(T)$, we first sort the edges in an ascending order by their $c(e)$ values and renumber them in the following sequence $\{e_{\alpha_i}\}$:

$$c(e_{\alpha_1}) \leq c(e_{\alpha_2}) \leq \cdots \leq c(e_{\alpha_n}) \tag{7.16}$$

Let $S(e) = |L(e)|(u(e) - w(e))$, and we then construct an auxiliary function

$$g(k) = \sum_{i=1}^{k} S(e_{\alpha_i}) \tag{7.17}$$

which represents the SRD added when updating the first k edges in the sequence $\{e_{\alpha_i}\}$. We use a binary search method to iteratively determine k^* such that $g(k^*) < \Delta D \leq g(k^* + 1)$ and give the following conclusion.

Theorem 7.10 ([83]) *If* $u(T) \geq D > w(T)$, *let* $\bar{E}_{\leq} := \{e_{\alpha_i} \in E \mid i = 1, 2, \ldots, k^* + 1\}$. *Then the weight vector*

$$\bar{w}(e) = \begin{cases} u(e), & e \in \bar{E}_{\leq} \\ w(e), & e \in E \backslash \bar{E}_{\leq} \end{cases} \tag{7.18}$$

is an optimal solution to ***(MCSRDIT$_{bH}$)***.

Based on Theorem 7.10, we present the following Algorithm 7.4 to solve the problem (**MCSRDIT**$_{bH}$).

Algorithm 7.4 $[\bar{w}, C(\bar{w})] :=$ MCSRDIT$_{bH}(T, n, w, u, c, D)$ [83]

Input: A rooted tree $T(V, E)$, the number n of edges, the weight vector w, an upper bound vector u, a cost vector c, and a given value D.
Output: An optimal vector $\bar{w}$ and the cost $C(\bar{w})$.
1: For each $e \in E$, calculate $F(e) := c(e)(u(e) - w(e))$. Let $\Delta D := D - w(T)$ and $u(T) := \sum_{t_k \in Y} u(P_k)$.
2: **if** $u(T) < D$ **then return** "The problem is infeasible" and $\bar{w} = [\ ]$, $C(\bar{w}) = +\infty$.
3: **else if** $w(T) \geq D$ **then return** $[w, 0]$.
4: **else**
5: Sort the cost vector $c(e)$ in a non-decreasing order in (7.16).
6: Set the initial values as $a := 1$, $b := n$, and $k^* := 0$.
7: **while** $b - a > 1$ and $k^* = 0$ **do**
8: $k := \lfloor \frac{a+b}{2} \rfloor$.
9: Calculate $g(k)$ and $g(k+1)$ using (7.17).
10: **if** $g(k) < \Delta D \leq g(k+1)$ **then** let $k^* := k$.
11: **else if** $g(k+1) < \Delta D$ **then** let $a := k$.
12: **else** let $b := k$.
13: **end if**
14: **end while**
15: **Return** $[\bar{w}, C(\bar{w})]$ by (7.18) and $C(\bar{w}) = c(e_{\alpha_{k^*+1}})$.
16: **end if**

Lemma 7.6 ([83]) *Algorithm 7.4 solves (**MCSRDIT**$_{bH}$) in $O(n \log n)$ time.*

7.2.7 ((BC-)Int, SRD-Trees, Bounded, Weighted Node Cost)

In this section, we prove that the problem (**SRDIT**$_N$) is $\mathcal{NP}$-hard by transforming it into a 0-1 knapsack problem, so is the problem (**MCSRDIT**$_N$).

If a node v is upgraded, then the lengths of the paths $P_{s,t_k}(t_k \in L(e), e \in A(v))$ decrease by $\Delta w(e) = \beta_0(e) - \beta_1(e)$.

Definition 7.3 ([151]) For any $v \in V$, let $B(v) = \sum_{e \in A(v)} Q(e)$ be the **total reduction amount** of the **node** v, which is the sum of the total reduction amount of the edges adjacent to the node v.

In a similar way, the problem (**MCSRDIT**$_N$) can also be proved to be equivalent to a 0-1 minimization knapsack problem.

Theorem 7.11 ([151]) *The problems (**SRDIT**$_N$) and (**MCSRDIT**$_N$) are $\mathcal{NP}$-hard.*

7.2.8 (BC-Int, SRD-Trees, Bounded, Unit Node Cost)

The problem (**SRDIT**$_N$) with unit cost, denoted by (**SRDIT**$_{uN}$), can be formulated as the following form.

$$
(\textbf{SRDIT}_{uN}) \quad \begin{aligned} &\max_z \sum_{v \in V} B(v) z(v) \\ &s.t. \ \sum_{v \in V} z(v) \leq K, \\ &z(v) = \begin{cases} 1, & \text{if } v \text{ is upgraded}, \\ 0, & \text{if } v \text{ is not upgraded}. \end{cases} \end{aligned} \tag{7.19}
$$

It is shown in model (7.19) that the problem (**SRDIT**$_{uN}$) aims to upgrade K nodes to maximize their relevant sum of $B(v)$-value. Thus we can sort the nodes by the values of $B(v)$ in non-increasing order and upgrade the first K largest $B(v)$-value nodes.

Sort the nodes by the values of $B(v)$ in non-increasing order as follows.

$$
B(v_{i_1}) \geq B(v_{i_2}) \geq \cdots \geq B(v_{i_K}) \geq \cdots \geq B(v_{i_{n+1}}).
$$

Theorem 7.12 ([151]) *Let $\bar{V}_K = \{v_{i_\tau} | \tau = 1, 2, \cdots, K\}$ be the set of the first K largest $B(v)$-value nodes in V. Then $\bar{w}(e) = \begin{cases} \beta_1(e), e \in A(v), v \in \bar{V}_K, \\ \beta_0(e), otherwise. \end{cases}$ is an optimal solution of the problem (**SRDIT**$_{uN}$).*

We can solve the problem (**SRDIT**$_{uN}$) similar to the problem (**SRDIT**$_{uH}$). We first find the K-th largest element q of an array B by the selection algorithm [33, 220-222] and then determine the sets V_1 and V_2 of nodes whose value is larger than and equal to q, respectively. Finally, we can obtain the set $S_K := V_1 \cup V_2(1 : K - |V_1|)$ of the first K largest $B(v)$-value edges in the array B. The aforementioned steps can both be completed in $O(n)$ time.

Algorithm 7.5 A greedy algorithm to solve the problem (**SRDIT**$_{uN}$) [151]

Input: A tree $T(V, E)$ rooted at s, the number n of edges, the set Y of leaf nodes, two edge weight vectors β_0, β_1, and the number K of upgrade nodes.

Output: The set S_K of upgrade nodes, an optimal solution $\bar{w}$, and the relevant value $d_{\bar{w}}(T)$.

1: **if** $K > n + 1$ **then return** "The problem has no feasible solution!"

2: **end if**

3: Calculate the set $L(e)$ of leaf nodes controlled by the edge $e \in E$ and the set $A(v)$ of adjacent edges of the node $v \in V$. Calculate $\Delta w(e) := \beta_0(e) - \beta_1(e)$ and $Q(e) := |L(e)| \cdot \Delta w(e)$ for any $e \in E$. Calculate $B(v) := \sum_{e \in A(v)} Q(e)$ for each node $v \in V$.

4: Call $q := Selection(B, K)$.

5: Let $V_1 := \{v_i | B(v_i) > q\}$, and $V_2 := \{v_i | B(v_i) = q\}$. Then $S_K := V_1 \cup V_2(1 : K - |V_1|)$.

6: The optimal solution is $\bar{w}(e) := \begin{cases} \beta_1(e), \ e \in A(v), v \in S_K, \\ \beta_0(e), \ otherwise. \end{cases}$ and the relevant value is $d_{\bar{w}}(T) := d_{\beta_0}(T) - \sum_{v \in S} B(v)$.

Theorem 7.13 ([151]) *Algorithm 7.5 can solve (**SRDIT**$_{uN}$) in $O(n)$ time.*

7.2.9 (Int, SRD-Trees, Bounded, Unit Node Cost)

The problem (**MCSRDIT**$_N$) with unit cost, denoted by (**MCSRDIT**$_{uN}$), can be formulated as follows:

$$\min_z \sum_{v \in V} z(v)$$

$$(\textbf{MCSRDIT}_{uN}) \quad s.t. \sum_{v \in V} B(v) z(v) \geq D^0, \tag{7.20}$$

$$z(v) = \begin{cases} 1, & \text{if } v \text{ is upgraded}, \\ 0, & \text{if } v \text{ is not upgraded}, \end{cases}$$

where $D^0 = d_{\beta_0}(T) - D$.

As shown in model (7.20), the problem **(MCSRDIT$_{uN}$)** aims to upgrade the least number of nodes such that the sum of the total reduction amount is no less than D^0. Obviously, we can first sort the nodes by the values of $B(v)$ non-increasingly and find the minimum number of nodes to be upgraded by a binary search such that the sum of the total reduction amount satisfies the constraint.

Theorem 7.14 ([151]) *Let $\bar{V}_k = \{v_{i_\tau} | \tau = 1, \cdots, k\}$ be the set with minimum number of nodes satisfying $\sum\limits_{\tau=1}^{k} B(v_{i_\tau}) \geq D^0$. Then for any $\tilde{V}_k = \{v_{\alpha_\tau} | \tau = 1, \cdots, k\}$ with $\sum\limits_{\tau=1}^{k} B(v_{\alpha_\tau}) \geq D^0$, $\bar{w}(e) = \begin{cases} \beta_1(e), \ e \in A(v), v \in \tilde{V}_k, \\ \beta_0(e), \ otherwise. \end{cases}$ is an optimal solution of* ***(MCSRDIT$_{uN}$)****.*

Algorithm 7.6 A greedy algorithm to solve the problem (**MCSRDIT$_{uN}$**) [151]

Input: A tree $T(V, E)$ rooted at s, the set Y of leaf nodes, two edge weight vectors β_0 and β_1, and the number K of upgrade nodes.

Output: The set S_{k^*} of upgrade nodes, the optimal solution $\bar{w}$, and the relevant value k^*.

Calculate the set $L(e)$ of leaf nodes controlled by the edge $e \in E$ and the set $A(v)$ of adjacent edges of the node $v \in V$. Calculate $\Delta w(e) := \beta_0(e) - \beta_1(e)$ and $Q(e) := |L(e)| \cdot \Delta w(e)$ for any $e \in E$. Calculate $B(v) := \sum_{e \in A(v)} Q(e), v \in V$. Calculate $D^0 = d_{\beta_0}(T) - D$.

if $D^0 > \sum_{v \in V} B(v)$ **then return** "The problem has no feasible solution!"

end if

Sort and relabel nodes in non-increasing order by their values of $B(v)$ as follows.

$$B(v_{i_1}) \geq B(v_{i_2}) \geq \cdots \geq B(v_{i_{n+1}}).$$

Initialization: $a := 1, b := m, k := \lfloor \frac{a+b}{2} \rfloor$, OPT='NO'.

while OPT='NO' **do**

  if $\sum\limits_{j=1}^{k} B(v_{i_j}) < D^0$ **then** update $a := k$.

  else if $\sum\limits_{j=1}^{k-1} B(v_{i_j}) < D^0$ **then** OPT:='YES', **return** $k^* := k$.

  else update $b := k$.

  end if

  Update $k := \lfloor \frac{a+b}{2} \rfloor$.

end while

13: The optimal solution is $\bar{w}(e) := \begin{cases} \beta_1(e), & e \in A(v), v \in S_{k^*}, \\ \beta_0(e), & otherwise. \end{cases}$ where $S_{k^*} := \{v_{i_\tau} | \tau = 1, \cdots, k^*\}$.

Corollary 7.1 ([151]) *The problem (**MCSRDIT**$_{uN}$) can be solved in $O(n \log n)$ time by Algorithm 7.6.*

7.3 The Sum of Root-Leaf Distance Interdiction Problem with Cardinality Constraint by Upgrading Edges on Trees

The Sum of Root-leaf Distance Interdiction Problem on Trees with Cardinality Constraint (**SRDITC**) aims to maximize the SRD by selecting an altered edge weight vector $\tilde{w}$. The objective is to adjust the edge weights such that the norm constraint, measured by $\|\tilde{w} - w\|$ in a specified norm, and the cardinality constraint on modified edges do not exceed predetermined limits K and N, respectively. This leads to the formulation of the mathematical model for the (**SRDITC**) as follows.

$$\begin{aligned} & \max_{\tilde{w}} \tilde{w}(T) := \sum_{t_k \in Y} \tilde{w}(P_k) \\ (\textbf{SRDITC}) \quad & s.t. \quad \|\tilde{w} - w\| \leq K, \\ & \sum_{e \in E} H(\tilde{w}(e), w(e)) \leq N, \\ & w(e) \leq \tilde{w}(e) \leq u(e), \quad e \in E. \end{aligned} \tag{7.21}$$

We also consider the corresponding minimum cost problem with a cardinality constraint (**MCSRDITC**) by interchanging its objective function and the norm constraint. The mathematical model for this problem is as follows:

$$\begin{aligned} & \min_{\tilde{w}} C(\tilde{w}) := \|\tilde{w} - w\| \\ (\textbf{MCSRDITC}) \quad & s.t. \sum_{t_k \in Y} \tilde{w}(P_k) \geq D, \\ & \sum_{e \in E} H(\tilde{w}(e), w(e)) \leq N, \\ & w(e) \leq \tilde{w}(e) \leq u(e), \quad e \in E. \end{aligned} \tag{7.22}$$

7.3.1 (BC-Int-Cardinality, SRD-Trees, Bounded, Weighted l_∞ Norm)

In this subsection, we considered the problem (SRDITC$_\infty$) under weighted l_∞ norm, in which the norm constraint in (7.21) can be described as

$$\max_{e \in E} c(e)(\bar{w}(e) - w(e)) \leq K.$$

Let

$$\bar{w}(e) = \min\left\{w(e) + \frac{K}{c(e)}, u(e)\right\} (e \in E) \tag{7.23}$$

$$\bar{S}(e) = |L(e)|(\bar{w}(e) - w(e))(e \in E) \tag{7.24}$$

Then, $\bar{S}(e)$ is the increment of SRD for e under $\bar{w}$. Let $\bar{S} := \{\bar{S}(e)|e \in E\}$. Then we call $\bar{S}^N :=$ **Selection**$(\bar{S}, N)$ to find the N-th largest element $\bar{S}^N$ in the set $\bar{S}$ by the selection algorithm in [33, pp.220-222].

Theorem 7.15 ([83]) *Let $\bar{E}_N := \{e \in E|\bar{S}(e) \geq \bar{S}^N\}$, which has exactly N values. Then the weight vector*

$$\tilde{w}(e) = \begin{cases} \bar{w}(e), & e \in \bar{E}_N \\ w(e), & \textit{otherwise} \end{cases} \tag{7.25}$$

is an optimal solution of the problem (**SRDITC$_\infty$**).

Specifically, when $N = n$, the cardinality constraint is trivially satisfied. Therefore, the solution to the problem $(SRDITC_\infty)$ for $N = n$ is identical to the solution of the problem $(SRDIT_\infty)$.

Algorithm 7.7 $[\tilde{w}, \tilde{w}(T)] :=$ SRDITC$_\infty(T, n, w, u, c, K, N)$ [83]

Input: A rooted tree $T(V, E)$, the number n of edges, the weight vector w, an upper bound vector u, a cost vector c, and given K and N.
Output: An optimal vector $\tilde{w}$ and the corresponding SRD $\tilde{w}(T)$.
1: For each e, calculate $\bar{w}(e)$ and $\bar{S}(e)$ by (7.23) and (7.24), respectively.
2: Call $\bar{S}^N :=$ **Selection**$(\bar{S}, N)$, $\bar{E}_N := \{e \in E|\bar{S}(e) \geq \bar{S}^N\}$.
3: Compute $\tilde{w}$ by (7.25) and $\tilde{w}(T) := \sum_{t_k \in Y} \tilde{w}(P_k)$.
4: **return** $[\tilde{w}, \tilde{w}(T)]$.

7.3.2 (Int-Cardinality, SRD-Trees, Bounded, Weighted l_∞ Norm)

Solving problem (**MCSRDITC**$_\infty$) directly is challenging, and therefore, we first focus on solving its relaxation problem (**MCSRDIT**$_\infty$), described as (7.11), without the cardinality constraint. If we can identify the set $\bar{E}^*$ of N edges that are updated in the optimal solution of (**MCSRDITC**$_\infty$), then it can be transformed into its relaxation problem (**MCSRDIT**$_\infty$).

For the problem (**MCSRDITC**$_\infty$), we first examine its feasibility. We define $S(e) := |L(e)|(u(e) - w(e))$ for each edge $e \in E$, representing the maximum possible increment in SRD that each edge can achieve. The edges are then sorted in non-increasing order of $S(e)$ and renumbered to form the following sequence:

$$S(e_{\beta_1}) \geq S(e_{\beta_2}) \geq \cdots \geq S(e_{\beta_{n-1}}) \geq S(e_{\beta_n}) \tag{7.26}$$

Given the cardinality constraint of N edges, the maximum total increment in SRD is $S_N = \sum_{i=1}^{N} S(e_{\beta_i})$.

Lemma 7.7 *When $S_N < \Delta D$, problem (**MCSRDITC**$_\infty$) is infeasible.*

Lemma 7.8 *Let $R = \sum_{e \in E} H(w(e), u(e))$. If $S_N \geq \Delta D$, then an optimal solution $\tilde{w}$ of problem (**MCSRDITC**$_\infty$) has the following property.*

$$\sum_{e \in E} H(\tilde{w}(e), w(e)) = \min\{R, N\}.$$

We next introduce an iterative method to determine the set $\bar{E}^{\tau^*}$ of N edges that are updated in the optimal solution $\tilde{w}$ of the problem (**MCSRDITC**$_\infty$), then we can find $\tilde{w}$ by solving a relaxation problem (**MCSRDIT**$_\infty$) with respect to a bound vector

$$u^{\tau^*}(e) = \begin{cases} u(e), e \in \bar{E}^{\tau^*}, \\ w(e), \text{otherwise.} \end{cases} \tag{7.27}$$

We first initialize $\tau := 0$ and the set $\bar{E}^\tau := \{e_{\beta_1}, e_{\beta_2}, \ldots, e_{\beta_N}\}$ and call $[\bar{w}^\tau, C(\bar{w}^\tau)]$:=**MCSRDIT**$_\infty(T, n, w, u^\tau, c, D)$.

Next, in each iteration, for each edge $e_i \in \bar{E}^\tau$, we check whether there is a better edge $e_j \in E \setminus \bar{E}^\tau$ with a lower cost, where j is traversed according to the sorted sequence of $\nu(e) := \frac{|L(e)|}{c(e)}$ by

$$\nu(e_{\gamma_1}) \geq \nu(e_{\gamma_2}) \geq \cdots \geq \nu(e_{\gamma_n}). \tag{7.28}$$

Let $S^\tau(e) := \min\{S(e), \nu(e)C(\bar{w}^\tau)\}$ denote the maximum SRD increment for edge e within $C(\bar{w}^\tau)$. Define $\bar{E}^\tau_= := \{e \in E \mid c(e)(\bar{w}^\tau(e) - w(e)) = C(\bar{w}^\tau)\}$ as the set of edges whose costs reach $C(\bar{w}^\tau)$. There are two replacement scenarios to obtain a better set $\bar{E}^{\tau+1} := \bar{E}^{\tau} \setminus \{e_i\} \cup \{e_j\}$: (1). $e_i \in \bar{E}^{\tau}_{=}$, $S^{\tau}(e_j) \geq S^{\tau}(e_i)$ and $\nu(e_i) < \nu(e_j)$; (2). $e_i \in \bar{E}^{\tau} \setminus \bar{E}^{\tau}_{=}$ and $S^{\tau}(e_j) > S^{\tau}(e_i)$.

Then, update $\tau := \tau + 1$ and continue iterating until there is no edge in $E \setminus \bar{E}^{\tau}$ better than any edge in the current set $\bar{E}^{\tau}$. Denote the resulting set of edges by $\bar{E}^{*}$.

Based on the earlier analysis, Li et al. [83] proposed Algorithm 7.8 to solve the problem (MCSRDITC$_{\infty}$).

Theorem 7.16 ([83]) *Problem (MCSRDITC$_{\infty}$) can be solved by Algorithm 7.8 in $O(Nn^2)$ time, which is bounded by $O(n^3)$.*

Algorithm 7.8 $\left[\tilde{w}, C(\tilde{w})\right] :=$ MCSRDITC$_{\infty}(T, n, w, u, c, D, N)$ [83]

Input: A rooted tree $T(V, E)$, the number n of edges, the weight vector w, an upper bound vector u, a cost vector c, and given value D and N.
Output: An optimal vector $\tilde{w}$ and the cost $C(\tilde{w})$.

For each $e \in E$, calculate $S(e) := |L(e)|(u(e) - w(e))$, and $\Delta D = D - w(T)$.
Sort the $S(e)$ in a non-increasing order by (7.26).
$S_N := \sum_{i=1}^{N} S(e_{\beta_i})$.
if $S_N < \Delta D$ **then**
  return "The problem is infeasible" and $\tilde{w} = [\ ]$, $C(\tilde{w}) = +\infty$.
else
  Sort $\nu(e) := \frac{|L(e)|}{c(e)}$, $e \in E$ in a non-increasing order by (7.28).
  $flg := 1, \tau := 0, \bar{E}^{\tau} := \left\{e_{\beta_1}, e_{\beta_2}, \ldots, e_{\beta_N}\right\}$.
  Let $S^{\tau}(e) := \min\{S(e), \nu(e)C(\tilde{w}^{\tau})\}$, $e \in E$.
  while $flg = 1$ **do**
    Let $flg := 0$ and $u^{\tau}(e) := \begin{cases} u(e), e \in \bar{E}^{\tau} \\ w(e), \text{otherwise} \end{cases}$.
    Call $[\tilde{w}^{\tau}, C(\tilde{w}^{\tau})]$=MCSRDIT$_{\infty}(T, n, w, u^{\tau}, c, D)$.
    $\bar{E}^{\tau}_{=} := \{e \in E | c(e)(\tilde{w}^{\tau}(e) - w(e)) = C(\tilde{w}^{\tau})\}$.
    for $e_i \in \bar{E}^{\tau}$ **do**
      for $e_j \in E \setminus \bar{E}^{\tau}$ in the order by (7.28) **do**
        if $e_i \in \bar{E}^{\tau}_{=}$, $S^{\tau}(e_j) \geq S^{\tau}(e_i)$ and $\nu(e_i) < \nu(e_j)$ **then**
          $\bar{E}^{\tau+1} := \bar{E}^{\tau} \setminus \{e_i\} \cup \left\{e_j\right\}$, $flg := 1$. **Break.**
        else if $e_i \in \bar{E}^{\tau} \setminus \bar{E}^{\tau}_{=}$, $S^{\tau}(e_j) > S^{\tau}(e_i)$ **then**
          $\bar{E}^{\tau+1} := \bar{E}^{\tau} \setminus \{e_i\} \cup \left\{e_j\right\}$, $flg := 1$. **Break.**
        end if
      end for
    end for
    Update $\tau := \tau + 1$.
  end while
end if
Update $\tau^* := \tau - 1$ and $\bar{E}^{*} := \bar{E}^{\tau^*}$.
Return $[\tilde{w} := \tilde{w}^{\tau^*}, C(\tilde{w}) := C(\tilde{w}^{\tau^*})]$.

7.3.3 (BC-Int-Cardinality, SRD-Trees, Bounded, Bottleneck Hamming)

When the bottleneck Hamming distance is applied to the upgrade cost, the problems (SRDITC_{bH}) and (MCSRDITC_{bH}) can be formulated as the models (7.21) and (7.22), respectively, by replacing

$$\|\tilde{w} - w\| := \max_{e \in E} c(e) H(\tilde{w}(e), w(e)).$$

In the next two subsections, we first provide an $O(n)$ algorithm for problem (**SRDITC**$_{bH}$). then we present an $O(n \log n)$ algorithm for the minimum cost problem (**MCSRDITC**$_{bH}$) by transforming it into its relaxation problem (**MCSRDIT**$_{bH}$).

Now we consider the problem (**SRDITC**$_{bH}$). Let $\tilde{E}_{\leq} := \{e \in E \mid c(e) \leq K\}$. Recall that $S(e) := |L(e)|(u(e) - w(e))$. Let $\tilde{S}_{\leq} := \left\{S(e) | e \in \tilde{E}_{\leq}\right\}$. When $|\tilde{E}_{\leq}| > N$, we first prioritize the first N largest edges in the set $\tilde{E}_{\leq}$ by calling algorithm $\tilde{S}_{\leq}^{N} := Selection(\tilde{S}_{\leq}, N)$[33] to select the N-th largest element $\tilde{S}_{\leq}^{N}$ from $\tilde{S}_{\leq}$.

Theorem 7.17 ([83]) *Let* $\tilde{E}_N := \begin{cases} \left\{e \in \tilde{E}_{\leq} | S(e) \geq \tilde{S}_{\leq}^{N}\right\}, & |\tilde{E}_{\leq}| > N \\ \tilde{E}_{\leq}, & |\tilde{E}_{\leq}| \leq N \end{cases}$. *Then the vector*

$$\tilde{w}(e) := \begin{cases} u(e), & e \in \tilde{E}_N \\ w(e), & e \in E \setminus \tilde{E}_N \end{cases} \tag{7.29}$$

*is an optimal solution to problem (**SRDITC**$_{bH}$).*

Specifically, when $N = n$, the cardinality constraint is trivially satisfied. Therefore, the solution to the problem $(SRDITC_{bH})$ for $N = n$ is identical to the solution of the problem $(SRDIT_{bH})$.

Based on Theorem 7.17, we propose Algorithm 7.9 for problem (**SRDITC**$_{bH}$).

Algorithm 7.9 $\left[\tilde{w}, \tilde{w}(T)\right] = \text{SRDITC}_{bH}(T, n, w, u, c, K, N)$ [83]

Input: A rooted tree $T(V, E)$, the number n of edges, the weight vector w, an upper bound vector u, a cost vector c, and given values K and N.
Output: An optimal vector $\tilde{w}$ and the SRD $\tilde{w}(T)$.
1: $\tilde{E}_{\leq} := \{e \in E \mid c(e) \leq K\}$
2: **if** $|\tilde{E}_{\leq}| > N$ **then**
3: $\quad S(e) := |L(e)|(w'(e) - w(e))$, $\tilde{S}_{\leq} := \left\{S(e) | e \in \tilde{E}_{\leq}\right\}$.
4: $\quad \tilde{S}_{\leq}^{N} := Selection(\tilde{S}_{\leq}, N)$, $\tilde{E}_N := \left\{e \in \tilde{E}_{\leq} | S(e) \geq \tilde{S}_{\leq}^{N}\right\}$.
5: **else** let $\tilde{E}_N := \tilde{E}_{\leq}$.
6: **end if**
7: **return** $[\tilde{w}, \tilde{w}(T)]$ by (7.29).

Lemma 7.9 ([83]) *Algorithm 7.9 can solve the problem (**SRDITC**$_{bH}$) in $O(n)$ time.*

7.3.4 (Int-Cardinality, SRD-Trees, Bounded, Bottleneck Hamming)

Now we can consider solving the cardinality problem (**MCSRDITC**$_{bH}$) under bottleneck Hamming distance. First, we need to determine the infeasibility by Lemma 7.7 similarly.

Lemma 7.10 ([83]) *When $S_N < \Delta D$, problem (**MCSRDITC**$_{bH}$) is infeasible.*

When $S_N \geq \Delta D$, we first call $[w', C(w')] = \text{MCSRDIT}_{bH}(T, n, w, u, c, D)$ to obtain the optimal solution to its relaxation problem **(MCSRDIT**$_{bH}$**)**.

Lemma 7.11 ([83]) *If $S_N \geq \Delta D$ and $\sum_{e \in E} H(w'(e), w(e)) \leq N$, then w' is an optimal solution to problem (**MCSRDITC**$_{bH}$).*

If $\sum_{e \in E} H(w'(e), w(e)) > N$, sort the edges $e \in E$ in a non-decreasing order according to their $c(e)$ values by (7.16). For any $N \leq k \leq n$, let $E_k := \{e_{\alpha_1}, e_{\alpha_2}, \ldots, e_{\alpha_k}\}$, $S_{E_k} := \{S(e) | e \in E_k\}$, call $S_k^N := Selection(S_{E_k}, N)$[33]. Let $E_k^N := \{e \in E_k | S(e) \geq S_k^N\}$, then

$$h(k) = \sum_{e \in E_k^N} S(e) \tag{7.30}$$

is the maximum SRD that can be increased by updating N edges among the first k edges in the sequence $\{e_{\alpha_i}\}$.

Next, we can use binary search for iteration to find the smallest k^* such that $h(k^*) \geq \Delta D$, and let the updated edge set at this point be $E_{k^*}^N$.

Theorem 7.18 ([83]) *If $\sum_{e \in E} H(w'(e), w(e)) > N$, the weight vector*

$$\tilde{w}(e) = \begin{cases} u(e), & e \in E_{k^*}^N \\ w(e), & e \in E \setminus E_{k^*}^N \end{cases} \tag{7.31}$$

*is an optimal solution to problem (**MCSRDITC**$_{bH}$).*

We present Algorithm 7.10 for solving **(MCSRDITC**$_{bH}$**)** based on Theorem 7.18.

Algorithm 7.10 $\left[\tilde{w}, C(\tilde{w})\right] := \text{MCSRDITC}_{bH}(T, n, w, u, c, D, N)$ [83]

Input: A rooted tree $T(V, E)$, the number n of edges, the weight vector w, an upper bound vector u, a cost vector c, and two given values D and N.
Output: An optimal vector $\tilde{w}$ and the cost $C(\tilde{w})$.

```
For each e, calculate S(e) := |L(e)|(u(e) − w(e)), and let ΔD := D − w(T).
Sort S(e) in a non-increasing order by (7.26).
Calculate S_N := Σ_{i=1}^{N} S(e_{β_i}).
if S_N < ΔD then
    return "The problem is infeasible" and w̃ = [ ], C(w̃) = +∞.
end if
[w', C(w')] :=MCSRDIT_{bH}(T, n, w, u, c, D).
if Σ_{e∈E} H(w'(e), w(e)) ≤ N then return [w', C(w')].
else sort c(e) in a non-increasing order by (7.16).
    Let a := N, b := n, k* := 0
    while a < b and k* = 0 do
        Let k := ⌊(a+b)/2⌋.
        Compute E_k^N, h(k) and h(k − 1) by (7.30).
        if h(k − 1) < ΔD ≤ h(k) then let k* := k, E_{k*}^N := E_k^N.
        else if h(k) < ΔD then let a := k.
        else let b := k.
        end if
    end while
    Compute w̃ by (7.31) and C(w̃) := c(e_{p_{k*}}).
    return [w̃, C(w̃)].
end if
```

Theorem 7.19 ([83]) *Algorithm 7.10 can solve (**MCSRDITC**$_{bH}$) in $O(n \log n)$ time.*

7.4 Double Interdiction Problem on the Sum of Root-Leaf Distances on Trees

Let $T = (V, E, w, u, c)$ be an edge-weighted tree rooted at s, where $V := \{s, v_1, v_2, \ldots, v_n\}$ and $E := \{e_1, e_2, \ldots, e_n\}$. Let $Y := \{t_1, t_2, \ldots, t_r\}$ be the set of leaf nodes.

Given two values K and M, the double interdiction problem on trees on the sum of root-leaf distance by upgrading edges (**DIT**) aims to maximize SRD by determining an edge weight vector $\tilde{w}$ such that the modification cost $\|\tilde{w} - w\|$ does not exceed K, and the Shortest Root-leaf Distance (**StRD**) must not be less than M. The mathematical representation of this problem can be articulated as follows.

$$\max_{\tilde{w}} \tilde{w}(T) := \sum_{t_k \in Y} \tilde{w}(P_k)$$

$$(\textbf{DIT}) \quad s.t. \quad \min_{t_k \in Y} \tilde{w}(P_k) \geq M, \tag{7.32}$$

$$\|\tilde{w} - w\| \le K,$$
$$w(e) \le \tilde{w}(e) \le u(e), \quad e \in E.$$

Its related minimum cost problem (MCDIT) by exchanging its objective function and the modification cost can be formulated as follows:

$$\begin{aligned} & \min_{\tilde{w}} C(\tilde{w}) := \|\tilde{w} - w\| \\ \textbf{(MCDIT)} \ & s.t. \ \min_{t_k \in Y} \tilde{w}(P_k) \ge M, \\ & \tilde{w}(T) \ge D, \\ & w(e) \le \tilde{w}(e) \le u(e), \quad e \in E. \end{aligned} \tag{7.33}$$

In practical applications, several norms are employed to quantify the cost of modifications, notably including the l_1 norm, l_∞ norm, bottleneck Hamming distance, and weighted sum Hamming distance. However, existing research has overlooked the scenario involving the simultaneous application of both l_∞ and weighted sum Hamming distances. This section aims to address this gap by measuring the upgrade cost with both l_∞ and weighted sum Hamming distances, as described as follows:

$$\begin{aligned} & \max_{\tilde{w}} \tilde{w}(T) := \sum_{t_k \in Y} \tilde{w}(P_k) \\ (\textbf{DIT}_{H\infty}) \ & s.t. \ \min_{t_k \in Y} \tilde{w}(P_k) \ge M, \\ & \max_{e \in E} c(e)(\tilde{w}(e) - w(e)) \le K, \\ & \sum_{e \in E} r(e) H(\tilde{w}(e), w(e)) \le N, \\ & w(e) \le \tilde{w}(e) \le u(e), \quad e \in E. \end{aligned} \tag{7.34}$$

where N is a given positive value.

Its related minimum cost problem obtained by exchanging the l_∞ norm cost and the SRD objective function can be written as

$$\begin{aligned} & \min_{\tilde{w}} C(\tilde{w}) := \max_{e \in E} c(e)(\tilde{w}(e) - w(e)) \\ (\textbf{MCDIT}_{H\infty}) \ & \text{s.t.} \ \min_{t_k \in Y} \tilde{w}(P_k) \ge M, \\ & \sum_{t_k \in Y} \tilde{w}(P_k) \ge D, \\ & \sum_{e \in E} r(e) H(\tilde{w}(e), w(e)) \le N, \\ & w(e) \le \tilde{w}(e) \le u(e), \quad e \in E. \end{aligned} \tag{7.35}$$

In the form of our uniform format of inverse optimizations, we denote the problems **(DIT$_{H\infty}$)** and **(MCDIT$_{H\infty}$)** by **((BC-)Double-Int, SRD-Trees, Bounded, wsH-l_∞)**.

7.4.1 *(BC-Double-Int, SRD-Trees, Bounded, wsH-l_∞)*

We first showed the $\mathcal{NP}$-Hardness of problem (DIT$_{H\infty}$), then proposed a binary search algorithm to solve the problem (DIT$_{H\infty}$), in which a dynamic programming algorithm is provided to solve a Combination Interdiction problem on Trees $(CIT^{\lambda}_{H\infty})$. Hence, we obtain a pseudo-polynomial time algorithm.

7.4.1.1 The $\mathcal{NP}$-Hardness of Problem (DIT$_{H\infty}$)

Note that the weighted sum Hamming distance is discrete, posing challenges in its treatment. To gain a clearer understanding of problem (DIT$_{H\infty}$), we initially examine its relaxation (DIT$_\infty$) by removing the constraint of weighted sum Hamming distance.

In the problem **(DIT$_\infty$)**, we can maximize the weight of each edge to the greatest extent possible under the constraint of cost K and the upper bound $u(e)$ as follows.

$$\bar{w}(e) = \min\left\{w(e) + \frac{K}{c(e)}, u(e)\right\} (e \in E). \tag{7.36}$$

If $\min_{t_k \in Y} \bar{w}(P_k) < M$, it implies problem **(DIT$_\infty$)** is infeasible. Otherwise, the solution $\bar{w}$ by (7.36) is optimal for the problem (DIT$_\infty$) and thus can be obtained in $O(n)$ time.

Building upon the discussion before, we understand that under weighted l_∞ norm, it is advantageous to upgrade an edge to its upper bound determined by (7.36). Consequently, the problem **(DIT$_{H\infty}$)** can be reformulated into a new 0-1 Integer Linear Programming model **(GDIT$_{H\infty}$)**, where the binary decision variable $x(e)$ is assigned to each edge $e \in E$ to denote e is considered upgraded or not.

$$\max \sum_{e\in E} In_{\bar{w}}(e)x(e)$$

$$\textbf{(GDIT}_{H\infty}\textbf{)} \text{ s.t.} \sum_{e\in P_k} (\bar{w}(e) - w(e))x(e) \geq M - w(P_k), \forall k = 1, \ldots, r. \tag{7.37a}$$

$$\sum_{e\in E} r(e)x(e) \leq N, \tag{7.37b}$$

$$x(e) \in \{0, 1\}, e \in E,$$

where

$$In_{\bar{w}}(e) := |L(e)|(\bar{w}(e) - w(e)). \tag{7.38}$$

Theorem 7.20 ([82]) *The problem* (DIT$_{H\infty}$) *is* $\mathcal{NP}$*-hard.*

7.4.1.2 Solve Problem (DIT$_{H\infty}$) When Upgrading One Edge

We denote the problem (DIT$_{H\infty}$) by (DIT$^1_{H\infty}$) when $N = 1$ and $r(e) = 1$ for all $e \in E$, which aims to modify a single critical edge to maximize SRD while ensuring compliance with the StRD constraint. To approach this task, we initially set forth some necessary definitions. Subsequently, we employ a greedy algorithm to efficiently address the problem, achieving a solution in $O(n)$ time complexity.

Definition 7.4 For any $e^* \in E$, we define the upgrade vector w_{e^*} as follows.

$$w_{e^*}(e) = \begin{cases} \bar{w}(e), & \text{if } e = e^* \\ w(e), & \text{otherwise.} \end{cases} \tag{7.39}$$

Theorem 7.21 ([82]) *If* $\min_{t_k \in Y} w(P_k) \geq M$, *let* $e^* := \arg\max_{e \in E} In_{w_e}(e)$, *the optimal solution of the problem (DITH$^1_\infty$) is* w_{e^*}.

Theorem 7.22 ([82]) *If* $\min_{t_k \in Y} w(P_k) < M$, *let* $k^* := \arg\min_{t_k \in Y} w(P_k)$, *then there must be an optimal solution* w_{e^*} *for some* $e^* \in P_{t_{k^*}}$.

Based on Theorem 7.22, we considered different cases to find the exact critical edge in each case. Therefore, we have Algorithm 7.11 to solve (DIT$^1_{H\infty}$).

Algorithm 7.11 A greedy algorithm to solve (DIT$^1_{H\infty}$) [82]

Input: A tree T rooted at s, two edge weight vectors w and u, two values K and M.
Output: The optimal value $w_{e^*}(T)$, the upgraded edge e^* and the optimal solution w_{e^*}.

1: Calculate $\bar{w}$ as in (7.36). Let $k^* := \arg\min_{t_k \in Y} w(P_k)$.
2: **if** $\min_{t_k \in Y} w(P_k) \geq M$ **then return** $e^* := \arg\max_{e \in E} In_{w_e}(e)$, $w_{e^*}(T)$ and w_{e^*}.
3: **else if** $\max_{e \in P_{k^*}} (\bar{w}(e) - w(e)) \geq M - w(P_{k^*})$ **then**
4: Let $Y_1 := \{t_k \in Y | w(P_k) < M\}$.
5: Let $B := \{e \in P_{k^*} | (\bar{w}(e) - w(e)) \geq M - w(P_{k^*}), Y_1 \subseteq L(e)\}$.
6: **if** $|Y_1| = 1$ **then return** $e^* := \arg\max_{e \in B} In_{w_e}(e)$, $w_{e^*}(T)$ and w_{e^*}.
7: **else if** $B \neq \emptyset$ **then return** $e^* := \arg\max_{e \in B} In_{w_e}(e)$, $w_{e^*}(T)$ and w_{e^*}.
8: **else return** No feasible solution.
9: **end if**
10: **else return** No feasible solution.
11: **end if**

Theorem 7.23 ([82]) *Algorithm 7.11 solves the problem (DIT$^1_{H\infty}$) in O(n) time.*

For the general form of problem ($\mathrm{DIT}_{H\infty}$), the complexity of the situation escalates significantly. This is due to the necessity to identify crucial edges that not only comply with the StRD constraint but also maximize the SRD value. To achieve a balance between the StRD and SRD, we introduce the Combination Interdiction problem on Trees ($CIT^{\lambda}_{H\infty}$), which employs a convex combination of these two factors and $0 \leq \lambda \leq 1$ is a parameter.

$$
(\mathbf{CIT}^{\lambda}_{H\infty}) \quad
\begin{aligned}
&\max_{\hat{w}} \left\{ (1-\lambda) \sum_{e \in E'} |L(e)| \hat{w}(e) + \lambda \min_{t_k \in \cup_{i=p}^{q} L(v_i)} \hat{w}(P(v, t_k)) \right\} \\
&s.t. \sum_{e \in E'} r(e) H(\hat{w}(e), w(e)) \leq N, \\
&\qquad w(e) \leq \hat{w}(e) \leq \bar{w}(e), \quad e \in E'.
\end{aligned}
\tag{7.40}
$$

To solve problem ($\mathrm{DIT}_{H\infty}$), we first propose a dynamic programming algorithm to solve ($CIT^{\lambda}_{H\infty}$), then develop a binary search algorithm for ($\mathrm{DIT}_{H\infty}$) based on their connection.

7.4.1.3 A Dynamic Programming Algorithm to Solve ($CIT^{\lambda}_{H\infty}$)

We begin by laying down foundational definitions (Fig. 7.1).

Definition 7.5 ([154]) Define $T_v = (V_v, E_v)$ as the subtree rooted at v for any $v \in V$. For any non-leaf node v, let $S(v) = \{v_{s_1}, \ldots, v_{s_{|S(v)|}}\}$ be the set of sons of v, where v_{s_q} is the q-th son. Define *the left q-subtree* of v as $T_v(q) := T_{v_{s_q}} \cup P(v, v_{s_q})$ and *the $p : q$-subtree* of v as $T_v(p : q) = \left(\bigcup_{i=p}^{q} T_v(i)\right) \cup \{v\}$. Specially, define $T_v(p : 0) = \emptyset$.

Definition 7.6 Define $h(v, p : q, N, \lambda)$ the optimal value and $\hat{w}_{(v,p:q,N,\lambda)}$ an optimal solution to the problem ($CIT^{\lambda}_{H\infty}$). Specifically, define $h(v, p : q, N, \lambda) = +\infty$, if $p > q$ or $N < 0$. And define the StRD and SRD under $h(v, p :$

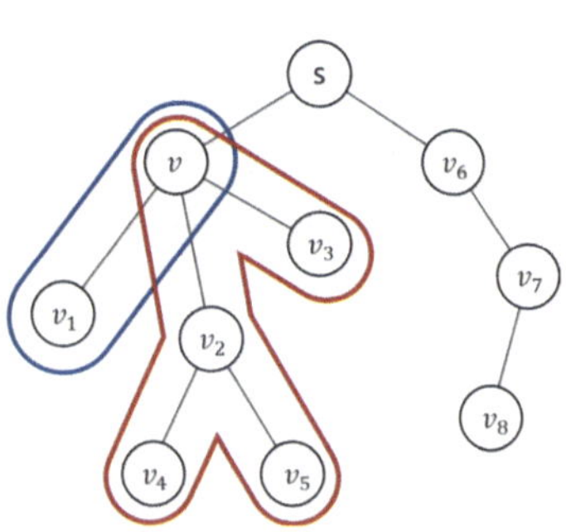

Fig. 7.1 The area labeled in blue and red are the subtrees $T_v(1)$ and $T_v(2 : 3)$, respectively [90]

$q, N, \lambda)$ as $SP_{(v,p:q,N,\lambda)} = \min_{t_k \in \cup_{i=p}^{q} L(v_{s_i})} \hat{w}(P(v, t_k))$ and $SRD_{(v,p:q,N,\lambda)} = \sum_{e \in E'} |L(e)|\hat{w}(e)$, respectively.

Through these definitions, we can perform dynamic programming methods to solve problem $(CIT_{H\infty}^{\lambda})$. Upon solving $(CIT_{H\infty}^{\lambda})$, we can demonstrate the existence of an optimal parameter λ^* such that the solution to $(CIT_{H\infty}^{\lambda})$ perfectly aligns with the solution to the general problem $(DIT_{H\infty})$.

Suppose $T_v(p : q) = (V', E')$. We define a transition equation based on the structure of left subtrees. We first show the state transition from tree $T_v(p, p)$ to its subtree $T_{v_{s_p}}(1 : |S(v_{s_p})|)$.

Theorem 7.24 ([82]) *For any non-leaf node* v *and any integer* $k \geq 1$*, let* $e_{s_p} := (v, v_{s_p})$*, then we have*

$$h(v, p : p, k, \lambda) = \max\{h\left(v_{s_p}, 1 : |S(v_{s_p})|, k, \lambda\right) + \lambda w(e_{s_p}) + (1-\lambda)In_w(e_{s_p}),$$
$$h\left(v_{s_p}, 1 : |S(v_{s_p})|, k - r(e_{s_p}), \lambda\right) + \lambda \bar{w}(e_{s_p}) + (1-\lambda)In_{\bar{w}}(e_{s_p})\} \tag{7.41}$$

Next, we show that the problem $(CIT_{H\infty}^{\lambda})$ defined on $T_v(p : q)$ can be divided into two sub-trees $T_v(p : l)$ and $T_v(l + 1 : q)$ with the following theorem.

Theorem 7.25 ([82]) *Suppose that* v *is a non-leaf node. For any child node index* p, l, q *satisfying* $p \leq l < r$*, then the optimal value of the problem* $(CIT_{H\infty}^{\lambda})$ *defined on* $T_v(p : q)$ *can be calculated by*

$$h(v, p : q, k, \lambda) = \max_{k_1+k_2 \leq k} \{\lambda \min\{SP_{(v,p:l,k_1,\lambda)}, SP_{(v,l+1:q,k_2,\lambda)}\}$$
$$+ (1-\lambda)(SRD_{(v,p:l,k_1,\lambda)} + SRD_{(v,l+1:q,k_2,\lambda)})\}. \tag{7.42}$$

Using the two theorems aforementioned, we are able to calculate function h in this two step.

$$h(v, 1 : p, N, \lambda) = \max_{0 \leq N_1 \leq N, N_1 \in \mathcal{Z}^+} \{\lambda \min\{SP_{(v,1:p-1,N_1,\lambda)}, SP_{(v,p:p,N-N_1,\lambda)}\}$$
$$+ (1-\lambda)(SRD_{(v,1:p-1,N_1,\lambda)} + SRD_{(v,p:p,N-N_1,\lambda)})\} \tag{7.43}$$

$$h(v, p : p, N - N_1, \lambda)$$
$$= \max\{h\left(v_{s_p}, 1 : |S(v_{s_p})|, N - N_1, \lambda\right) + \lambda w(e_{s_p}) + (1-\lambda)|L(e_{s_p})|w(e_{s_p}), \tag{7.44}$$
$$h\left(v_{s_p}, 1 : |S(v_{s_p})|, N - N_1 - r(e_{s_p}), \lambda\right) + \lambda \bar{w}(e_{s_p}) + (1-\lambda)|L(e_{s_p})|\bar{w}(e_{s_p})\}.$$

Throughout the discussion earlier, we propose Algorithm 7.12 to solve the problem $(CIT_{H\infty}^{\lambda})$, where we need to call Depth-First Search algorithm **DFS**(s, $1 : |S(s)|, N, \lambda$) recursively [33].

Algorithm 7.12 A dynamic programming algorithm to solve $(CIT^{\lambda}_{H\infty})$ [82]

Input: A tree T rooted at s, two values N, K, two weight vectors w and u, a parameter λ.
Output: The optimal value $h(s, 1 : |S(s)|, N, \lambda)$, an optimal solution $\hat{w}_{(s,1:|S(s)|,N,\lambda)}$.

Call **DFS**$(s, 1 : |S(s)|, N, \lambda)$,
Return $h(s, 1 : |S(s)|, N, \lambda)$, $\hat{w}_{(s,1:|S(s)|,N,\lambda)}$.
function DFS$(v, p : q, N, \lambda)$
  if $v \in L(s)$ or $q < p$ or $N < 0$ **then**
    Return $SP_{(v,p:p,N,\lambda)} := 0$, $SRD_{(v,p:p,N,\lambda)} := 0$, $h(v, p : p, N, \lambda) := 0$, $E_{(v,p:p,N,\lambda)} := \emptyset$, $\hat{w}_{(v,p:p,N,\lambda)} := w$.
  end if
  if $p = q$ **then**
    Calculate **DFS**$(v, p : p, N, \lambda)$, **DFS**$(v_{s_p}, p : p, N - r(e_{s_p}), \lambda)$.
    Let $h_1 := h\left(v_{s_p}, 1 : |S(v_{s_p})|, N - N_1, \lambda\right) + \lambda w(e_{s_p}) + (1-\lambda)|L(e_{s_p})|w(e_{s_p})$, $h_2 := h\left(v_{s_p}, 1 : |S(v_{s_p})|, N - N_1 - r(e_{s_p}), \lambda\right) + \lambda \bar{w}(e_{s_p}) + (1-\lambda)|L(e_{s_p})|\bar{w}(e_{s_p})$.
    if $h_1 < h_2$ **then return** $SP_{(v,p:p,N,\lambda)} := SP_{(v_{s_p},1:|S(v_{s_p})|,N-r(e_{s_p}),\lambda)} + \bar{w}(e_{s_p})$,
      $SRD_{(v,p:p,N,\lambda)} := SRD_{(v_{s_p},1:|S(v_{s_p})|,N-r(e_{s_p}),\lambda)} + |L(e_{s_p})|\bar{w}(e_{s_p})$,
      $h(v, p : p, N, \lambda) := h_2$, $E_{(v,p:p,N,\lambda)} := E_{(v_{s_p},1:|S(v_{s_p})|,N-r(e_{s_p}),\lambda)} \cup \{e_{s_p}\}$,
      $\hat{w}_{(v,p:p,N,\lambda)}(e) := \begin{cases} \bar{w}(e), e \in E_{(v,p:p,N,\lambda)}, \\ w(e), otherwise. \end{cases}$
    else return $SP_{(v,p:p,N,\lambda)} := SP_{(v_{s_p},1:|S(v_{s_p})|,N,\lambda)} + w(e_{s_p})$,
      $SRD_{(v,p:p,N,\lambda)} := SRD_{(v_{s_p},1:|S(v_{s_p})|,N,\lambda)} + |L(e_{s_p})|w(e_{s_p})$,
      $h(v, p : p, N, \lambda) := h_1$, $E_{(v,p:p,N,\lambda)} := E_{(v_{s_p},1:|S(v_{s_p})|,N,\lambda)}$,
      $\hat{w}_{(v,p:p,N,\lambda)}(e) := \begin{cases} \bar{w}(e), e \in E_{(v,p:p,N,\lambda)}, \\ w(e), otherwise. \end{cases}$
    end if
  else
    for $N_1 = 0$ to N **do**
      Calculate **DFS**$(v, 1 : p-1, N_1, \lambda)$, **DFS**$(v, p : p, N - N_1, \lambda)$.
    end for
    Calculate $h(v, 1 : p, N, \lambda)$ by (7.43) whose maximum achieved at $N_1 := N_1^*$.
    Return $SP_{(v,1:p,N,\lambda)} := \min\{SP_{(v,1:p-1,N_1^*,\lambda)}, SP_{(v,p:p,N-N_1^*,\lambda)}\}$,
    $SRD_{(v,1:p,N,\lambda)} := SRD_{(v,1:p-1,N_1^*,\lambda)} + SRD_{(v,p:p,N-N_1^*,\lambda)}$,
    $h(v, 1 : p, N, \lambda)$, $E_{(v,1:p,N,\lambda)} := E_{(v,1:p-1,N_1^*,\lambda)} \cup E_{(v,p:p,N-N_1^*,\lambda)}$,
    $\hat{w}_{(v,1:p,N,\lambda)}(e) := \begin{cases} \bar{w}(e), e \in E_{(v,1:p,N,\lambda)}, \\ w(e), otherwise. \end{cases}$
  end if
end function

Theorem 7.26 ([82]) *Algorithm 7.12 can solve problem* $(CIT^{\lambda}_{H\infty})$ *in* $O(nN^2)$ *time, which incurs pseudo-polynomial time complexity.*

7.4.1.4 A Binary Search Algorithm to Solve Problem (DIT$_{H\infty}$)

To solve problem (DIT$_{H\infty}$), we first analyse the connection between problem $(CIT^{\lambda}_{H\infty})$ and (DIT$_{H\infty}$), then propose the monotonicity theorem of problem $(CIT^{\lambda}_{H\infty})$, based on which, we develop a binary search algorithm.

Note that by setting $\lambda = 1$ or 0, we are able to derive optimal solutions for two planning problems: maximizing the StRD and maximizing SRD, subject to upper bound and weighted sum Hamming distance constraints, respectively. Then the following theorem concerning infeasibility can be obtained.

Theorem 7.27 ([82]) *If $h(s, 1 : |S(s)|, N, 1) < M$, then ($DIT_{H\infty}$) is infeasible.*

Analysing problem ($\text{DIT}_{H\infty}$), we seek to establish a lower bound for the StRD while simultaneously maximizing SRD. Given that the objective function h, comprises two components, we observe that as the parameter λ varies from 1 to 0, the contribution of the StRD to h steadily diminishes, while the contribution of the SRD progressively increases. Despite the inherent discontinuity imposed by the weighted sum Hamming distance constraints, as λ changes, the algorithm becomes biased in the selection of update edges, which allows us to draw the following theorems.

Theorem 7.28 ([82]) *If $1 \geq \lambda_1 \geq \lambda_2 \geq 0$, then $SP_{(s,1:|S(s)|,N,\lambda_1)} \geq SP_{(s,1:|S(s)|,N,\lambda_2)}$, $SRD_{(s,1:|S(s)|,N,\lambda_1)} \leq SRD_{(s,1:|S(s)|,N,\lambda_2)}$.*

Theorem 7.29 ([82]) *If $h(s, 1 : |S(s)|, N, 1) \geq M$, then the optimal solution of ($DIT_{H\infty}$) is given by $\hat{w}_{(s,1:|S(s)|,N,\lambda^*)}$, where λ^* is the critical value with $0 \leq \lambda^* \leq 1$.*

Finally, we present Algorithm 7.13 to solve ($\text{DIT}_{H\infty}$), encapsulating our findings. This algorithm leverages a binary search method to determine the critical value λ^* corresponding to Theorem 7.29. Then we call Algorithm 7.12 to solve problem ($\text{CIT}_{H\infty}^{\lambda^*}$), in which **DFS**$(s, 1 : |S(s)|, N, \lambda^*)$ is utilized.

Algorithm 7.13 A binary search algorithm to solve ($\text{DIT}_{H\infty}$) [82]

Input: A tree T rooted at s, two given values N, K, two edge weight vectors w and u.
Output: An optimal solution $\tilde{w}$, the optimal value $\tilde{w}(T)$.

Let $U := \min_{t_k \in Y} u(P_k) + \sum_{e \in E} |L(e)|u(e)$.
Calculate $\bar{w}$ by (7.36), let $p^1 := SP_{(s,1:|S(s)|,N,1)}$ by calling **DFS**$(s, 1 : |S(s)|, N, 1)$ in Algorithm 7.12.
if $p^1 < M$ **then return** "No feasible solution."
else
    Let $lr := 0, rr := 1$.
    while $rr - lr > \frac{1}{U^2}$ **do**
        Let $mid := \frac{(lr+rr)}{2}$, $p^{mid} := SP_{(s,1:|S(s)|,N,mid)}$ by **DFS**$(s, 1 : |S(s)|, N, mid)$ in Algorithm 7.12.
        if $p^{mid} > M$ **then** let $rr := mid$
        else let $lr := mid$
        end if
    end while
    Let $\lambda^* := rr$. Call **DFS**$(s, 1 : |S(s)|, N, \lambda^*)$.
    Return $\tilde{w} := \hat{w}_{(s,1:|S(s)|,N,\lambda^*)}$, $\tilde{w}(T) := SRD_{(s,1:|S(s)|,N,\lambda^*)}$.
end if

Theorem 7.30 ([82]) *Let $U := \min_{t_k \in Y} u(P_k) + \sum_{e \in E} |L(e)|u(e)$. When $N \geq 1$, Algorithm 7.13 can solve problem ($DIT_{H\infty}$) within $O(nN^2 \log_2 U)$ time.*

7.4.2 *(Double-Int, SRD-Trees, Bounded, wsH-l_∞)*

7.4.2.1 The $\mathcal{NP}$-Hardness of (MCDITH) under the Weighted l_∞ Norm

Next, we consider its related minimum cost problem (MCDIT$_{H\infty}$) by exchanging the objective function and the l_∞ norm of problem (DIT$_{H\infty}$), which is formulated in (7.35). To prove (MCDIT$_{H\infty}$) is $\mathcal{NP}$-hard, it suffices to show that its decision version (DMCDIT$_{H\infty}$) is $\mathcal{NP}$-complete. Typically, proving a decision problem is $\mathcal{NP}$-complete involves two steps: first, demonstrating that the decision version is in $\mathcal{NP}$; second, showing that the decision version can be reduced from a problem already proven to be $\mathcal{NP}$-complete. For our purposes, we choose the decision version of the 0-1 knapsack problem as the original problem [74]. This leads to the following theorem.

Theorem 7.31 ([82]) *The problem (MCDIT$_{H\infty}$) is $\mathcal{NP}$-hard.*

7.4.2.2 A Pseudo-Polynomial Time Algorithm to Solve Problem (MCDIT$_{H\infty}$)

For problem (MCDIT$_{H\infty}$), the objective function, which represents the maximum cost under the l_∞ norm, is actually constrained within the interval $[K^1, K^2]$, where $K^1 = 0$, and $K^2 = \max_{e \in E} c(e)(u(e) - w(e))$. Upon investigating the interconnection between problems (DIT$_{H\infty}$) and (MCDIT$_{H\infty}$), it is discovered that the latter can be addressed by sequentially solving (DIT$_{H\infty}$) to ascertain the minimum K^* such that $\tilde{w}(T) > D$. Moreover, in problem (DIT$_{H\infty}$), it is observed that as the value of K ascends, the optimal SRD value, $\tilde{w}(T)$ also exhibits a monotonic increase.

To find optimal K^* within the defined interval $[K^1, K^2]$, a binary search Algorithm is employed. Each phase of the iteration leverages Algorithm 7.13 for the computation of $\tilde{w}(T)$, thus iteratively pinpointing K^*. Consequently, we have the following theorem.

Theorem 7.32 ([82]) *Problem (MCDIT$_{H\infty}$) can be solved by Algorithm 7.14 within $O(nN^2 \log_2 K \log_2 U)$ time.*

Algorithm 7.14 A pseudo-polynomial time algorithm for (MCDIT$_{H\infty}$) [82]

Input: A tree T rooted at s, a given value N, two edge weight vectors w and u.
Output: Minimal cost value K^*, a set of upgrade edges E^*, and the optimal solution $\tilde{w}$.
Let $K^1 := 0$ and $K^2 := \max_{e \in E} c(e)(u(e) - w(e))$.
Let y the optimal SRD value of DIT$_{H\infty}(T, N, K^2, w, u)$
if $y < D$ **then**
  Return "No feasible solution."
else
  while $K^2 - K^1 > 1$ **do**
    Let $mid := \frac{(K^2+K^1)}{2}$, SRD^{mid} the SRD returned by DIT$_{H\infty}(T, N, mid, w, u)$.
    if $SRD^{mid} > D$ **then** let $K^2 := mid$

else let $K^1 := mid$
end if
end while
Return K^2, the upgrade edge set E^* returned by $\text{DIT}_{H\infty}(T, N, K^2, w, u)$, the optimal solution $\tilde{w}$ returned by $\text{DIT}_{H\infty}(T, N, K^2, w, u)$.
end if

7.5 Conclusions

This chapter concludes by addressing the sum of root-leaf distance interdiction problem on trees (SRDITC) and its associated minimal cost problem (MCSRDITC) under the l_∞ norm and bottleneck Hamming distance. Two binary search algorithms are proposed, both operating in $O(n \log n)$ time, for solving the SRDITC under the weighted l_∞ norm and bottleneck Hamming distance, respectively. For the (MCSRDITC), two binary search algorithms are introduced, with time complexities of $O(Nn^2)$ and $O(n \log n)$ for the weighted l_∞ norm and bottleneck Hamming distance, respectively.

Subsequently, a shortest path constraint is incorporated, leading to the formulation of the double interdiction problem ($\text{DIT}_{H\infty}$) and its minimal cost counterpart. These extended problems are proven to be $\mathcal{NP}$-hard. However, by employing a combination of dynamic programming and binary search techniques, we successfully devise solutions that operate in pseudo-polynomial time. This approach provides a robust method for navigating and addressing complex interdiction challenges in tree-structured data.

Chapter 8
Restricted Inverse Optimal Value Problem on Shortest Path Under Weighted l_1 Norm on Trees

Abstract We introduce the restricted inverse shortest path problems under weighted l_1 norm on trees. It aims at adjusting the weights of some edges to minimize the total cost under weighted l_1 norm on the premise that the length of the shortest root-leaf path of the tree is lower-bounded by a given value D, which is just the restriction on the length of a given root-leaf path P_0. We solve this problem in $O(n^2)$ time through a series of subproblems based on searching for a minimum cost cut.

Keywords Restricted inverse optimal value problem · Shortest path · Weighted l_1 norm · Trees · Minimum cost cut

8.1 (RIOV, SP-Trees, Bounded, Weighted l_1 Norm)

Let $T = (V, E, w)$ be an edge-weighted tree rooted at v_1, where $V := \{v_1, v_2, \cdots, v_n\}$ and $E := \{e_2, e_3, \cdots, e_n\}$ are the sets of nodes and edges, respectively. Let $Y := \{t_0, t_1, \cdots, t_{r-1}\}$ be the set of leaves. Let $w(e)$, $l(e)$, and $u(e)$ be the original weight, the lower and upper bounds of the adjusted weight for an edge $e \in E$, respectively, where $l(e) \leq w(e) \leq u(e)$. Let $c(e) > 0$ be a cost to adjust the weight of edge $e \in E$. Denoted by P_{v_i,v_j} the unique path from v_i to v_j on the tree T and record $P_k := P_{v_1,t_k}$ as the "root-leaf path" of the leaf node $t_k (k = 0, 1, 2, \cdots, r-1)$ in brief. The restricted inverse optimal value problem on shortest path on trees **(RIOVSPT)** aims at adjusting some weights of edges to minimize the total cost under some measurement such that the length of the shortest root-leaf path is lower-bounded by a given value D, which is just the restriction of the length of the given path P_0. The mathematical model can be stated as follows:

$$\begin{aligned} &\min_{\bar{w}} \; \|\bar{w} - w\| \\ \textbf{(RIOVSPT)} \quad &s.t. \quad \bar{w}(P_k) \geq D, t_k \in Y \setminus \{t_0\}, \end{aligned}$$

Qiao Zhang and Xiucui Guan have equally contributed to this chapter.

X. Guan et al., *Inverse Combinatorial Optimization Problems*, Springer Optimization and Its Applications 225, https://doi.org/10.1007/978-3-031-91175-0_8

$$\bar{w}(P_0) = D,$$
$$l(e) \leq \bar{w}(e) \leq u(e), e \in E.$$

The problem **(RIOVSPT$_1$)** under weighted l_1 norm can be formulated as follows.

$$\text{(RIOVSPT}_1\text{)} \quad \begin{aligned} \min_{\bar{w}} \; & C(\bar{w}) := \sum_{e \in E} c(e)|\bar{w}(e) - w(e)| \\ s.t. \; & \bar{w}(P_k) \geq D, t_k \in Y \setminus \{t_0\}, \\ & \bar{w}(P_0) = D, \\ & l(e) \leq \bar{w}(e) \leq u(e), e \in E. \end{aligned} \tag{8.1}$$

8.2 Some Important Properties of the Problem (RIOVSPT$_1$)

We present a sufficient and necessary condition to check the feasibility of the problem **(RIOVSPT$_1$)**.

Theorem 8.1 ([152, Theorem 2]) *The problem* ***(RIOVSPT$_1$)*** *is feasible if and only if **1)** $l(P_0) \leq D \leq \min_{t_k \in Y} u(P_k)$ and **2)** $l(P_0) \leq \min_{t_k \in Y}(l(P_k \cap P_0) + u(P_k \backslash P_0))$ hold.*

Denote by $x(e) \geq 0$ and $y(e) \geq 0$ the increament and the decrement of an edge $e \in E$, respectively. Notice that the length of an edge e cannot increase and decrease at the same time, then $x(e) > 0$, $y(e) > 0$ cannot hold simultaneously. Hence, $\bar{w}(e) - w(e) = x(e) - y(e)$, and $|\bar{w}(e) - w(e)| = x(e) + y(e)$ for $e \in E$. Therefore, the model (8.1) is equivalent to the following model (8.2).

$$\text{(RIOVSPT}_1\text{)} \quad \begin{aligned} \min_{x,y} \; & \sum_{e \in E} c(e)\big(x(e) + y(e)\big) \\ s.t. \; & x(P_k) - y(P_k) \geq D - w(P_k), t_k \in Y \backslash \{t_0\}, \\ & x(P_0) - y(P_0) = D - w(P_0), \\ & 0 \leq x(e) \leq \Delta u(e), e \in E, \\ & 0 \leq y(e) \leq -\Delta l(e), e \in E. \end{aligned} \tag{8.2}$$

In Sect. 6.3, we developed a primal-dual algorithm in $O(n^2)$ time to solve the shortest path interdiction problem **(Int-SPT$_1$)** by upgrading edges under l_1 norm on trees, in which we ignore the restriction on the path P_0 in the problem **(RIOVSPT$_1$)**. The mathematical model of the problem **(Int-SPT$_1$)** with respect to $\tilde{w}$ can be stated as follows.

$$\begin{aligned}
&\min_{\tilde{w}} \ C(\tilde{w}) := \sum_{e\in E} c(e)|\tilde{w}(e) - w(e)| \\
\textbf{(Int} - \textbf{SPT}_1\textbf{)} \quad & s.t. \quad \tilde{w}(P_k) \geq D, t_k \in Y, \\
& \qquad w(e) \leq \tilde{w}(e) \leq u(e), e \in E.
\end{aligned} \tag{8.3}$$

To ensure the uniqueness of the solution, we define a canonical optimal solution[152, Definition 6-7,Theorem 6] of the problem **(Int-SPT$_1$)**, as well as the problem **(RIOVSPT$_1$)**, as the one whose edge-weight reduction or increment occurs as close to the root v_1 as possible in the optimal solutions.

For any edge $e_j \in E$, we define $T_{e_j} := P_{v_1,v_j} \cup T_{v_j}$. Let $\ddot{e} = (v_1, \ddot{v}) \in P_0$ be the **gate** edge, which is the first edge from the root v_1 on the path P_0. Denote by $\ddot{T} := T_{\ddot{e}} := \{\ddot{e}\} \cup T_{\ddot{v}}$ in brief. Then the tree $T = (T\backslash\ddot{T}) \cup \ddot{T}$ can be divided into two parts. Meanwhile, the problem **RIOVSPT$_1$**(T) can also be divided into two independent subproblems **RIOVSPT$_1$**$(T\backslash\ddot{T})$ and **RIOVSPT$_1$**$(\ddot{T})$, where the optimal solution of the former subproblem can be obtained from an optimal solution $\tilde{w}$ of the problem **Int-SPT$_1$**$(T\backslash\ddot{T})$. Then an an optimal solution of the problem **(RIOVSPT$_1$)** can be generated by the optimal solutions of the aforementioned two subproblems[152, Theorem 5].

we propose an important property of the optimal solutions of the problems **(Int-SPT$_1$)** and **RIOVSPT$_1$**$(\ddot{T})$.

Theorem 8.2 ([152, Theorem 4,11]) *Let $\tilde{w}$ be an optimal solution of the problem **(Int-SPT$_1$)** and $\ddot{w}$ be a canonical optimal solution of the problem **RIOVSPT$_1$**$(\ddot{T})$.*

*(1) If $\tilde{w}(P_0) = D$, then $\bar{w} := \tilde{w}$ is also an optimal solution of the problem **(RIOVSPT$_1$)**.*

*(2) If $\tilde{w}(P_0) > D$, then **1)** $\ddot{w}(e) \leq \tilde{w}(e)$ for $e \in P_0$ and **2)** $\ddot{w}(e) \geq \tilde{w}(e)$ for $e \in \ddot{T}\backslash P_0$.*

Suppose $\tilde{w}(P_0) > D$. Let

$$\Delta\tilde{u}(e) := \begin{cases} +\infty, & e \in P_0 \\ u(e) - \tilde{w}(e), & e \notin P_0 \end{cases}, \quad \Delta\tilde{l}(e) := \begin{cases} +\infty, & e \notin P_0 \\ l(e) - \tilde{w}(e), & e \in P_0 \end{cases},$$

$$E_l := \{e \in P_0 | \Delta l(e) = 0\}, \quad E_u := \{e \notin P_0 | \Delta\tilde{u}(e) = 0\},$$

$$\tilde{c}(e) := \begin{cases} +\infty, & e \in E_l \cup E_u \\ c(e), & otherwise \end{cases}.$$

Then the model (8.2) of the problem **RIOVSPT$_1$**$(\ddot{T})$ is equivalent to the following form based on **Theorem 8.2**.

$$\begin{aligned}
&\min_{x,y} \sum_{e\notin P_0} \tilde{c}(e)x(e) + \sum_{e\in P_0} \tilde{c}(e)y(e) \\
&s.t. \quad x(P_k\backslash P_0) - y(P_k \cap P_0) \geq D - \tilde{w}(P_k), t_k \subset Y\backslash\{t_0\},
\end{aligned}$$

$$\textbf{RIOVSPT}_1(\ddot{T}) \qquad -y(P_0) = D - \tilde{w}(P_0), \tag{8.4}$$
$$0 \leq x(e) \leq \Delta\tilde{u}(e), \quad e \notin P_0,$$
$$0 \leq y(e) \leq -\Delta\tilde{l}(e), \quad e \in P_0.$$

8.3 Construct a Slack Problem ($\mathbf{RIOVSPT_1^s}$)

Then we reconstruct the subtree $\ddot{T}$ by the following double-edge technique.

Double-edge technique: [152] Let $\tilde{P}_0 = \{e_b \in P_0 | \tilde{w}(e_b) > w(e_b)\}$. Replace each edge $e_b = (v_a, v_b) \in \tilde{P}_0$ by two edges $e_b^1 := (v_a, v_b^1)$ and $e_b^2 := (v_b^1, v_b)$. Then let $\tilde{P}_0^1 = \{e_b^1 | e_b \in \tilde{P}_0\}$ and $\tilde{P}_0^2 = \{e_b^2 | e_b \in \tilde{P}_0\}$. Denote the resulting paths by $\breve{P}_k$ for $t_k \in \breve{Y} := Y \cap \ddot{T}$ and **the auxiliary tree** by $\breve{T}$. Let

$$\breve{w}(e) := \begin{cases} \tilde{w}(e) - w(e), & e \in \tilde{P}_0^1 \\ w(e), & e \in \tilde{P}_0^2 \\ \tilde{w}(e), & otherwise \end{cases}, \quad \breve{c}(e) := \begin{cases} -\tilde{c}(e), & e \in \tilde{P}_0^1 \\ \tilde{c}(e), & otherwise \end{cases},$$

$$\Delta\breve{u}(e) := \begin{cases} +\infty, & e \in \tilde{P}_0^1 \cup \tilde{P}_0^2 \\ \Delta\tilde{u}(e), & otherwise \end{cases}, \quad \Delta\breve{l}(e) := \begin{cases} w(e) - \tilde{w}(e), & e \in \tilde{P}_0^1 \\ \Delta\tilde{l}(e), & otherwise \end{cases}.$$

Now we consider a slack problem ($\mathbf{RIOVSPT_1^s}$) defined on **the auxiliary tree** $\breve{T}$, where the length of the path $\breve{P}_0$ is greater than D.

$$\min_{\breve{x},\breve{y}} \sum_{e \notin \breve{P}_0} \breve{c}(e)\breve{x}(e) + \sum_{e \in \breve{P}_0} \breve{c}(e)\breve{y}(e)$$
$$(\mathbf{RIOVSPT_1^s}) \quad s.t. \quad \breve{x}(\breve{P}_k \backslash \breve{P}_0) - \breve{y}(\breve{P}_k \cap \breve{P}_0) \geq D - \breve{w}(\breve{P}_k), t_k \in \breve{Y} \backslash \{t_0\},$$
$$-\breve{y}(\breve{P}_0) \geq D - \breve{w}(\breve{P}_0),$$
$$0 \leq \breve{x}(e) \leq \Delta\breve{u}(e), \quad e \notin \breve{P}_0,$$
$$0 \leq \breve{y}(e) \leq -\Delta\breve{l}(e), e \in \breve{P}_0.$$

If the problem ($\mathbf{RIOVSPT_1}$) is feasible, so is the problem ($\mathbf{RIOVSPT_1^s}$). For any $\hat{e} \in \breve{P}_0$ with $\Delta\breve{l}(\hat{e}) < 0$, denote by $Y_{\hat{e}}^{>} := \{t_k \in Q(\hat{e}) | \breve{w}(\breve{P}_k) > D\}$ the set of leaf nodes whose root-leaf paths passing $\hat{e}$ and their lengths are larger than D, where $Q(e) := \{t_k | e \in P_{v_1,t_k}, k = 0, 1, \cdots, r-1\}$ is the set of leaf nodes whose root-leaf paths pass through e[152, Definition 8]. Then we call the sub-algorithm **Mincut** in [153, Algorithm 4] on $\breve{T}_{\hat{e}}$ with the length $D + \varepsilon_0(\hat{e})$ to obtain the set $\underline{E_{\hat{e}}}$ of a relevant minimum cost cut, where $\varepsilon_0(\hat{e}) := \min\left\{\min_{t_k \in Y_{\hat{e}}^{>}} \{\breve{w}(\breve{P}_k) - D\}, -\Delta\breve{l}(\hat{e})\right\}$.

Let $\varepsilon_1(\hat{e}) := \min\left\{\min_{e\in \underline{E}_{\hat{e}}} \Delta\breve{u}(e), \varepsilon_0(\hat{e})\right\}$, then

$$\breve{x}(e) := \begin{cases} \varepsilon_1(\hat{e}), & e \in \underline{E}_{\hat{e}} \\ 0, & otherwise \end{cases}, \quad \breve{y}(e) := \begin{cases} \varepsilon_1(\hat{e}), & e = \hat{e} \\ 0, & otherwise \end{cases}. \tag{8.5}$$

is a feasible solution of the problem (**RIOVSPT$_1^s$**)[152, Lemma 12].

8.4 Construct and Solve the Subproblems (RIOVSPT$_1^i$)

Next we show that we can obtain the feasible solutions of the slack problem (**RIOVSPT$_1^s$**) gradually approaching to an optimal solution of the problem (**RIOVSPT$_1$**) by solving a series of subproblems (**RIOVSPT$_1^i$**). Let $\breve{P}_0^i := \{e \in \breve{P}_0 | \Delta\bar{l}^i(e) < 0\}$. Then the subproblem (**RIOVSPT$_1^i$**) aims at determining the only weight-decreasing edge $\hat{e} \in \breve{P}_0^i$ with the minimum total upgrade cost $C_{\hat{e}}^i$ under $\bar{c}^i$ so that the lengths of all the root-leaf paths are no less than D. Furthermore, in each subproblem, we require either the length of one of the paths $P_k(t_k \in Y_{\hat{e}}^{>})$ equals D or the weight of at least one edge reaches to its upper or lower bound. Let E_l^i (E_u^i, resp.) be the sets of the edges (not, resp.) in $\breve{P}_0$ achieving its lower bound (upper bound, resp.). Then we describe the latter case in the aforementioned requirement by $|E_l^i| + |E_u^i| \geq |E_l^{i-1}| + |E_u^{i-1}| + 1$.

Initialization: $i := 1$, $\bar{w}^1 := \breve{w}$, $\Delta\bar{l}^1 := \Delta\breve{l}$, $\Delta\bar{u}^1 := \Delta\breve{u}$, $\bar{c}^1 := \breve{c}$, $E_l^0 := \{e \in \breve{P}_0 | \Delta\bar{l}^1(e) = 0\}$ and $E_u^0 := \{e \notin \breve{P}_0 | \Delta\bar{u}^1(e) = 0\}$. Notice that $\hat{e}$ is the only weight-decreasing edge in $\breve{P}_0^i$. Since $0 < \bar{y}^i(\hat{e}) \leq -\Delta\bar{l}^i(\hat{e})$ and $\breve{w}(\breve{P}_0) - \bar{y}^i(\hat{e}) \geq D$, then $0 < \bar{y}^i(\hat{e}) \leq \min\{\bar{w}^i(\breve{P}_0) - D, -\Delta\bar{l}^i(\hat{e})\}$, for $\hat{e} \in \breve{P}_0^i$. Then we call **Mincut** in [153, Algorithm 4] on $\breve{T}_{\hat{e}}$ with the length $D + \varepsilon_0^i(\hat{e})$ to obtain the set $\underline{E}_{\hat{e}}^i$ of a relevant minimum cost cut, where $\varepsilon_0^i(\hat{e}) := \min\left\{\min_{t_k\in Y_{\hat{e}}^{>}} \{\bar{w}^i(\breve{P}_k) - D\}, -\Delta\bar{l}^i(\hat{e})\right\}$. Then we have the following subproblem (**RIOVSPT$_1^i$**).

$$\begin{aligned}
&\min_{\hat{e}\in\breve{P}_0^i} \quad C_{\hat{e}}^i := \sum_{e\in\underline{E}_{\hat{e}}^i} \bar{c}^i(e)\bar{x}^i(e) + \bar{c}^i(\hat{e})\bar{y}^i(\hat{e}) \\
&(\mathbf{RIOVSPT_1^i}) \quad s.t. \quad \sum_{e\in\breve{P}_k\cap\underline{E}_{\hat{e}}^i} \bar{x}^i(e) - \bar{y}^i(\hat{e}) \geq D - \bar{w}^i(\breve{P}_k), t_k \in Q(\hat{e})\backslash\{t_0\}, \\
&\qquad 0 \leq \bar{x}^i(e) \leq \Delta\bar{u}^i(e), \quad e \in \underline{E}_{\hat{e}}^i, \\
&\qquad 0 < \bar{y}^i(\hat{e}) \leq \min\{\bar{w}^i(\breve{P}_0) - D, -\Delta\bar{l}^i(\hat{e})\}, \quad \hat{e} \in \breve{P}_0^i, \\
&|E_l^i| + |E_u^i| \geq |E_l^{i-1}| + |E_u^{i-1}| + 1 \text{ or } \exists t_k \in Y_{\hat{e}}^{>}, w^{i+1}(\breve{P}_k) = D.
\end{aligned}$$

Let $\bar{w}^{i+1} := \bar{w}^i + \bar{x}^i - \bar{y}^i$, $\Delta\bar{u}^{i+1} := \Delta\bar{u}^i - \bar{x}^i + \bar{y}^i$, $\Delta\bar{l}^{i+1} := \Delta\bar{l}^i - \bar{x}^i + \bar{y}^i$. Let $E_l^i := \{e \in \breve{P}_0 | \Delta\bar{l}^{i+1}(e) = 0\}$ and $E_u^i := \{e \notin \breve{P}_0 | \Delta\bar{u}^{i+1}(e) = 0\}$. Then $\bar{c}^{i+1}(e) := +\infty$ for $e \in E_l^i \cup E_u^i$ and $\bar{c}^{i+1}(e) := \check{c}(e)$, otherwise.

Theorem 8.3 ([152, Theorem 13]) *Let* $e^* := \arg\min_{\hat{e} \in \breve{P}_0^i} C_{\hat{e}}^i$, $\varepsilon_1^i(e^*) := \min\{\varepsilon_0^i(e^*), \min_{e \in \underline{E}_{e^*}^i} \Delta\breve{u}^i(e)\}$. *Then* $(\bar{x}^i, \bar{y}^i)$ *is an optimal solution of the problem* (**RIOVSPT$_1^i$**), *where* $\bar{x}^i(e) := \begin{cases} \varepsilon_1^i(e^*), & e \in \underline{E}_{e^*}^i \\ 0, & otherwise \end{cases}$, $\bar{y}^i(e) := \begin{cases} \varepsilon_1^i(e^*), & e = e^* \\ 0, & otherwise \end{cases}$.

Then we present a linear time algorithm to solve the subproblem (**RIOVSPT$_1^i$**) [152, Algorithm 1]. Notice that an optimal solution of the problem (**RIOVSPT$_1^i$**) is a feasible solution of the problem (**RIOVSPT$_1^s$**), but may not be a feasible solution of the problem (**RIOVSPT$_1$**) unless the length of the path P_0 equals D.

8.5 An $O(n^2)$ Time Algorithm to Solve the Problem (RIOVSPT$_1$)

Based on the relationship of the optimal solutions between (**Int-SPT$_1$**) and (**RIOVSPT$_1$**) in **Theorem 8.2**, we first take the optimal solution of the problem (**Int-SPT$_1$**) as an initial infeasible solution of problem (**RIOVSPT$_1$**). Then we consider a slack problem (**RIOVSPT$_1^s$**), where the length of the path P_0 is greater than D. We obtain its feasible solutions gradually approaching to an optimal solution of the problem (**RIOVSPT$_1$**) by solving a series of subproblems (**RIOVSPT$_1^i$**).

Suppose the problem (**RIOVSPT$_1^i$**) is solved totally j^* times with the optimal solutions $(\bar{x}^i, \bar{y}^i)(i := 1, 2, \cdots, j^*)$ until the length of the path P_0 equals D. Then we can obtain the optimal solution of the problem (**RIOVSPT$_1$**) by the following recover edge technique.

Recover edge technique: [152] Recombine each edge $e_b = (v_a, v_b) \in \tilde{P}_0$ from two edges $e_b^1 = (v_a, v_b^1)$ and $e_b^2 = (v_b^1, v_b)$ with

$$\bar{x}^i(e) := 0, \bar{y}^i(e) := \bar{y}^i(e_b^1) + \bar{y}^i(e_b^2), i := 1, 2, \cdots, j^*. \tag{8.6}$$

Theorem 8.4 ([152, Theorem 15]) *Let* $x^* := \sum_{i=1}^{j^*} \bar{x}^i$, $y^* := \sum_{i=1}^{j^*} \bar{y}^i$. *Then* $\bar{w} := \tilde{w} + x^* - y^*$ *is a cononical optimal solution of the problem* (***RIOVSPT$_1$***) *based on the optimal solution* $\tilde{w}$ *of the problem* (***MCSPIT$_1$***).

We devise Algorithm 8.1 to solve the problem **(RIOVSPT$_1$)**.

Algorithm 8.1 $[\bar{w}, C(\bar{w})] :=$ **RIOVSPT**$_1(T, Y, w, l, u, c, D, t_0)$ [152]

Input: The tree $T(V, E)$ rooted at v_1, the set Y of leaf nodes; three weight vectors w, l, u and the cost vector c; the given leaf node t_0 and D.

Output: The optimal solution $\bar{w}$ and its relevant objective value $C(\bar{w})$.

if $D < l(P_0)$ or $D > \min_{t_k \in Y} u(P_k)$ or $l(P_0) > \min_{t_k \in Y}(l(P_k \cap P_0) + u(P_k \setminus P_0))$ **then return** "The problem is infeasible."
end if
Determine the sets CD, CA, χ, LN of the tree T and the subtree $\ddot{T}$.
Call $[x^0, C(x^0)] :=$ **Int − SPT**$_1(T, Y, CD, CA, \chi, LN, w, u, c, D)$ [Algorithm 6.4].
if $w(P_0) + x^0(P_0) = D$ **then return** $\bar{w} := w + x^0, C(\bar{w}) := C(x^0)$.
end if
Initialize $x^*(e) := 0, y^*(e) := 0, e \in T \setminus \ddot{T}$.
Construct the tree $\breve{T}(\breve{w}, \breve{u}, \breve{l}, \breve{c}, D, t_0)$ by the **double-edge technique**.
Initialization: $i := 1, \bar{w}^1 := \breve{w}, \Delta\bar{l}^1 := \Delta\breve{l}, \Delta\bar{u}^1 := \Delta\breve{u}$.
$\bar{c}^1(e) := \begin{cases} +\infty, & e \in E_l^0 \cup E_u^0 \\ \breve{c}(e), & otherwise \end{cases}$, where $E_l^0 := \{e \in \breve{P}_0 | \Delta\bar{l}^1(e) = 0\}$ and $E_u^0 := \{e \notin \breve{P}_0 | \Delta\bar{u}^1(e) = 0\}$.
while $\bar{w}^i(\breve{P}_0) > D$ **do**
    Call $[\bar{x}^i, \bar{y}^i] :=$ **RIOVSPT$^{\mathbf{i}}_{\mathbf{1}}$**$(\breve{T}, S_{\breve{T}}, \bar{w}^i, \Delta\bar{l}^i, \Delta\bar{u}^i, \bar{c}^i, D, t_0)$[152, Algorithm 1].
    Update $\bar{w}^{i+1} := \bar{w}^i + \bar{x}^i - \bar{y}^i, \Delta\bar{u}^{i+1} := \Delta\bar{u}^i - \bar{x}^i + \bar{y}^i, \Delta\bar{l}^{i+1} := \Delta\bar{l}^i - \bar{x}^i + \bar{y}^i$.
    Update $E_l^i := \{e \in \breve{P}_0 | \Delta\bar{l}^{i+1}(e) = 0\}$ and $E_u^i := \{e \notin \breve{P}_0 | \Delta\bar{u}^{i+1}(e) = 0\}$. Update $\bar{c}^{i+1}(e) := +\infty$ for $e \in E_l^i \cup E_u^i$ and $\bar{c}^{i+1}(e) := \breve{c}(e)$, otherwise.
    Update $i := i + 1$.
end while
Let $j^* := i - 1$. For $e \in \breve{P}_0$, update the solution $\bar{x}^i(e), \bar{y}^i(e), i := 1, 2, \cdots, j^*$ by (8.6) in the **recover edge technique**. Let $x^*(e) := \sum_{i=1}^{j^*} \bar{x}^i(e)$ and $y^*(e) := \sum_{i=1}^{j^*} \bar{y}^i(e)$ for $e \in \ddot{T}$.
Return $\bar{w} := w + x^0 + x^* - y^*$ and $C(\bar{w}) := C(x^0) + \sum_{e \in E} c(e)(x^*(e) + y^*(e))$.

Theorem 8.5 ([152, Theorem 16]) *We can solve the problem **(RIOVSPT$_1$)** by Algorithm 8.1 in $O(n^2)$ time.*

8.6 Conclusions

In this chapter, we introduce restricted the inverse shortest path problems under weighted l_1 norm on trees(**RIOVSPT$_1$**). It aims at adjusting some edge weights to minimize the total cost under weighted l_1 norm on the premise that the length of the shortest root-leaf path of the tree is lower-bounded by a given value D, which is just the restriction on the length of a given root-leaf path P_0. If we ignore the restriction on the path P_0, then we obtain the shortest path interdiction problem on trees (**Int-SPT$_1$**). We analyze some properties of the problem (**RIOVSPT$_1$**) and explore the relationship of the optimal solutions between (**Int-SPT$_1$**) and (**RIOVSPT$_1$**). We first take the optimal solution of the problem (**Int-SPT$_1$**) as an initial infeasible solution of problem (**RIOVSPT$_1$**). Then we consider a slack problem (**RIOVSPT$_1^s$**), where the length of the path P_0 is greater than D. We obtain its feasible solutions gradually approaching to an optimal solution of the problem (**RIOVSPT$_1$**) by solving a series of subproblems (**RIOVSPT$_1^i$**). It aims at determining the only weight-decreasing edge on the path P_0 with the minimum cost so that the length of the shortest root-leaf path is no less than D. The subproblem can be solved by searching for a minimum cost cut in $O(n)$ time. The iterations continue until the length of the path P_0 equals D. Consequently, the time complexity of the algorithm is $O(n^2)$.

Part III
Generalized Inverse Spanning Tree Problems

This part delves into the complexities and nuances of generalized inverse spanning tree problems, presenting a compendium of theoretical insights and algorithmic advancements. This review synthesizes the core contributions of the four chapters: Inverse Minimum Spanning Tree (IMST), Inverse Max+Sum Spanning Tree (IMSST), Restricted Inverse Optimal Value Problem on Minimum Spanning Tree (RIOVMST), and Partial Inverse Minimum Spanning Tree (PIMST) problems. Each chapter meticulously addresses distinct facets of inverse combinatorial optimization, providing a robust framework for both academic and practical applications.

Chapter 9, focusing on the (IMST) problems, challenges conventional paradigms by inverting the optimization process. It explores the transformation of inverse problems into classical optimization models, such as maximum-weight flow and unbalanced assignment problems, offering polynomial-time solutions grounded in rigorous mathematical proofs.

Chapter 10 tackles the (IMSST) problems, introducing a novel class of inverse problems with combined minimax-minsum objectives. The chapter innovates by formulating the problem into a linear programming model and proposing a column generation algorithm, demonstrating the versatility of the approach in handling complex real-world scenarios.

Chapter 11 presents a study on the (RIOVMST) problems under l_∞ and l_1 norms. It provides optimality conditions and proposes efficient algorithms, highlighting the development of a minimum cost flow-based approach to solve the problem, achieving a time complexity of $O(m^2n^2 \log n \log(nC))$.

Lastly, Chap. 12 introduces the (PIMST) problems, aiming to find a new weight function that minimally differs from the original while ensuring the given edge set's inclusion in some minimum spanning tree. The chapter presents exact algorithms for tractable cases and approximation algorithms for $\mathcal{NP}$-hard scenarios, underlining the computational complexities and future research directions.

We list the research results on generalized inverse (MST/MSST) problems on Table III, where in problem ($PIMST^-$) the weights can only be decreased and in ($PInvMST^+$) the weights can only be increased.

Table III Research results on generalized inverse MST/MSST problems

Problem type	Norm	Time complexity	Reference
Inv-MST	(Bounded, unit l_1 norm)	$O(m^8 \log m)$	Sect. 9.2.1 [148]
	(Unbounded, unit l_1 norm)	$O(n^3)$	Sect. 9.2.2 [116]
	(Unbounded, unit l_1 norm)	$O(n^2 \log n)$	Sect. 9.2.2 [6]
	(Unbounded, weighted l_1 norm)	$O(n^2 m \log(nC))$	Sect. 9.2.3 [116]
	(Unbounded, unit l_∞ norm)	$O(n^2)$	Sect. 9.2.4 [116]
	(Bounded, weighted sum HD)	$O(mn^3)$	Sect. 9.3.1 [62]
	(Bounded, bottleneck HD)	$O(n^2)$	Sect. 9.3.2 [37]
Inv-MSST modifying c	(Unbounded, weighted l_1 norm)	Column generation alg.	Sect. 10.3 [51]
	(Bounded, weighted l_∞ norm)	$O(m^2 \log n)$	Sect. 10.2.2 [53]
	Bottleneck HD	$O(m \log^2 n)$	Sect. 10.4 [50]
	Sum-type HD	$\mathcal{NP}$-hard	Sect. 10.4 [50]
Inv-MSST modifying w	(Bounded, weighted l_∞ norm)	$O(m^2 \log n)$	Sect. 10.5 [67]
RIOV-MST	(Bounded, weighted l_∞ norm)	$O(m^2 \log n)$	Sect. 11.2 [65]
	(Bounded, unit l_∞ norm)	$O(mn)$	Sect. 11.2 [65]
	(Bounded, weighted l_1 norm)	$O(m^2 n^2 \log n \log(nC))$	Sect. 11.3 [66]
	Bottleneck HD	$mn \log m$	Sect. 11.4 [128]
PIMST$^-$	(Bounded, l_p norm)	$O(m + n \log n + k^2)$	Sect. 12.2.1 [23]
PIMST$^+$	(Bounded, unit l_1 norm)	$\mathcal{NP}$-Hard	Sect. 12.2.3 [86]
PIMST		k-approx.	Sect. 12.3 [87]
	(Bounded, weighted sum Ham.)	$\mathcal{NP}$-Hard	Sect. 12.2.4,
		k-approx.	Sect. 12.2.5 [84]
			Sect. 12.3 [87]
	(Bounded, Bottleneck Hamming)	$O(\log n(m + \log n))$	Sect. 12.2.7 [84]
PIMST$^+$	(Bounded, weighted l_∞ norm)	$O(m + n \log n + (m - n)k)$	Sect. 12.2.2 [81]
PIMST	(Bounded, weighted l_∞ norm)	$O(mn \log n)$	Sect. 12.2.8 [85]

In conclusion, the part offers a profound exploration into the theoretical and practical aspects of inverse spanning tree problems, marking a significant contribution to the field of combinatorial optimization. The proposed algorithms and theoretical findings not only advance the academic discourse but also offer a blueprint for solving real-world network optimization problems.

Chapter 9
Inverse Minimum Spanning Tree Problems

Abstract In this chapter, we delve into the inverse minimum spanning tree (IMST) problems under various norm constraints, including the l_1, l_∞ norms and the weighted Hamming distance. We present mathematical formulations and propose efficient algorithms tailored to each scenario. The bounded IMST problem under the unit l_1 norm is reformulated as a maximum flow problem, which can be effectively solved within $O(m^8 \log m)$. In contrast, the unbounded IMST problem under the unit l_1 norm is transformed into an unbalanced assignment problem, solvable in $O(n^2 m)$ time. Furthermore, this problem is also cast as a minimum cost flow problem, solvable in $O(n^2 \log n)$ time. When considering the unbounded IMST problem under the weighted l_1 norm, it is mapped onto an unbalanced transportation problem, with a solution time of $O(n^2 m \log(nc))$, where $C = \max\{|c_j| : a_j \in A\}$. The unbounded IMST problem under the unit l_∞ norm is more straightforward, admitting a solution in $O(n^2)$ time. For the bounded IMST problem under the weighted sum Hamming distance, we have transformed it into a minimum-weight node cover problem of a bipartite graph, which can be addressed in $O(n^3 m)$ time. Employing the binary method, the constrained bounded IMST problem under the weighted bottleneck Hamming distance is also transformed into a minimum-weight node cover problem of a bipartite graph, with a solution time of $O(n^3 m \log m)$.

Keywords Inverse minimum spanning tree problems · Maximum flow problem · Unbalanced assignment problem · Unbalanced transportation problem

9.1 The Inverse Minimum Spanning Tree Problem

Let $G = (V, E)$ be a graph with a vertex set $V = \{v_1, v_2, \ldots, v_n\}$ and an edge set $E = \{e_1, e_2, \ldots, e_m\}$. Let $\boldsymbol{w} = (w_1, \ldots, w_m)$ and $\boldsymbol{c} = (c_1, \ldots, c_m)$ be the weight and cost vectors, respectively, defined on E. Additionally, let $\boldsymbol{l} = (l_1, \ldots, l_m)$ and

Binwu Zhang, Junhua Jia, and Xiucui Guan have equally contributed to this chapter.

X. Guan et al., *Inverse Combinatorial Optimization Problems*, Springer Optimization and Its Applications 225, https://doi.org/10.1007/978-3-031-91175-0_9

$\boldsymbol{u} = (u_1, \ldots, u_m)$ denote the lower and upper bound vectors for a modified weight vector $\bar{\boldsymbol{w}}$, respectively. Given a set of spanning trees $\{T_1, \ldots, T_r\}$ in G, the inverse minimum spanning tree problem **(Inv, MST, Bounded)** aims to identify a weight vector $w^* \geq 0$ such that:

1. Each of the trees $T_1, \ldots, T_r$ becomes a minimum spanning tree of G under w^*.
2. For each, $j = 1, \ldots, m, l_j \leq w_j^* \leq u_j$.
3. The objective is to minimize the distance $\|\boldsymbol{w}^* - \boldsymbol{w}\|$.

9.2 The Inverse Minimum Spanning Tree Problem Under l_1/l_∞ Norm

We explore the inverse MST problem under l_1 norm, presenting various forms of the problem, both bounded and unbounded, and proposing efficient algorithms to solve them and demonstrate the transformation of these problems into well-known optimization problems like maximum flow and unbalanced assignment problems, leading to polynomial-time algorithms.

9.2.1 *(Inv, MST, Bounded, l_1 Norm)*

Zhang and Ma [148] studied the following problem **(Inv, MST, Bounded, l_1 Norm)**:

$$
\begin{aligned}
&\min \sum_{e_i \in E} |w_i^* - w_i| \\
&\text{s.t. } T_1, \ldots, T_r \text{ all become minimum spanning trees under } w^*, \\
&\qquad l_j \leq w_j^* \leq u_j, \; j = 1, 2, \cdots, m.
\end{aligned} \tag{9.1}
$$

It is well known that T_k is a minimum spanning tree under weight $\hat{w}$ if and only if for each $e_j \notin T_k, e_j \cup \{T_k\}$ contains a unique cycle denoted by $\mathscr{C}_j(T_k)$, and $\hat{w}_j \geq \hat{w}_i, \quad \forall e_i \in \mathscr{C}_j(T_k)$ [106].

Letting $\tilde{E}_1 = \{(e_j, e_i) \mid$, there exists a T_k such that $e_j \notin T_k,\ e_i \notin \mathscr{C}_j(T_k) \cap T_k\}$, and letting $\hat{w}_i = w_i + \alpha_i - \beta_i$ with $\alpha_i, \beta_i \geq 0, i = 1, \ldots m,$, then Zhang and Ma [148] formulated (9.1) as

$$
\begin{aligned}
&\min \sum_{e_i \in E} \alpha_i + \sum_{e_i \in E} \beta_i \\
(LP1)\ &\text{s.t. } w_j + \alpha_j - \beta_j \geq w_i + \alpha_i - \beta_i, \quad \forall (e_j, e_i) \in \tilde{E}_1, \\
&\qquad 0 \leq \alpha_j \leq u_j - c_j, \quad 0 \leq \beta_j \leq c_j + l_j, \quad j = 1, \ldots, m.
\end{aligned}
$$

Zhang and Ma [148] obtained the following dual problem of (LP1):

$$
(DP2)\quad
\begin{aligned}
\max\ & \sum_{(e_j,e_i)\in\tilde{E}_1} (w_i - w_j)\, y_{ji} - \sum_{j=1}^{m} (u_j - c_j)x_j - \sum_{j=1}^{m} (c_j - l_j) z_j \\
\text{s.t.}\ & \sum_{\{e_i|(e_j,e_i)\in\tilde{E}_1\}} y_{ji} - \sum_{\{e_k|(e_k,e_j)\in\tilde{E}_1\}} y_{kj} - x_j \le 1, \quad j = 1,\dots,m, \\
- & \sum_{\{e_i|(e_j,e_i)\in\tilde{E}_1\}} y_{ji} + \sum_{\{e_k|(e_k,e_j)\in\tilde{E}_1\}} y_{kj} - z_j \le 1, \quad j = 1,\dots,m, \\
& y_{ji} \ge 0, \quad \forall x_j \ge 0, \quad \forall z_j \ge 0.
\end{aligned}
$$

In order to solve the problem by a strongly polynomial method, Zhang and Ma [148] constructed a directed graph $\tilde{G} = (\tilde{V}, \tilde{E})$ as follows:

$$\tilde{V} = \{s\} \cup \{t\} \cup E, \quad \tilde{E} = \tilde{E}_1 \cup \tilde{E}_2,$$

$$\tilde{E}_2 = \bigcup_{e_j\in E} (s, e_j) \bigcup_{e_j\in E} (s, e_j)' \bigcup_{e_j\in E} (e_j, t) \bigcup_{e_j\in E} (e_j, t)'.$$

For each arc in $\tilde{G}$, they assigned a weight and a capacity range and constructed the following maximum flow problem:

Maximum Flow Problem (MWF1)
In graph $\tilde{G}$, let s be the only source, t be the unique sink, and other vertices be transshipment points where the flows in and out are equal. We consider the maximum flow problem in the graph under the data given in Table 9.1.

For problems (DP2) and (MWF1), Zhang and Ma [148] obtained the next two lemmas:

Lemma 9.1 ([148]) *Each optimal solution of problem (DP2) corresponds to a feasible flow of problem (MWF1) with equal objective value.*

Lemma 9.2 ([148]) *Each maximum solution of problem (MWF1) corresponds to a feasible solution of problem (DP2) with equal objective value.*

By the above two lemmas, [148] obtain the next main theorem:

Table 9.1 The data in maximum flow problem (MWF1)

Arc	Weight	Capacity range	Flow
$(e_j, e_j) \in \tilde{E}_1$	$w_i - w_j$	$[0, \infty)$	y_{ji}
$(s, e_j) \in \tilde{E}_2$		$[0, 1]$	y_{sj}
$(s, e_j)' \in \tilde{E}_2$	$c_j - u_j$	$[0, \infty)$	x_j
$(e_j, t) \in \tilde{E}_2$	0	$[0, 1]$	y_{jt}
$(e_j, t)' \in \tilde{E}_2$	$l_j - c_j$	$[0, \infty)$	z_j

Theorem 9.1 ([148]) *Solving the maximum flow problem (MWF1) is equivalent to solving the dual LP problem (DP2), and the inverse minimum spanning tree problem under unit l_1 norm problem (LP1) can be solved in $O\left(|E|^8 \log |E|\right)$ time.*

9.2.2 (Inv, MST, Unbounded, l_1 Norm)

When $r = 1$ in the general inverse problem (9.1), Sokkalingam et al. [116] considered the following problem **(Inv, MST, Unbounded l_1 Norm)**:

$$\min \sum_{e_i \in E} |w_i^* - w_i|$$
$$\text{s.t. } T^0 \text{ is a minimum spanning tree under } w^*. \tag{9.2}$$

Let $T^0 = \{e_1, e_2, \cdots, e_{n-1}\}$ be the given spanning tree. We refer to the arcs in T^0 as tree arcs and the arcs in $E \setminus T^0$ as nontree arcs. In any spanning tree, there is a unique tree path between any two nodes. For each nontree arc $e_j \in E$, let P_j be the indices of all the tree arcs between the two endpoints of the arc e_j. We introduce the following path optimality conditions for the minimum spanning tree:

Theorem 9.2 ([4]) *A tree T^0 is a minimum spanning tree with respect to the cost vector w^* if and only if $w_i^* \le w_j^*$ for each $e_i \in P_j$, $j = n, \ldots, m$.*

Letting $\alpha_i = w_i^* - w_i$ for $e_i \in E$, Sokkalingam et al. [116] formulated (9.2) as the following mathematical program by Theorem 9.2:

$$\min \sum_{j=1}^{m} |\alpha_j|$$
$$s.t.\ w_i + \alpha_i \le w_j + \alpha_j \text{ for each } e_i \in P_j \text{ and } j = n, \ldots, m. \tag{9.3}$$

Lemma 9.3 ([116]) *In every optimal solution α of (9.3), we have $\alpha_i \le 0$ for each $e_i \in T^0$ and $\alpha_j \ge 0$ for each $e_j \in E \setminus T^0$.*

Then we have the following model equivalent to (9.3):

$$\min \sum_{j=n}^{m} \alpha_j - \sum_{i=1}^{n-1} \alpha_i,$$
$$\text{s.t } w_i + \alpha_i \le w_j + \alpha_j \text{ for each } e_i \in P_j \text{ and } e_j \in E \setminus T^0, \tag{9.4}$$
$$\alpha_i \le 0 \text{ for each } e_i \in T^0,$$
$$\alpha_j \ge 0 \text{ for each } E \setminus T^0.$$

Construct a network $G^0 = \left(N^0, A^0\right)$, where $N^0 = N^1 \cup N^2$ satisfies $N^1 = \{i | e_i \in T^0\}$, $N^2 = \{j | e_j \in E \setminus T^0\}$, and $A^0 = \left\{(i, j) : e_i \in P_j, j \in N_2\right\}$. Letting $f_{ij} = w_j - w_i$ for each $(i, j) \in A_0$ and associating the dual variable x_{ij} with the arc $(i, j) \in A^0$, then the dual of problem (9.4) can be stated as the following problem:

$$\min \sum_{(i,j)\in A^0} f_{ij} x_{ij}, \tag{9.5a}$$

$$(DP3)\ \text{s.t} \sum_{\{j:(i,j)\in A^0\}} x_{ij} \le 1 \quad \text{for each node } i \in N^1, \tag{9.5b}$$

$$\sum_{\{i:(i,j)\in A^0\}} x_{ij} \le 1 \quad \text{for each node } j \in N^2, \tag{9.5c}$$

$$x_{ij} \ge 0 \quad \text{for each arc } (i, j) \in A^0.$$

The problem (DP3) is a variant of the well-known assignment problem. In the standard assignment problem, the first two constraints (9.5b) and (9.5c) are in the equality form and $\left|N^1\right| = \left|N^2\right|$. But in problem (DP3), we have inequalities and possibly have $\left|N^1\right| \ll \left|N^2\right|$. We call the problem (DP3) as *the unbalanced assignment problem.*

Consider any feasible assignment x of (DP1) that is a 0-1 vector. For this assignment, the nodes that satisfy the constraints (9.5b) or (9.5c) with equality are called *matched node* and *unmatched nodes* otherwise.

Sokkalingam et al. [116] have elegantly transformed the unbalanced assignment problem (DP3) on $G^0 = (N^1 \cup N^2, A^0)$ into a minimum cost flow problem (MCF) on a bipartite network $G' = (N', A')$ defined as follows:

1. Introduce a source node s with supply $b(s) = n$, and for each $i \in N^1$, add an arc (s, i) with cost $f_{si} = 0$ and capacity $u_{si} = 1$.
2. Introduce a sink node t with demand $b(t) = -n$, and for each $j \in N^2$, add an arc (j, t) with cost $f_{jt} = 0$ and capacity $u_{jt} = 1$.
3. Include an additional arc (s, t) with cost $f_{st} = 0$ and capacity $u_{st} = (n + 1)$.
4. Set the supply/demand $b(i)$ of each node $i \in N^0$ to zero.
5. For each arc $(i, j) \in A^0$, set the capacity to n and assign the cost f_{ij}.

The set of arcs emanating from node i in A' is denoted by $A'(i)$. The following property elucidates the connection between the minimum cost flow problem in G' and the assignment problem in G^0.

Lemma 9.4 ([116]) *There is a one-to-one correspondence between feasible flows in G' and feasible assignments in G^0, and the costs of the flows and assignments are the same.*

Letting x denote an optimal flow of G' and π denote an optimal dual solution of the minimum cost flow problem (MCF), then Sokkalingam et al. [116] obtained the next lemma.

Lemma 9.5 ([116]) *Suppose that $\pi(t) = 0$. Any optimal primal solution x together with the optimal dual solution π of the minimum cost flow problem satisfies the following conditions:*

(a) $\pi(s) = 0$ and $\pi(t) = 0$.
(b) For each arc (s, i), if $x_{si} = 1$, then $\pi(i) \le 0$; if $x_{si} = 0$, then $\pi(i) \ge 0$.
(c) For each arc (j, t), if $x_{jt} = 1$, then $\pi(j) \ge 0$; if $x_{jt} = 0$, then $\pi(j) \le 0$.

By the above two lemmas, Sokkalingam et al. [116] showed that the solutions x and π can be used to obtain an optimal solution of the inverse spanning tree problem (9.2) by the next theorem.

Theorem 9.3 ([116]) *The vector α defined by (9.6) is an optimal solution of (9.2):*

$$\alpha_i = \begin{cases} \pi(i) & \text{for every matched node } i \text{ in } x, \\ 0 & \text{for every unmatched node } i \text{ in } x. \end{cases} \tag{9.6}$$

Sokkalingam et al. [116] used the successive shortest path algorithm (see [4]) to solve the minimum cost flow problem (MCF) in G', and the running time of the algorithm is $O(n^2 m)$.

When $n^2 \le (m' + n' logn)$ in G' for the minimum cost flow problem (MCF), Sokkalingam et al. [116] designed the Dijkstra's algorithm, which is applied on the residual network $G'(x)$ with arc lengths $f_{ij}^{\pi} = f_{ij} - \pi(i) + \pi(j)$ to find a shortest path from node s to node i, maintaining a distance label $\mathbf{d}(i)$ with each node $i \in N'$. If the distance label $\mathbf{d}(i)$ is finite, then it denotes the length of some directed path from node s to node i; otherwise, it implies that a directed path to node i is yet to be discovered. A finite distance label $\mathbf{d}(i)$ is permanent if it is guaranteed to be the shortest path length to node i and is temporary otherwise. In each iteration, Dijkstra's algorithm selects a minimum temporary distance label $\mathbf{d}(i)$, makes it permanent, and examines each arc $(i, j) \in A'(i)$ to update $\mathbf{d}(j) = \min\left\{\mathbf{d}(j), \mathbf{d}(i) + f_{ij}^{\pi}\right\}$. The algorithm terminates when all distance labels become permanent.

By using the Dijkstra's algorithm, Sokkalingam et al. [116] designed successive shortest path algorithm to solve the unbalanced assignment problem (DP1) in $O(n^3)$. Thus, the inverse spanning tree problem (9.2) can be solved in $O(n^3)$ if $n^2 \le (m' + n' \log n)$ in G'.

Furthermore, Ahuja and Orlin [6] also considered the **(Inv, MST, Unbounded, l_1 Norm)** problem by solving its dual problem (DP3) more efficiently, which is called as the bipartite node weighted matching problem on the path graph. Ahuja and Orlin [6] first transformed the bipartite node weighted matching problem into a specially structured minimum cost flow problem on the path graph and developed an $O(n^3)$ algorithm. They next used the special structure of the minimum cost flow problem more effectively and developed an $O(n^2 \log n)$ time algorithm. This result improved the $O(n^3)$ time algorithm obtained by Sokkalingam et al. [116].

9.2.3 (Inv, MST, Unbounded, Weighted l_1 Norm)

Sokkalingam et al. [116] also considered the following problem **(Inv, MST, Unbounded, Weighted l_1 Norm)**:

$$\min \sum_{e_i \in E} c_i |w_i^* - w_i|$$

$$\text{s.t. } T^0 \text{ is a minimum spanning tree under } w^*. \tag{9.7}$$

Sokkalingam et al. [116] reformulated the problem (9.7) as

$$\begin{aligned} \text{Min } & \sum_{j=n}^{m} c_j \alpha_j - \sum_{i=1}^{n-1} c_i \alpha_i \\ \text{s.t } & w_i + \alpha_i \le w_j + \alpha_j \text{ for each } e_i \in P_j \text{ and } e_j \in E \setminus T^0, \\ & \alpha_i \le 0 \text{ for each } e_i \in T^0, \\ & \alpha_j \ge 0 \text{ for each } E \setminus T^0. \end{aligned} \tag{9.8}$$

The dual of problem (9.8) is as follows:

$$\min \sum_{(i,j)\in A^0} f_{ij} x_{ij}, \tag{9.9a}$$

$$(DP4) \text{ s.t } \sum_{\{j:(i,j)\in A^0\}} x_{ij} \le c_{ij} \quad \text{for each node } i \in N^1, \tag{9.9b}$$

$$\sum_{\{i:(i,j)\in A^0\}} x_{ij} \le c_{ij} \quad \text{for each node } j \in N^2, \tag{9.9c}$$

$$x_{ij} \ge 0 \quad \text{for each arc } (i, j) \in A^0.$$

The problem (DP4) is called as *unbalanced transportation problem.* Using the cost scaling algorithm (see Goldberg and Tarjan [47]) for bipartite networks, the unbalanced transportation problem was solved in $O(n^2 m \log(nC))$ time, where $C = \max\{|c_j| : a_j \in A\}$.

9.2.4 (Inv, MST, Unbounded, l_∞ Norm)

Sokkalingam et al. [116] considered the following problem **(Inv, MST, Unbounded, l_∞ Norm)**:

$$\begin{aligned} &\min\ \max\{|w_i^* - w_i| : e_i \in E\} \\ &\text{s.t. } T^0 \text{ is a minimum spanning tree under } w^*. \end{aligned} \tag{9.10}$$

Letting $\alpha_i = w_i^* - w_i$ for $e_i \in E$, Sokkalingam et al. [116] formulated (9.10) as the following mathematical program:

$$\begin{aligned} \min\ & \max_{1\le j\le m} \alpha_j \\ \text{s.t } & w_i + \alpha_i \le w_j + \alpha_j \text{ for each } e_i \in P_j \text{ and } e_j \in E \setminus T^0, \\ & \alpha_i \le 0 \text{ for each } e_i \in T^0, \\ & \alpha_j \ge 0 \text{ for each } E \setminus T^0. \end{aligned} \tag{9.11}$$

Theorem 9.4 ([116]) *Letting* $\delta = \max\{c_i - c_j :$ *for each* $e_i \in P_j, j = n, \cdots, m\}$, $\alpha_i = \min\{0, -\delta/2\}$ *for each* i, $1 \le i \le (n-1)$, *and* $\alpha_j = \max\{0, \delta/2\}$ *for each* $j, n \le j \le m$, *then* $\alpha = (\alpha_1, \alpha_2, \ldots, \alpha_m)$ *is an optimal solution of the problem (9.11).*

By Theorem 9.4, Sokkalingam et al. [116] showed that **(Inv, MST, Unbounded, l_∞ Norm)** can be solved in $O(n^2)$ time.

9.3 The Inverse Minimum Spanning Tree Problem Under Hamming Distance

This section addresses the inverse MST problem under Hamming distance, focusing on both weighted sum Hamming and bottleneck Hamming distances. It transforms these problems into minimum-weight node cover problems of bipartite graphs and applies the binary method to solve constrained problems efficiently.

9.3.1 (Inv, MST, Bounded, Weighted Sum Hamming)

Supposing T^0 is a given spanning tree, He et al. [62] considered the problem **(Inv, MST, Bounded, wsH)** below:

$$\begin{aligned} &\min \textstyle\sum_{i=1}^{m} c_i H(w_i, w_i^*) \\ &\text{s.t. } \textstyle\sum_{e_i \in T^0} w_i^* \le \sum_{e_j \in T} w_j^*, \quad \text{for any spanning tree } T \text{ of } G, \\ &\qquad l_i \le w_i^* \le u_i \qquad\qquad 1 \le i \le m. \end{aligned} \tag{9.12}$$

For the spanning tree T^0, denote by E^0 the edge set consisting of edges in T^0 as *tree edge set* and $\overline{E^0}$ the set of other edges as *nontree edge set*. For each $e_j \in \overline{E^0}$, $E^0 \cup \{e_j\}$ contains a unique cycle. Denote by P_j the edge set consisting of all edges in this cycle except e_j. By letting $\alpha_i = |w_i^* - w_i|, i = 1, 2, \cdots, m$, He et al. [62] transformed (9.12) to the following problem:

$$\begin{aligned} &\min \textstyle\sum_{i=1}^{m} c_i H(\alpha_i, 0) \\ &\text{s.t. } w_i - \alpha_i \le w_j + \alpha_j, \qquad \text{for each } e_j \in \overline{E^0} \text{ and } e_i \in P_j', \\ &\qquad 0 \le \alpha_i \le w_i - l_i, \qquad\qquad e_i \in E^0, \\ &\qquad 0 \le \alpha_j \le u_j - w_j m \qquad\qquad e_j \in \overline{E^0}. \end{aligned} \tag{9.13}$$

To solve (9.13) in polynomial time, He et al. [62] construct a bipartite graph G' with respect to the tree T^0 as follows: $G' = (N', A) = (N_1' \cup N_2', A)$, where the node set $N' = N_1' \cup N_2'$ satisfies $N_1' = E^0$ and $N_2' = \overline{E^0}$ and the edge set $A = \{(e_i, e_j)|\, e_j \in \overline{E^0} \text{ and } e_i \in P_j'\}$. Define the weight of each node e_i of G' as c_i.

Lemma 9.6 ([62]) *Problem (9.13) has a feasible solution if and only if $l_i \le u_j$ for each $(e_i, e_j) \in A$.*

Let $Z = \{e_i \,|\alpha_i \neq 0\}$ in every feasible solution $\alpha = (\alpha_1, \alpha_2, \ldots, \alpha_m)\}$. Next lemma tell us how to determine Z.

Lemma 9.7 ([62]) *Suppose problem (9.13) is feasible. Then*

1. *For each $e_i \in E^0$, $e_i \in Z$ if and only if there exists e_j such that $(e_i, e_j) \in A$ and $w_i > u_j$.*
2. *For each $e_j \in \overline{E^0}$, $e_j \in Z$ if and only if there exists e_i such that $(e_i, e_j) \in A$ and $w_j < l_i$.*

Building upon the aforementioned lemma, He and Zhang [62] further normalize the bipartite graph G' through the following procedure: Initialize $Z = \emptyset$ and examine each edge (e_i, e_j) in G' to determine whether Cases 1 and 2 from Lemma 9.7 are applicable. If so, update Z by adding e_i, e_j, or both, and subsequently remove the edge (e_i, e_j) from G'. Thereafter, for each $e_i \in Z$, eliminate all edges in G' incident to node e_i. The resulting bipartite network is denoted by $G'' = (N_1' \cup N_2', A')$, where

$A' = \{(e_i, e_j) \mid (e_i, e_j) \in A, e_i \notin Z, \text{ and } e_j \notin Z\}$, with N_1' and N_2' remaining unaltered.

He and Zhang [62] then address (9.13) by computing the minimum-weight node cover of the bipartite graph G'', which is derived by removing Z and all edges incident to nodes in Z. The minimum-weight node cover problem, when confined to bipartite graphs, is solvable in strongly polynomial time through a reduction to the maximum flow problem, as demonstrated by Lawler [80]. The reduction is executed by constructing a directed network from G'' as follows:

- Introduce a source s into G'' with an arc leading to each node in N_1', where the capacity of each arc equals the weight of the respective node.
- Introduce a sink t with an arc incoming from each node in N_2', analogously.
- Direct each arc in A' from N_1' to N_2' and assign an infinite capacity to these arcs.
- Denote the resulting network as GN.

The minimum (s, t)-cut in GN can be determined through a maximum flow computation. It is crucial to note that the minimum cut must be finite, given that the net flow out of the source is finite, thus precluding any edge in A' from being part of that cut. Letting

$$C' = \{a_j \mid a_j \neq s, t, \text{ and } a_j \text{ is a node of the edge in the minimum } (s, t)\text{-cut}\},$$

then C' constitutes a node cover for G'', and the capacity of the cut is precisely equivalent to the weight of this node cover.

Theorem 9.5 ([62]) *Suppose the problem (9.13) has feasible solutions. Letting C' be the minimum-weight node cover for G'' and $\alpha = (\alpha_1, \alpha_2, \ldots, \alpha_m)$ satisfy*

$$\alpha_i = \begin{cases} w_i - l_i, & \text{if } e_i \in (C' \cup Z) \cap E^0, \\ u_i - w_i, & \text{if } e_i \in (C' \cup Z) \cap \overline{E^0}, \\ 0, & \text{otherwise,} \end{cases}$$

then α is an optimal solution for (9.13) with value $w(C' \cup Z)$.

Algorithm 9.1 An algorithm to solve problem (9.13) [62]

Input: A network $G = (V, E)$ with a weight vector w, a cost vector c, the lower bounded vector l, the upper bounded vector u, and a spanning tree T^0.
Output: The optimal solution α.
1: For each $e_j \in \overline{E^0}$, determine P_j and then determine P_j'.
2: Construct a bipartite graph $G' = (E^0 \cup \overline{E^0}, A)$, where $A = \{(e_i, e_j) \mid e_j \in \overline{E^0}, \text{ and } e_i \in P_j'\}$, and define the weight of each node e_i to be c_i.
3: **if** there exists some $(e_i, e_j) \in A$ such that $l_i > u_j$ **then** return "the problem is infeasible" and stop.

4: **else** Compute G'' and Z by normalizing G'.
5: Construct a directed network GN from G'', find its minimum (s,t)−cut, and transform it to a minimum-weight node cover C' for G''.
6: Output an optimal solution $\alpha = (\alpha_1, \alpha_2, \ldots, \alpha_m)$ where

$$\alpha_i = \begin{cases} w_i - l_i, & \text{if} \quad e_i \in (C' \cup Z) \cap E^0, \\ u_i - w_i, & \text{if} \quad e_i \in (C' \cup Z) \cap \overline{E^0}, \\ 0, & \text{otherwise.} \end{cases}$$

7: **end if**

9.3.2 *(Inv, MST, Bounded, Bottleneck Hamming)*

For a given spanning tree T^0, Zhang et al. [139] considered the following constrained inverse problem *(Constrained-Inv, MST, Bounded, Bottleneck Hamming)*:

$$\begin{aligned} \min \quad & \max_{i=1,\cdots,m} c_i H(w_i^*, w_i) \\ \text{s.t.} \quad & \sum_{e_i \in T^0} w_i^* \le \sum_{e_j \in T} w_j^*, \qquad \text{for any spanning tree } T \text{ of } G, \qquad (9.14) \\ & l_i \le w_i^* \le u_i, \qquad 1 \le i \le m, \\ & \sum_{i=1}^{m} c_i H(w_i^*, w_i) \le M. \end{aligned}$$

For a given spanning tree T^0, the definitions of E^0, $\overline{E^0}$, and P_j are the same as in Sect. 9.3.1. Let $\alpha_i = |w_i^* - w_i|, i = 1, 2, \cdots, m$. Define $\phi(\alpha) = \max_{i=1,\cdots,m} c_i H(\alpha_i, 0)$, and $\eta(\alpha) = \sum_{i=1}^{m} c_i H(\alpha_i, 0)$. For a given subset S of edges, define $c^s(S) = \sum_{e_i \in S} c_i$ and $c^b(S) = \max\{c_i | e_i \in S\}$ (letters s and b here stand for "sum" and "bottleneck," respectively). Zhang et al. [139] transformed (9.14) to the following problem:

$$\begin{aligned} \min_\alpha \quad & \phi(\alpha) \\ \text{s.t.} \quad & w_i - \alpha_i \le w_j + \alpha_j, \qquad \text{for each } e_j \in \overline{E^0} \text{ and } e_i \in P_j'; \qquad (9.15) \\ & 0 \le \alpha_i \le w_i - l_i, \qquad e_i \in E^0; \\ & 0 \le \alpha_j \le u_j - w_j, \qquad e_j \in \overline{E^0}; \\ & \eta(\alpha) \le M. \end{aligned}$$

In order to solve problem (9.15), Zhang et al. [139] first considered the next standard problem:

$$
\begin{aligned}
\min_{\alpha}\ & \phi(\alpha) \\
\text{s.t.}\ & w_i - \alpha_i \le w_j + \alpha_j, && \text{for each } e_j \in \overline{E^0} \text{ and } e_i \in P'_j; \\
& 0 \le \alpha_i \le w_i - l_i, && e_i \in E^0; \\
& 0 \le \alpha_j \le u_j - w_j, && e_j \in \overline{E^0}.
\end{aligned}
\tag{9.16}
$$

The constructions of G', A, Z, G'', and A' are the same as in Sect. 9.3.1. A node set C^* is a minimum bottleneck-weight node cover of G'' such that $c^b(C^*) = \min\{c^b(C)\ |C \text{ is a node cover of } G''\}$. Zhang et al. [139] gave the next Algorithm 9.2 to solve a minimum bottleneck-weight node cover of bipartite graph G'', which runs in $O(mn)$ time.

Algorithm 9.2 An algorithm to find a minimum bottleneck-weight node cover [139]

Input: A bipartite graph G'' in Sect. 9.3.1.
Output: A minimum bottleneck-weight node cover C^*.
1: Initialize $C := \emptyset$. Sort the nodes of the current graph (it is G'' initially) according to the nondecreasing order of node weights.
2: Find the node e_i with the minimum weight and a positive degree in the current graph, set $C := C \cup \{e_i\}$, and delete the node e_i and all edges incident to it in the current graph.
3: Repeat the process in Step 2 until the edge set of the current graph is empty. Output $C^* := C$.

Theorem 9.6 ([139]) *Suppose problem (9.15) is feasible. Let C^* be a minimum bottleneck-weight node cover of G''. Then $\alpha' = (\alpha'_1, \alpha'_2, \ldots, \alpha'_m)$ defined below is an optimal solution of problem (9.16) with the minimum objective value $c^b(C^* \cup Z)$.*

$$
\alpha'_i = \begin{cases} w_i - l_i, & \text{if } e_i \in (C^* \cup Z) \cap E^0, \\ u_i - w_i, & \text{if } e_i \in (C^* \cup Z) \cap \overline{E^0}, \\ 0, & \text{if } e_i \notin (C^* \cup Z). \end{cases} \tag{9.17}
$$

Based on Theorem 9.6 and Algorithm 9.2, Zhang et al. [139] gave the following algorithm for solving problem (9.16) in $O(mn)$ time:

Algorithm 9.3 An $O(mn)$ algorithm to solve problem (9.16) [139]

Input: A network $G = (V, E)$ with a weight vector w, a cost vector c, the lower bounded vector l, the upper bounded vector u, and a spanning tree T^0.
Output: An optimal solution α' of (9.16).

For each $e_j \in \overline{E^0}$, determine P_j and then determine $P_j^{'}$.
Construct a bipartite graph $G' = (E^0 \cup \overline{E^0}, A)$, where $A = \{(e_i, e_j) \mid e_j \in \overline{E^0}$ and $e_i \in P_j^{'}\}$, and define the weight of each node e_i as c_i.
if there exists some $(e_i, e_j) \in A$ such that $l_i > u_j$ **then**, return "the problem is infeasible".
else compute Z by Lemma 9.7 and construct G'' by normalizing G'.
    Run Algorithm 9.2 to get a minimum bottleneck-weight node cover C^* of G''.
    Output an optimal solution $\alpha' = (\alpha_1', \alpha_2', \ldots, \alpha_m')$ by (9.17).
end if

By using the binary method and Algorithm 9.3, Zhang et al. [139] obtained the following algorithm for solving problem (9.15):

Algorithm 9.4 An algorithm to solve problem (9.15) [139]

Input: A network $G = (V, E)$ with a weight vector w, a cost vector c, the lower bounded vector l, the upper bounded vector u, and a spanning tree T^0.
Output: An optimal solution $\bar{\alpha}$.

Let $c_0 := -1$, and rearrange the costs $c_0, c_1, c_2, \cdots, c_m$ in an increasing order by $-1 =: c_{j_1} < c_{j_2} < \cdots < c_{j_k}$.
Call Algorithm 9.3 to solve problem (9.16).
if the problem (9.16) is infeasible **then** output that problem (9.15) is infeasible.
else let α' be an optimal solution of (9.16), and find the index $\underline{t}$ such that $c_{j_{\underline{t}+1}} = \phi(\alpha')$.
    Let $\bar{\alpha} = (\bar{\alpha}_1, \bar{\alpha}_2, \ldots, \bar{\alpha}_m)$ be an optimal solution of problem (9.13) with objective value $\eta(\bar{\alpha})$.
    if $\eta(\bar{\alpha}) > M$ **then** output that problem (9.15) is infeasible.
    else $\bar{\alpha}$ is a feasible solution of problem (9.15) with the objective value $\phi(\bar{\alpha})$. Find the index t such that $c_{j_t} = \phi(\bar{\alpha})$.
        while $t - \underline{t} > 1$ **do**
            Let $t' := \lceil \frac{t+\underline{t}}{2} \rceil$. Solve problem (9.13) with the restriction that

$$l_i = 0, \text{ if } e_i \in E^0 \text{ and } c_i > c_{j_{t'}}, \tag{9.18}$$

$$u_i = 0, \text{ if } e_i \in \overline{E^0} \text{ and } c_i > c_{j_{t'}}. \tag{9.19}$$

if the restricted problem (9.13) is infeasible **then** set $\underline{t} := t'$.
else let $\hat{\alpha}$ be an optimal solution of the restricted problem (9.13) with minimum value $\eta(\hat{\alpha}) = \sum_{i=1}^{m} c_i H(\hat{\alpha}_i, 0)$.
if $\eta(\hat{\alpha}) \leq M$ **then** update $t := t', \bar{\alpha} := \hat{\alpha}, \phi(\bar{\alpha}) := \phi(\hat{\alpha})$.
else let $\underline{t} := t'$.
end if
end if
end while
if $t - \underline{t} = 1$ **then** output an optimal solution $\bar{\alpha}$ of problem (9.15) with minimum cost $\phi(\bar{\alpha})$.
end if
end if
end if

The time complexity of Algorithm 9.4 for problem (9.15) is $O(mn^3 \log m)$.

For problem (9.16), Duin et al. [37] designed the following Algorithm 9.5, whose time complexity is $O(n^2)$.

Algorithm 9.5 An $O(n^2)$ algorithm to solve problem (9.16) [37]

Input: A network $G = (V, E)$ with a weight vector w, a cost vector c, the lower bounded vector l, the upper bounded vector u, and a spanning tree T^0.
Output: an optimal solution $\boldsymbol{\alpha}$ with optimal value c^*.
Initialize $\alpha := \mathbf{0}, i := 2, k := 1, c^* := 0$.
while $i < n$ **do**
for all chords e_j incident to v_i, which is violating to $w + \alpha$ **do**
if $\alpha_i = w_i - l_i$ and $\alpha_j = u_j + w_j$ **then** return "the instance is infeasible."
else if $c_j > c_k$ or $\alpha_j = u_j + w_j$ **then** update $\alpha_k := w_k - l_k, c^* := c_k, k := k + 1$.
else let $\alpha_j := u_j + w_j, c^* := \max\{c^*, c_j\}$
end if
end for
Update $i := i + 1$.
end while
Output an optimal solution α with optimal value c^*.

9.4 Conclusions

The chapter offers a profound exploration into the intricacies of the minimum spanning tree context. It is not merely a collection of problems but a systematic exposition of the theoretical underpinnings and practical algorithms that are pivotal to solving these complex problems efficiently. The inverse MST problem challenges traditional optimization by seeking to determine input parameters that yield a given solution, rather than finding an optimal solution from a set of possible outcomes.

Section 9.2 delves into the inverse MST problem under the l_1 norm, presenting a spectrum of problems ranging from bounded to unbounded scenarios. We adeptly demonstrate the transformation of these inverse problems into well-known optimization problems such as the maximum flow problem and the unbalanced assignment problem. These transformations are more than mere theoretical curiosities. We provide a pathway to developing efficient algorithms that can solve complex, real-world problems in polynomial time. The solutions presented are grounded in a rigorous mathematical framework, ensuring both their correctness and their practical applicability.

Section 9.3 extends the discussion to the inverse MST problem under weighted sum Hamming and bottleneck Hamming distances. We transformed these problems into minimum-weight node cover problems of bipartite graphs. The application of the binary method to solve constrained problems is a testament to the innovative approach.

Throughout the chapter, the authors maintain a clear and rigorous mathematical narrative, ensuring that the complexity of the problems and the solutions is fully explored. The use of dual problems, the construction of directed graphs, and the application of maximum flow techniques are all explained with meticulous detail, allowing readers to follow the logical progression of the arguments and understand the nuances of the solutions. It provides not only a comprehensive treatment of the inverse MST problems but also a blueprint for approaching and solving similar inverse combinatorial optimization problems.

Chapter 10
Inverse Max+Sum Spanning Tree Problems

Abstract In this chapter, the inverse max+sum spanning tree problem, a novel class of inverse optimization problems with combined minimax-minsum objectives, has been thoroughly examined. By focusing on the modification of the sum-cost vector under both weighted l_1 and l_∞ norms, as well as under weighted Hamming distance, we have developed mathematical models and efficient algorithms. The l_1 norm problem has been transformed into a linear programming problem, addressed with a column generation algorithm, revealing a fascinating connection to the max+sum spanning tree problem, solvable in $O(m \log n)$ time. The l_∞ norm problem has been proven to be a linear fractional combinatorial optimization problem, tackled with a discrete Newton method within $O(m^2 \log m)$ iterations. For the Hamming distance, a binary search algorithm has been designed for the bottleneck case with a complexity of $O(m \log^2 n)$, while the sum case has been identified as NP-hard. Modifying the max weight vector under the l_∞ norm led to a strongly polynomial time algorithm with a runtime of $O(m^2 \log n)$.

Keywords Inverse max+sum spanning tree problems · Minimax-minsum objective · Column generation algorithm · Linear fractional combinatorial optimization problem · Discrete Newton method; NP-hard

10.1 Inverse Max+Sum Spanning Tree Problems

The *max+sum spanning tree (MSST) problem* is to find a *CS* $S^* \in \mathcal{S}$ (S^* is a spanning tree and $\mathcal{S}$ is the set of all spanning trees) whose max+sum weight $\boldsymbol{w}(S^*) = \max_{e \in S^*} w(e) + \sum_{e \in S^*} c(e)$ is minimum on the network $G(V, E, c, w)$, which can be formulated as (10.1):

$$(MSST) \quad \min_{S \in \mathcal{S}} \max_{e \in S} w(e) + \sum_{e \in S} c(e) \tag{10.1}$$

Junhua Jia and Xiucui Guan have equally contributed to this chapter.

X. Guan et al., *Inverse Combinatorial Optimization Problems*, Springer Optimization and Its Applications 225, https://doi.org/10.1007/978-3-031-91175-0_10

Let S^0 be a given *CS*. The inverse max+sum spanning tree (IMSST) problems aim to modify the sum-cost vector c or the max weight vector w such that S^0 becomes an optimal solution in the new network $G(V, E, \bar{c}, w)$ or $G(V, E, c, \bar{w})$ under different norms.

For convenience, let $T_0 = S^0$, $\Gamma = \mathcal{S}$, $c^s(T) := \sum_{e\in T} c(e)$, $w^b(T) := \max_{e\in T} w(e)$, and $f(w, c, T) := w^b(T) + c^s(T)$ for any cost vector c, weight vector w, and tree $T \in \Gamma$.

10.2 (Inv-sum-cost, MSST, (Un)Bounded, Weighted l_∞ Norm)

In this section, we introduce the inverse max+sum spanning tree problem by modifying the sum-cost vector c under weighted l_∞ norm [53]. Its mathematical model can be stated as follows:

$$\begin{aligned} &\min \max_{e\in E} q(e)|\bar{c}(e) - c(e)| \\ (IMSST^b_\infty)\ \ &s.t.\ \max_{e\in T_0} w(e) + \sum_{e\in T_0} \bar{c}(e) \le \max_{e\in T} w(e) + \sum_{e\in T} \bar{c}(e), \forall T \in \Gamma, \\ &c(e) - l(e) \le \bar{c} \le c(e) + u(e). \end{aligned}$$

Let

$$\beta(e) = \begin{cases} c(e) - \bar{c}(e), & \text{if } e \in T_0, \\ \bar{c}(e) - c(e), & \text{if } e \notin T_0. \end{cases} \tag{10.2}$$

Lemma 10.1 ([53, Lemma 1]) *There is an optimal solution β^* of the problem $(IMSST^b_\infty)$ such that $\beta^*(e) \ge 0$.*

Using Lemma 10.1, the problem (IMSST^b_∞) can be reformed as follows:

$$\begin{aligned} &\min \max_{e\in E} q(e)\beta(e) \\ (IMSST^b_\infty)\ \ &s.t.\ \sum_{e\in T\setminus T_0} \beta(e) + \sum_{e\in T_0\setminus T} \beta(e) \ge f(w, c, T_0) - f(w, c, T), \forall T \in \Gamma, \\ &0 \le \beta(e) \le l(e), \text{ if } e \in T_0, \\ &0 \le \beta(e) \le u(e), \text{ if } e \notin T_0. \end{aligned} \tag{10.3}$$

10.2.1 (Inv-sum-cost, MSST, Unbounded, Weighted l_∞ Norm)

Denote by (IMSST^u_∞) the unbounded inverse MSST problem, where $l(e) = u(e) = +\infty$ for all $e \in E$. Letting Δ^* be its optimal value, then $\beta^*(e) = \frac{\Delta^*}{q(e)}$ is its optimal solution. It follows from the constraint condition in (10.3) that

$$\Delta^* \geq \Delta(T) := \frac{f(w, c, T_0) - f(w, c, T)}{\sum_{e \in T \setminus T_0} \frac{1}{q(e)} + \sum_{e \in T_0 \setminus T} \frac{1}{q(e)}},$$

which means $\Delta^* = \max\{\Delta(T) | T \in \Gamma\}$.

Theorem 10.1 ([53, Theorem 2]) *The problem (IMSST^u_∞) is a linear fractional combinatorial optimization problem.*

By Theorem 10.1, we can design a discrete type Newton method to solve the problem (IMSST^u_∞). Consider the parametric version of problem (IMSST^u_∞) below:

$$\begin{aligned} &\min \Delta \\ (IMSST_\infty)\ s.t.\ & f(w, c, T_0) - f(w, c, T) - \sum_{e \in T \setminus T_0} \frac{\Delta}{q(e)} - \sum_{e \in T_0 \setminus T} \frac{\Delta}{q(e)} \leq 0, \forall T \in \Gamma. \end{aligned}$$

Define

$$h(\Delta) := \max\{f(w, c, T_0) - f(w, c, T) - \sum_{e \in T \setminus T_0} \frac{\Delta}{q(e)} - \sum_{e \in T_0 \setminus T} \frac{\Delta}{q(e)} | T \in \Gamma\}.$$

Thus, the another equivalent formulation of the problem (IMSST^u_∞) is as follows:

$$h(\Delta) = 0.$$

Lemma 10.2 ([53, Lemma 3]) *The problem $h(\Delta) = 0$ is equivalent to finding a minimum Δ such that $\min_{T \in \Gamma} f(w, c_\Delta, T) = f(w, c_\Delta, T_0)$, where*

$$c_\Delta(e) := \begin{cases} c(e) - \frac{\Delta}{q(e)}, & \text{if } e \in T_0, \\ c(e) + \frac{\Delta}{q(e)}, & \text{if } e \notin T_0. \end{cases} \tag{10.4}$$

Algorithm 10.1 A Newton algorithm to solve (IMSST^u_∞) [53, Algorithm 4]

Input: A network $G = (V, E)$ with a weight vector w, a cost vector c, and a vector q on E.
Output: The optimal value $\Delta^* := \Delta_k$ and an optimal solution $\bar{c}$.
1: Let $\Delta_0 := 0$ and initialize $\Delta := \Delta_0$ and $k := 0$.
2: **while** $k = 0$ or $\Delta_k > \Delta$ **do**

Update $\Delta := \Delta_k$ and $k := k + 1$.
Solve the max+sum spanning tree problem on the network $G(V, E, c_\Delta, w)$, where c_Δ is defined as in (10.4). Let the optimal MSST be T_k.
if $f(w, c_{\Delta,T_k}) < f(w, c_{\Delta,T_0})$ **then** let $\Delta_k := \Delta(T_k)$.
else let $\Delta_k := \Delta$.
end if
end while
Return The optimal value $\Delta^* := \Delta_k$ and the optimal solution $\bar{c}$ below:

$$\bar{c}(e) := \begin{cases} c(e) - \frac{\Delta^*}{q(e)}, & \text{if } e \in T_0, \\ c(e) + \frac{\Delta^*}{q(e)}, & \text{if } e \notin T_0. \end{cases} \tag{10.5}$$

Theorem 10.2 ([53, Theorem 12]) *Algorithm 10.1 can solve the problem (IMSST$^u_\infty$) in $O(m^2 \log n)$ time.*

10.2.2 (Inv-sum-cost, MSST, Bounded, Weighted l_∞ Norm)

Next, we consider how to solve the bounded inverse MSST problem. Let

$$\tilde{c}(e) := \begin{cases} c(e) - l(e), & \text{if } e \in T_0, \\ c(e) + u(e), & \text{if } e \notin T_0. \end{cases} \tag{10.6}$$

Then the next lemma contributes to check the feasibility of the bounded inverse MSST problem (IMSST$^b_\infty$).

Lemma 10.3 ([53, Lemma 15]) *Let $\tilde{T}$ be an optimal max+sum spanning tree on the network $G(V, E, \tilde{c}, w)$. If $f(w, \tilde{c}, T_0) = f(w, \tilde{c}, \tilde{T})$, then the problem (IMSST$^b_\infty$) is feasible. If $f(w, \tilde{c}, T_0) > f(w, \tilde{c}, \tilde{T})$, then it is infeasible.*

Let

$$Q(e) := \begin{cases} q(e)l(e), & \text{if } e \in T_0, \\ q(e)u(e), & \text{if } e \notin T_0. \end{cases} \tag{10.7}$$

Sort the values $\{Q(e) | e \in E\}$ in a strictly ascending order. Suppose $Q(e_{i_1}) < Q(e_{i_1}) < \cdots < Q(e_{i_\tau})$. If the problem (IMSST$^b_\infty$) is feasible, then its optimal value Q^* must be in an interval $(Q(e_{i_k}), Q(e_{i_{k+1}})]$ for some index k. Define $E^* := \{e \in E | Q(e) \leq Q^*\}$ and

$$\beta^*(e) := \begin{cases} l(e), & \text{if } e \in E^* \cap T_0, \\ u(e), & \text{if } e \in E^* \setminus T_0, \\ \frac{Q^*}{q(e)}, & \text{if } e \notin E^*. \end{cases} \tag{10.8}$$

Theorem 10.3 ([53, Theorem 16]) *Let Q^* be the optimal value of the problem (IMSST$^b_\infty$). Then β^* defined in (10.8) is an optimal solution of the problem (IMSST$^b_\infty$).*

Let $\bar{\Delta}(T, E^*) := \dfrac{f(w,c,T_0)-f(w,c,T)-\sum_{e\in T\cap E^*\setminus T_0} u(e)-\sum_{e\in T_0\cap E^*\setminus T} l(e)}{\sum_{e\in T\setminus E^*\setminus T_0}\frac{1}{q(e)}+\sum_{e\in T_0\setminus E^*\setminus T}\frac{1}{q(e)}}$.

Then $Q^* = \max\{\bar{\Delta}(T, E^*)|T \in \Gamma\}$.

Clearly, this equality motivates the following main idea to solve the problem (IMSST$^b_\infty$). First, find an interval $(Q(e_{i_k}), Q(e_{i_{k+1}})]$ that Q^* belongs to by the binary search method, and then find the exact value of Q^* in this interval by solving an unbounded inverse MSST problem defined below:

$$\begin{aligned}
&\min \Delta\\
&s.t.\ f(w,c,T_0) - f(w,c,T) - \sum_{e\in T\cap E_k\setminus T_0} u(e) - \sum_{e\in T_0\cap E_k\setminus T} l(e)\\
&\quad - \sum_{e\in T\setminus E_k\setminus T_0}\frac{\Delta}{q(e)} - \sum_{e\in T_0\setminus E_k\setminus T}\frac{\Delta}{q(e)} \le 0, \forall T \in \Gamma.
\end{aligned}\tag{10.9}$$

where $E_k := \{e \in E|Q(e) \le Q(e_{i_k})\}$.

Lemma 10.4 ([53, Lemma 17]) *The problem (10.9) is equivalent to finding a minimum Δ such that* $\min_{T\in\Gamma} f(w, \tilde{\Delta}, T) = f(w, \tilde{\Delta}, T_0)$*, where*

$$\tilde{c}_\Delta(e) := \begin{cases} c(e) - l(e), & if\ e \in T_0 \cap E_k,\\ c(e) - \frac{\Delta}{q(e)}, & if\ e \in T_0 \setminus E_k,\\ c(e) + u(e), & if\ e \in E_k \setminus T_0,\\ c(e) + \frac{\Delta}{q(e)}, & if\ e \notin E_k \cap T_0. \end{cases}\tag{10.10}$$

Algorithm 10.2 A binary search method to solve (IMSST$^b_\infty$) [53, Algorithm 19]

Input: A network $G = (V, E)$ with a weight vector w, a cost vector c, and three vectors q, l, and u on E.

Output: the optimal value $Q^* := \bar{\Delta}_{\bar{k}}$ and an optimal solution $\bar{c}^*$.

Step 1: Check the feasibility of the instance.

1: Let $\tilde{T}$ be an optimal max+sum spanning tree on the network $G(V, E, \tilde{c}, w)$, where $\tilde{c}$ is defined as in (10.6). If $f(w, \tilde{c}, T_0) = f(w, \tilde{c}, T)$, then the instance is feasible; otherwise, it is infeasible.

Step 2: Initialization.

2: Sort the values $\{Q(e)|e \in E\}$ in a strictly ascending order, where $Q(e)$ is as defined in (10.7). Assume $Q(e_{i_1}) < Q(e_{i_2}) < \cdots < Q(e_{i_\tau})$.

3: Let $a := 1$ and $b := \tau$.

Step 3: Find the interval that the optimal value belongs to.

4: **while** $b - a > 1$ **do**

Let $k := \lfloor \frac{a+b}{2} \rfloor$, $E_k : \{e \in E | Q(e) \leq Q(e_{i_k})\}$, and $\Delta_k := Q(e_{i_k})$.
Find an optimal max+sum spanning tree $\tilde{T}_k$ on network $G(V, E, \tilde{c}_{\Delta_k}, w)$, where $\tilde{c}_{\Delta_k}$ is defined as in (10.10).
if $f(w, \tilde{c}_{\Delta_k}, \tilde{T}_k) < f(w, \tilde{c}_{\Delta_k}, T_0)$ **then** update $a := k$.
else update $b := k$.
end if
end while
Step 4: Find the optimal value.
Let $\bar{\Delta} := Q(e_{i_a})$, $\bar{k} := 0$, and $\bar{\Delta}_{\bar{k}} := \bar{\Delta}$.
while $\bar{k} = 0$ or $\bar{\Delta}_{\bar{k}} > \bar{\Delta}$ **do**
Update $\bar{\Delta} := \bar{\Delta}_{\bar{k}}$ and $\bar{k} := \bar{k} + 1$.
Find an optimal max+sum spanning tree $\bar{T}_{\bar{k}}$ on network $G(V, E, \tilde{c}_{\bar{\Delta}}, w)$, where $\tilde{c}_{\bar{\Delta}}$ is defined as in (10.10) by replacing E_k with E_a.
if $f(w, \tilde{c}_{\bar{\Delta}}, \bar{T}_{\bar{k}}) < f(w, \tilde{c}_{\bar{\Delta}}, T_0)$ **then** update $\bar{\Delta}_{\bar{k}} := \bar{\Delta}(\bar{T}_{\bar{k}}, E_a)$.
else update $\bar{\Delta}_{\bar{k}} := \bar{\Delta}$.
end if
end while
Return the optimal value $Q^* := \bar{\Delta}_{\bar{k}}$ and the optimal solution $\bar{c}^*$ below:

$$\bar{c}^*(e) := \begin{cases} c(e) - l(e), & \text{if } e \in T_0 \cap E_a, \\ c(e) - \frac{Q^*}{q(e)}, & \text{if } e \in T_0 \setminus E_a, \\ c(e) + u(e), & \text{if } e \in E_a \setminus T_0, \\ c(e) + \frac{Q^*}{q(e)}, & \text{if } e \notin E_a \cup T_0. \end{cases} \tag{10.11}$$

Theorem 10.4 ([53, Theorem 20]) *The problem (IMSST$^b_\infty$) can be solved by Algorithm in $O(m^2 \log n)$ operations.*

10.3 (Inv-sum-cost, MSST, Unbounded, Weighted l_1 Norm)

In this section, we introduce the unbounded inverse max+sum spanning tree problem by modifying the sum-cost vector c under weighted l_1 norm [51]. Its mathematical model can be stated as follows:

$$\min \sum_{e \in E} q(e)|\bar{c}(e) - c(e)|$$

$$(IMSST^u_1) \; s.t. \; \max_{e \in T_0} w(e) + \sum_{e \in T_0} \bar{c}(e) \leq \max_{e \in T} w(e) + \sum_{e \in T} \bar{c}(e), \forall T \in \Gamma.$$

Let

$$\pi(e)=\begin{cases}\bar{c}(e)-c(e), & \text{if } e\in T_0,\\ c(e)-\bar{c}(e), & \text{if } e\notin T_0.\end{cases} \tag{10.12}$$

Lemma 10.5 ([51, Lemma 1]) *There is an optimal solution π of the problem (IMSST$_1^u$) such that $\pi(e)\le 0$ for any $e\in E$.*

Then the problem (IMSST$_1^u$) can be reformulated as follows by Lemma 10.5:

$$\begin{aligned} &\max \sum_{e\in E} q(e)\pi(e)\\ (IMSST_1^u)\ \ &s.t. \sum_{e\in T_0\setminus T}\pi(e)+\sum_{e\in T\setminus T_0}\pi(e)\le \sum_{e\in T\setminus T_0}c(e)-\sum_{e\in T_0\setminus T}c(e)+w^b(T)\\ &\quad -w^b(T_0),\ \forall T\in\Gamma,\\ &\quad \pi(e)\le 0,\ \forall e\in E.\end{aligned}$$

For any spanning tree T_j, let $d_j=\sum_{e\in T_j\setminus T_0}c(e)-\sum_{e\in T_0\setminus T_j}c(e)+w^b(T_j)-w^b(T_0)$, $d=(d_1,d_2,\ldots,d_{|\Gamma|})^T$ and $A=(a_{ij})$ be an $m\times|\Gamma|$ matrix, where

$$a_{ij}=\begin{cases}1, & \text{if } e\in T_0\setminus T_j,\\ 1, & \text{if } e\in T_j\setminus T_0,\\ 0, & otherwise.\end{cases} \tag{10.13}$$

Then the matrix form of the problem (IMSST$_1^u$) can be stated as follows:

$$\begin{aligned} &\max q^T\pi\\ (IMSST_1^u)\ \ &s.t.\ A^T\pi\le d\\ &\quad \pi\le 0.\end{aligned} \tag{10.14}$$

Denote by t_j the dual variable corresponding to constraint (10.14). Then $t=(t_1,t_2,\cdots,t_{|\Gamma|})^T$ is a $|\Gamma|$-dimensional column vector. Then the dual problem of (IMSST$_1^u$) is as follows:

$$\begin{aligned} &\min d^Tt\\ (DIMSST_1^u)\ \ &s.t.\ At\le q,\\ &\quad t\ge 0.\end{aligned}$$

The standard form of the problem (DIMSST$_1^u$) can be formulated below:

$$\begin{aligned} &\min\ d^T t + \theta^T x \\ (SDIMSST_1^u)\ \ &s.t.\ \ At + Ix = q, \\ &\qquad t, x \geq 0. \end{aligned}$$

where θ is an m-dimensional column vector whose elements are all zeros and I is an $m \times m$ identity matrix.

Let

$$\tilde{c}(e) = \begin{cases} c(e) - \pi(e), & \text{if } e \notin T_0, \\ c(e) + \pi(e), & \text{if } e \in T_0. \end{cases} \tag{10.15}$$

Theorem 10.5 ([51, Theorem 2]) *When the column generation method is applied to solve the dual problem ($SDIMSST_1^u$), the entering variable selection in each iteration can be determined in $O(m \log n)$ time by solving a max+sum spanning tree problem on the network $G(V, E, \tilde{c}, w)$, where $\tilde{c}$ is defined as in (10.15).*

Algorithm 10.3 The column generation algorithm to solve the problem (IMSST$_1^u$) [51, Algorithm 3]

Input: A network $G = (V, E)$ with a weight vector w, a cost vector c, and a vector q on E.

Output: The optimal value and solution.

Step 1: Initialization.

Let the initial basic feasible solution be $t^0 := 0$, $x_i^0 := q_i$, $i = 1, \cdots, m$ and the initial simplex multiplier be $\pi^0 := \{0, 0, \cdots, 0\}^T$. Then the initial basis matrix is $B_0 := I$, and the initial cost coefficient vector with respect to B_0 is $f_{B_0} := (0, 0, \cdots, 0)^T$. Initialize k, the number of iterations, to 0.

Step 2: Calculate the reduced cost $\bar{f}_{\bar{j}}^k$.

Solve the max+sum spanning tree problem on the network $G(V, E, \tilde{c}^k, w)$, where $\tilde{c}^k$ is defined as in (10.15). Let the optimal max+sum spanning tree be T_{j^*}.

Let $\bar{f}_{T_{j^*}}^k := w^b(T_{j^*}) + \sum_{e \in T_{j^*}} \tilde{c}^k(e) - \sum_{e \in T_0} \tilde{c}^k(e) - w^b(T_0)$.

Let $\bar{f}_{e_{i^*}}^k := \min_{j=|\Gamma|+i \notin J_B, i=1,\cdots,m}\{-\pi^k(e_i)\} = -\pi^k(e_{i^*})$.

Calculate $\bar{f}_{\bar{j}}^k := \min\{\bar{f}_{T_{j^*}}^k, \bar{f}_{e_{i^*}}^k\}$.

Step 3: Check the optimality of the dual problem (SDIMSST$_1^u$).

if $\bar{f}_{\bar{j}}^k \geq 0$ **then**

  The current solution π^k is an optimal solution of the problem (IMSST$_1^u$).

end if

Step 3.2

if $\bar{f}_{\bar{j}}^k = \bar{f}_{e_{i*}}^k < 0$ **then**

The column vector entering the basis is $H_{\bar{j}} := I_{i*}$, whose i^*-th element is 1 and all the other elements are 0, and the corresponding cost coefficient is $f_{\bar{j}} := 0$.

end if

Step 3.3

if $\bar{f}_{\bar{j}}^k = \bar{f}_{T_{j*}}^k < 0$ **then**

The column vector entering the basis is $H_{\bar{j}} := A_{j*}$, which can be calculated by (10.13), and the corresponding cost coefficient is $f_{\bar{j}} := d_{j*} = c^s(T_{j*}) + w^b(T_{j*}) - c^s(T_0) - w^b(T_0)$.

end if

Step 4: Perform a simplex iteration.

Calculate the new column vector $\bar{H}_{\bar{j}} := B_k^{-1} H_{\bar{j}}$ entering the basis matrix and the new right-hand side vector $\bar{q} := B_k^{-1} q$ in the current simplex tabular.

Follow the lexicographical rule to find the pivot index r and the pivot element $\bar{h}_{r\bar{j}}$, where $(B_k^{-1})_i$ is the i-th row vector of the current inverse matrix B_k^{-1}:

$$lex\ \min \left\{ \left(\frac{\bar{q}_i}{\bar{h}_{i\bar{j}}}, \frac{(B_k^{-1})_i}{\bar{h}_{i\bar{j}}} \right) | \bar{h}_{i\bar{j}} > 0 \right\} = \left(\frac{\bar{q}_r}{\bar{h}_{r\bar{j}}}, \frac{(B_k^{-1})_r}{\bar{h}_{r\bar{j}}} \right)$$

Let

$$I_k := \begin{bmatrix} 1 & & -\bar{h}_{1\bar{j}}/\bar{h}_{r\bar{j}} & & \\ & \ddots & \vdots & & \\ & & 1/\bar{h}_{r\bar{j}} & & \\ & & \vdots & \ddots & \\ & & -\bar{h}_{m\bar{j}}/\bar{h}_{r\bar{j}} & & 1 \end{bmatrix}$$

and update $B_{k+1}^{-1} := I_k B_k^{-1}$.

Obtain the new cost coefficient vector $f_{B_{k+1}}$ by changing the r-th element in f_{B_k} to 0 if Step 3.2 is implemented or with d_{j*} if Step 3.3 is implemented.

Calculate the new simplex multiplier $\pi^{k+1} := f_{B_{k+1}} B_{k+1}^{-1}$.

Update $k := k + 1$. Go to Step 2.

10.4 (Inv-sum-cost, MSST, Bounded, Hamming Distance)

In this section, we introduce the bounded inverse max+sum spanning tree problem by modifying the sum-cost vector c under bottleneck Hamming distance (IMSST$_{bH}^b$) and unbounded inverse MSST problem by modifying the sum-cost vector c under weighted sum Hamming distance (IMSST$^u_{wsH}$) [50]. Their mathematical models can be stated as follows:

$$(IMSST^b_{bH})\quad \begin{aligned} &\min \max_{e\in E} q(e)H(\bar{c}(e), c(e)) \\ &s.t.\ \max_{e\in T_0} w(e) + \sum_{e\in T_0} \bar{c}(e) \leq \max_{e\in T} w(e) + \sum_{e\in T} \bar{c}(e), \forall T \in \Gamma. \\ &c(e) - l(e) \leq \bar{c} \leq c(e) + u(e), \forall e \in E. \end{aligned}$$

$$(IMSST^u_{wsH})\quad \begin{aligned} &\min \sum_{e\in E} H(\bar{c}(e), c(e)) \\ &s.t.\ \max_{e\in T_0} w(e) + \sum_{e\in T_0} \bar{c}(e) \leq \max_{e\in T} w(e) + \sum_{e\in T} \bar{c}(e), \forall T \in \Gamma. \end{aligned}$$

Let

$$\beta(e) = \begin{cases} c(e) - \bar{c}(e), & \text{if } e \in T_0, \\ \bar{c}(e) - c(e), & \text{if } e \notin T_0. \end{cases} \tag{10.16}$$

Lemma 10.6 ([50, Lemma 1]) *There is an optimal solution β^* of the problem (IMSST$^b_{bH}$) such that $\beta^*(e) \geq 0$.*

Using Lemma 10.6, the problem (IMSST$^b_{bH}$) can be reformed as follows:

$$(IMSST^b_{bH})\quad \begin{aligned} &\min \max_{e\in E} q(e)H(\beta(e), 0) \\ &s.t.\ \sum_{e\in T\setminus T_0} \beta(e) + \sum_{e\in T_0\setminus T} \beta(e) \geq f(w, c, T_0) - f(w, c, T), \forall T \in \Gamma, \\ &0 \leq \beta(e) \leq l(e), \forall e \in T_0. \\ &0 \leq \beta(e) \leq u(e). \forall e \notin T_0. \end{aligned}$$

Let

$$\tilde{c}(e) := \begin{cases} c(e) - l(e), & \text{if } e \in T_0, \\ c(e) + u(e), & \text{if } e \notin T_0. \end{cases} \tag{10.17}$$

Lemma 10.7 ([50, Lemma 2]) *Let $\tilde{T}$ be an optimal max+sum spanning tree of network $G(V, E, \tilde{c}, w)$. If $f(w, \tilde{c}, T_0) = f(w, \tilde{c}, \tilde{T})$, then the inverse problems (IMSST$^b_{bH}$) and (IMSST$^b_{wsH}$) are feasible. If $f(w, \tilde{c}, T_0) > f(w, \tilde{c}, \tilde{T})$, then the inverse problems (IMSST$^b_{bH}$) and (IMSST$^b_{wsH}$) are infeasible.*

First, sort the values $\{q(e)|e \in E\}$ in a strictly increasing order. Assume

$$q(e_{i_1}) < q(e_{i_2}) < \cdots < q(e_{i_\tau}).$$

It follows from the bottleneck Hamming distance in the objective function of the problem (IMSST$^b_{bH}$), and there is an index $k \in \{1, 2, \cdots, \tau\}$ such that the optimal objective value is $q(e_{i_k})$. Let $E_k = \{e \in E | q(e) \leq q(e_{i_k})\}$. Then we define a new vector:

$$\bar{c}_k(e) := \begin{cases} c(e) - l(e), & \text{if } e \in T_0 \cap E_k, \\ c(e) + u(e), & \text{if } e \in E_k \setminus T_0, \\ c(e), & \text{if } e \notin E_k. \end{cases} \tag{10.18}$$

Lemma 10.8 ([50, Lemma 3]) *Let $\bar{T}_k$ be an optimal max+sum spanning tree of network $G(V, E, \bar{c}_k, w)$. If $f(w, \bar{c}_k, T_0) = f(w, \bar{c}_k, \bar{T}_k)$, then the optimal value Q^* satisfies the inequality $Q^* \leq q(e_{i_k})$. If $f(w, \bar{c}_k, T_0) > f(w, \bar{c}_k, \bar{T}_k)$, then the optimal value Q^* satisfies the inequality $Q^* > q(e_{i_k})$.*

Algorithm 10.4 A binary search method to solve (IMSST$^b_{bH}$) [50, Algorithm 4]

Input: A network $G = (V, E)$ with a weight vector w, a cost vector c, and three vectors q, l, and u on E.
Output: The optimal value $q(e_{i_b})$ and an optimal solution $\bar{c}^*$.

Step 1: Check feasibility.

Let $\tilde{T}$ be an optimal max+sum spanning tree on the network $G(V, E, \tilde{c}, w)$, where $\tilde{c}$ is defined as in (10.17). If $f(w, \tilde{c}, T_0) = f(w, \tilde{c}, \tilde{T})$, then the instance is feasible; otherwise, it is infeasible.

Step 2: Sort the values.

Sort the values $\{q(e)|e \in E\}$ in a strictly ascending order. Assume $q(e_{i_1}) < q(e_{i_2}) < \cdots < q(e_{i_\tau})$. Initialize $a := 1$ and $b := \tau$.

Step 3: Find the optimal objective value.

while $b - a > 1$ **do**
  Let $k := \lfloor \frac{a+b}{2} \rfloor$, $E_k := \{e \in E | q(e) \leq q(e_{i_k})\}$.
  Find an optimal max+sum spanning tree $\bar{T}_k$ on network $G(V, E, \bar{c}_k, w)$, where $\bar{c}_k$ is defined as in (10.18).
  if $f(w, \bar{c}_k, \bar{T}_k) = f(w, \bar{c}_k, T_0)$ **then** update $b := k$.
  else update $a := k$.
  end if
end while

Step 4: Output the optimal value and optimal solution.

Return The optimal value $q(e_{i_b})$ and an optimal solution below:

$$\bar{c}^*(e) := \begin{cases} c(e) - l(e), & \text{if } e \in E_b \cap T_0, \\ c(e) + u(e), & \text{if } e \in E_b \setminus T_0, \\ c(e), & \text{if } e \notin E_b. \end{cases} \tag{10.19}$$

Theorem 10.6 ([50, Theorem 6]) *The problem (IMSST$^b_{bH}$) can be solved by Algorithm 10.4 in $O(m\log^2 n)$ operations.*

Next, we study the problem (IMSST$^u_{wsH}$). Using Lemma 10.6, the problem (IMSST$^u_{wsH}$) can be reformed as follows:

$$\begin{aligned}&\min\ \|\beta\|_0\\ &s.t.\ \sum_{e\in T\setminus T_0}\beta(e)+\sum_{e\in T_0\setminus T}\beta(e)\geq w^b(T_0)+\sum_{e\in T_0\setminus T}c(e)-w^b(T)-\sum_{e\in T\setminus T_0}c(e),\ T\in\Gamma\\ &\quad\ \beta(e)\geq 0,\ \forall\, e\in T_0.\end{aligned}$$

The matrix form of the problem (IMSST$^u_{wsH}$) can be stated as follows by (10.14):

$$\begin{aligned}&\min\ \|\beta\|_0\\ (IMSST^u_{wsH})\ \ &s.t.\ A\beta\geq d,\\ &\quad\ \ \beta\geq 0.\end{aligned}\tag{10.20}$$

Theorem 10.7 ([50, Lemma 11]) *The problem (IMSST$^u_{wsH}$) is $\mathcal{NP}$-hard.*

10.5 (Inv-max-weight, MSST, Bounded, Weighted l_∞ Norm)

In this section, we introduce the inverse max+sum spanning tree problem by modifying the max weight vector w under weighted l_∞ norm [67]. Its mathematical model can be stated as follows:

$$\begin{aligned}&\min\ \max_{e\in E} q(e)|\bar w(e)-w(e)|\\ (IMSST^b_\infty)\ \ &s.t.\ \max_{e\in T_0}\bar w(e)+\sum_{e\in T_0}c(e)\leq\max_{e\in T}\bar w(e)+\sum_{e\in T}c(e),\ \forall T\in\Gamma,\\ &\quad\ \ w(e)-l(e)\leq\bar w(e)\leq w(e)+u(e),\ \forall e\in E.\end{aligned}$$

For a given value $Q\geq 0$, let

$$\Delta(Q)=\max_{e\in T^0}\left\{w(e)-\min\left\{l(e),\frac{Q}{q(e)}\right\}\right\}.\tag{10.21}$$

Construct a vector w_Q as follows:

$$w_Q(e)=\begin{cases}\Delta(Q), & if\ e=e^w_{T_0},\\ w(e)+\min\{\frac{Q}{q(e)},\Delta(Q)-w(e),u(e)\}, & if\ e\in T_0\setminus\{e^w_{T_0}\},\\ w(e)+\min\{\frac{Q}{q(e)},u(e)\}, & if\ e\in E\setminus T_0.\end{cases}\tag{10.22}$$

where $e^w_{T_0} = \arg\max_{e\in T_0} w(e)$. Then the vector w_Q has the following properties:

Theorem 10.8 ([67, Cor 10]) *Let Q^* be the optimal objective value and $T_Q = \arg\min_{T\in\Gamma} f(w_Q, c, T)$. Then the following two results hold:*

(1) $Q \geq Q^*$ *if and only if* $f(w_Q, c, T_Q) = f(w_Q, c, T^0)$.
(2) $Q < Q^*$ *if and only if* $f(w_Q, c, T_Q) < f(w_Q, c, T^0)$.

Based on Theorem 10.8, we can use a binary search method to determine the interval $(\lambda_\tau, \lambda_{\tau+1}]$ where the optimal value Q^* lies in, where $\lambda_{\tau+1} \geq Q^*$, $\lambda_\tau < Q^*$. When the interval $(\lambda_\tau, \lambda_{\tau+1}]$ is small enough, we can find the exact value Q^*. However, we can hardly ensure that the number of iterations in the binary search method is upper bounded by a polynomial function $P(m, n)$ of the numbers m, n of edges and vertices, since it depends on the precision and the maximum value

$$\bar{Q} := \max\left\{\max_{e\in T^0}\{q(e)l(e)\}, \max_{e\in E}\{q(e)u(e)\}\right\}. \tag{10.23}$$

In order to find the optimal value Q^* in polynomial time, we need to perform a binary search Algorithm 10.5 on a set of $O(m^2)$ values to find the interval $(\lambda_\tau, \lambda_{\tau+1}]$ where the optimal value Q^* lies in.

Algorithm 10.5 $(\lambda_\tau, \lambda_{\tau+1}] :=$ **BinarySearch**$(\Lambda, T^0, \boldsymbol{w}, \boldsymbol{l}, \boldsymbol{u}, \boldsymbol{c}, \boldsymbol{q})$ [67, Alg. 1]

Input: A network $G = (V, E)$, a spanning tree T^0, a weight vector $\boldsymbol{w}$, a lower bound vector $\boldsymbol{l}$, an upper bound vector $\boldsymbol{u}$, a cost vector $\boldsymbol{c}$, a modification cost vector $\boldsymbol{q}$, and the set $\Lambda := \{\lambda_j\}$ of values.
Output: The interval $(\lambda_\tau, \lambda_{\tau+1}]$.

```
Sort the different values in Λ increasingly.
Initialize a := 1, b := |Λ|.
if b ≤ 2 then output (λ_τ, λ_{τ+1}] := (min Λ, max Λ].
else
    while b − a > 1 do
        Let k := ⌊(b + a)/2⌋. Construct w_{λ_k} by (10.22).
        Calculate the optimal MSST T_{λ_k} under w_{λ_k}.
        if f(w_{λ_k}, c, T^0) = f(w_{λ_k}, c, T_{λ_k}) then update b := k.
        else update a := k.
        end if
    end while
    Output (λ_τ, λ_{τ+1}] := (λ_a, λ_b].
end if
```

Next, we propose Lemma 10.9 to judge the feasibility of the problem (**IMSST$_\infty$**). Define

$$\theta = \max_{e \in T^0}\{w(e) - l(e)\}. \tag{10.24}$$

Based on the relationship between θ and $\Delta(Q)$, we have

$$\Delta(Q) = \max\left\{\theta, \max_{e \in T^0}\left\{w(e) - \frac{Q}{q(e)}\right\}\right\} \tag{10.25}$$

.

Lemma 10.9 ([67, Lemma 11]) *Construct w_0 as follows:*

$$w_0(e) = \begin{cases} \theta, & if\ e = e^w_{T^0}, \\ \min\{w(e) + u(e), \theta\}, & if\ e \in T^0 \setminus \{e^w_{T^0}\}, \\ w(e) + u(e), & if\ e \in E \setminus T^0. \end{cases} \tag{10.26}$$

*Let $f(w_0, c, T_{w_0}) = \min_{T \in \Gamma} f(w_0, c, T)$. The problem (**IMSST$_\infty$**) is feasible if and only if $f(w_0, c, T^0) = f(w_0, c, T_{w_0})$.*

In the following discussion, we assume that the problem (**IMSST$_\infty$**) is feasible.

Definition 10.1 ([67, Definition 1]) Suppose $Q_l, Q_r \in [Q_1, Q_2]$ and $Q_l < Q_r$. Let w_{Q_l}, w_{Q_r} be constructed by (10.22). Let $E' = \{e_{i_1}, e_{i_2}, \cdots, e_{i_s}\} \subset E$. We call the interval $[Q_1, Q_2]$ has an order invariance with respect to E'; if $w_{Q_l}(e_{i_1}) \le w_{Q_l}(e_{i_2}) \le \cdots \le w_{Q_l}(e_{i_s})$ holds, then $w_{Q_r}(e_{i_1}) \le w_{Q_r}(e_{i_2}) \le \cdots \le w_{Q_r}(e_{i_s})$ holds.

Lemma 10.10 ([67, Lemma 12]) *Suppose $Y = \{y_i = a_i x_i + b_i | i = 1, 2, \ldots, N_1\}$ is a family of different lines and $X = [X_1, X_2, \ldots, X_{N_2}]$ is the set of $x-$coordinates of intersection points of all lines in Y satisfying $X_j < X_{j+1}$, $j = 1, 2, \ldots, N_2 - 1$. For any $x_j \in [X_j, X_{j+1}]$, $j = 1, 2, \ldots, N_2 - 1$, letting $\hat{x}_j = \frac{X_j + X_{j+1}}{2}$, if $y_{i_1}(\hat{x}_j) < y_{i_2}(\hat{x}_j) < \cdots < y_{i_{N_1}}(\hat{x}_j)$, then $y_{i_1}(x_j) \le y_{i_2}(x_j) \le \cdots \le y_{i_{N_1}}(x_j)$.*

For $\theta = \max_{e \in T^0}\{w(e) - l(e)\}$, we define

$$Q_\theta = \max_{e \in T^0}\{q(e)(w(e) - \theta)\}, \tag{10.27}$$

$$e_\theta = \arg\max_{e \in T^0}\{q(e)(w(e) - \theta)\}. \tag{10.28}$$

Lemma 10.11 ([67, Lemma 13]) *(1) If $Q_\theta < Q^*$, then we have $\Delta(Q) = \theta$ for any $Q \in [Q_\theta, Q^*]$. (2) If $Q_\theta \ge Q^*$ and then there is a lower bound Q^0_τ satisfying $Q^0_\tau < Q^*$ and a tree edge $e_0 \in T^0$, we have $\Delta(Q) = w(e_0) - \frac{Q}{q(e_0)}$ for any $Q \in [Q^0_\tau, Q^*]$.*

In the next discussion, we assume $Q_\theta < Q^*$, and then we have $\Delta(Q) = \theta$ for any $Q \in [Q_\theta, Q^*]$ by Lemma 10.11. Firstly, we propose a lemma to calculate the three subsets

$$T_\theta^0 = \{e \in T^0 | w_Q(e) = \theta\},$$
$$E_u = \{e \in E | w_Q(e) = w(e) + u(e)\}, \text{ and}$$
$$E_Q = \left\{e \in E | w_Q(e) = w(e) + \frac{Q}{q(e)}\right\}$$

and a lower bound $\underline{Q}$ of the optimal value. Secondly, we raise the lower bound $\underline{Q}$ to Q_τ^1 such that the interval $[Q_\tau^1, Q^*]$ has the order invariance with respect to E. Finally, we propose an algorithm to calculate the three subsets T_θ^0, E_u, and E_Q and the two values Q_τ^1 and $Q_{\tau+1}^1$ satisfying $Q_\tau^1 < Q^* \le Q_{\tau+1}^1$.

To describe the three subsets T_θ^0, E_u, and E_Q precisely, we first define some subsets of edges. Let

$$T_\theta^{01} = \{e_i \in T^0 | w(e_i) \ge \theta\},$$
$$T_\theta^{02} = \{e_i \in T^0 | w(e_i) < \theta, (\theta - w(e_i))q(e_i) < Q^*, w(e_i) + u(e_i) \ge \theta\},$$
$$\overline{T}_u^0 = \{e_i \in E \setminus T^0 | q(e_i)u(e_i) < Q^*\},$$
$$T_u^{01} = \{e_i \in T^0 | w(e_i) < \theta, (\theta - w(e_i))q(e_i) \ge Q^*, q(e_i)u(e_i) < Q^*\},$$
$$T_u^{02} = \{e_i \in T^0 | w(e_i) < \theta, (\theta - w(e_i))q(e_i) < Q^*, w(e_i) + u(e_i) < \theta\},$$
$$\overline{T}_Q^0 = \{e_i \in E \setminus T^0 | q(e_i)u(e_i) \ge Q^*\},$$
$$T_Q^0 = \{e_i \in T^0 | w(e_i) < \theta, (\theta - w(e_i))q(e_i) \ge Q^*, q(e_i)u(e_i) \ge Q^*\},$$

$$T_\theta^0 = T_\theta^{01} \cup T_\theta^{02}, \quad E_u = \overline{T}_u^0 \cup T_u^{01} \cup T_u^{02}, \quad E_Q = \overline{T}_Q^0 \cup T_Q^0, \tag{10.29}$$

$$\underline{Q} = \max\left\{Q_\theta, \max_{e_i \in T_\theta^{02} \cup T_u^{02}} \{(\theta - w(e_i))q(e_i)\}, \max_{e_i \in \overline{T}_u^0 \cup T_u^{01}} \{q(e_i)u(e_i)\}\right\}. \tag{10.30}$$

Lemma 10.12 ([67, Lemma 14]) *For any $Q \in [\underline{Q}, Q^*]$, w_Q defined as in (10.22) can be calculated by*

$$w_Q(e) = \begin{cases} \theta, & if\ e \in T_\theta^0, \\ w(e) + u(e), & if\ e \in E_u, \\ w(e) + \frac{Q}{q(e)}, & if\ e \in E_Q. \end{cases} \tag{10.31}$$

Next, we devise Algorithm 10.6 to find the three subsets T^0_θ, E_u, and E_Q and the two values $Q^1_\tau and Q^1_{\tau+1}$ satisfying $\underline{Q} \le Q^1_\tau < Q^* \le Q^1_{\tau+1}$, even if we did not know the optimal value Q^*. The main theoretical basis is that we can judge the feasibility of a given value Q by whether the equality $f(w_Q, c, T^0) = f(w_Q, c, T_{w_Q})$ holds or not, where $T_Q = \arg\min_{T\in\Gamma} f(w_Q, c, T)$. The main idea is as follows: For any nontree edges $e_i \in E \setminus T^0$, if $Q_i := q(e_i)u(e_i)$ is a feasible objective value, then $e_i \in \overline{T}^0_Q \subset E_Q$; otherwise, $e_i \in \overline{T}^0_u \subset E_u$. For any tree edges $e_i \in T^0$, if $w(e_i) \ge \theta$, then $e_i \in T^{01}_\theta \subset T^0_\theta$; otherwise, if $Q'_i := (\theta - w(e_i))q(e_i)$ and $Q_i := q(e_i)u(e_i)$ are both feasible objective values, then $e_i \in T^0_Q \subset E_Q$; if Q'_i is but Q_i is not a feasible objective value, then $e_i \in T^{01}_u \subset E_u$; if Q'_i is not feasible and $w(e_i) + u(e_i) < \theta$, then $e_i \in T^{02}_u \subset E_u$; if Q'_i is not feasible and $w(e_i) + u(e_i) \ge \theta$, then $e_i \in T^{02}_\theta \subset T^0_\theta$.

Algorithm 10.6 $(T^0_\theta, E_u, E_Q, Q^1_\tau, Q^1_{\tau+1}) :=$ **Findsets1**$(V, E, T^0, \boldsymbol{w}, \boldsymbol{c}, \boldsymbol{q}, \boldsymbol{l}, \boldsymbol{u})$ [67, Algorithm 2]

Input: A network $G = (V, E)$, a spanning tree T^0, a weight vector $\boldsymbol{w}$, a lower bound vector $\boldsymbol{l}$, an upper bound vector $\boldsymbol{u}$, a cost vector $\boldsymbol{c}$ and a modification cost vector $\boldsymbol{q}$.
Output: $T^0_\theta, E_u, E_Q, Q^1_\tau, Q^1_{\tau+1}$.

Let $T^{01}_\theta, T^{02}_\theta, \overline{T}^0_u, T^{01}_u, T^{02}_u, \overline{T}^0_Q, T^0_Q := \emptyset$.
Calculate $\bar{Q}, \theta, Q_\theta$ by (10.23), (10.24), (10.27), respectively.
for $i = 1$ to m **do**
  if $e_i \in E \setminus T^0$ **then** $Q_i := q(e_i)u(e_i)$, construct w_{Q_i} by (10.22), and calculate the optimal **MSST** $T_{w_{Q_i}}$ under w_{Q_i}.
    if $f(w_{Q_i}, c, T^0) = f(w_{Q_i}, c, T_{w_{Q_i}})$ **then** $\overline{T}^0_Q := \overline{T}^0_Q \cup \{e_i\}$.
    else $\overline{T}^0_u := \overline{T}^0_u \cup \{e_i\}$.
    end if
  else if $w(e_i) \ge \theta$ **then** $T^{01}_\theta := T^{01}_\theta \cup \{e_i\}$.
  else $Q'_i := (\theta - w(e_i))q(e_i)$, construct $w_{Q'_i}$ by (10.22), and calculate the optimal **MSST** $T_{w_{Q'_i}}$ under $w_{Q'_i}$.
    if $f(w_{Q'_i}, c, T^0) = f(w_{Q'_i}, c, T_{w_{Q'_i}})$ **then** $Q_i := q(e_i)u(e_i)$, construct w_{Q_i} by (10.22), and calculate the optimal **MSST** $T_{w_{Q_i}}$ under w_{Q_i}.
      if $f(w_{Q_i}, c, T^0) = f(w_{Q_i}, c, T_{w_{Q_i}})$ **then** $T^0_Q := T^0_Q \cup \{e_i\}$.
      else $T^{01}_u := T^{01}_u \cup \{e_i\}$.
      end if
    else if $w(e_i) + u(e_i) < \theta$ **then** $T^{02}_u := T^{02}_u \cup \{e_i\}$.
    else $T^{02}_\theta := T^{02}_\theta \cup \{e_i\}$.
    end if
  end if
end for

19: Let $T_\theta^0 := T_\theta^{01} \cup T_\theta^{02}$, $E_u := \overline{T}_u^0 \cup T_u^{01} \cup T_u^{02}$, $E_Q := \overline{T}_Q^0 \cup T_Q^0$.
20: Calculate $\underline{Q}$ by (10.30).
21: Let $Q_{s_1} := \{Q_{ij} | Q_{ij} = \frac{(w(e_i)-w(e_j))q(e_i)q(e_j)}{q(e_i)-q(e_j)}, q(e_i) \neq q(e_j), e_i, e_j \in E_Q\}$.
22: $Q_{s_2} := \{Q_{ij} | Q_{ij} = q(e_j)(w(e_i) + u(e_i) - w(e_j)), e_i \in E_u, e_j \in E_Q\}$.
23: $Q_{s_3} := \{Q_i | Q_i = (\theta - w(e_i))q(e_i), e_i \in E_Q\}$.
24: $Q_s^1 := Q_{s_1} \cup Q_{s_2} \cup Q_{s_3} \cup \{\underline{Q}, \bar{Q}\}$.
25: $(Q_\tau^1, Q_{\tau+1}^1] :=$ BinarySearch$(Q_s, T^0, \boldsymbol{w}, \boldsymbol{l}, \boldsymbol{u}, \boldsymbol{c}, \boldsymbol{q})$.

Lemma 10.13 ([67, Lemma 15]) *The three subsets T_θ^0, E_u, and E_Q and the two values Q_τ^1 and $Q_{\tau+1}^1$ calculated by Algorithm 10.6 satisfy the following properties:*

(1) T_θ^0, E_u, E_Q remain unchanged for any $Q \in [Q_\tau^1, Q^*]$.
(2) $\underline{Q} \leq Q_\tau^1 < Q^* \leq Q_{\tau+1}^1$.
(3) The interval $[Q_\tau^1, Q^*]$ has the order invariance with respect to E.

In the next discussion, assuming $Q_\theta \geq Q^*$ and then there is a lower bound Q_τ^0 satisfying $Q_\tau^0 < Q^*$ and a tree edge $e_0 \in T^0$, we have $\Delta(Q) = w(e_0) - \frac{Q}{q(e_0)}$ for any $Q \in [Q_\tau^0, Q^*]$ by Lemma 10.11. We first propose a lemma to calculate the three subsets

$$T_\Delta^0 = \{e \in T^0 | w_Q(e) = \Delta(Q)\},$$
$$E_u = \{e \in E | w_Q(e) = w(e) + u(e)\}, \text{and}$$
$$E_Q = \left\{e \in E | w_Q(e) = w(e) + \frac{Q}{q(e)}\right\}$$

and a lower bound $\underline{Q}_0$ of the optimal value. Secondly, we raise the lower bound $\underline{Q}_0$ to Q_τ^2 such that the interval $[Q_\tau^2, Q^*]$ has the order invariance with respect to E. Finally, we propose an algorithm to calculate the three subsets $T_\Delta^0, E_u, and E_Q$, a tree edge e_0, and the two values Q_τ^2 and $Q_{\tau+1}^2$ satisfying $Q_\tau^2 < Q^* \leq Q_{\tau+1}^2$.

To describe the three subsets $T_\Delta^0, E_u, and E_Q$ precisely, we first define some subsets of edges. For any $e_i \in T^0$, let $Q_i := \frac{(w(e_0)-w(e_i))q(e_0)q(e_i)}{q(e_0)+q(e_i)}$, $Q_i' := (w(e_0) - w(e_i) - u(e_i))q(e_0)$,

$$T_\Delta^{01} = \{e_i \in T^0 | w(e_i) \geq w_{Q_\tau^0}(e_0)\},$$
$$T_\Delta^{02} = \{e_i \in T^0 | w(e_i) < w_{Q_\tau^0}(e_0), Q_i < Q^*, Q_i' < Q^*\},$$
$$\overline{T}_u^0 = \{e_i \in E \setminus T^0 | q(e_i)u(e_i) < Q^*\},$$
$$T_u^{01} = \{e_i \in T^0 | w(e_i) < w_{Q_\tau^0}(e_0), Q_i \geq Q^*, q(e_i)u(e_i) < Q^*\},$$
$$T_u^{02} = \{e_i \in T^0 | w(e_0) < w_{Q_\tau^0}(e_0), Q_i < Q^* \leq Q_i'\},$$

$$\overline{T}^0_Q = \{e_i \in E \setminus T^0 | q(e_i)u(e_i) \geq Q^*\},$$

$$T^0_Q = \{e_i \in T^0 | w(e_i) < w_{Q^0_\tau}(e_0), Q_i \geq Q^*, q(e_i)u(e_i) \geq Q^*\},$$

$$T^0_\Delta = T^{01}_\Delta \cup T^{02}_\Delta, \tag{10.32}$$

$$E_u = \overline{T}^0_u \cup T^{01}_u \cup T^{02}_u, \tag{10.33}$$

$$E_Q = \overline{T}^0_Q \cup T^0_Q, \tag{10.34}$$

$$\underline{Q}_0 = \max\left\{Q^0_\tau, \max_{e_i \in \overline{T}^0_u \cup T^{01}_u} \{q(e_i)u(e_i)\}, \max_{e_i \in T^{02}_\Delta \cup T^{02}_u} \{Q_i\}, \max_{e_i \in T^{02}_\Delta} \{Q'_i\}\right\}. \tag{10.35}$$

Lemma 10.14 ([67, Lemma 17]) *For any $Q \in [\underline{Q}_0, Q^*]$, w_Q defined as in (10.22) can be calculated by*

$$w_Q(e) = \begin{cases} \Delta(Q), & if\ e \in T^0_\Delta \\ w(e) + u(e), & if\ e \in E_u \\ w(e) + \frac{Q}{q(e)}, & if\ e \in E_Q \end{cases} \tag{10.36}$$

Next, we devise Algorithm 10.7 to find the three subsets $T^0_\Delta, and E_u, E_Q$, a tree edge e_0, and the two values $Q^2_\tau and Q^2_{\tau+1}$ satisfying $\underline{Q}_0 \leq Q^2_\tau < Q^* \leq Q^2_{\tau+1}$, even if we did not know the optimal value Q^*. The main idea is as follows: For any nontree edges $e_i \in E \setminus T^0$, if $Q''_i := q(e_i)u(e_i)$ is a feasible objective value, then $e_i \in \overline{T}^0_Q$; otherwise, $e_i \in \overline{T}^0_u$. For any tree edges $e_i \in T^0$, if $w(e_i) \geq w_{Q^0_\tau}(e_0)$, then $e_i \in T^{01}_\Delta$; otherwise, if Q_i and Q'_i are not feasible objective values, then $e_i \in T^{02}_\Delta$; if Q_i is a feasible objective value but Q''_i is not, then $e_i \in T^{01}_u$; if Q_i is not a feasible objective value but Q'_i is, then $e_i \in T^{02}_u$; if Q_i and Q''_i are both feasible objective values, then $e_i \in T^0_Q$.

Algorithm 10.7
$(e_0, T^0_\Delta, E_u, E_Q, Q^2_\tau, Q^2_{\tau+1}) :=$ **Findsets2**$(V, E, T^0, \boldsymbol{w}, \boldsymbol{c}, \boldsymbol{q}, \boldsymbol{l}, \boldsymbol{u})$ [67, Algorithm 3]

Input: A network $G = (V, E)$, a spanning tree T^0, a weight vector $\boldsymbol{w}$, a lower bound vector $\boldsymbol{l}$, an upper bound vector $\boldsymbol{u}$, a cost vector $\boldsymbol{c}$, and a modification cost vector $\boldsymbol{q}$.
Output: $e_0, T^0_\Delta, E_u, E_Q, Q^2_\tau, Q^2_{\tau+1}$.
1: Calculate θ, Q_θ by (10.24) and (10.27), respectively.
2: $\Lambda^0 := \{Q_{ij} = \frac{(w(e_i)-w(e_j))q(e_i)q(e_j)}{q(e_j)-q(e_i)} | q(e_i) \neq q(e_j), e_i, e_j \in T^0\} \cup \{0, Q_\theta\}$.
3: $(Q^0_\tau, Q^0_{\tau+1}] =$**BinarySearch**$(\Lambda^0, T^0, \boldsymbol{w}, \boldsymbol{l}, \boldsymbol{u}, \boldsymbol{c}, \boldsymbol{q})$.
4: Let $Q_0 := \frac{Q^0_\tau + Q^0_{\tau+1}}{2}$.

$e_0 := \arg\max_{e_i \in T_0}\{w(e_i) - \frac{Q_0}{q(e_i)}\}$.
Construct $w^0_{Q_\tau}$ by (10.22).
Let $T^{01}_\Delta, T^{02}_\Delta, \overline{T}^0_u, T^{01}_u, T^{02}_u, \overline{T}^0_Q, T^0_Q := \emptyset$.
for $i = 1$ to m **do**
  if $e_i \in E \setminus T^0$ **then** $Q_i := q(e_i)u(e_i)$, construct w_{Q_i} by (10.22), and calculate the optimal **MSST** $T_{w_{Q_i}}$ under w_{Q_i}.
    if $f(w_{Q_i}, c, T^0) = f(w_{Q_i}, c, T_{w_{Q_i}})$ **then** $\overline{T}^0_Q := \overline{T}^0_Q \cup \{e_i\}$.
    else $\overline{T}^0_u := \overline{T}^0_u \cup \{e_i\}$.
    end if
  else if $w(e_i) \geq w^0_{Q_\tau}(e_0)$ **then** $T^{01}_\Delta := T^{01}_\Delta \cup \{e_i\}$.
  else Let $Q_i := \frac{(w(e_0)-w(e_i))q(e_0)q(e_i)}{q(e_0)+q(e_i)}$, $Q'_i := (w(e_0) - w(e_i) - u(e_i))q(e_0)$, $Q''_i := q(e_i)u(e_i)$, construct $w_{Q_i}, w_{Q'_i}, w_{Q''_i}$ by (10.22), and calculate the optimal **MSST** $T_{w_{Q_i}}, T_{w_{Q'_i}}, T_{w_{Q''_i}}$ under $w_{Q_i}, w_{Q'_i}, w_{Q''_i}$.
    if $f(w_{Q_i}, c, T_{w_{Q_i}}) < f(w_{Q_i}, c, T_0)$ **then**
      if $f(w_{Q'_i}, c, T_{w_{Q'_i}}) < f(w_{Q'_i}, c, T_0)$ **then** $T^{02}_\Delta := T^{02}_\Delta \cup \{e_i\}$.
      else $T^{02}_u := T^{02}_u \cup \{e_i\}$.
      end if
    else if $f(w_{Q''_i}, c, T_{w_{Q''_i}}) < f(w_{Q''_i}, c, T_0)$ **then** $T^{01}_u := T^{01}_u \cup \{e_i\}$.
    else $T^0_Q := T^0_Q \cup \{e_i\}$.
    end if
  end if
end for
Let $T^0_\Delta := T^{01}_\Delta \cup T^{02}_\Delta$, $E_u := \overline{T}^0_u \cup T^{01}_u \cup T^{02}_u$, $E_Q := \overline{T}^0_Q \cup T^0_Q$.
Calculate $\underline{Q}_0$ by (10.35).
Let $Q_{s_1} := \{Q_{ij} | Q_{ij} = \frac{(w(e_i)-w(e_j))q(e_i)q(e_j)}{q(e_i)-q(e_j)}, q(e_i) \neq q(e_j), e_i, e_j \in E_Q\}$
$Q_{s_2} := \{Q_i | Q_i = (w(e_0) - w(e_i) - u(e_i))q(e_0), e_i \in E_u\}$
$Q_{s_3} := \{Q_i = \frac{(w(e_0)-w(e_i))q(e_0)q(e_i)}{q(e_0)+q(e_i)} | e_i \in E_Q\}$.
$Q_{s_4} := \{Q_{ij} = (w(e_i) + u(e_i) - w(e_j))q(e_j) | e_i \in E_u, e_j \in E_Q\}$
$Q^2_s := Q_{s_1} \cup Q_{s_2} \cup Q_{s_3} \cup Q_{s_4} \cup \{\underline{Q}_0, Q_\theta\}$.
$(Q^2_\tau, Q^2_{\tau+1}] :=$ **BinarySearch** $(Q^2_s, T^0, \boldsymbol{w}, \boldsymbol{l}, \boldsymbol{u}, \boldsymbol{c}, \boldsymbol{q})$.

Lemma 10.15 ([67, Lemma 18]) *The three sets T^0_Δ, E_u, and E_Q, the two values Q^2_τ, $Q^2_{\tau+1}$, and the tree edge e_0 outputted by Algorithm 10.7 satisfy the following properties:*

(1) For any $Q \in [Q^2_\tau, Q^*]$, we have $\Delta(Q) = w(e_0) - \frac{Q}{q(e_0)}$.
(2) T^0_Δ, E_u, and E_Q remain unchanged for any $Q \in [Q^2_\tau, Q^*]$.
(3) $Q^2_\tau < Q^* \leq Q^2_{\tau+1}$.
(4) The interval $[Q^2_\tau, Q^*]$ has the order invariance with respect to E.

Finally, we devise an algorithm to solve the problem (**IMSST**$_\infty$). The main idea is to start iterating from Q_τ^1 or Q_τ^2 according to $Q_\theta < Q^*$ or $Q_\theta \geq Q^*$. In each iteration, we need to determine a minimum increment Q_λ such that $Q + Q_\lambda \leq Q^*$ for any infeasible objective value $Q \in [Q_\tau^1, Q^*)$ or $Q \in [Q_\tau^2, Q^*)$.

For simplicity, we denote the optimal solution w_{Q^*} by w^*. To determine the minimum increment Q_λ, we propose two lemmas:

Lemma 10.16 ([67, Lemma 20]) *Suppose* $Q_\theta < Q^*$ *and* $(T_\theta^0, E_u, E_Q, Q_\tau^1, Q_{\tau+1}^1) :=$ ***Findsets1***$(V, E, T^0, \boldsymbol{w}, \boldsymbol{c}, \boldsymbol{q}, \boldsymbol{l}, \boldsymbol{u})$. *For any* $Q \in [Q_\tau^1, Q^*)$, *letting* $\lambda_Q = f(w_Q, c, T^0) - f(w_Q, c, T_Q) > 0$, $Q_\lambda = \lambda_Q \cdot q(e_Q)$ *and* $\tilde{Q} = Q + Q_\lambda$, *then* $\tilde{Q} \leq Q^*$ *and* $f(w_{\tilde{Q}}, c, T_0) = f(w_{\tilde{Q}}, c, T_Q)$, *where* $T_Q = \arg\min_{T\in\Gamma} f(w_Q, c, T)$ *and* $e_Q = \arg\max_{e\in T_Q} w_Q(e)$.

Lemma 10.17 ([67, Lemma 21]) *Suppose* $Q_\theta \geq Q^*$ *and* $(e_0, T_\Delta^0, E_u, E_Q, Q_\tau^2, Q_{\tau+1}^2) :=$ ***Findsets2***$(V, E, T^0, \boldsymbol{w}, \boldsymbol{c}, \boldsymbol{q}, \boldsymbol{l}, \boldsymbol{u})$. *For any* $Q \in [Q_\tau^2, Q^*)$, *let* $\lambda_Q = f(w_Q, c, T^0) - f(w_Q, c, T_Q) > 0$, *where* $T_Q = \arg\min_{T\in\Gamma} f(w_Q, c, T)$, $e_Q = \arg\max_{e\in T_Q} w_Q(e)$. *Let* $\tilde{Q} = Q + Q_\lambda$, *where*

$$Q_\lambda = \begin{cases} \lambda_Q q(e_0), & if\ e_Q \in E_u, \\ \frac{\lambda_Q q(e_0) q(e_Q)}{q(e_0)+q(e_Q)}, & if\ e_Q \in E_Q. \end{cases} \tag{10.37}$$

Then $\tilde{Q} \leq Q^*$ *and* $f(w_{\tilde{Q}}, c, T_0) = f(w_{\tilde{Q}}, c, T_Q)$.

Now, we present a strongly polynomial time algorithm for the problem (**IMSST**$_\infty$).

Algorithm 10.8 An $O(m^2 \log n)$ algorithm for (**IMSST**$_\infty$) [67, Algorithm 4]

Input: A network $G = (V, E)$, a spanning tree T^0, a weight vector $\boldsymbol{w}$, a lower bound vector $\boldsymbol{l}$, an upper bound vector $\boldsymbol{u}$, a cost vector $\boldsymbol{c}$, and a modification cost vector $\boldsymbol{q}$.

Output: An optimal solution $\boldsymbol{w^*}$ and its objective value $\boldsymbol{Q^*}$.

1: Calculate θ by (10.24). Construct w_0 by (10.26). Calculate the optimal **MSST** T_{w_0} and T_w under w_0 and w, respectively.
2: **if** $f(w_0, c, T^0) > f(w_0, c, T_{w_0})$ **then** the instance is infeasible and stop.
3: **else if** $f(w, c, T^0) = f(w, c, T_w)$ **then** T^0 is an optimal **MSST** and stop.
4: **else**
5: Calculate Q_θ by (10.27). Construct w_{Q_θ} by (10.22). Calculate the optimal **MSST** T_{Q_θ} under w_{Q_θ}.
6: **if** $f(w_{Q_\theta}, c, T^0) > f(w_{Q_\theta}, c, T_{Q_\theta})$ **then** call $(T_\theta^0, E_u, E_Q, Q_\tau^1, Q_{\tau+1}^1) =$ **Findsets1**$(V, E, T^0, w, c, q, l, u)$. Let $Q' := \frac{Q_\tau^1 + Q_{\tau+1}^1}{2}$, and construct $w_{Q'}$ by (10.31).
7: Let $k := 0$, $Q_k := Q_\tau^1$, construct w_{Q_k} by (10.31), and calculate the optimal **MSST** T_{Q_k} under w_{Q_k}.

while $f(w_{Q_k}, c, T^0) > f(w_{Q_k}, c, T_{Q_k})$ **do**
$\lambda_k := f(w_{Q_k}, c, T^0) - f(w_{Q_k}, c, T_{Q_k})$, $e_{Q_k} := \arg\max_{e_i \in T_{Q_k}} w_{Q'}(e_i)$.
$Q_{\lambda_k} := q(e_{Q_k})\lambda_k$, $Q_{k+1} := Q_k + Q_{\lambda_k}$.
Calculate the optimal **MSST** $T_{Q_{k+1}}$ under $w_{Q_{k+1}}$ by (10.31).
Update $k := k + 1$.
end while
Let $Q^* := Q_k$, $w^* := w_{Q_k}$.
end if
if $f(w_{Q_\theta}, c, T^0) = f(w_{Q_\theta}, c, T_{Q_\theta})$ **then** call $(e_0, T^0_\Delta, E_u, E_Q, Q^2_\tau, Q^2_{\tau+1})=$ **Findsets2** $(V, E, T^0, w, c, q, l, u)$, Let $Q' := \frac{Q^2_\tau + Q^2_{\tau+1}}{2}$, and construct $w_{Q'}$ by (10.36).
Let $k := 0$, $Q_k := Q^2_\tau$, and construct w_{Q_k} by (10.36).
Calculate the optimal **MSST** T_{Q_k} under w_{Q_k}.
while $f(w_{Q_k}, c, T^0) > f(w_{Q_k}, c, T_{Q_k})$ **do**
$\lambda_k := f(w_{Q_k}, c, T^0) - f(w_{Q_k}, c, T_{Q_k})$, $e_{Q_k} := \arg\max_{e_i \in T_{Q_k}} w_{Q'}(e_i)$.
$Q_{\lambda_k} := \begin{cases} q(e_0)\lambda_k, & \text{if } e_{Q_k} \in E_u. \\ \frac{q(e_0)q(e_k)\lambda_k}{q(e_0)+q(e_k)}, & \text{if } e_{Q_k} \in E_Q. \end{cases}$ $Q_{k+1} := Q_k + Q_{\lambda_k}$.
Calculate the optimal **MSST** $T_{Q_{k+1}}$ under $w_{Q_{k+1}}$ by (10.36).
Update $k := k + 1$.
end while
Let $Q^* := Q^k$, $w^* := w_{Q_k}$.
end if
end if

Theorem 10.9 ([67, Theorem 25]) *The problem (**IMSST**$_\infty$) can be solved by Algorithm 10.8 in $O(m^2 \log n)$ time.*

10.6 Conclusions

In this chapter, an inverse max+sum spanning tree problem is considered. It is the first class of inverse problems on optimization problems with combined minimax-minsum objective functions. As there are two parameter vectors in the max+sum spanning tree (MSST) problem, when we only modify the sum-cost vector under weighted l_1 norm, to make a given spanning tree become an optimal MSST on the network with respect to the new sum-cost vector, we formulate the inverse max+sum spanning tree problem under weighted l_1 norm into a linear programming problem with exponential number of constraints. Then we design a column generation algorithm to solve its dual problem. It is extremely interesting that in each iteration the choice of entering variable can be transformed into solving a max+sum spanning tree problem on the original network structure with a new sum-cost vector and the old max-weight vector, which can be solved in $O(m \log n)$ time.

When we only modify the sum-cost vector under weighted l_∞ norm, we show that the unbounded inverse MSST problem under weighted l_∞ norm is a linear fractional combinatorial optimization problem, which can be solved by a discrete type Newton method in $O\left(m^2 \log m\right)$ iterations. Furthermore, we prove an $O(m)$ bound on the number of iterations for the Newton method. Therefore, both the unbounded and bounded inverse problems can be solved by $O(m)$ MSST problems with respect to a new sum-cost vector and the original max-weight vector, and the total time complexity is $O(m^2 \log n)$.

When we only modify the sum-cost vector under weighted Hamming distance, firstly, we present a mathematical model for the inverse MSST problem and a method to check its feasibility. Then, under the weighted bottleneck Hamming distance, we design a binary search algorithm whose time complexity is $O(m \log^2 n)$. Next, under the unit sum Hamming distance, we show that it is NP-hard.

When we only modify the max weight vector under weighted l_∞ norm, firstly, we provide some optimality conditions of optimal solutions and then propose a strongly polynomial time algorithm with running time $O(m^2 \log n)$. The main idea is to start iteration from Q, where $Q < Q^*$ and Q^* is the optimal value. Each iteration will cause an increment Q_λ, where $Q + Q_\lambda \leq Q^*$. Furthermore, we prove that the number of iterations is at most $O(m)$ and the algorithm needs to solve an MSST problem under a new max weight vector in each iteration.

As a future research topic, we will study the inverse max+sum spanning tree problem when modifying the max weight vector under weighted l_1 norm and sum Hamming distance, which is more challenging.

Chapter 11
Restricted Inverse Optimal Value Problem on Minimum Spanning Tree

Abstract In this chapter, we address the restricted inverse optimal value problem on minimum spanning tree under various norms. Under the weighted l_∞ norm with bounds, we establish optimality conditions and develop two efficient algorithms operating in $O(m^2n)$ and $O(m^2 \log n)$, where m and n are the number of edges and nodes of a graph G. Additionally, an $O(mn)$ algorithm is introduced for the (RIOVMST$_\infty$) problem with unit norm and bounds. For the l_1 norm case, we formulate the problem as a linear program, derive a dual subproblem, and calculate a critical value z^* using binary search, ultimately solving the problem with a complexity of $O(m^2n^2 \log n \log(nC))$. Under the bottleneck Hamming distance, three binary search algorithms are developed with a uniform time complexity of $O(mn \log n)$. An open problem is posed regarding a strongly polynomial time algorithm for the (RIOVMST$_1$) problem.

Keywords Restricted inverse optimal value problem · Minimum spanning tree · Binary search method · Dual theory · Polynomial time algorithm · Open problem

11.1 An Introduction to Restricted Inverse Optimal Value Problem on MST

In this chapter, we study the restricted inverse optimal value problem on minimum spanning tree (RIOVMST) under some norms.

In a connected undirected network $G = (V, E, \boldsymbol{w})$, we are given a spanning tree T^0, a weight vector $\boldsymbol{w}$, a decrement vector $\boldsymbol{l}$, an increment vector $\boldsymbol{u}$, a cost vector $\boldsymbol{c}$, and a constant K. We aim to obtain a new weight vector $\bar{\boldsymbol{w}}$ satisfying the lower and upper bounds such that T^0 is a minimum spanning tree under the vector $\bar{\boldsymbol{w}}$ with weight K. The objective is to minimize the modification cost $\|\bar{\boldsymbol{w}} - \boldsymbol{w}\|$ under some

Junhua Jia, Xiucui Guan, and Binwu Zhang have equally contributed to this chapter.

X. Guan et al., *Inverse Combinatorial Optimization Problems*, Springer Optimization and Its Applications 225, https://doi.org/10.1007/978-3-031-91175-0_11

norms. Hence, its mathematical model can be stated as follows:

$$\begin{aligned} & \min \|w - \bar{w}\| \\ (RIOVMST)\ & s.t.\ \bar{w}(T^0) = K, \\ & \bar{w}(T) \geq K,\ \forall T \in \Gamma, \\ & w - \boldsymbol{l} \leq \bar{\boldsymbol{w}} \leq w + \boldsymbol{u}. \end{aligned}$$

We call the edges in (out of) the given tree T^0 tree (nontree) edges. For each nontree edge $e_j \notin T^0$, denote by P_j the only path connecting the two endpoints of e_j. For any tree edge e_i, define $\Omega_i = \{e_j \notin T^0 | e_i \in P_j\}$ as the set of nontree edges e_j belonging to at least one common fundamental cycle with e_i. Let $\Omega^0 = \{e_i \in T^0 | \Omega_i = \emptyset\}$. If $\Omega^0 \neq \emptyset$, then each edge in Ω^0 belongs to every spanning tree of G, and hence Ω^0 is also the set of cut edges of G. If $\Omega^0 = \emptyset$, then there is no cut edge, and each edge in T^0 belongs to at least one fundamental cycle.

Let

$$U_i := \begin{cases} \min\{\min\{w_j + u_j | e_j \in \Omega_i\}, w_i + u_i\}, & for\ e_i \in T^0 \setminus \Omega^0, \\ w_i + u_i, & for\ e_i \in \Omega^0 \cup (E \setminus T^0). \end{cases} \tag{11.1}$$

$$L = w - l. \tag{11.2}$$

Lemma 11.1 ([65, Lemma 2.3]) *The problem (RIOVMST) is feasible if and only if $L(T^0) \leq K \leq U(T^0)$ and $L_i \leq U_i$ for each $e_i \in T^0$.*

Lemma 11.2 ([65, Lemma 2.1]) *A tree T^0 is a minimum spanning tree with respect to the weight vector $\bar{\boldsymbol{w}}$ if and only if one of the following optimality conditions is satisfied:*

$$\bar{w}_i \leq \bar{w}_j,\ e_j \notin T^0\ and\ e_i \in P_j. \tag{11.3a}$$

$$\bar{w}_i \leq \bar{w}_j,\ for\ each\ e_i \in (T^0 \setminus \Omega^0)\ and\ e_j \in \Omega_i. \tag{11.3b}$$

11.2 (RIOV, MST, Bounded, Weighted l_∞ Norm)

In this section, we study the restricted inverse optimal value problem on minimum spanning tree under weighted l_∞ norm (RIOVMST$_\infty$) [65]. Its mathematical model can be stated as follows:

$$\begin{aligned} & \min_{\bar{\boldsymbol{w}}} \max_{e_i \in E} c_i |\bar{w}_i - w_i| \\ (RIOVMST_\infty)\ & s.t.\ \bar{w}(T^0) = K, \end{aligned} \tag{11.4}$$

$$\bar{w}(T) \geq K, \ \forall T \in \Gamma, \tag{11.5}$$

$$w - \boldsymbol{l} \leq \bar{\boldsymbol{w}} \leq w + \boldsymbol{u}. \tag{11.6}$$

Lemma 11.3 ([65, Lemma 2.2]) *If $\bar{\boldsymbol{w}}$ is a feasible solution of problem (**RIOVMST**$_\infty$) with objective value $\bar{\lambda}$, then $\hat{\boldsymbol{w}}$ is also a feasible solution of problem (**RIOVMST**$_\infty$) with objective value $\hat{\lambda} \leq \bar{\lambda}$, where*

$$\hat{w}_j = \begin{cases} \bar{w}_j, & \text{if } e_j \in T^0, \\ \max\{\max_{e_t \in P_j} \bar{w}_t, w_j\}, & \text{if } e_j \notin T^0. \end{cases} \tag{11.7}$$

By Lemmas 11.2 and 11.3 and notation Ω_i for $e_i \in (T^0 \setminus \Omega^0)$, the problem (**RIOVMST**$_\infty$) can be reformulated as follows:

$$\min_{\bar{\boldsymbol{w}}} \max_{e_i \in E} c_i |\bar{w}_i - w_i|$$

$$(RIOVMST_\infty) \ s.t. \ \bar{w}_i \leq \bar{w}_j, \ \text{for each } e_i \in (T^0 \setminus \Omega^0) \text{ and } e_j \in \Omega_i, \tag{11.8}$$

$$w_i - l_i \leq \bar{w}_i \leq w_i + u_i, \ \text{for each } e_i \in T^0, \tag{11.9}$$

$$w_j \leq \bar{w}_j \leq w_j + u_j, \ \text{for each } e_j \notin T^0, \tag{11.10}$$

$$\bar{w}(T^0) = K. \tag{11.11}$$

Lemma 11.4 ([65, Lemma 2.4]) *If $\bar{\boldsymbol{w}}$ is a feasible solution of problem (RIOVMST$_\infty$) with objective value $\bar{\lambda}$, then $\bar{\lambda} \geq \alpha$, where*

$$\alpha = \max_{e_i \in T^0 \setminus \Omega^0} \max_{e_k \in \Omega_i} \max \left\{ \frac{c_i c_k (w_i - w_k)}{c_i + c_k}, c_k (L_i - w_k), c_i (w_i - U_k), 0 \right\}. \tag{11.12}$$

Theorem 11.1 ([65, Theorem 2.1]) *If $L(T^0) = K$, then w^* defined as in (11.13) is an optimal solution of the problem (RIOVMST$_\infty$) with objective value*

$$\lambda^* = \max \left\{ \max_{e_i \in T^0} \{c_i(w_i - L_i)\}, \max_{e_j \notin T^0} \left\{ c_j \left(\max_{e_t \in P_j} L_t - w_j \right) \right\} \right\}.$$

$$w_i^* = \begin{cases} L_i, & \text{if } e_i \in T^0, \\ \max\{\max_{e_t \in P_i} L_t, w_i\}, & \text{if } e_i \notin T^0. \end{cases} \tag{11.13}$$

Theorem 11.2 ([65, Theorem 2.2]) *If $U(T^0) = K$, then w^* defined as in (11.14) is an optimal solution of the problem (RIOVMST$_\infty$) with objective value*

$$\lambda^* = \max \left\{ \max_{e_i \in T^0} \{c_i |w_i - U_i|\}, \max_{e_j \notin T^0} \left\{ c_j \left(\max_{e_t \in P_j} U_t - w_j \right) \right\} \right\}.$$

$$w_i^* = \begin{cases} U_i, & \text{if } e_i \in T^0, \\ \max\{\max_{e_t \in P_i} U_t, w_i\}, & \text{if } e_i \notin T^0. \end{cases} \tag{11.14}$$

Based on Lemma 11.1, Theorem 11.1, and Theorem 11.2, we assume $L(T^0) < K < U(T^0)$ and $L_i \le U_i$ for each $e_i \in T^0$ in the next discussion.

We first analyze some properties of the feasible solutions of (**RIOVMST**$_\infty$). Then we propose an algorithm to solve the problem. Finally, we show its time complexity is $O(m^2 n)$.

Theorem 11.3 ([65, Theorem 3.1]) *Suppose $L(T^0) < K < U(T^0)$ and $L_i \le U_i$ for each $e_i \in T^0$. For any $\tilde{\lambda} \ge \alpha$, let*

$$\tilde{w}_i = \begin{cases} \max\left\{w_i - \frac{\tilde{\lambda}}{c_i}, L_i\right\}, & \text{if } e_i \in T^0, \\ \min\left\{w_i + \frac{\tilde{\lambda}}{c_i}, U_i\right\}, & \text{if } e_i \notin T^0. \end{cases} \tag{11.15}$$

$$\tilde{w}_{j_i} = \begin{cases} \min\limits_{e_k \in \Omega_i} \tilde{w}_k, & \text{if } e_i \in T^0 \setminus \Omega^0, \\ U_i, & \text{if } e_i \in \Omega^0. \end{cases} \tag{11.16}$$

Then $\tilde{w}_i \le \tilde{w}_{j_i}$ for any $e_i \in T^0$.

Theorem 11.4 ([65, Theorem 3.2]) *Suppose $L(T^0) < K < U(T^0)$ and $L_i \le U_i$ for each $e_i \in T^0$. For any $\tilde{\lambda} \ge \alpha$, if $\tilde{w}(T^0) = K$, then $\tilde{w}$ defined as in (11.15) is a feasible solution with objective value $\max_{e_i \in E} c_i |\tilde{w}_i - w_i| \le \tilde{\lambda}$.*

We now introduce a robust polynomial time algorithm for addressing the (**RIOVMST**$_\infty$) challenge, complemented by an examination of its optimality and computational efficiency.

The algorithm's core concept encompasses five distinct scenarios: **(1)** (Refer to lines 1–7). We assess the viability of the (**RIOVMST**$_\infty$) challenge and yield the optimal solution via equations (11.13) and (11.14) when $L(T^0) = K$ or $U(T^0) = K$, respectively. **(2)** (Refer to lines 9–12.) Initially, we set $\Delta^0 := K - w^0(T^0)$, $k := 0$, with w^0 defined as per equation (11.15) at $\tilde{\lambda} = \alpha$. Should $\Delta^0 = 0$, it implies that $w^0(T^0) = K$ and w^0 constitutes an optimal solution with the objective value α, as per Theorem 11.4. **(3)** (Refer to lines 13–19.) If $\Delta^0 < 0$, indicating that $w^0(T^0) > K$, the optimal objective value λ^* exceeds α. We iteratively diminish the weights $w_i^{k+1} := w_i^k - \frac{\lambda_D^k}{c_i}$ for edges $e_i \in E_L^k := \{e_i \in T^0 | w_i^k > L_i\}$ in a greedy manner until $w^{k+1}(T^0) = K$, where λ_D^k denotes the least cost required to ensure $w^{k+1}(T^0)$ meets the lower bound $w_i^{k+1} \ge L_i$ for each $e_i \in E_L^k$. **(4)** (Refer to lines 21–29.) If $\Delta^0 > 0$, suggesting that $w^0(T^0) < K$, we initially verify whether α could represent the optimal objective value. This is achieved by incrementing the weights from w_i^0 to $\tilde{w}_i := \min\{w_i + \frac{\alpha}{c_i}, U_i, w_{j_i}^0\}$ for each $e_i \in T^0$ and computing the cumulative increments $\tilde{\Delta}_k := \tilde{\Delta}_{k-1} + \tilde{w}_i - w_i^0$. If $\tilde{\Delta}_k \ge \Delta^0$, then α is confirmed as the optimal objective value. **(5)** (Refer to lines 31–36.) When $\tilde{w}(T^0) < K$, it signifies that λ^* is greater than α. We iteratively augment the weights $w_i^{k+1} := \min\{w_i + \frac{\lambda^{k+1}}{c_i}, U_i, w_{ji}^{k+1}\}$ for edges $e_i \in E_U^k := \{e_i \in T^0 | w_i^k < U_i\}$ in a greedy fashion until $w^{k+1}(T^0) = K$.

Algorithm 11.1 An $O(m^2 n)$ algorithm for (**RIOVMST**$_\infty$) [65, Algorithm 3.1]

Input: A network $G = (V, E)$, a spanning tree T^0, a weight vector $\boldsymbol{w}$, a decrement vector $\boldsymbol{l}$, an increment vector $\boldsymbol{u}$, a cost vector $\boldsymbol{c}$, and a value K.

Output: An optimal solution $\boldsymbol{w}^*$ and its objective value $\boldsymbol{\lambda}^*$.

Calculate P_j for each $e_j \notin T^0$; calculate U_i, Ω_i for each $e_i \in T^0$, Ω^0, and α.
if $L(T^0) > K$ or $U(T^0) < K$ or $\{e_i \in T^0 | L_i > U_i\} \neq \emptyset$ **then**
    the problem (2.3) is infeasible, and stop.
else if $L(T^0) = K$ **then**
    Output an optimal solution w^* by (11.13) and its objective value λ^*, stop.
else if $U(T^0) = K$ **then**
    Output an optimal solution w^* by (11.14) and its objective value λ^*, stop.
else
    Initialize $k := 0$, $\lambda^k := \alpha$.
    Let $w_i^k := \begin{cases} \max\{w_i - \frac{\alpha}{c_i}, L_i\}, e_i \in T^0, \\ \min\{w_i + \frac{\alpha}{c_i}, U_i\}, e_i \notin T^0. \end{cases}$,

$w_{ji}^k := \begin{cases} \min\limits_{e_j \in \Omega_i} w_j^k, \; if \; e_i \in T^0 \setminus \Omega^0, \\ U_i, \quad if \; e_i \in \Omega^0. \end{cases}$ and $\Delta^k := K - w^k(T^0)$.

    if $\Delta^k = 0$ **then**
        output an optimal solution $w^* := w^k$ and its objective value $\lambda^* := \alpha$.
    else if $\Delta^k < 0$ **then**
        Let $E_L^k := \{e_i \in T^0 | w_i^k > L_i\}$.
        while $\Delta^k < 0$ **do**
            $\lambda_D^k := \min \left\{ \frac{-\Delta^k}{\sum\limits_{e_i \in E_L^k} \frac{1}{c_i}}, \min\limits_{e_i \in E_l^k} c_i(w_i^k - L_i) \right\}$.
            Update $w_i^{k+1} := w_i^k - \frac{\lambda_D^k}{c_i}$ if $e_i \in E_L^k$, $\Delta^{k+1} := \Delta^k + \sum\limits_{e_i \in E_L^k} \frac{\lambda_D^k}{c_i}$,

$\lambda^{k+1} := \lambda^k + \lambda_D^k$, $E_L^{k+1} := \{e_i \in E_L^k | w_i^{k+1} > L_i\}$, $k := k + 1$.

        end while
        Output $\lambda^* := \lambda^{k+1}$ and an optimal solution $w_i^* := w_i^{k+1}$ for $e_i \in T^0$ and $w_i^* := \min\{w_i + \frac{\lambda^*}{c_i}, U_i\}$ for $e_i \notin T^0$, stop.
    else
        Initialize $\tilde{\Delta}_k := 0$, $\tilde{T}^0 := \emptyset$.
        while $\tilde{\Delta}_k < \Delta^0$ and $T^0 \setminus \tilde{T}^0 \neq \emptyset$ **do**
            Choose $e_i \in T^0 \setminus \tilde{T}^0$.
            Let $\tilde{w}_i := \min\{w_i + \frac{\alpha}{c_i}, U_i, w_{ji}^0\}$.
            Let $\tilde{\Delta}_{k+1} := \tilde{\Delta}_k + \tilde{w}_i - w_i^0$, $\tilde{T}^0 := \tilde{T}^0 \cup \{e_i\}$, $e_{i^*} := e_i$, $k := k + 1$.

27: **end while**
28: **if** $\tilde{\Delta}_k \geq \Delta^0$ **then**
29: Let $w_{i*}^* := \Delta^0 - \tilde{\Delta}_{k-1} + w_{i*}^0$.
30: Output an optimal solution

$$w_i^* := \begin{cases} w_i^0, & if\ e_i \notin T^0, \\ \tilde{w}_i, & if\ e_i \in \tilde{T}^0 \setminus \{e_{i^*}\}, \\ w_i^0, & if\ e_i \in T^0 \setminus \tilde{T}^0, \\ w_{i*}^*, & if\ e_i = e_{i^*}. \end{cases} \tag{11.17}$$

and its value $\lambda^* := \alpha$, stop.
31: **else**
32: Update $w_i^k := \tilde{w}_i$ for $e_i \in \tilde{T}^0$ and $w_i^k := w_i^0$ for other e_i, $\Delta^k := K - w^k(T^0)$ and $\lambda^k := \alpha$.
33: **while** $\Delta^k > 0$ **do**
34: Let $E_1^k := \{e_i \in T^0 | w_i^k < U_i, w_i^k = w_i + \frac{\lambda^k}{c_i}\}$, $E_2^k := \{e_i \in T^0 |\ w_i^k < U_i, w_i^k = w_{j_i}^k\}$, $\lambda_U^k := \dfrac{\Delta^k}{\sum_{e_i \in E_1^k} \frac{1}{c_i} + \sum_{e_i \in E_2^k} \frac{1}{c_{j_i}}}$.
35: Update $\lambda^{k+1} := \lambda^k + \lambda_U^k$, $w_i^{k+1} := \min\{w_i + \frac{\lambda^{k+1}}{c_i}, U_i\}, e_i \notin T^0$, $w_{j_i}^{k+1} := \begin{cases} \min_{e_j \in \Omega_i} w_j^{k+1}, & if\ e_i \in T^0 \setminus \Omega^0, \\ U_i, & if\ e_i \in \Omega^0. \end{cases}$ $w_i^{k+1} := \min\{w_i + \frac{\lambda^{k+1}}{c_i}, U_i, w_{j_i}^{k+1}\}$, $e_i \in T^0$, $\Delta^{k+1} := K - w^{k+1}(T^0)$, $k := k + 1$.
36: **end while**
37: Output $\lambda^* := \lambda^{k+1}$ and an optimal solution $w^* := w^{k+1}$, stop.
38: **end if**
39: **end if**
40: **end if**

Theorem 11.5 ([65, Cor 3.1]) *The time complexity of Algorithm 11.1 is $O(m^2n)$.*

The main computation in Algorithm 11.1 is $O(m^2n)$ when $\tilde{w}(T^0) < K$ in lines 32–35 where

$$\tilde{w}_i := \begin{cases} \min\{w_i + \frac{\alpha}{c_i}, U_i\}, & e_i \notin T^0, \\ \min\{w_i + \frac{\alpha}{c_i}, U_i, \tilde{w}_{j_i}\}, & e_i \in T^0, \end{cases} \quad \tilde{w}_{j_i} := \begin{cases} \min_{e_j \in \Omega_i} \tilde{w}_j, & e_i \in T^0 \setminus \Omega^0, \\ U_i, & e_i \in \Omega^0. \end{cases} \tag{11.18}$$

The number of iterations is $O(m)$ in the while loop in lines 32–35. Moreover, in each iteration, the main computation is to calculate $w_{j_i}^{k+1}$ for each $e_i \in T^0$ in line 34, which can be done in $O(mn)$ time. For simplicity, we denote by $e_{j_i}^k := \arg\min_{e_j \in \Omega_i} w_j^k$ the minimum value edge with respect to a feasible solution w^k.

In the subsequent discourse, we elaborate on a sophisticated algorithm, boasting an enhanced time complexity of $O(m^2 \log n)$, tailored to address the (**RIOVMST**$_\infty$) challenge under the condition $\tilde{w}(T^0) < K$. The crux of the approach involves executing a pair of binary search methodologies (Algorithm 11.2) to pinpoint the precise range $(\bar{\lambda}_\tau, \bar{\lambda}_{\tau+1}]$ where the optimal value λ^* resides. Subsequently, we are able to accurately identify the edge with the minimum value $e^*_{j_i}$ in relation to an optimal solution w^*, even without prior knowledge of the precise solution w^*.

Algorithm 11.2 $(\lambda_\tau, \lambda_{\tau+1}] :=$ **BinarySearch**$(\Lambda, T^0, K, \boldsymbol{w}, \boldsymbol{l}, \boldsymbol{U}, \boldsymbol{c})$ [65, Alg. 4.1]

Input: A network $G = (V, E)$, a spanning tree T^0, a weight vector $\boldsymbol{w}$, a decrement vector $\boldsymbol{l}$, an increment vector $\boldsymbol{U}$, a cost vector $\boldsymbol{c}$, a values K, and the set $\Lambda := \{\lambda_j\}$ of sorted values.

Output: The interval $(\lambda_\tau, \lambda_{\tau+1}]$.

1: Initialize $a := 1, b := |\Lambda|$.
2: **if** $b \le 2$ **then** output $(\lambda_\tau, \lambda_{\tau+1}] := (\min \Lambda, \max \Lambda]$.
3: **end if**
4: **while** $b - a > 1$ **do**
5: Let $k := \lfloor (b+a)/2 \rfloor$. Let

$$\underline{w}_i^k := \begin{cases} \min\{w_i + \frac{\lambda_k}{c_i}, U_i\}, \; e_i \notin T^0, \\ \max\{w_i - \frac{\lambda_k}{c_i}, L_i\}, \; e_i \in T^0, \end{cases} \tag{11.19}$$

$$\begin{cases} \bar{w}_i^k := \min\{w_i + \frac{\lambda_k}{c_i}, U_i\}, & e_i \notin T^0, \\ \bar{w}_{j_i}^k := \min_{e_j \in \Omega_i} \bar{w}_j^k, & e_i \in T^0 \setminus \Omega^0, \\ \bar{w}_{j_i}^k := U_i, & e_i \in \Omega^0, \\ \bar{w}_i^k := \min\{w_i + \frac{\lambda_k}{c_i}, U_i, \bar{w}_{j_i}^k\}, & e_i \in T^0, \end{cases} \tag{11.20}$$

6: **if** $\underline{w}^k(T^0) \le K \le \bar{w}^k(T^0)$ **then** update $b := k$.
7: **else** update $a := k$.
8: **end if**
9: **end while**
10: Output $(\lambda_\tau, \lambda_{\tau+1}] := (\lambda_a, \lambda_b]$.

Algorithm 11.3 An algorithm for (**RIOVMST**$_\infty$) when $\tilde{w}(T^0) < K$ [65, Alg. 4.2]

Input: A network $G = (V, E)$, a spanning tree T^0, a weight vector $\boldsymbol{w}$, Ω_i for each $e_i \in T^0$, Ω^0, a decrement vector $\boldsymbol{l}$, an increment vector $\boldsymbol{U}$, a cost vector $\boldsymbol{c}$, and three values K, α, and ε.

Output: The optimal objective value λ^* and an optimal solution w^*.

1: Initialize $\underline{\lambda} := \alpha$, $\bar{\lambda} := \max\left\{\max\limits_{e_j \in E} c_j(u_j - w_j), \max\limits_{e_i \in T^0} c_i(w_i - L_i)\right\}$, $\Lambda^0 :=$ $\{\lambda_j := c_j(u_j - w_j) | \lambda_j > \underline{\lambda}, e_j \notin T^0\}$.

2: Let $\Lambda^1 := \{\underline{\lambda}, \bar{\lambda}\} \cup \Lambda^0$. Sort the values in Λ^1 in nondecreasing order.

3: Call $(\lambda_\tau, \lambda_{\tau+1}] :=$ **BinarySearch**$(\Lambda^1, T^0, K, \boldsymbol{w}, \boldsymbol{l}, \boldsymbol{U}, \boldsymbol{c})$.

4: Let $\bar{T}^0 := \{e_j \notin T^0 | \lambda_j \geq \lambda_{\tau+1}\}$.

5: **for** $i = 1$ to $m - n + 1$ **do**

6: **for** $j = i + 1$ to $m - n + 1$ **do**

7: **if** $e_i \in \bar{T}^0, e_j \in \bar{T}^0$ and $c_i \neq c_j$ **then** let $\bar{\lambda}_{ij} := \frac{(w_i - w_j)c_i c_j}{c_i - c_j}$.

8: **else if** $e_i \in \bar{T}^0$ and $e_j \notin \bar{T}^0$ **then** let $\bar{\lambda}_{ij} := c_i(U_j - w_i)$.

9: **else if** $e_i \notin \bar{T}^0$ and $e_j \in \bar{T}^0$ **then** let $\bar{\lambda}_{ij} := c_j(U_i - w_j)$.

10: **end if**

11: **end for**

12: **end for**

13: Let $\Lambda^2 := \{\lambda_\tau, \lambda_{\tau+1}\} \cup \{\bar{\lambda}_{ij} | \lambda_\tau < \bar{\lambda}_{ij} \leq \lambda_{\tau+1}, e_i, e_j \notin T^0\}$. Sort the values in Λ^2 in nondecreasing order.

14: Call $(\bar{\lambda}_\tau, \bar{\lambda}_{\tau+1}] = BinarySearch(\Lambda^2, T^0, K, \boldsymbol{w}, \boldsymbol{l}, \boldsymbol{U}, \boldsymbol{c})$.

15: Let $\hat{\lambda} := \bar{\lambda}_{\tau+1} - \varepsilon$.

16: Calculate $\underline{\hat{w}}$, $\hat{\bar{w}}$ by (11.19)-(11.20), where λ_k is replaced with $\hat{\lambda}$.

17: **if** $\underline{\hat{w}}(T^0) > K$ or $K > \hat{\bar{w}}(T^0)$ or $\hat{\lambda} < \bar{\lambda}_\tau$ **then**

18: Output $\lambda^* := \bar{\lambda}_{\tau+1}$ and $w^* := \bar{w}^{\tau+1}$ by (11.20), where λ_k is replaced with $\bar{\lambda}_{\tau+1}$.

19: **else**

20: Let $\hat{w}_j := \begin{cases} w_j + \frac{\hat{\lambda}}{c_j}, & e_j \in \bar{T}^0, \\ U_j, & e_j \notin (T^0 \cup \bar{T}^0). \end{cases}$

21: For $e_i \in T^0 \setminus \Omega^0$, let $\hat{e}_{j_i} := \arg\min_{e_j \in \Omega_i} \hat{w}_j$.

22: Initialize $k := 0$. Let $\hat{\lambda}^k := \bar{\lambda}_\tau$,

$$\hat{w}_i^k := \begin{cases} \min\{w_i + \frac{\hat{\lambda}^k}{c_i}, U_i\}, & e_i \in \Omega^0 \cup (E \setminus T^0), \\ \min\{w_i + \frac{\hat{\lambda}^k}{c_i}, U_i, \hat{w}^k(\hat{e}_{j_i})\}, & e_i \in T^0 \setminus \Omega^0, \end{cases} \tag{11.21}$$

and $\Delta^k := K - \hat{w}^k(T^0)$.

23: **while** $\Delta^k > 0$ **do**

24: Let $E_1^k := \{e_i \in \Omega^0 | \hat{w}_i^k < U_i\} \cup \{e_i \in T^0 \setminus \Omega^0 | \hat{w}_i^k < \hat{w}^k(\hat{e}_{j_i}), \hat{w}_i^k < U_i\}$ and $E_2^k := \{e_i \in T^0 \setminus \Omega^0 | \hat{e}_{j_i} \in \bar{T}^0, \hat{w}_i^k = \hat{w}^k(\hat{e}_{j_i}) < U_i\}$.

Let $\lambda_U^k := \frac{\Delta^k}{\sum_{e_i \in E_1^k} \frac{1}{c_i} + \sum_{e_i \in E_2^k} \frac{1}{c(\hat{e}_{j_i})}}$. Update $\hat{\lambda}^{k+1} := \hat{\lambda}^k + \lambda_U^k$ and $\hat{w}^{k+1}$ by (11.21). Update $\Delta^{k+1} := K - \hat{w}^{k+1}(T^0)$ and $k := k + 1$.

end while

Output $\lambda^* := \hat{\lambda}^k$ and $w^* := \hat{w}^k$.

end if

Remark 11.1 (Algorithm 11.4 [65, Algorithm 4.3]) We can obtain **Algorithm 11.4** by replacing lines 31–36 in Algorithm 11.1 with Algorithm 11.3.

Theorem 11.6 ([65, Theorem 4.2]) *Algorithm 11.4 can solve problem (**RIOVMST**$_\infty$) in $O(m^2 \log n)$ time.*

Additionally, Zhang et al. [136, 138] studied the unbounded and the lower bounded RIOVMST problem under unit l_∞ norm. They presented two $O(mn)$ algorithms to solve them.

11.3 (RIOV, MST, Bounded, Weighted l_1 Norm)

In this section, we introduce the restricted inverse optimal value problem on minimum spanning tree under weighted l_1 norm (RIOVMST$_1$) [66]. Its mathematical model can be stated as follows:

$$\min_{\bar{w}} \sum_{e_i \in E} c_i |\bar{w}_i - w_i|$$

$$(RIOVMST_1) \quad s.t. \quad \bar{w}(T^0) = K, \tag{11.22}$$

$$\bar{w}_i \le \bar{w}_j, \quad \text{for each } e_j \notin T^0 \text{ and } e_i \in P_j, \tag{11.23}$$

$$w_i - l_i \le \bar{w}_i \le w_i + u_i, \quad \text{for each } e_i \in E. \tag{11.24}$$

Lemma 11.5 ([66, Lemma 2.2]) *Supposing w^* is an optimal solution of the problem (RIOVMST$_1$), then $w_j^* \ge w_j$ for any $e_j \in E \setminus T^0$.*

Let $M^1 = \{1, 2, \ldots, n-1\}$, $M^2 = \{n, \ldots, m\}$, $M = M^1 \cup M^2$, $E^0 = \{(i, j) | e_i \in P_j, n \le j \le m\}$. Let $\bar{\alpha}_i$ and $\bar{\beta}_i$ represent the increasing and decreasing amount for $i \in M^1$, respectively. Let $\bar{\alpha}_j$ represent the increasing amount for $j \in M^2$. By Lemma 11.5, the problem (*RIOVMST*$_1$) can be reformulated as follows:

$$\min \sum_{i \in M^1} c_i(\bar{\alpha}_i + \bar{\beta}_i) + \sum_{j \in M^2} c_j \bar{\alpha}_j$$

$$s.t. \ w_i + \bar{\alpha}_i - \bar{\beta}_i \le w_j + \bar{\alpha}_j, \ \forall (i, j) \in E^0,$$

$$(P) \qquad \sum_{i \in M^1} (\bar{\alpha}_i - \bar{\beta}_i) = K - w(T_0),$$
$$0 \le \bar{\alpha}_i \le u_i, \forall i \in M^1,$$
$$0 \le \bar{\beta}_i \le l_i, \forall i \in M^1,$$
$$0 \le \bar{\alpha}_j \le u_j, \forall j \in M^2.$$

Letting $\omega_{ij} = w_j - w_i$, then we can get its dual problem:

$$\max \sum_{(i,j)\in E^0} \omega_{ij} x_{ij} + (K - w(T_0))z + \sum_{i\in M^1} (u_i \pi_i + l_i \gamma_i) + \sum_{j \in M^2} u_j \pi_j$$
$$s.t. \sum_{j:(i,j)\in E^0} x_{ij} + z + \pi_i \le c_i, \ \forall i \in M^1,$$
$$- \sum_{j:(i,j)\in E^0} x_{ij} - z + \gamma_i \le c_i, \ \forall i \in M^1,$$
$$- \sum_{i:(i,j)\in E^0} x_{ij} + \pi_j \le c_j, \ \forall j \in M^2,$$
$$x_{ij} \le 0, \ \forall (i, j) \in E^0,$$
$$\pi_i, \gamma_i \le 0, \ \forall i \in M^1,$$
$$\pi_j \le 0, \ \forall j \in M^2.$$

When we substitute $x_{ij}, \pi_i, \gamma_i, \pi_j$ by $-x_{ij}, -\pi_i, -\gamma_i, -\pi_j$, then we can reformulate the dual problem as follows:

$$\max - \sum_{(i,j)\in E^0} \omega_{ij} x_{ij} + (K - w(T_0))z - \sum_{i\in M^1} (u_i \pi_i + l_i \gamma_i) - \sum_{j \in M^2} u_j \pi_j$$
$$s.t. - \sum_{j:(i,j)\in E^0} x_{ij} - \pi_i \le c_i - z, \ \forall i \in M^1,$$
$$(D) \qquad \sum_{j:(i,j)\in E^0} x_{ij} - \gamma_i \le c_i + z, \ \forall i \in M^1,$$
$$\sum_{i:(i,j)\in E^0} x_{ij} - \pi_j \le c_j, \ \forall j \in M^2,$$
$$x_{ij} \ge 0, \ \forall (i, j) \in E^0,$$
$$\pi_i, \gamma_i \ge 0, \ \forall i \in M^1,$$
$$\pi_j \ge 0, \ \forall j \in M^2.$$

Next, we first focus on a subproblem (D^z) below of (D) for a given $z \in \mathbb{R}$, where we delete $(K - w(T_0))z$ from the objective function of the dual problem (D). By solving a series of problems (D^z) corresponding to different z, we can successfully solve the problem (D):

$$
(D^z)\quad
\begin{aligned}
\phi(z) := \max\ & -\sum_{(i,j)\in E^0} \omega_{ij} x_{ij} - \sum_{i\in M^1}(u_i\pi_i + l_i\gamma_i) - \sum_{j\in M^2} u_j\pi_j \\
s.t.\quad & -\sum_{j:(i,j)\in E^0} x_{ij} - \pi_i \le c_i - z,\ \forall i \in M^1, && (11.25)\\
& \sum_{j:(i,j)\in E^0} x_{ij} - \gamma_i \le c_i + z,\ \forall i \in M^1, && (11.26)\\
& \sum_{i:(i,j)\in E^0} x_{ij} - \pi_j \le c_j,\ \forall j \in M^2,, && (11.27)\\
& x_{ij} \ge 0,\ \forall (i,j) \in E^0,\\
& \pi_i, \gamma_i \ge 0,\ \forall i \in M^1,\\
& \pi_j \ge 0,\ \forall j \in M^2.
\end{aligned}
$$

We can obtain **the dual problem** (D_d^z) of (D^z):

$$
(D_d^z)\quad
\begin{aligned}
\psi(z) := \min & \sum_{i\in M^1}(-\alpha_i + \beta_i)z + \sum_{i\in M^1} c_i(\alpha_i + \beta_i) + \sum_{j\in M^2} c_j\alpha_j \\
s.t.\quad & -\alpha_i + \beta_i + \alpha_j \ge -\omega_{ij},\ \forall (i,j) \in E^0, && (11.28)\\
& 0 \le \alpha_i \le u_i,\ \forall i \in M^1, && (11.29)\\
& 0 \le \beta_i \le l_i,\ \forall i \in M^1,, && (11.30)\\
& 0 \le \alpha_j \le u_j,\ \forall j \in M^2. && (11.31)
\end{aligned}
$$

Lemma 11.6 ([66, Lemma 3.1]) *The function $\psi(z)$ has the following properties:*

(P1) $\psi(z)$ is a continuous function.
(P2) $\psi(z)$ is a piecewise linear function of z.
(P3) $\psi(z)$ is a concave function of z.

Assumption 11.1 ([66, Assumption 1]) The problem (RIOVMST_1) is feasible.

Assumption 11.2 ([66, Assumption 2]) The cost vector $c \in \mathbb{Z}^m$ is a positive integer vector.

Lemma 11.7 ([66, Lemma 3.2]) *Let* $\Delta = \{0, 1, -1\}$ *and* $TP = \{\frac{\rho}{\varsigma} | \varsigma \in \mathbb{Z}, |\varsigma| \leq n-1, \rho = \sum_{1\leq i\leq m} \vartheta_i c_i + 2 \sum_{1\leq k\leq n-1} \vartheta_k c_k, \vartheta_i, \vartheta_k \in \Delta\}$. *If* $\frac{\rho_1}{\varsigma_1} \neq \frac{\rho_2}{\varsigma_2}$ *and* $\frac{\rho_1}{\varsigma_1}, \frac{\rho_2}{\varsigma_2} \in TP$, *then* $|\frac{\rho_1}{\varsigma_1} - \frac{\rho_2}{\varsigma_2}| \geq \frac{1}{(n-1)^2}$.

Lemma 11.8 ([66, Lemma 3.3]) *Supposing* $z_k and z_{k+1}$ *are two adjacent turning coordinates, then* $|z_k - z_{k+1}| \geq \frac{1}{(n-1)^2}$.

Lemma 11.9 ([66, Lemma 3.4]) *Suppose* (α^1, β^1) *is a feasible solution of the problem* (D_d^z) *with the objective value* $h(z, \alpha^1, \beta^1)$. *Then* (α^2, β^2) *is also a feasible solution of the problem* (D_d^z) *and* $h(z, \alpha^2, \beta^2) \leq h(z, \alpha^1, \beta^1)$*:*

$$\begin{cases} \alpha_i^2 = 0, \beta_i^2 = -\alpha_i^1 + \beta_i^1, & \text{if } -\alpha_i^1 + \beta_i^1 \geq 0, i \in M^1, \\ \alpha_i^2 = \alpha_i^1 - \beta_i^1, \beta_i^2 = 0, & \text{if } -\alpha_i^1 + \beta_i^1 < 0, i \in M^1, \\ \alpha_j^2 = \alpha_j^1, & \text{if } j \in M^2. \end{cases} \tag{11.32}$$

Definition 11.1 ([66, Definition 3.2]) We call a feasible solution (α^2, β^2) of the problem (D_d^z) in the form of (11.32) *a standard feasible solution.*

Assumption 11.3 ([66, Assumption 3]) Next, we consider a standard feasible solution of the problem (D_d^z). Then at least one of α_i^2 and β_i^2 is 0 for any $i \in M^1$. □

Let

$$\bar{z} = \sum_{j\in M^2} c_j + \max_{i\in M^1} c_i + 1,$$

$$\underline{z} = -\max_{i\in M^1} c_i - 1.$$

For any $i \in M^1$, let

$$M_i^2 = \{j \in M^2 | (i, j) \in E^0\},$$

$$L_i^1 = \max\{-\omega_{ij} - u_j | j \in M_i^2\},$$

$$L_i^2 = \max\{L_i^1, -u_i\}.$$

For any $j \in M^2$, let

$$M_j^1 = \{i \in M^1 | (i, j) \in E^0\},$$

$$U_j^1 = \max\{-\omega_{ij} - L_i^2 | i \in M_j^1\},$$

$$U_j^2 = \max\{U_j^1, 0\}.$$

Theorem 11.7 ([66, Theorem 3.1]) *The vector (α^*, β^*) defined in (11.33) is an optimal solution of the problem (D_d^z) for any $z \geq \bar{z}$.*

$$\begin{cases} \alpha_i^* = 0, \beta_i^* = L_i^2, & \text{if } i \in M^1, L_i^2 \geq 0, \\ \alpha_i^* = -L_i^2, \beta_i^* = 0, & \text{if } i \in M^1, L_i^2 < 0, \\ \alpha_j^* = U_j^2, & \text{if } j \in M^2. \end{cases} \tag{11.33}$$

Based on Theorem 11.7, we know that for any $z \geq \bar{z}$, the linear function $\psi(z) = \sum_{i\in M^1}(-\alpha_i^*+\beta_i^*)z + \sum_{i\in M^1} c_i(\alpha_i^*+\beta_i^*) + \sum_{j\in M^2} c_j\alpha_j^*$ keeps unchanged. Furthermore, the slope of $\psi(z)$ is $k_{\bar{z}} = \sum_{i\in M^1}(-\alpha_i^* + \beta_i^*)$ at point $(\bar{z}, \psi(\bar{z}))$. The following properties obviously hold:

Lemma 11.10 ([66, Lemma 3.6]) *For any $z \geq \bar{z}$, we have $w(T^0) - K \geq k_{\bar{z}}$.*

For any $j \in M^2$, let $L_j^1 = \max\{-\omega_{ij} - l_i | i \in M_j^1\}$ and $L_j^2 = \max\{L_j^1, 0\}$.

Theorem 11.8 ([66, Theorem 3.2]) *The vector (α^*, β^*) defined in (11.34) is an optimal solution of the problem (D_d^z) for any $z \leq \underline{z}$:*

$$\begin{cases} \alpha_i^* = 0, \beta_i^* = l_i, & \text{if } i \in M^1, \\ \alpha_j^* = L_j^2, & \text{if } j \in M^2. \end{cases} \tag{11.34}$$

Based on Theorem 11.8, we know that for any $z \leq \underline{z}$ the linear function $\psi(z) = \sum_{i\in M^1}(-\alpha_i^*+\beta_i^*)z + \sum_{i\in M^1} c_i(\alpha_i^*+\beta_i^*) + \sum_{j\in M^2} c_j\alpha_j^*$ remains unchanged. Furthermore, the slope of $\psi(z)$ is $k_{\underline{z}} = \sum_{i\in M^1}(-\alpha_i^* + \beta_i^*)$ at point $(\underline{z}, \psi(\underline{z}))$. We can obtain the following property:

Lemma 11.11 *[66, Lemma 3.8] For any $z \leq \underline{z}$, we have $w(T^0) - K \leq k_{\underline{z}}$.*

In the forthcoming analysis, we will study how to solve the problem (D_d^z) for a given value $z \in \mathbb{R}$ with $\underline{z} < z < \bar{z}$.

Let $M^{11} = \{i \in M^1 | -c_i \leq z \leq c_i\}$, $M^{12} = \{i \in M^1 | z < -c_i\}$, and $M^{13} = \{i \in M^1 | z > c_i\}$. Therefore, (11.25) holds for $i \in M^{11} \cup M^{12}$ since the left side is non-positive while $c_i - z \geq 0$ for $i \in M^{11}$ and $c_i - z > 2c_i > 0$ for $i \in M^{12}$. Furthermore, in an optimal solution of the problem (D^z), we have $\pi_i^* = 0$ for $i \in M^{11} \cup M^{12}$. Then we can obtain the constraint condition (11.37) from (11.25) for $i \in M^{13}$ by subtracting a surplus variable ξ_i. The equality condition of (11.26) by adding a slack variable η_i can be divided into three conditions (11.35), (11.36), and (11.38), where (11.38) can be obtained by subtracting the condition (11.37) from (11.26) for $i \in M^{13}$. Hence, the standard form (D_s^z) of the problem (D^z) can be stated below.

$$
\begin{aligned}
&\max -\sum_{(i,j)\in E^0} \omega_{ij}x_{ij} - \sum_{i\in M^{11}} l_i\gamma_i - \sum_{i\in M^{12}} l_i\gamma_i \\
&\qquad - \sum_{i\in M^{13}} (u_i\pi_i + l_i\gamma_i) - \sum_{j\in M^2} u_j\pi_j
\end{aligned}
$$

$$
(D_s^z)\quad
\begin{aligned}
\text{s.t.}\ & \sum_{j:(i,j)\in E^0} x_{ij} + \eta_i = c_i + z + \gamma_i, \quad i \in M^{11}, && (11.35)\\
& \sum_{j:(i,j)\in E^0} x_{ij} + \eta_i + (-c_i - z) = \gamma_i, \quad i \in M^{12}, && (11.36)\\
& \sum_{j:(i,j)\in E^0} x_{ij} + \pi_i = z - c_i + \xi_i, \quad i \in M^{13}, && (11.37)\\
& \eta_i + \xi_i = 2c_i + \gamma_i + \pi_i, \quad i \in M^{13}, && (11.38)\\
& \sum_{i:(i,j)\in E^0} x_{ij} + \varrho_j = c_j + \pi_j, \quad j \in M^2, && (11.39)\\
& x_{ij} \geq 0, \quad (i,j) \in E^0,\\
& \gamma_i, \eta_i \geq 0, \quad i \in M^1,\\
& \pi_j, \varrho_j \geq 0, \quad j \in M^2,\\
& \pi_i, \xi_i \geq 0, \quad i \in M^{13}.
\end{aligned}
$$

Next, we will show that the problem (D_s^z) can be solved by a minimum cost flow problem on **an auxiliary graph** $\bar{G}^z(\bar{V}, \bar{E}, \bar{b}^z, \bar{c})$, where $\bar{b}^z$ and $\bar{c}$ are the capacity vector and cost vector, respectively. For convenience, let $\bar{\lambda} = \max\{\max\{l_i, u_i\} | i \in M^1\} + 1$ and $S = \{w, l, u, c, \omega, M^1, M^2, E^0, \bar{\lambda}\}$. Algorithm 11.4 shows the construction process of auxiliary graph $\bar{G}^z(\bar{V}, \bar{E}, \bar{b}^z, \bar{c})$.

Algorithm 11.4 $\bar{G}^z(\bar{V}, \bar{E}, \bar{b}^z, \bar{c}) =$ **AG1**(S, z) [66, Algorithm 1]

Input: The parameter set S and a real number z.

Output: An auxiliary graph $\bar{G}^z(\bar{V}, \bar{E}, \bar{b}^z, \bar{c})$, where $\bar{b}^z$ and $\bar{c}$ are the capacity vector and cost vector, respectively.

Let $M^{11} := \{i \in M^1 | -c_i \leq z \leq c_i\}$, $M^{12} := \{i \in M^1 | z < -c_i\}$, $M^{13} := \{i \in M^1 | z > c_i\}$.

Initialize $\bar{V} := M^1 \cup M^2 \cup \{s, t\}$, $\bar{E} := E^0 \cup (s, t)$, $\bar{c}_{st} := 0$, $\bar{b}_{st}^z := +\infty$.

for $(i, j) \in E^0$ **do**

  $\bar{c}_{ij} := \omega_{ij}$, $\bar{b}_{ij}^z := +\infty$.

end for

for $j \in M^2$ **do**

Let $\bar{V} := \bar{V} \cup \{j_1, j_2\}$, $\bar{E} := \bar{E} \cup \{(j, j_1), (j, j_2), (j_1, t), (j_2, t), (s, j)\}$, $\bar{c}_{j,j_1} := -\bar{\lambda}$, $\bar{c}_{j,j_2} := u_j$, $\bar{c}_{j_1,t} := 0$, $\bar{c}_{j_2,t} := 0$, $\bar{c}_{s,j} := 0$, $\bar{b}^z_{j,j_1} := c_j$, $\bar{b}^z_{j,j_2} := +\infty$, $\bar{b}^z_{j_1,t} := +\infty$, $\bar{b}^z_{j_2,t} := +\infty$, and $\bar{b}^z_{s,j} := +\infty$.
end for
for $i \in M^{11}$ **do**
Let $\bar{V} := \bar{V} \cup \{i_1, i_2\}$, $\bar{E} := \bar{E} \cup \{(i_1, i), (i_2, i), (s, i_1), (s, i_2), (i, t)\}$, $\bar{c}_{i_1,i} := l_i$, $\bar{c}_{i_2,i} := -\bar{\lambda}$, $\bar{c}_{s,i_1} := 0$, $\bar{c}_{s,i_2} := 0$, $\bar{c}_{i,t} := 0$, $\bar{b}^z_{i_1,i} := +\infty$, $\bar{b}^z_{i_2,i} := c_i + z$, $\bar{b}^z_{s,i_1} := +\infty$, $\bar{b}^z_{s,i_2} := +\infty$, and $\bar{b}^z_{i,t} := +\infty$.
end for
for $i \in M^{12}$ **do**
Let $\bar{V} := \bar{V} \cup \{i_1\}$, $\bar{E} := \bar{E} \cup \{(i, i_1), (i_1, t), (i, t), (s, i)\}$, $\bar{c}_{si} := l_i$, $\bar{c}_{i,i_1} := -\bar{\lambda}$, $\bar{c}_{i_1,t} := 0$, $\bar{c}_{i,t} := 0$, $\bar{b}^z_{si} := +\infty$, $\bar{b}^z_{i,i_1} := -c_i - z$, $\bar{b}^z_{i_1,t} := +\infty$, and $\bar{b}^z_{i,t} := +\infty$.
end for
for $i \in M^{13}$ **do**
Let $\bar{V} := \bar{V} \cup \{i_1, i_2, i_3, i_4, i_5\}$, $\bar{E} := \bar{E} \cup \{(s, i_1), (i_1, i), (i, i_5), (i_5, i_2), (i_2, i), (i_2, t), (i_3, i_2), (i_4, i_2), (s, i_3), (s, i_4)\}$, $\bar{c}_{i_1,i} = \bar{c}_{i_4,i_2} := -\bar{\lambda}$, $\bar{c}_{i,i_5} := u_i$, $\bar{c}_{i_3,i_2} := l_i$, $\bar{c}_{s,i_1} = \bar{c}_{i_2,i} = \bar{c}_{i_5,i_2} = \bar{c}_{i_2,t} = \bar{c}_{s,i_3} = \bar{c}_{s,i_4} := 0$, $\bar{b}^z_{i_1,i} = z - c_i$, $\bar{b}^z_{i_4,i_2} := 2c_i$, and $\bar{b}^z_{s,i_1} = \bar{b}^z_{i_2,i} = \bar{b}^z_{i_5,i_2} = \bar{b}^z_{i_2,t} = \bar{b}^z_{s,i_3} = \bar{b}^z_{s,i_4} = \bar{b}^z_{i,i_5} = \bar{b}^z_{i_3,i_2} := +\infty$.
end for
return $\bar{G}^z(\bar{V}, \bar{E}, \bar{b}^z, \bar{c})$.

We next give an example to demonstrate how to construct $\bar{G}^z(\bar{V}, \bar{E}, \bar{b}^z, \bar{c})$. Let $M^1 = \{1, 2, 3\}$, $M^2 = \{4, 5\}$, $E^0 = \{(1, 4), (1, 5), (2, 4), (2, 5), (3, 4)\}$, $M^{11} = \{1\}$, $M^{12} = \{2\}$, and $M^{13} = \{3\}$. The left graph in Fig. 11.1 is a bipartite graph $G_b(M^1, M^2, E^0)$. The right graph in Fig. 11.1 is the auxiliary graph $\bar{G}^z(\bar{V}, \bar{E}, \bar{b}^z, \bar{c})$ based on the left bipartite graph G_b.

Let $\bar{A}$ be the node-arc incidence matrix of graph $\bar{G}^z$ and ν^z be the value of a flow. Next, we consider the following problem $(\bar{P}^z)$ defined on $\bar{G}^z$:

$$\min \varpi(f) := \sum_{(i,j)\in E^0} \omega_{ij} f_{ij} + \sum_{i\in M^{11}} (l_i f_{i_1,i} - \bar{\lambda} f_{i_2,i}) + \sum_{i\in M^{12}} (l_i f_{si} - \bar{\lambda} f_{i,i_1})$$

$$+ \sum_{i\in M^{13}} [l_i f_{i_3,i_2} - \bar{\lambda}(f_{i_1,i} + f_{i_4,i_2}) + u_i f_{i,i_5}] + \sum_{j\in M^2} (u_j f_{j,j_2} - \bar{\lambda} f_{j,j_1})$$

$$(\bar{P}^z) \text{ s.t. } \bar{A} f = (\nu^z, -\nu^z, 0, 0, .., 0)^T,$$

$$f \le \bar{b}^z,$$

$$f \ge 0,$$

$$\nu^z \ge 0.$$

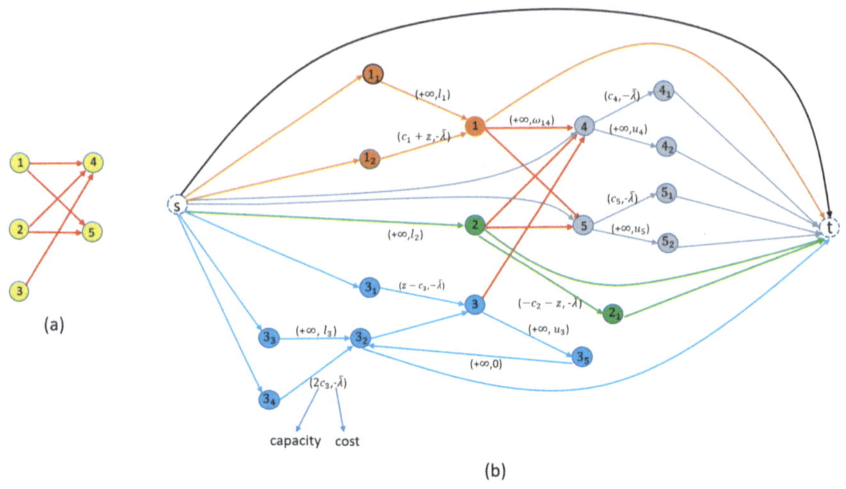

Fig. 11.1 A demonstration to construct the auxiliary graph $\bar{G}^z$

Lemma 11.12 ([66, Lemma 3.9]) *For a given value z, if f^* is an optimal solution of the problem ($\bar{P}^z$), then f^* satisfies the following conditions:*

(1) $f^*_{i_2,i} = c_i + z, i \in M^{11}$.
(2) $f^*_{i,i_1} = -c_i - z, i \in M^{12}$.
(3) $f^*_{i_4,i_2} = 2c_i, f^*_{i_1,i} = z - c_i, f^*_{i,i_5} = f^*_{i_5,i_2}, i \in M^{13}$.
(4) $f^*_{j,j_1} = c_j, j \in M^2$.

Theorem 11.9 ([66, Theorem 3.3]) *Suppose f^* is an optimal solution of the problem ($\bar{P}^z$). Let*

$$\begin{cases} x^*_{ij} = f^*_{ij}, & (i,j) \in E^0, \\ \eta^*_i = f^*_{it}, \gamma^*_i = f^*_{i_1,i}, & i \in M^{11}, \\ \eta^*_i = f^*_{i,t}, \gamma^*_i = f^*_{si}, & i \in M^{12}, \\ \pi^*_i = f^*_{i,i_5}, \xi^*_i = f^*_{i_2,i}, \eta^*_i = f^*_{i_2,t}, \gamma^*_i = f^*_{i_3,i_2}, & i \in M^{13}, \\ \varrho^*_j = f^*_{s,j}, \pi^*_j = f^*_{j,j_2}, & j \in M^2. \end{cases} \tag{11.40}$$

Then $(x^*, \eta^*, \pi^*, \xi^*, \varrho^*, \gamma^*)$ is an optimal solution of the problem (D^z_s).

Based on Theorem 11.9, we know that solving the problem (D^z_s) is equivalent to solve the problem ($\bar{P}^z$). Next, we solve the problem ($\bar{P}^z$) by a minimum cost flow problem on $\bar{G}^z$.

Definition 11.2 ([106]) Given a feasible flow f in a weighted network $\bar{G}^z$, define **the residual network** $\bar{G}'^z(f)$ with respect to f as follows: $\bar{G}'^z$ has the same node set as $\bar{G}^z$. For each arc (i, j) in $\bar{G}^z$ with flow f_{ij}, capacity $\bar{b}_{ij}$, and cost $\bar{c}_{ij}$, put two arcs in $\bar{G}'^z$: one arc (i, j) with capacity $\bar{b}_{ij} - f_{ij} \geq 0$ and cost $\bar{c}_{ij}$ and another arc (j, i) with capacity $f_{ij} \geq 0$ and cost $-\bar{c}_{ij}$. Omit all arcs with zero capacity.

Theorem 11.10 (An Optimality Condition of Minimum Cost Flow [106]) *A feasible flow is a minimum cost flow if and only if the residual network has no negative cost circle.*

Let $E^z_{\bar{\lambda}} = \{(i_2, i)|i \in M^{11}\} \cup \{(i, i_1)|i \in M^{12}\} \cup \{(i_1, i), (i_4, i_2)|i \in M^{13}\} \cup \{(j, j_1)|j \in M^2\}$ and

$$v^z_0 = \sum_{i \in M^{11}} (c_i + z) + \sum_{i \in M^{12}} (-c_i - z) + \sum_{i \in M^{13}} (c_i + z) + \sum_{j \in M^2} c_j. \tag{11.41}$$

Suppose f is a feasible flow of the problem $(\bar{P}^z)$ and P_0 is a directed path from s to t. Let $f(P_0) = \min_{e \in P_0} f(e)$, $\bar{c}(P_0) = \sum_{e \in P_0} \bar{c}(e)$, and $\overleftarrow{P_0}$ be a reverse path in which every arc is a reverse arc of that in P_0.

Theorem 11.11 ([66, Theorem 3.5]) *The problem $(\bar{P}^z)$ is equivalent to the following minimum cost flow problem $(\bar{P}^z_1)$ on $\bar{G}^z$, where v^z_0 is defined in (11.41):*

$$\begin{aligned}
&\min \sum_{(i,j) \in E^0} \omega_{ij} f_{ij} + \sum_{i \in M^{11}} (l_i f_{i_1,i} - \bar{\lambda} f_{i_2,i}) + \sum_{i \in M^{12}} (l_i f_{si} - \bar{\lambda} f_{i,i_1}) \\
&\quad + \sum_{i \in M^{13}} [l_i f_{i_3,i_2} - \bar{\lambda}(f_{i_1,i} + f_{i_4,i_2}) + u_i f_{i,i_5}] + \sum_{j \in M^2} (u_j f_{j,j_2} - \bar{\lambda} f_{j,j_1}) \\
&(\bar{P}^z_1) \text{ s.t. } \bar{A} f = (v^z_0 + 1, -v^z_0 - 1, 0, 0, .., 0)^T, \\
&\qquad\qquad f \leq \bar{b}^z, \\
&\qquad\qquad f \geq 0.
\end{aligned}$$

Note that the only difference between the problems $(\bar{P}^z_1)$ and $(\bar{P}^z)$ is the value of the minimum cost flow, which is a given value $v^z_0 + 1$ in $(\bar{P}^z_1)$ and a variable v^z in $(\bar{P}^z)$.

It is easy to verify that we can limit the infinity capacity to $v^z_0 + 1$. However, some costs of edges may be negative, which is different from the classical minimum cost flow problem. Fortunately, we can transform it into a classical minimum cost flow problem by an arc reversal technology [4].

Definition 11.3 (An Arc Reversal Transformation [4]) The arc reversal transformation is typically used to remove arcs with negative costs. Let $\bar{b}^z_{ij}$ denote the capacity of the arc (i, j). We replace the variable f_{ij} by $\bar{b}^z_{ij} - f_{ji}$. In this transformation, we replace the arc (i, j) with a cost $\bar{c}_{ij}$ by the arc (j, i) with an associated cost $-\bar{c}_{ij}$ and the same capacity. As shown in Fig. 11.2, the transformation has the following network interpretation. We first sent $\bar{b}^z_{ij}$ units of

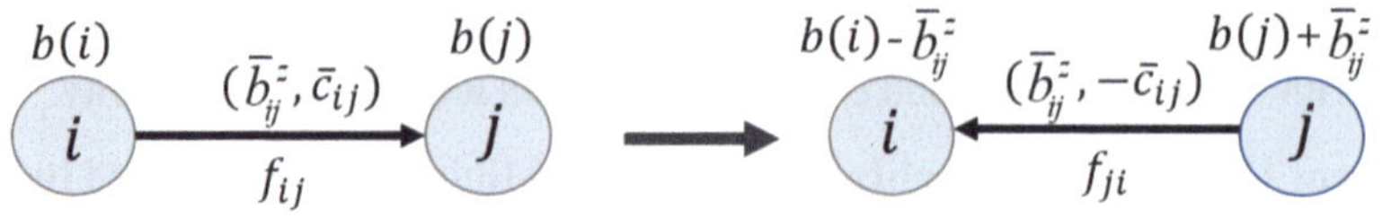

Fig. 11.2 An arc reversal transformation

flow on the arc (i, j), which decrease $b(i)$ by $\bar{b}^z_{ij}$ units and increase $b(j)$ by $\bar{b}^z_{ij}$ units, and then we replace arc (i, j) by arc (j, i) with cost $-\bar{c}_{ij}$. The new flow f_{ji} measures the amount of flow we "remove" from the "full-capacity" flow of $\bar{b}^z_{ij}$.

By the *arc reversal transformation*, we can construct a new auxiliary network $\tilde{G}^z(\tilde{V}, \tilde{E}, \tilde{b}^z, \tilde{c}, \tilde{d}^z)$ from $\bar{G}^z(\bar{V}, \bar{E}, \bar{b}^z, \bar{c})$. Then an optimal solution of the problem (D^z_s) can be obtained by solving a classical minimum cost flow problem on $\tilde{G}^z$. Let $E^0_< = \{(i, j) \in E^0 | \omega_{ij} < 0\}$. Algorithm 11.5 gives its construction process.

Algorithm 11.5 $\tilde{G}^z(\tilde{V}, \tilde{E}, \tilde{b}^z, \tilde{c}, \tilde{d}^z)$:= **AG2**$(\bar{G}^z(\bar{V}, \bar{E}, \bar{b}^z, \bar{c}), E^0_<, M^1, M^2, z)$ [66, Algorithm 2]

Input: The auxiliary graph $\bar{G}^z(\bar{V}, \bar{E}, \bar{b}^z, \bar{c})$ and the sets $E^0_<$, M^1, and M^2 and a real number z.

Output: A new auxiliary graph $\tilde{G}^z(\tilde{V}, \tilde{E}, \tilde{b}^z, \tilde{c}, \tilde{d}^z)$, where $\tilde{d}^z$ represents the supply or demand value of every node.

Let $M^{11} := \{i \in M^1 | -c_i \le z \le c_i\}$, $M^{12} := \{i \in M^1 | z < -c_i\}$, $M^{13} := \{i \in M^1 | z > c_i\}$. $E^z_{\bar{\lambda}} = \{(i_2, i) | i \in M^{11}\} \cup \{(i, i_1) | i \in M^{12}\} \cup \{(i_1, i), (i_4, i_2) | i \in M^{13}\} \cup \{(j, j_1) | j \in M^2\}$.
Let $v^z_0 := \sum_{i \in M^{11}} (c_i + z) + \sum_{i \in M^{12}} (-c_i - z) + \sum_{i \in M^{13}} (c_i + z) + \sum_{j \in M^2} c_j$.
Initialize $\tilde{G}^z(\tilde{V}, \tilde{E}, \tilde{b}^z, \tilde{c}, \tilde{d}^z) := \bar{G}^z(\bar{V}, \bar{E}, \bar{b}^z, \bar{c})$ and $\tilde{d}^z := (v^z_0 + 1, -v^z_0 - 1, 0, \ldots, 0)$.
for $(i, j) \in \tilde{E}$ **do**
    if $\tilde{b}^z_{ij} = +\infty$ **then** let $\tilde{b}^z_{ij} := v^z_0 + 1$.
    end if
end for
for $(i, j) \in E^0_<$ **do**
    Reverse this edge to (j, i) and $\tilde{d}^z(i) := \tilde{d}^z(i) - v^z_0 - 1$, $\tilde{d}^z(j) := \tilde{d}^z(j) + v^z_0 + 1$, $\tilde{c}_{ji} := -\tilde{c}_{ij}$, $\tilde{b}^z_{ji} = v^z_0 + 1$.
end for
for $(i, j) \in E^z_{\bar{\lambda}}$ **do**
    Reverse this edge to (j, i) and $\tilde{d}^z(i) := \tilde{d}^z(i) - \tilde{b}^z_{ij}$, $\tilde{d}^z(j) := \tilde{d}^z(j) + \tilde{b}^z_{ij}$, $\tilde{c}_{ji} := \bar{\lambda}$, $\tilde{b}^z_{ji} = \tilde{b}^z_{ij}$.
end for
return $\tilde{G}^z(\tilde{V}, \tilde{E}, \tilde{b}^z, \tilde{c}, \tilde{d}^z)$.

If $\tilde{f}^*$ is a minimum cost flow on network $\tilde{G}^z(\tilde{V}, \tilde{E}, \tilde{b}^z, \tilde{c}, \tilde{d}^z)$, then we can get an optimal solution of the problem (D_s^z) by (11.42):

$$\begin{cases} x_{ij}^* = \tilde{f}_{ij}^*, & (i,j) \in E^0 \setminus E_<^0, \\ x_{ij}^* = v_0^z + 1 - \tilde{f}_{ji}^*, & (i,j) \in E_<^0, \\ \eta_i^* = \tilde{f}_{it}^*, \gamma_i^* = \tilde{f}_{i_1,i}^*, \pi_i^* = 0, & i \in M^{11}, \\ \eta_i^* = \tilde{f}_{i,t}^*, \gamma_i^* = \tilde{f}_{si}^*, \pi_i^* = 0, & i \in M^{12}, \\ \pi_i^* = \tilde{f}_{i,i_5}^*, \xi_i^* = \tilde{f}_{i_2,i}^*, \eta_i^* = \tilde{f}_{i_2,t}^*, \gamma_i^* = \tilde{f}_{i_3,i_2}^*, & i \in M^{13}, \\ \varrho_j^* = \tilde{f}_{s,j}^*, \pi_j^* = \tilde{f}_{j,j_2}^*, & j \in M^2. \end{cases} \tag{11.42}$$

Let $E^* = \{(i,j) \in E^0 | x_{ij}^* > 0\}$, $M_\pi^1 = \{i \in M^1 | \pi_i^* > 0\}$, $M_\gamma^1 = \{i \in M^1 | \gamma_i^* > 0\}$, $M_\pi^2 = \{j \in M^2 | \pi_j^* > 0\}$, $M_1^1 = \{i \in M^1 | -\sum_{j:(i,j) \in E^0} x_{ij}^* + z - \pi_i^* < c_i\}$, $M_2^1 = \{i \in M^1 | \sum_{j:(i,j) \in E^0} x_{ij}^* - z - \gamma_i^* < c_i\}$, and $M_1^2 = \{j \in M^2 | \sum_{i:(i,j) \in E^0} x_{ij}^* - \pi_j^* < c_j, \}$. Then we can determine an optimal solution of the problem (D_d^z) by solving the following system (11.43) of linear equations based on the dual feasibility and **theorem of complementary slackness** [106].

$$\begin{cases} -\alpha_i^* + \beta_i^* + \alpha_j^* = -\omega_{ij}, & (i,j) \in E^* \\ \alpha_i^* = u_i, & i \in M_\pi^1 \\ \beta_i^* = l_i, & i \in M_\gamma^1 \\ \alpha_j^* = u_j, & i \in M_\pi^2 \\ \alpha_i^* = 0, & i \in M_1^1 \\ \beta_i^* = 0, & i \in M_2^1 \\ \alpha_j^* = 0, & j \in M_1^2 \end{cases} \tag{11.43}$$

In the next discussion, we will propose Algorithm 11.6 to calculate the function value $\psi(z)$ and its slope at point $(z, \psi(z))$. We first demonstrate how to calculate the optimal solution (α^*, β^*) of problem (D_d^z) from the optimal solution (x^*, π^*, γ^*) of problem (D^z) based on the *theorem of complementary slackness* [106].

There are two cases: (1) If $z \geq \bar{z}$ or $z \leq \bar{z}$, then (α^*, β^*) defined by (11.33) or (11.34) is an optimal solution of problem (D_d^z). (2) If $\underline{z} < z < \bar{z}$, then we can get the optimal solution (x^*, π^*, γ^*) of problem (D^z) by calculating a minimum cost flow on network $\tilde{G}^z(\tilde{V}, \tilde{E}, \tilde{b}^z, \tilde{c}, \tilde{d}^z)$. Furthermore, we can obtain an optimal solution (α^*, β^*) of problem (D_d^z) by (11.43). Hence, $\psi(z) = \sum_{i \in M^1}(-\alpha_i^* + \beta_i^*)z + \sum_{i \in M^1} c_i(\alpha_i^* + \beta_i^*) + \sum_{j \in M^2} c_j \alpha_j^*$ and its slope $k_z = \sum_{i \in M^1}(-\alpha_i^* + \beta_i^*)$ at point $(z, \psi(z))$.

Algorithm 11.6 $(k_z, \psi(z)) =$ **Slope**$(S, E^0_<, \bar{z}, \underline{z}, z)$ [66, Algorithm 3]

Input: The parameter set S and the set $E^0_<$ and the real numbers $\bar{z}, \underline{z}, z$.
Output: The function value of $\psi(z)$ and its slope at an point $(z, \psi(z))$.

if $z \geq \bar{z}$ **then**
  Calculate (α^*, β^*) by (11.33).
else if $z \leq \underline{z}$ **then**
  Calculate (α^*, β^*) by (11.34).
else
  Call $\bar{G}^z(\bar{V}, \bar{E}, \bar{b}^z, \bar{c})$=**AG1**$(S, z)$.
  Call $\tilde{G}^z(\tilde{V}, \tilde{E}, \tilde{b}^z, \tilde{c}, \tilde{d}^z)$=**AG2**$(\bar{G}^z(\bar{V}, \bar{E}, \bar{b}^z, \bar{c}), E^0_<, M^1, M^2, z)$.
  Calculate a minimum cost flow $\tilde{f}^*$ on network $\tilde{G}^z(\tilde{V}, \tilde{E}, \tilde{b}^z, \tilde{c}, \tilde{d}^z)$.
  Calculate the optimal solution (x^*, π^*, γ^*) of model (D^z) by (11.42).
  Solve the equations (11.43) to calculate (α^*, β^*).
end if
Return $k_z := \sum_{i \in M^1} (-\alpha_i^* + \beta_i^*)$, $\psi(z) := k_z z + \sum_{i \in M^1} c_i(\alpha_i^* + \beta_i^*) + \sum_{j \in M^2} c_j \alpha_j^*$.

Remark 11.2 Notice that $\psi(z)$ is a piecewise linear and continuous function. Hence, if z is a turning coordinate, then k_z is uncertain. We should avoid this situation when call Algorithm 11.6.

In the next discussion, we aim to construct an optimality condition for solving the problem (D). For convenience, we introduce the following definition:

Definition 11.4 ([66, Definition 4.1]) Supposing $(x^*, \pi^*, \gamma^*, z^*)$ is an optimal solution of problem (D), then we call z^* *a critical value*.

Obviously, if we can determine **a critical value** z^*, then we can solve the subproblem (D^{z^*}) and obtain its optimal solution (x^*, π^*, γ^*), which renders an optimal solution $(x^*, \pi^*, \gamma^*, z^*)$ of problem (D). Hence, the main issue now is how to determine z^*. Next, we propose an optimality condition on z^*.

Theorem 11.12 ([66, Theorem 4.1])

(1) If $k_{\bar{z}} = w(T^0) - K$*, then* $\bar{z}$ *is* ***a critical value.***
(2) If $k_{\underline{z}} = w(T^0) - K$*, then* $\underline{z}$ *is* ***a critical value.***
(3) Suppose $(z^*, \psi(z^*))$ *is an intersection point of two adjacent lines* L_1*and*L_2 *in* $\psi(z)$*, where* $L_i : \psi(z) = k_i z + b_i, i = 1, 2$*. If* $k_2 < w(T_0) - K \leq k_1$*, then* z^* *is* ***a critical value.***

Next, we will design Algorithm 11.7 to solve problem (P). There are two main steps:
(1) We first determine **the critical value** z^*. The main idea is as follows: First of all, we calculate $k_{\bar{z}}$ and $k_{\underline{z}}$. If $k_{\bar{z}} = w(T^0) - K$ or $k_{\underline{z}} = w(T^0) - K$, then $z^* = \bar{z}$ or $z^* = \underline{z}$ by Theorem 11.12. Next, if $k_{\bar{z}} < w(T^0) - K < k_{\underline{z}}$, then we need to

find the intersection point $(z^*, \psi(z^*))$ in the case (3) of Theorem 11.12, which can be realized by a binary search method. Iterations will terminate if the length of an interval in a binary search is less than $\frac{1}{(n-1)^2}$. In this case, there is only a **a turning coordinate** in this interval by Lemma 11.8. Therefore, we can get z^* by calculating the intersection point of the two lines.
(2) When z^* is determined, we can get the optimal solution (x^*, π^*, γ^*) of the problem (D^{z^*}) by solving a minimum cost flow on network $\tilde{G}^{z^*}(\tilde{V}, \tilde{E}, \tilde{b}^{z^*}, \tilde{c}, \tilde{d}^{z^*})$. Thus, $(x^*, \pi^*, \gamma^*, z^*)$ is an optimal solution of problem (D). Moreover, we can determine the optimal solution of the problem (P) by solving the linear equations (11.44) based on *theorem of complementary slackness* [106]:

$$\begin{cases} w_i + \bar{\alpha}_i - \bar{\beta}_i = w_j + \bar{\alpha}_j, & (i, j) \in E^* \\ \bar{\alpha}_i = u_i, & i \in M_\pi^1 \\ \bar{\beta}_i = l_i, & i \in M_\gamma^1 \\ \bar{\alpha}_j = u_j, & i \in M_\pi^2 \\ \bar{\alpha}_i = 0, & i \in M_1^1 \\ \bar{\beta}_i = 0, & i \in M_2^1 \\ \bar{\alpha}_j = 0, & j \in M_1^2 \\ \sum\limits_{i \in M^1} (\bar{\alpha}_i - \bar{\beta}_i) = K - w(T_0) \end{cases} \tag{11.44}$$

Algorithm 11.7 $(\bar{\alpha}^*, \bar{\beta}^*) = \textbf{RIOVMST}_1(G(V, E), T^0, w, l, u, c, K)$ [66, Alg. 4]

Input: A network $G = (V, E)$, a spanning tree T^0, a weight vector w, a reduction vector l, an increment vector u, a cost vector c, and a real number K
Output: An optimal solution (α^*, β^*) of problem (P).

```
Calculate Ū, L̲.
if L̲(T^0) > K or Ū(T^0) < K then
    The problem (RIOVMST_1) is infeasible and stop.
end if
Calculate z̲, z̄ by (11.33), (11.33) and ω, M^1, M^2, E^0, λ̄, E^0_<.
Let a := z̲, b := z̄ and flag := 0.
Call (k_a, ψ(a))=Slope(S, E^0_<, z̄, z̲, a).
Call (k_b, ψ(b))=Slope(S, E^0_<, z̄, z̲, b).
if k_a = w(T^0) − K then let z* := a.
else if k_b = w(T^0) − K then let z* := b.
else
    while b − a > 1/(n−1)^2 do
        Let z := (a+b)/2.
        if z ∈ TP then
            Let z_l := z − 1/(2(n−1)^2) and z_r := z + 1/(2(n−1)^2).
            Call (k_{z_l}, ψ(z_l))=Slope(S, E^0_<, z̄, z̲, z_l).
            Call (k_{z_r}, ψ(z_r))=Slope(S, E^0_<, z̄, z̲, z_r).
```

18: **if** $z_l \geq w(T^0) - K$ and $z_r < w(T^0) - K$ **then** let $z^* := z$ and $flag := 1$. **break.**
19: **else if** $z_l \geq w(T^0) - K$ and $z_r \geq w(T^0) - K$ **then** let $a := z_r$.
20: **else** let $b := z_l$.
21: **end if**
22: **else**
23: Call $(k_z, \psi(z))$=**Slope**$(S, E^0_<, \bar{z}, \underline{z}, z)$.
24: **if** $k_z \geq w(T^0) - K$ **then** update $a := z$.
25: **else** update $b := z$.
26: **end if**
27: **end if**
28: **end while**
29: **if** $flag = 0$ **then**
30: Call $(k_a, \psi(a))$=**Slope**$(S, E^0_<, \bar{z}, \underline{z}, a)$.
31: Call $(k_b, \psi(b))$=**Slope**$(S, E^0_<, \bar{z}, \underline{z}, b)$.
32: Calculate $z^* := \frac{\psi(b)-\psi(a)-k_b b+k_a a}{k_a-k_b}$.
33: **end if**
34: **end if**
35: Call $\bar{G}^{z^*}(\bar{V}, \bar{E}, \bar{b}^{z^*}, \bar{c})$=**AG1**$(S, z^*)$.
36: Call $\tilde{G}^{z^*}(\tilde{V}, \tilde{E}, \tilde{b}^{z^*}, \tilde{c}, \tilde{d}^{z^*})$=**AG2**$(\bar{G}^{z^*}(\bar{V}, \bar{E}, \bar{b}^{z^*}, \bar{c}), E^0_<, M^1, M^2, z^*)$.
37: Calculate a minimum cost flow $\tilde{f}^*$ on network $\tilde{G}^{z^*}(\tilde{V}, \tilde{E}, \tilde{b}^{z^*}, \tilde{c}, \tilde{d}^{z^*})$.
38: Calculate an optimal solution (x^*, π^*, γ^*) of model (D^{z^*}) by (11.42).
39: **Return** $(\bar{\alpha}^*, \bar{\beta}^*)$ by (11.44).

Remark 11.3 Notice that any $z \in TP$ is in the form of $\frac{\rho}{\varsigma}$ by Lemma 11.7, where ρ and $|\varsigma| \leq n-1$ are integers. Hence, we can use another way to judge whether z belongs to TP in line 16 even if we do not know the set TP beforehand. Suppose $z = \frac{q}{p}$ and p, q are relatively prime. If $|p| \leq n-1$, then $z \in TP$. Furthermore, if $z \in TP$, then $z_l, z_r \notin TP$ by Lemma 11.7.

There are many different algorithms to solve the minimum cost flow problem. To our knowledge, the fastest algorithm for minimum cost flow problem is proposed by Orlin [103] with time complexity $O(m \log m(m + n \log n))$, where $n = |V(G)|$ and $m = |E(G)|$. **It is an open problem** that whether there is a strongly polynomial time algorithm with better complexity to find a minimum cost flow problem on the special network $\tilde{G}^z(\tilde{V}, \tilde{E}, \tilde{b}^z, \tilde{c}, \tilde{d}^z)$. Hence, we can get the time complexity of Algorithm 11.7.

Theorem 11.13 ([66, Theorem 4.2]) *The time complexity of Algorithm 11.7 is* $O(m^2 n^2 \log n \log(n\,C))$, *where* $C = \max c$.

11.4 (RIOV, MST, Bounded, Bottleneck Hamming Distance)

In this section, we introduce the restricted inverse optimal value problem on minimum spanning tree under bottleneck Hamming distance (**RIOVMST**$_{bH}$) [128]. Its mathematical model can be stated as follows:

$$\textbf{(RIOVMST}_{bH}\textbf{)} \quad \begin{aligned} z(\bar{w}) = {} & \min_{\bar{w}} \max_{e_i \in E} c_i^0 H(\bar{w}_i, w_i) \\ \text{s.t.} \quad & \bar{w}(T^0) = K, \\ & \bar{w}(T) \geq K, \quad \forall T \in \Gamma, \\ & l \leq \bar{w} \leq u. \end{aligned} \tag{11.45}$$

Define a vector $c = (c_1, \cdots, c_\sigma)$, where $c_i \neq c_j$ for any i, j, which is obtained by deleting the same elements from c^0. For a given c_k $(k = 1, 2, \cdots, \sigma)$, we define a subproblem (**RIOVMST**$_{bH}(c_k)$) of (**RIOVMST**$_{bH}$), whose objective value $z(\bar{w})$ is no more than c_k.

In this subsection, we first introduce some notations and then analyze some properties of a feasible solution of the problem (**RIOVMST**$_{bH}$).

For a given c_k $(k = 1, 2, \cdots, \sigma)$, let

$$T^0_{\leq c_k} = \{e_i \in T^0 | c_i^0 \leq c_k\}, \quad T^0_{>c_k} = \{e_i \in T^0 | c_i^0 > c_k\}$$

be the sets of edges in T^0 satisfying $c_i^0 \leq c_k$ and $c_i^0 > c_k$, respectively. Let

$$\overline{T}^0_{\leq c_k} = \{e_i \notin T^0 | c_i^0 \leq c_k\}, \quad \overline{T}^0_{>c_k} = \{e_i \notin T^0 | c_i^0 > c_k\}$$

be the sets of edges not in T^0 satisfying $c_i^0 \leq c_k$ and $c_i^0 > c_k$, respectively.

For each $e_j \notin T^0$, $T^0 \cup \{e_j\}$ contains a **fundamental cycle**, and we denote by P_j the **fundamental path** containing all the edges in this cycle except e_j. For each $e_i \in T^0$, define $\Omega_i = \{e_j \notin T^0 | e_i \in P_j\}$ as the set of non-tree edges e_j such that e_i is in the **fundamental path** P_j. Let $\Omega^0 = \{e_i \in T^0 | \Omega_i = \emptyset\}$ be the set of **isolated edges** in T^0, which belong to every spanning tree of G and define $\Omega^0_{\leq c_k} = \Omega^0 \cap T^0_{\leq c_k}$. If $\Omega^0 \neq \emptyset$, then Ω^0 is also the set of cut edges of G. If $\Omega^0 = \emptyset$, then each edge in T^0 belongs to at least one fundamental cycle.

Define $\widetilde{T}^0_{\leq c_k} = \{e_i \in T^0_{\leq c_k} \setminus \Omega^0 | \Omega_i \subset \overline{T}^0_{\leq c_k}\}$ and

$$\Phi^0_{\leq c_k} = \Omega^0_{\leq c_k} \cup \widetilde{T}^0_{\leq c_k}. \tag{11.46}$$

Let $\Gamma^0_{\leq c_k} = T^0_{\leq c_k} \setminus \Phi^0_{\leq c_k}$. For each $e_i \in \Gamma^0_{\leq c_k}$, define

$$w_{j_i} = \min_{e_j \in \Omega_i \cap \overline{T}^0_{>c_k}} w_j, \tag{11.47}$$

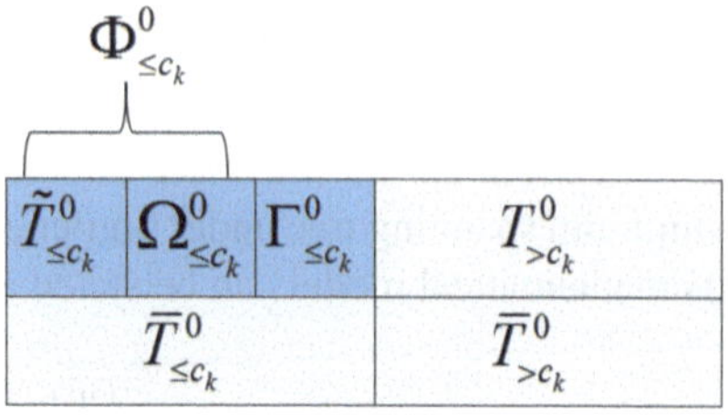

Fig. 11.3 The relationships among some notations

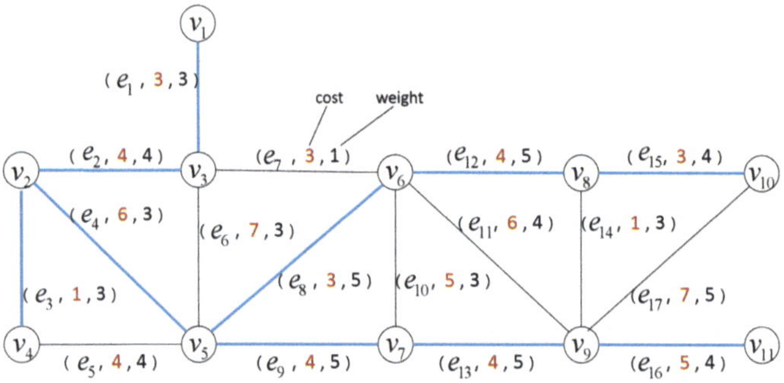

Fig. 11.4 An example to show the meanings of notations

where the minimum value is obtained on the edge e_{j_i}. And define

$$w_j(\Gamma^0_{\le c_k}) = \sum_{e_i \in \Gamma^0_{\le c_k}} w_{j_i}. \tag{11.48}$$

Obviously, by the definitions of $T^0_{\le c_k}$, $\widetilde{T}^0_{\le c_k}$, $\Omega^0_{\le c_k}$, $\Phi^0_{\le c_k}$, and $\Gamma^0_{\le c_k}$, we can conclude that

Lemma 11.13 ([128, Lemma 1]) $T^0_{\le c_k} = \widetilde{T}^0_{\le c_k} \cup \Omega^0_{\le c_k} \cup \Gamma^0_{\le c_k} = \Phi^0_{\le c_k} \cup \Gamma^0_{\le c_k}$.

Figure 11.3 shows the relationships among $T^0_{\le c_k}$, $\widetilde{T}^0_{\le c_k}$, $\Omega^0_{\le c_k}$, $\Phi^0_{\le c_k}$, $\Gamma^0_{\le c_k}$, $\overline{T}^0_{\le c_k}$, $T^0_{>c_k}$, and $\overline{T}^0_{>c_k}$, in which $T^0_{\le c_k}$ is shown in blue part.

We give **Example 11.1** to explain the meaning of the above notations.

Example 11.1 As shown in Fig. 11.4, let $V = \{v_1, v_2, \ldots, v_{11}\}$, $E = \{e_1, e_2, \ldots, e_{17}\}$, and

$$w = (3, 4, 3, 3, 4, 3, 1, 5, 5, 3, 4, 5, 5, 3, 4, 4, 5),$$
$$c^0 = (3, 4, 1, 6, 4, 7, 3, 3, 4, 5, 6, 4, 4, 1, 3, 5, 7).$$

Suppose $c_k = 4$, $T^0 = \{e_1, e_2, e_3, e_4, e_8, e_9, e_{12}, e_{13}, e_{15}, e_{16}\}$ is denoted by blue lines in Fig.11.4.

In **Example 11.1** shown in Fig.11.4,

(1) For each $e_j \notin T^0$, $P_5 = \{e_3, e_4\}$, $P_6 = \{e_2, e_4\}$, $P_7 = \{e_2, e_4, e_8\}$, $P_{10} = \{e_8, e_9\}$, $P_{11} = \{e_8, e_9, e_{13}\}$, $P_{14} = \{e_8, e_9, e_{12}, e_{13}\}$, and $P_{17} = \{e_8, e_9, e_{12}, e_{13}, e_{15}\}$.
(2) For each $e_i \in T^0$, $\Omega_1 = \Omega_{16} = \emptyset$, $\Omega_2 = \{e_6, e_7\}$, $\Omega_3 = \{e_5\}$, $\Omega_4 = \{e_5, e_6, e_7\}$, $\Omega_8 = \{e_7, e_{10}, e_{11}, e_{14}, e_{17}\}$, $\Omega_9 = \{e_{10}, e_{11}, e_{14}, e_{17}\}$, $\Omega_{12} = \{e_{14}, e_{17}\}$, $\Omega_{13} = \{e_{11}, e_{14}, e_{17}\}$, and $\Omega_{15} = \{e_{17}\}$.
(3) $T^0_{\leq c_k} = \{e_1, e_2, e_3, e_8, e_9, e_{12}, e_{13}, e_{15}\}$, $T^0_{>c_k} = \{e_4, e_{16}\}$, $\overline{T}^0_{\leq c_k} = \{e_5, e_7, e_{14}\}$, $\overline{T}^0_{>c_k} = \{e_6, e_{10}, e_{11}, e_{17}\}$. $\Omega^0 = \{e_1, e_{16}\}$, $\Omega^0_{\leq c_k} = \{e_1\}$, $\widetilde{T}^0_{\leq c_k} = \{e_3\}$, $\Phi^0_{\leq c_k} = \{e_1, e_3\}$, and $\Gamma^0_{\leq c_k} = \{e_2, e_8, e_9, e_{12}, e_{13}, e_{15}\}$.
(4) For each $e_i \in \Gamma^0_{\leq c_k}$, $w_{j_2} = w_6 = 3$, $w_{j_8} = w_{j_9} = w_{10} = 3$, $w_{j_{13}} = w_{11} = 4$, and $w_{j_{12}} = w_{j_{15}} = w_{17} = 5$.

Next, we analyze some properties of a feasible solution of (**RIOVMST**$_{bH}$).

Lemma 11.14 ([128, Lemma 3]) *If $\bar{w}$ is a feasible solution of (**RIOVMST**$_{bH}(c_k)$), then $\bar{w}_i \leq w_{j_i}$ for each edge $e_i \in \Gamma^0_{\leq c_k}$, where w_{j_i} is defined as in (11.47).*

Suppose $l > -\infty$ and $u < +\infty$. By using the notation of P_j for $e_j \notin T^0$, the problem (11.45) is equivalent to

$$
(\mathbf{RIOVMST}_{bH})\quad \begin{aligned} z(\bar{w}) = &\min \max_{e_i \in E} c_i^0 H(\bar{w}_i, w_i) \\ \text{s.t.}\quad & \bar{w}_i \leq \bar{w}_j, \quad e_j \notin T^0 \text{ and } e_i \in P_j, \\ & \bar{w}(T^0) = K, \\ & l \leq \bar{w} \leq u. \end{aligned} \tag{11.49}
$$

For a given c_k $(k = 1, 2, \cdots, \sigma)$, let (**RIOVMST**$_{bH}(c_k)$) be a subproblem of problem (**RIOVMST**$_{bH}$), whose objective value $z(\bar{w})$ is no more than c_k. In this section, we first analyze some properties of problem (**RIOVMST**$_{bH}(c_k)$). Then we present an algorithm to solve it. Finally, we analyze the time complexity of the algorithm.

In the next discussion, we analyze properties for the problem (**RIOVMST**$_{bH}(c_k)$) and either find a feasible solution of the problem (**RIOVMST**$_{bH}(c_k)$) or determine that the problem is infeasible.

Lemma 11.15 ([128, Lemma 16]) *If the problem (11.49) is feasible, then we have $l(T^0) \leq K \leq u(T^0)$.*

Lemma 11.16 ([128, Lemma 17]) *If $\bar{w}$ is a feasible solution of problem (11.49) with a given value $\bar{c}_k$, then $\tilde{w}$ is also a feasible solution of (11.49) with objective value $\tilde{c}_k = c_k$, where*

$$
\tilde{w}_i = \begin{cases} \bar{w}_i, & \text{if } e_i \in T^0 \cup \overline{T}^0_{>\bar{c}_k}, \\ \bar{w}_i + \alpha, 0 \leq \alpha \leq u_i - w_i, & \text{if } e_i \in \overline{T}^0_{\leq \bar{c}_k}. \end{cases}
$$

Firstly, we introduce some notations.

Given an edge $e_j \in \overline{T}^0_{\leq c_k}$, if there is an edge $e_i \in P_j \cap T^0_{\geq c_k}$ with $w_i > u_j$, then we add the pair (e_i, e_j) of edges to **the third-class infeasible set** Ψ_k^3. Define

$$\Psi_k^3 = \left\{(e_i, e_j) | e_j \in \overline{T}^0_{\leq c_k}, e_i \in P_j \cap T^0_{>c_k}, w_i > u_j\right\}. \tag{11.50}$$

If $\Psi_k^3 = \emptyset$, then $w_i \leq u_j$ for all $e_j \in \overline{T}^0_{\leq c_k}$ and $e_i \in P_j \cap T^0_{>c_k}$.

Given an edge $e_i \in \widetilde{T}^0_{\leq c_k} \cup \Gamma^0_{\leq c_k}$, let

$$u_{j_i} = \begin{cases} \min_{e_j \in \Omega_i \cap \overline{T}^0_{\leq c_k}} u_j, & \text{if } \Omega_i \cap \overline{T}^0_{\leq c_k} \neq \emptyset, \\ w_{j_i} = \min_{e_j \in \Omega_i \cap \overline{T}^0_{>c_k}} w_j, & \text{if } \Omega_i \cap \overline{T}^0_{\leq c_k} = \emptyset. \end{cases} \tag{11.51}$$

which is obtained on e_{j_i}. If there is an edge $e_{j_i} \in \Omega_i$ with $l_i > u_{j_i}$, then we add the pair (e_i, e_{j_i}) of edges to **the fourth-class infeasible set** Ψ_k^4, where

$$\Psi_k^4 = \left\{(e_i, e_{j_i}) | e_i \in \widetilde{T}^0_{\leq c_k} \cup \Gamma^0_{\leq c_k}, e_{j_i} \in \Omega_i, l_i > u_{j_i}\right\}. \tag{11.52}$$

If $\Psi_k^4 = \emptyset$, then $l_i \leq u_{j_i}$ for all $e_i \in \widetilde{T}^0_{\leq c_k} \cup \Gamma^0_{\leq c_k}$ and $e_{j_i} \in \Omega_i$.

We can conclude that

Lemma 11.17 ([128, Lemma 18]) *If at least one of the four infeasible sets is nonempty, that is, $\Psi_k^1 \neq \emptyset$ or $\Psi_k^2 \neq \emptyset$ or $\Psi_k^3 \neq \emptyset$ or $\Psi_k^4 \neq \emptyset$, then problem* (**RIOVMST**$_{bH}(c_k)$) *has no feasible solution.*

Lemma 11.18 ([128, Lemma 19]) *Suppose the four infeasible sets $\Psi_k^1 = \emptyset$, $\Psi_k^2 = \emptyset$, $\Psi_k^3 = \emptyset$, $\Psi_k^4 = \emptyset$, and $T^0_{\leq c_k} = \emptyset$ $(T^0 = T^0_{>c_k})$. If $w(T^0_{>c_k}) \neq K$, then* (**RIOVMST**$_{bH}(c_k)$) *has no feasible solution. If $w(T^0_{>c_k}) = K$, then*

$$\bar{w}_i = \begin{cases} w_i, \text{ if } e_i \in T^0_{>c_k} \cup \overline{T}^0_{>c_k}, \\ u_i, \text{ if } e_i \in \overline{T}^0_{\leq c_k} \end{cases} \tag{11.53}$$

is a feasible solution of (**RIOVMST**$_{bH}(c_k)$).

Lemma 11.19 ([128, Lemma 20]) *If $\bar{w}$ is a feasible solution of* (**RIOVMST**$_{bH}(c_k)$), *then*

$$l_i \leq \bar{w}_i \leq \tilde{u}_{j_i} \leq u_i, \quad \forall e_i \in \widetilde{T}^0_{\leq c_k}, \tag{11.54}$$

$$\text{and } l_i \leq \bar{w}_i \leq \widetilde{w}_{j_i} \leq u_i, \quad \forall e_i \in \Gamma^0_{\leq c_k}, \tag{11.55}$$

where

$$\widetilde{u}_{j_i} = \min\{u_i, u_{j_i}\} \tag{11.56}$$

$$and \widetilde{w}_{j_i} = \min\{u_i, u_{j_i}, w_{j_i}\}. \tag{11.57}$$

Define

$$\tilde{u}_j(\widetilde{T}^0_{\le c_k}) = \sum_{e_i \in \widetilde{T}^0_{\le c_k}} \tilde{u}_{j_i} \tag{11.58}$$

$$and \widetilde{w}_j(\Gamma^0_{\le c_k}) = \sum_{e_i \in \Gamma^0_{\le c_k}} \widetilde{w}_{j_i}. \tag{11.59}$$

Next, we propose Lemma 11.20 to solve the problem (**RIOVMST**$_{bH}(c_k)$) when $\Psi^1_k = \emptyset$, $\Psi^2_k = \emptyset$, $\Psi^3_k = \emptyset$, $\Psi^4_k = \emptyset$, and $T^0_{\le c_k} \ne \emptyset$. For edge $e_i \in T^0_{\le c_k}$, define

$$w^k_i = \begin{cases} \tilde{u}_{j_i}, & \text{if } e_i \in \widetilde{T}^0_{\le c_k}, \\ u_i, & \text{if } e_i \in \Omega^0_{\le c_k}, \\ \widetilde{w}_{j_i}, & \text{if } e_i \in \Gamma^0_{\le c_k}. \end{cases} \tag{11.60}$$

Suppose $T^0_{\le c_k} = \{e_{\bar{1}}, e_{\bar{2}}, \ldots, e_{\bar{p}}\}$ satisfying $w^k_{\bar{1}} \ge w^k_{\bar{2}} \ge \ldots \ge w^k_{\bar{p}}$.

For $e_i \in T^0_{\le c_k}$, define

$$\bar{s}^* = \min\left\{\bar{s} \,\middle|\, \sum_{\bar{t}=\bar{1}}^{\bar{s}} w^k_{\bar{t}} + \sum_{\bar{t}=\bar{s}+1}^{\bar{p}} l_{\bar{t}} \ge K - w(T^0_{>c_k})\right\}, \tag{11.61}$$

and define $T^1_{\le c_k} = \{e_{\bar{1}}, e_{\bar{2}}, \ldots, e_{\bar{s}^*-1}\}$, $T^2_{\le c_k} = \{e_{\bar{s}^*+1}, e_{\bar{s}^*+2}, \ldots, e_{\bar{p}}\}$.

By the definitions of $\bar{s}^*$, $T^1_{\le c_k}$, and $T^2_{\le c_k}$, we can obtain the following result:

$$T^0_{\le c_k} = T^1_{\le c_k} \cup T^2_{\le c_k} \cup e_{\bar{s}^*}, \tag{11.62}$$

$$w^k(T^1_{\le c_k}) + l_{\bar{s}^*} + l(T^2_{\le c_k}) < K - w(T^0_{>c_k}) \le w^k(T^1_{\le c_k}) + w^k_{\bar{s}^*} + l(T^2_{\le c_k}). \tag{11.63}$$

Lemma 11.20 ([128, Lemma 21]) *Suppose the four infeasible sets* $\Psi^1_k = \emptyset$, $\Psi^2_k = \emptyset$, $\Psi^3_k = \emptyset$, $\Psi^4_k = \emptyset$, *and* $T^0_{\le c_k} \ne \emptyset$.

(1) If $l(T^0_{\le c_k}) + w(T^0_{>c_k}) > K$ *or* $\tilde{u}_j(\widetilde{T}^0_{\le c_k}) + u(\Omega^0_{\le c_k}) + \widetilde{w}_j(\Gamma^0_{\le c_k}) + w(T^0_{>c_k}) < K$, *then (**RIOVMST**$_{bH}(c_k)$) has no feasible solution, where* $\tilde{u}_j(\widetilde{T}^0_{\le c_k})$ *and* $\widetilde{w}_j(\Gamma^0_{\le c_k})$ *are defined as in (11.58) and (11.59), respectively.*

(2) *If* $l(T^0_{\le c_k}) + w(T^0_{>c_k}) = K$, *then*

$$\bar{w}_i = \begin{cases} w_i, & \text{if } e_i \in T^0_{>c_k} \cup \overline{T}^0_{>c_k}, \\ l_i, & \text{if } e_i \in T^0_{\le c_k}, \\ u_i, & \text{if } e_i \in \overline{T}^0_{\le c_k} \end{cases} \tag{11.64}$$

is a feasible solution of (**RIOVMST**$_{bH}(c_k)$).

(3) *If* $\tilde{u}_j(\widetilde{T}^0_{\le c_k}) + u(\Omega^0_{\le c_k}) + \widetilde{w}_j(\Gamma^0_{\le c_k}) + w(T^0_{>c_k}) = K$, *then*

$$\bar{w}_i = \begin{cases} w_i, & \text{if } e_i \in T^0_{>c_k} \cup \overline{T}^0_{>c_k}, \\ w_i^k, & \text{if } e_i \in T^0_{\le c_k}, \\ u_i, & \text{if } e_i \in \overline{T}^0_{\le c_k} \end{cases} \tag{11.65}$$

is a feasible solution of (**RIOVMST**$_{bH}(c_k)$).

(4) *If* $l(T^0_{\le c_k})+w(T^0_{>c_k}) < K$ *and* $\tilde{u}_j(\widetilde{T}^0_{\le c_k})+u(\Omega^0_{\le c_k})+\widetilde{w}_j(\Gamma^0_{\le c_k})+w(T^0_{>c_k}) > K$, *then*

$$\bar{w}_i = \begin{cases} w_i, & \text{if } e_i \in T^0_{>c_k} \cup \overline{T}^0_{>c_k}, \\ w_i^k, & \text{if } e_i \in T^1_{\le c_k}, \\ K - w(T^0_{>c_k}) - w^k(T^1_{\le c_k}) - l(T^2_{\le c_k}), & \text{if } e_i = e_{\bar{s}^*}, \\ l_i, & \text{if } e_i \in T^2_{\le c_k}, \\ u_i, & \text{if } e_i \in \overline{T}^0_{\le c_k} \end{cases} \tag{11.66}$$

is a feasible solution of (**RIOVMST**$_{bH}(c_k)$).

Next, we propose an algorithm to solve (**RIOVMST**$_{bH}$). The main idea of the algorithm is to use the binary search for the cost vector c to find the optimal cost c_{k^*}. In each iteration, we either find a feasible solution of the subproblem (**RIOVMST**$_{bH}(c_k)$) or determine its infeasibility.

Algorithm 11.8 A binary search algorithm for (**RIOVMST**$_{bH}$) [128, Algorithm 3]

Input: A network $G = (V, E, w, c^0, l, u)$, a given spanning tree T^0 of G, a given number K.

Output: The optimal value c_{k^*}, an optimal solution w^*, or "no feasible solution."

1: Determine the sets P_j for each $e_j \notin T^0$, Ω_i for each $e_i \in T^0$, and Ω^0;

2: Sort different c_i^0 in ascending order. Suppose $c_1 < c_2 < \cdots < c_\sigma$. Let $c := (c_1, c_2, \cdots, c_\sigma)$.

3: Let $a := 1$, $b := \sigma$, and sign:=0.

4: **while** $b - a > 1$ **do**

5: $\quad k := \lfloor \frac{a+b}{2} \rfloor$.

6: $\quad$ Determine $T^0_{\le c_k}$, $T^0_{>c_k}$, and $\overline{T}^0_{\le c_k}$,$\overline{T}^0_{>c_k}$, $\Omega^0_{\le c_k}$, and $\widetilde{T}^0_{\le c_k} := \{e_i \in T^0_{\le c_k} \setminus \Omega^0 | \Omega_i \subset \overline{T}^0_{\le c_k}\}$, $\Phi^0_{\le c_k} := \widetilde{T}^0_{\le c_k} \cup \Omega^0_{\le c_k}$, $\Gamma^0_{\le c_k} := T^0_{\le c_k} \setminus \Phi^0_{\le c_k}$. Calculate w_{j_i} for each $e_i \in \Gamma^0_{\le c_k}$ and u_{j_i} by (11.51) for each $e_i \in \Gamma^0_{\le c_k} \cup \widetilde{T}^0_{\le c_k}$. Determine Ψ^1_k, Ψ^2_k, Ψ^3_k by (11.50) and Ψ^4_k by (11.52), where $\Psi^1_k = \{(e_i, e_j) | e_j \in \overline{T}^0_{>c_k}, e_i \in P_j \cap T^0_{>c_k}, w_i > w_j\}$, and $\Psi^2_k = \{(e_i, e_{j_i}) | e_i \in \Gamma^0_{\le c_k}, l_i > w_{j_i}\}$.

  if $\Psi^1_k \ne \emptyset$ or $\Psi^2_k \ne \emptyset$ or $\Psi^3_k \ne \emptyset$ or $\Psi^4_k \ne \emptyset$ **then** let $a := k$.
    if $b - a = 1$ **then** let $k := b$.
    end if
  else
    if $T^0_{\le c_k} = \emptyset$ **then**
      if $w(T^0_{>c_k}) \ne K$ **then** let $a := k$
        if $b - a = 1$ **then** let $k := b$.
        end if
      else let $b := k$, sign:=1 and output a feasible solution $\bar{w}$ by (11.53).
      end if
    else
      Let $K' := K - w(T^0_{>c_k})$. Calculate $\tilde{u}_{j_i}$ by (11.56) for each $e_i \in \widetilde{T}^0_{\le c_k}$ and $\widetilde{w}_{j_i}$ by (11.57) for each $e_i \in \Gamma^0_{\le c_k}$. Determine w^k by (11.60).
      if $l(T^0_{\le c_k}) > K'$ or $w^k(T^0_{\le c_k}) < K'$ **then** let $a := k$.
        if $b - a = 1$ **then** let $k := b$.
        end if
      else
        if $l(T^0_{\le c_k}) = K'$ **then**
          Let $b := k$, sign:=1 and output a solution $\bar{w}$ by (11.64).
        else if $w^k(T^0_{\le c_k}) = K'$ **then**
          Let $b := k$, sign:=1 and output a solution $\bar{w}$ by (11.65).
        else
          Determine $T^1_{\le c_k} := \{e_{\bar{1}}, e_{\bar{2}}, \ldots, e_{\bar{s}^*-1}\}$, $T^2_{\le c_k} := \{e_{\bar{s}^*+1}, e_{\bar{s}^*+2}, \ldots, e_{\bar{p}}\}$, where $\bar{s}^*$ is defined as in (11.61).
          Let $b := k$, sign:=1 and output a solution $\bar{w}$ by (11.66).
        end if
      end if
    end if
  end if
end while
if sign=0 **then** output "no feasible solution."
else output the optimal value $c_{k^*} := c_b$ and an optimal solution $w^* := \bar{w}$.
end if

Finally, we can similarly analyze the time complexity of Algorithm 11.8.

Theorem 11.14 ([128, Theorem 22]) *The problem (**RIOVMST**$_{bH}$) can be solved in $O(mn \log m)$ time by Algorithm 11.8.*

11.5 Conclusions

In this chapter, we study the restricted inverse optimal value problem on minimum spanning tree. Under weighted l_∞ norm with both lower and upper bounds, we provide some optimality conditions of optimal solutions of the problem and then propose two strongly polynomial time algorithms with running time $O(m^2 n)$ and $O(m^2 \log n)$. As a byproduct, we obtain an $O(mn)$ algorithm to solve the (**RIOVMST**$_\infty$) problem under unit l_∞ norm with both lower and upper bounds.

Under weighted l_1 norm with both lower and upper bounds, we study the restricted inverse optimal value problem on minimum spanning tree. We first construct its mathematical model and convert it into a linear programming problem (P). Then we construct a subproblem (D^z) of the dual problem (D) of (P) by omitting the item $(K - w(T^0))z$ from the objective function of (D). The main issue is to calculate **the critical value** z^* in the optimal solution of (D) by a binary search method, in which a subproblem (D^z) is solved in each iteration. Then solving the problem (D) is equivalent to solving the subproblem (D^{z^*}). It is interesting that the subproblem (D^z) can be transformed into a minimum cost flow problem on an auxiliary network $\tilde{G}^z$. As a conclusion, we prove that the problem **RIOVMST**$_1$ can be solved by $\log(nC)$ subproblems (D^z), which render the total time complexity $O(m^2 n^2 \log n \log(nC))$. We also provide **an open problem**, that is, if there is a strongly polynomial time algorithm for the problem **RIOVMST**$_1$.

Under bottleneck Hamming distance, we consider the restricted inverse optimal value problem on minimum spanning tree. We develop three binary search algorithms. The time complexities for solving the unbounded, the lower bounded, and the bounded problems are all $O(|V||E| \log |E|)$ time.

As a future research topic, we will study the inverse optimal value problem on minimum spanning tree under weighted sum Hamming distance, which is more challenging.

Chapter 12
Partial Inverse Minimum Spanning Tree Problems

Abstract This chapter introduces the partial inverse minimum spanning tree problem (*PInvMST*). Given an edge-weighted graph and an acyclic edge set, the target of (*PInvMST*) is to get a new weight vector such that the given edge set is included in some minimum spanning tree with respect to the new weight vector, and the difference between the two vectors is minimum. Firstly, we show the computational complexities of (*PInvMST*) in some special cases and general case under different norms and provide exact algorithms for polynomial time solvable cases. Then, for some cases that have been proven to be $\mathcal{NP}$-hard, we introduce their approximation algorithms and analyze their approximation ratios. Finally, we summarize the conclusion and look forward to future works.

Keywords Partial inverse minimum spanning tree problems · Polynomial time algorithms · $\mathcal{NP}$-hardness · Fitted cut · Approximation algorithms

12.1 An Introduction to Partial Inverse Minimum Spanning Tree Problems

In this section, we give a mathematical model of *partial inverse minimum spanning tree problem (PInvMST)* and some useful results.

12.1.1 A Mathematical Model

Given an edge-weighted graph $G = (V, E)$ with weight vector $w : E \to \mathcal{R}$, a cost function $c : E \to \mathcal{R}^+$, an acyclic edge subset F, and decreasing and increasing

---

Xianyue Li has contributed to this chapter.

X. Guan et al., *Inverse Combinatorial Optimization Problems*, Springer Optimization and Its Applications 225, https://doi.org/10.1007/978-3-031-91175-0_12

bound functions l and u, the aim of (*PInvMST*) is to find a new weight vector $\bar{w}$ satisfying the following:

1. F can be extended to a minimum spanning tree T^0 with respect to $\bar{w}$.
2. $-l(e) \leq \bar{w}(e) - w(e) \leq u(e)$ for each edge $e \in E$.
3. $\| \bar{w} - w \|$ is minimum under some norm.

A weight vector $\bar{w}$ is *feasible* if it satisfies the first two conditions. Similarly, a minimum spanning tree T^0 with respect to a feasible solution containing F is called a *feasible tree*. As above, its mathematical model is given below, where $\mathcal{T}$ is the set of all spanning trees of G:

$$\begin{aligned}
& \min \|\bar{w} - w\| \\
(PInvMST) \quad & \text{s.t.} \sum_{e \in T^0} \bar{w}(e) \leq \sum_{e \in T} \bar{w}(e), \quad \text{for some } F \subseteq T^0 \in \mathcal{T} \text{ and } \forall T \in \mathcal{T}, \\
& w - l \leq \bar{w} \leq w + u.
\end{aligned}$$

We denote (*PInvMST*) with $u \equiv 0$ (the weights can only be decreased) by (*PInvMST*$^-$) and $l \equiv 0$ (the weights can only be increased) by (*PInvMST*$^+$), respectively.

12.1.2 Preliminary Results

Let T be a spanning tree of a graph $G = (V, E)$. For any edge $e \in E(G) \setminus E(T)$, $T + e$ has an unique cycle, which is called the *fundamental cycle* with respect to T and e and denoted by $C(T, e)$. It is easy to see that for any edge $\bar{e} \in C(T, e)$, $T + e - \bar{e}$ is also a spanning tree of G. On the other hand, for any edge $\bar{e} \in E(T)$, $T - \bar{e}$ has exactly two components. The set of edges of G connecting these two components is called the *fundamental cut* with respect to T and $\bar{e}$ and denoted by $K(T, \bar{e})$. Similarly, for any edge $e \in K(T, \bar{e})$, $T - \bar{e} + e$ is also a spanning tree of G.

The following two necessary and sufficient conditions are given for a spanning tree T of G to be minimum (see Exercises 6.2.2 and 6.2.3 in [16]):

$$\text{For any edge } e \in E(T), \ w(e) = \min\{w(y) : y \in K(T, e)\}. \tag{12.1}$$

$$\text{For any edge } e \in E(G) \setminus E(T), \ w(e) = \max\{w(y) : y \in C(T, e)\}. \tag{12.2}$$

The following two theorems generalize the above results to (*PInvMST*):

Theorem 12.1 ([87, Theorem 1]) *Let $I = (G, w, c, F, l, u)$ be an instance of (PInvMST). A new weight vector $\bar{w}$ is feasible for I if and only if*

1. *There exists a spanning tree T of G containing F such that $\bar{w}(e') = \min\{\bar{w}(y) : y \in K(T, e')\}$ for any edge $e' \in F$.*
2. *$-l(e) \leq \bar{w}(e) - w(e) \leq u(e)$, for any edge $e \in E$.*

Theorem 12.2 ([87, Theorem 2]) *Let $I = (G, w, c, F, l, u)$ be an instance of (PInvMST). A new weight vector $\bar{w}$ is feasible for I if and only if*

1. *There exists a spanning tree T of G containing F such that $C(T, e) \cap F = \emptyset$ or $\bar{w}(e) \geq \max\{\bar{w}(y) : y \in C(T, e) \cap F\}$ for any edge $e \notin E(T)$.*
2. *$-l(e) \leq \bar{w}(e) - w(e) \leq u(e)$, for any edge $e \in E$.*

It can be seen that the above two theorems simplify the determination of feasible solutions of (*PInvMST*). Furthermore, an optimal solution of (*PInvMST*) has a *separable property*.

Lemma 12.1 ([84, Lemma 1]) *Let I be an instance of (PInvMST). If I has feasible solutions, then there exists an optimal solution w^* of I, such that*

$$w^*(e) \begin{cases} \leq w(e), & e \in F; \\ \geq w(e), & e \notin F. \end{cases}$$

12.2 Computational Complexities of (*PInvMST*)

In this section, we shall discuss the computational complexity or exact algorithm of *PInvMST* under different norms. According to the chronological order of the research, we will introduce in the order of (*PInvMST*$^-$), (*PInvMST*$^+$), and (*PInvMST*).

12.2.1 A Polynomial Time Algorithm for (PInvMST$^-$)

Recall that for the problem (*PInvMST*$^-$), the weight vector can only be decreased ($u \equiv 0$), that is, $\bar{w}(e) \leq w(e)$ for each edge $e \in E$. Thus, (*PInvMST*$^-$) can be formulated as follows:

$$\begin{aligned} &\min \ \|\bar{w} - w\| \\ (PInvMST^-) \quad &\text{s.t.} \sum_{e \in T^0} \bar{w}(e) \leq \sum_{e \in T} \bar{w}(e), \quad \text{for some } F \subseteq T^0 \in \mathcal{T} \text{ and } \forall T \in \mathcal{T}, \end{aligned}$$

$$w(e) - l(e) \leq \bar{w}(e) \leq w(e), \forall e \in E.$$

For any instance $I = (G, w, c, F, l, u \equiv 0)$ of (*PInvMST*$^-$), if I has feasible solutions, by *separable property*, there is an optimal solution w^* of I, such that the difference between w^* and w lies entirely in the partial solution F, and $w^*(e) \leq w(e)$, for each edge $e \in F$. Now, we define a new weight vector $\bar{w}$ as follows:

$$\bar{w}(e) = \begin{cases} \min\limits_{P \in \mathcal{P}(e)} \max\{w(e') \mid e' \in P \setminus (F \setminus e)\}, & \forall e \in F, \\ w(e), & \forall e \notin F, \end{cases} \tag{12.3}$$

where $\mathcal{P}(e)$ is the set of all paths joining the end vertices of edge e. Note that e is also a path in $\mathcal{P}(e)$, so $\bar{w}(e) \leq w(e)$ for each edge $e \in F$. In the following, we shall show that $\bar{w}$ is an optimal solution.

Theorem 12.3 ([23, Theorem 2]) *If $\bar{w}(e) \geq w(e) - l(e)$ for each edge $e \in F$, then $\bar{w}$ defined as in* (12.3) *is a feasible solution of (PInvMST$^-$).*

Theorem 12.4 ([23, Theorem 1]) *Let w' be a feasible solution. Then, $w'(e) \leq \bar{w}(e)$ for each edge $e \in E$.*

Based on Theorem 12.3 and Theorem 12.4, we can obtain the following solution:

Corollary 12.1 ([23, Corollary 2]) *Let I be an instance of (PInvMST$^-$) under the monotone norm. Then, I has a feasible solution if and only if $\bar{w}(e) \geq w(e) - l(e)$ for each edge $e \in F$; in that case, the weight vector $\bar{w}$ is optimal.*

Remark 12.1 This conclusion does not depend on the cost function c.

Although the scale of $\mathcal{P}(e)$ may be exponential for a given edge e, Corollary 12.1 yields the following polynomial time algorithm based on the characteristics of the minimum spanning tree.

Algorithm 12.1 An algorithm to solve $(PInvMST^-)$ under the monotone norm

---

Input: An instance $I = (G, w, c, F, l, u \equiv 0)$ of (*PInvMST*$^-$) under the monotone norm.

Output: An optimal solution w^* or claim that I has no feasible solutions.

1: Find a minimum spanning tree T' of G with respect to the weight vector w' by

$$w'(e) = \begin{cases} -1 + \min\limits_{e \in E} w(e), & \forall e \in F, \\ w(e), & \forall e \notin F. \end{cases} \tag{12.4}$$

2: Contract every component of $T' - F$ into a single vertex and obtain resulting graph G'.

3: Find a minimum spanning tree T'' of G' with respect to w.

4: Sort the edges of $T'' \setminus F$ to nondecreasing order, say $T'' \setminus F = \{e_1, e_2, \ldots, e_t\}$ with $w(e_1) \leq w(e_2) \leq \cdots \leq w(e_t)$.

```
Set R := F, w* := w, and i := 1.
while R ≠ ∅ and i ≤ t do
    for each edge e ∈ C'(F, e_i) ∩ R do
        Set w*(e) := min{w(e), w(e_i)}, R := R \ e, and i := i + 1.
        if w(e) − w*(e) > l(e) then
            return "No Feasible Solutions!".
        end if
    end for
end while
return w*.
```

Theorem 12.5 ([23, Theorem 3]) *The problem (PInvMST$^-$) can be solved in time* $O(|E| + |V|\log|V| + |F|^2\})$.

Notice that spanning tree is a special type of matroid. In 2016, Zhang et al. [155] extended the above results to the matroid and presented two polynomial time exact algorithms for partial inverse problem on minimum base of matroid, which weight vector can only be decreased. In fact, in that paper, they studied partial inverse problem on maximum base of matroid, which weight vector can only be increased. These two problems are equivalent.

12.2.2 A Polynomial Time Algorithm for (PInvMST$^+_\infty$)

Recall that for (*PInvMST*$^+$), the weight vector can only be increased ($l \equiv 0$), that is, $\bar{w}(e) \geq w(e)$ for each edge $e \in E$. Thus, (*PInvMST*$^+$) can be formulated as follows:

$$\begin{aligned} & \min \|\bar{w} - w\| \\ (\mathit{PInvMST}^+) \quad & \text{s.t.} \sum_{e\in T^0} \bar{w}(e) \leq \sum_{e\in T} \bar{w}(e), \ \text{for some } F \subseteq T^0 \in \mathcal{T} \text{and } \forall T \in \mathcal{T}, \\ & w(e) \leq \bar{w}(e) \leq w(e) + u(e), \forall e \in E. \end{aligned}$$

By *separable property*, for any instance I of (*PInvMST*$^+$), there is an optimal solution w^* of I, such that the difference between w^* and w lies only in the edges outside the partial solution F and $w^*(e) \geq w(e)$ for each edge $e \notin F$.

In 2016, Li et al. [81] firstly considered (*PInvMST*$^+$) under the weighted l_∞ norm, denoted by (*PInvMST*$^+_\infty$). In fact, in that paper, they studied partial inverse problem on maximum base of matroid under the weighted l_∞ norm, which weight vector can only be decreased. This problem includes (*PInvMST*$^+_\infty$).

Similar to the method for ($PInvMST^-$), we first define the following weight vector

$$\bar{w}(e) = \begin{cases} w(e), & \forall e \in T'; \\ \max\{w(y) : y \in C(T', e)\}, & \forall e \in E \setminus T', \end{cases} \tag{12.5}$$

where T' is a minimum spanning tree G with respect to w' as (12.4). Based on the new weight vector, we can obtain the following algorithm and theorem:

Algorithm 12.2 An algorithm to solve ($PInvMST^+_\infty$) [81, Algorithm 1]

Input: An instance $I = (G, w, c, F, l \equiv 0, u)$ of ($PInvMST^+_\infty$).
Output: An optimal solution w^* or claim that I has no feasible solutions.

```
Find a minimum spanning tree T' of G with respect to the weight vector w' as
  (12.4).
Set w* := w.
for each edge e ∈ E \ T' do
    Set w*(e) := max{w(y) : y ∈ C(T', e)}.
    if w*(e) − w(e) > u(e) then
        return "No Feasible Solutions!".
    end if
end for
return w*.
```

Theorem 12.6 ([81, Theorem 3.2]) *The problem ($PInvMST^+_\infty$) can be solved in time* $O(\max\{|E| + |V|\log|V|, (|E| - |V|)|V|\})$.

Remark 12.2 If the contraction technique (used in ($PInvMST^-$)) is utilized, the running time of the above algorithm can be reduced to $O(\max\{|E| + |V|\log|V|, (|E| - |V|)|F|\})$.

A natural idea is whether the method for solving ($PInvMST^+_\infty$) can be applied to ($PInvMST^+$) under other norms. Unfortunately, it is easy to see from Fig. 12.1 that this method is not applicable. In that figure, (a) is an instance I of ($PInvMST^+$) with $c \equiv 1$ and $u \equiv \infty$. The partial solution is the two dashed edges, and the number on each edge is its weight. (b) is the output of Algorithm 12.2. The weights on edges constitute a feasible solution w^* of I, and the dark edges form a feasible tree with respect to w^*. (c) is another feasible solution w' of I, and the dark edges form the feasible tree with respect to w'. By Theorem 12.6, w^* is an optimal solution of ($PInvMST^+_\infty$). But when I is an instance of ($PInvMST^+$) under the weighted l_1 norm, $\|w' - w\|_1 = 8 < 25 = \|w^* - w\|$.

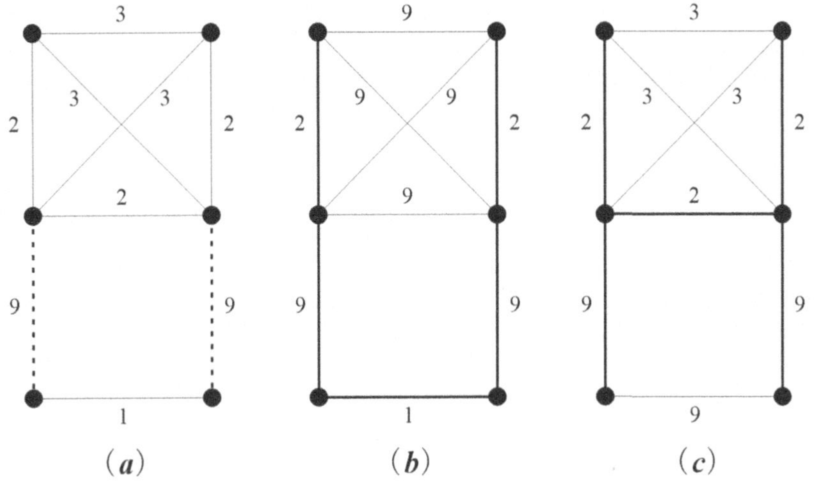

Fig. 12.1 An instance of (*PInvMST*$^+$) and its different solutions

12.2.3 $\mathcal{NP}$-Hardness of (PInvMST$_p^+$)

In 2018, Li et al. [86] studied the computational complexity of (*PInvMST*$^+$) under the weighted l_p norm, denoted by (*PInvMST*$_p^+$). In fact, in that paper, they studied partial inverse problem on maximum spanning tree under the weighted l_p norm, which weight vector can only be decreased. This problem is equivalent to (*PInvMST*$_p^+$). For the convenience of readers, in this part, we only discuss the case of l_1 norm, and the case of l_p norm can be naturally extended from the case of l_1 norm.

Firstly, we show that (*PInvMST*$_1^+$) is $\mathcal{NP}$-hard by reduction from *multiterminal k-cut problem (MTkC)*. Given an edge-weighted graph $G = (V, E)$ with weight vector w and a terminal set $S = \{s_1, s_2, \ldots, s_k\}$, an edge subset E' is a *multiterminal k-cut* if for any two terminals, s_i and s_j, they are disconnected in $G - E'$. The decision version of (*MTkC*) is to decide whether there is a multiterminal k-cut E' with $w(E') \leq K$ for a given real number K. Clearly, the edges that connect any two terminals in S must be in any multiterminal cut, so we can make an assumption that S is an independent set in G. (*MTkC*) is $\mathcal{NP}$*-hard* for any fixed integer $k \geq 3$ [34], even if w is unit.

Now, for any instance, $I' = (G', w' \equiv 1, S)$ of decision version of (*MTkC*) with real number K, we construct an instance $I = (G, w, c, F, l \equiv 0, u)$ of the decision version of (*PInvMST*$_1^+$) with the same number K, where $F = \{s_1s_2, s_2s_3, \ldots, s_{k-1}s_k\}$, G is obtained from G' by adding F, $c \equiv 1$, $u \equiv \infty$, and $w(e) = 1$ for any edge $e \in F$ and $w(e) = 0$ for any edge $e \in E(G')$.

Let E_{kc} be a minimal multiterminal k-cut of instance I' with $|E_{kc}| \leq K$. Then $G' - E_{kc}$ has exactly k components $C_1, C_2, \ldots, C_k$ with $s_i \in C_i$. Now, we define a new weight vector $\bar{w}$ of I as follows: For any edge $e \in E_{kc}$, $\bar{w}(e) = 1$, and for any other edge e, $\bar{w}(e) = w(e)$. Obviously, $\bar{w} \geq w$ and $\|\bar{w} - w\|_1 = |E_{kc}| \leq K$. Let T_i be any spanning tree of $G[C_i]$ and $\bar{T} = (\cup_{1 \leq i \leq k} T_i) \cup F$. Then, $\bar{T}$ is a spanning tree of G containing F. For any edge $e \in F$, the fundamental cut $K(\bar{T}, e) \subseteq (E_{kc} \cup \{e\})$; it implies that the weights of all edges in $K(\bar{T}, e)$ are equal to 1 and $\bar{w}(e) = \min\{\bar{w}(y) : y \in K(\bar{T}, e)\}$. Hence, by Theorem 12.1, $\bar{w}$ is a feasible solution of I with $\|w' - w\|_1 = |E_{kc}| \leq K$.

Conversely, let $\bar{w}$ be a feasible solution of (*PInvMST*$_1^+$) with $\|w' - w\|_1 \leq K$ and T be a feasible tree with respect to $\bar{w}$. Then, $T - F$ has exactly k components, and each of them contains exactly one terminal in S. For any edge $\bar{e} \in F$ and any edge $e \in K(T, \bar{e}) \setminus \{\bar{e}\}$,

$$\bar{w}(e) \geq \bar{w}(\bar{e}) \geq w(\bar{e}) \geq 1,$$

since T is minimum. Let $E_{kc} = (\cup_{e \in F} K(T, e)) \setminus F$. Clearly, E_{kc} is a feasible solution of I', and

$$w'(E_{kc}) = |E_{kc}| \leq \|w' - w\|_1 \leq K.$$

Based on above results, we can obtain the following result:

Theorem 12.7 *For any fixed integer $k \geq 2$, (PInvMST$_1^+$) is $\mathcal{NP}$-hard for $|F| = k$, even if restricted to cases in which (i) F induces a path, (ii) $u \equiv \infty$, (iii) c is unit, and (iv) the weight vector w is binary.*

Remark 12.3 Actually, by *l-reduction* from (*MTkC*), (*PInvMST*$_1^+$) can be proved as APX-hard for any fixed integer $k \geq 2$ with $|F| = k$ [86, Theorem 1].

In the following, we consider the rest case $|F| = 1$ of (*PInvMST*$_1^+$), that is, $F = \{\bar{e} = st\}$.

Lemma 12.2 *Let $I = (G, w, c, \bar{e}, 0, u)$ be an instance of (PInvMST$_1^+$). For any feasible solution $\bar{w}$ of I and for any feasible tree $\bar{T}$ with respect to $\bar{w}$,*

$$\|\bar{w} - w\|_1 \geq \sum_{e \in K^-(\bar{T}, \bar{e})} c(e)(w(\bar{e}) - w(e)), \tag{12.6}$$

where $K^-(\bar{T}, \bar{e})$ is a subset of fundamental cut $K(\bar{T}, \bar{e})$ that includes all edges in $K(\bar{T}, \bar{e})$ with weights less than $w(\bar{e})$.

According to (12.6), it can be seen that in order to obtain a better feasible solution, the following two conditions need to be met:

1. Minimize the value at the right of the (12.6).
2. The value on the left of the (12.6) should be as close as possible to the right.

Since $\sum_{e\in K^-(\bar{T},\bar{e})} c(e)(w(\bar{e}) - w(e)) = \sum_{e\in K(\bar{T},\bar{e})} c(e)\cdot \max\{w(\bar{e}) - w(e), 0\}$ and every fundamental cut $K(\bar{T}, \bar{e})$ is an s-t cut, the approach to achieving condition 1 is to find an s-t cut K of G such that

$$\sum_{e\in K(\bar{T},\bar{e})} c(e)\cdot \max\{w(\bar{e}) - w(e), 0\}$$

is minimum. Based on this, we define a new weight vector w' with

$$w'(e) = \begin{cases} 0, & w(e) \geq w(\bar{e}); \\ c(e)(w(\bar{e}) - w(e)), & w(e) < w(\bar{e}) \leq w(e) + u(e); \\ \infty, & w(\bar{e}) > w(e) + u(e). \end{cases} \tag{12.7}$$

Therefore, condition 1 can be achieved by finding the minimum s-t cut in G with respect to w'. And the purpose of the last item in (12.7) is to ensure that after the weights are changed, $\bar{e}$ can become the edge with the smallest weight in this cut.

To achieve condition 2, we need to only adjust the weights of the edges in $K^-(\bar{T}, \bar{e})$ and make them equal to $w(\bar{e})$.

Based on the above analysis, we obtain Algorithm 12.3 for solving ($PInvMST_1^+$) with $|F| = 1$.

Algorithm 12.3 An algorithm to solve ($PInvMST_1^+$) with $|F| = 1$ [86, Algorithm 2]

Input: An instance $I = (G, w, c, \bar{e}, l \equiv 0, u)$ of ($PInvMST_1^+$) with $\bar{e} = st$.
Output: An optimal solution w^* or claim that I has no feasible solutions.

```
Calculate a new weight vector w' as (12.7).
Use a min-cut algorithm on (G, s, t, w') to find a minimum cut K*.
if w'(K*) < ∞ then
    Set w* := w.
    for each edge e ∈ K* do
        Set w*(e) := max{w(e), w(ē)}.
    end for
else
    return "No Feasible Solutions!".
end if
return w*.
```

Theorem 12.8 *($PInvMST_1^+$) with $|F| = 1$ can be solved in time $O(CUT(|V|, |E|))$, where $CUT(|V|, |E|)$ is the time needed to solve minimum cut problem on graph with $|V|$ vertices and $|E|$ edges.*

12.2.4 $\mathcal{NP}$-Hardness of ($PInvMST^+_{wsH}$)

In this part, we study ($PInvMST^+$) under the weighted sum Hamming distance, denoted by ($PInvMST^+_{wsH}$), and the bottleneck Hamming distance, denoted by ($PInvMST^+_{bH}$).

For ($PInvMST^+_{wsH}$), based on the similarity between the weighted sum Hamming distance and weighted l_1 norm, Li et al. [84] proved that it has similar computational complexity to ($PInvMST^+_1$). Specifically, when the partial solution contains at least two edges, ($PInvMST^+_{wsH}$) is $\mathcal{NP}$-hard; when the partial solution contains only one edge, ($PInvMST^+_{wsH}$) can be solved in polynomial time.

For the $\mathcal{NP}$-hardness of ($PInvMST^+_{wsH}$), we can prove it by reducing from (*MTkC*) and constructing the same instance as ($PInvMST^+_1$).

Theorem 12.9 ([84, Theorem 1]) *For any fixed integer $k \geq 2$, ($PInvMST^+_{wsH}$) is $\mathcal{NP}$-hard for $|F| = k$, even if restricted to cases in which (i) F induces a path, (ii) $u \equiv \infty$, (iii) c is unit, and (iv) the weight vector w is binary.*

Remark 12.4 Actually, Theorem 1 in [84] showed that ($PInvMST^+_{wsH}$) is APX-hard for any fixed integer $k \geq 2$ with $|F| = k$.

For the rest case $|F| = 1$, let $F = \{\bar{e} = st\}$. As the result of ($PInvMST^+_1$), we can get the following Lemma:

Lemma 12.3 *Let $I = (G, w, c, \bar{e}, 0, u)$ be an instance of ($PInvMST^+_{wsH}$). For any feasible solution $\bar{w}$ of I and for any feasible tree $\bar{T}$ with respect to $\bar{w}$,*

$$\|\bar{w} - w\|_{wsH} \geq \sum_{e \in K^-(\bar{T}, \bar{e})} c(e). \tag{12.8}$$

Based on (12.8), we can modify w' in (12.7) to the following form:

$$w'(e) = \begin{cases} 0, & w(e) \geq w(\bar{e}); \\ c(e), & w(e) < w(\bar{e}) \leq w(e) + u(e); \\ \infty, & w(\bar{e}) > w(e) + u(e). \end{cases} \tag{12.9}$$

Furthermore, using this new weight, an algorithm for solving ($PInvMST^+_{wsH}$) with $|F| = 1$ is presented as follows:

Algorithm 12.4 An algorithm for ($PInvMST^+_{wsH}$) with $|F| = 1$ [84, Algorithm 1]

Input: An instance $I = (G, w, c, \bar{e}, l \equiv 0, u)$ of ($PInvMST^+_{wsH}$) with $\bar{e} = st$.
Output: An optimal solution w^* or claim that I has no feasible solutions.
1: Calculate a new weight vector w' as (12.9).
2: Use a min-cut algorithm on (G, s, t, w') to find a minimum cut K^*.

```
3: if w'(K*) < ∞ then
4:     Set w* := w.
5:     for each edge e ∈ K* do
6:         Set w*(e) := max{w(e), w(ē)}.
7:     end for
8: else
9:     return "No Feasible Solutions!".
10: end if
11: return w*.
```

Remark 12.5 The difference between Algorithm 12.3 and Algorithm 12.4 is only in their line 1, where the definitions of w' are different.

Theorem 12.10 *($PInvMST^+_{wsH}$) can be solved in time $CUT(|V|, |E|)$ when $|F| = 1$.*

For ($PInvMST^+_{bH}$), as it is a special case ($l \equiv 0$) of ($PInvMST_{bH}$), we will introduce it in Sect. 12.2.7.

12.2.5 $\mathcal{NP}$-Hardness of ($PInvMST_1$) and ($PInvMST_{wsH}$)

Since ($PInvMST^+_1$) and ($PInvMST^+_{wsH}$) are special cases ($l \equiv 0$) of ($PInvMST_1$) and ($PInvMST_{wsH}$), respectively, we can determine their computational complexities directly.

Theorem 12.11 *For any fixed integer $k \geq 2$, ($PInvMST_1$) and ($PInvMST_{wsH}$) are $\mathcal{NP}$-hard for $|F| = k$, even if restricted to cases in which (i) F induces a path, (ii) c is unit, and (iii) the weight vector w is binary.*

Remark 12.6 For the special case $l = u \equiv \infty$ of (*PInvMST*), if we set $c(e) = \infty$ for each edge $e \in F$, it implies that the weights of all edges in F cannot be changed. Moreover, by *separable property*, the current problem is equivalent to ($PInvMST^+$). Hence, we can conclude that both of ($PInvMST_1$) and ($PInvMST_{wsH}$) are $\mathcal{NP}$-hard, even if $l = u \equiv \infty$.

12.2.6 Polynomial Time Algorithms for ($PInvMST_1$) and ($PInvMST_{wsH}$) with $|F| = 1$

Now, we consider the rest case, that is, $F = \{\bar{e} = st\}$. By *separable property*, except for the possible decrease in weight of $\bar{e}$, the weights of other edges will not decrease. This implies that when the new weight of $\bar{e}$ is determined, the current problem is equivalent to ($PInvMST^+$) with $|F| = 1$. In this way, existing polynomial time algorithms of (*PInvMST*$^+$) can be used to solve the current problem. In summary, the key to solving the rest case is how to determine the new weight of $\bar{e}$.

For any instance $I = (G, w, c, \bar{e}, l, u)$ of ($PInvMST_{wsH}$) and any feasible solution $\bar{w}$, according to the definition of Hamming distance, if the weight of $\bar{e}$ changes, its corresponding cost $c(e)$ is calculated into the objective value. This implies that if $\bar{w}(\bar{e}) \neq w(\bar{e})$, then change the weight of $\bar{e}$ to $w(\bar{e}) - l(\bar{e})$, the new weight remains feasible, and the objective value remains unchanged. Hence, executing Algorithm 12.4 twice can solve ($PInvMST_{wsH}$) with $|F| = 1$.

Algorithm 12.5 An algorithm for ($PInvMST_{wsH}$) with $|F| = 1$ [84, Algorithm 2]

Input: An instance $I = (G, w, c, \bar{e}, l, u)$ of ($PInvMST_{wsH}$) with $\bar{e} = st$.
Output: An optimal solution w^* or claim that I has no feasible solutions.

Execute Algorithm 12.4 on $I = (G, w, c, F, l' \equiv 0, u)$ to obtain solution w_1.
Set $w' := w$.
Set $w'(\bar{e}) := w(\bar{e}) - l(\bar{e})$.
Execute Algorithm 12.4 on $I' = (G, w', c, F, l' \equiv 0, u)$ to obtain solution w_2.
if Both w_1 and w_2 are "No Feasible Solutions!" **then**
    return "No Feasible Solutions!"
else if $\|w_1 - w\|_{wsH} \leq \|w_2 - w\|_{wsH}$ **then**
    Set $w^* := w_1$.
else
    Set $w^* := w_2$.
end if
return w^*.

Theorem 12.12 *The problem ($PInvMST_{wsH}$) can be solved in time $CUT(|V|, |E|)$ when $|F| = 1$.*

Due to the difference between the weighted l_1 norm and the weighted sum Hamming distance, Algorithm 12.5 cannot be directly applied to ($PInvMST_1$) with $|F| = 1$. Fortunately, by exploring the characteristics of the optimal solution to ($PInvMST_1$) with $|F| = 1$, we can still obtain a polynomial time algorithm for solving it. The following result gives a useful property of optimal solution:

Lemma 12.4 *Let $I = (G, w, c, \bar{e}, l, u)$ be an instance of ($PInvMST_1$). For any optimal solution w^* of I and any feasible tree T with respect to w^*,*

1. $w^*(\bar{e}) \leq w(\bar{e})$.
2. $w^*(e) = \max\{w(e), w^*(\bar{e})\}$, *for each edge* $e \in K(T, \bar{e}) \setminus \{\bar{e}\}$.
3. $w^*(e) = w(e)$, *for each edge* $e \notin K(T, \bar{e})$.

From Lemma 12.4, we can obtain that the only edges which affect the objective function are $\bar{e}$ and the edges in $K(T^*, \bar{e})$ with original weights less than $w^*(\bar{e})$. Let T be a spanning tree and $\bar{e}$ be an edge in T. For two edge weight vectors w and w', define

$$K^-(T, \bar{e}; w, w') = \{e \in K(T, \bar{e}) \mid w(e) < w'(\bar{e})\}.$$

The following results can be obtained directly from Lemma 12.4:

Corollary 12.2 *Let w^* be an optimal solution of I and T be a feasible tree with respect to w^*.*

$$\|w^* - w\|_1 = c(\bar{e})(w(\bar{e}) - w^*(\bar{e})) + \sum_{e \in K^-(T,\bar{e};w,w^*)} c(e)(w^*(\bar{e}) - w(e)). \quad (12.10)$$

Remark 12.7 Lemma 12.4 and Corollary 12.2 can both be directly extended to weighted l_p norms. At this point, (12.10) should be changed to

$$\|w^* - w\|_p^p = c(\bar{e})(w(\bar{e}) - w^*(\bar{e}))^p + \sum_{e \in K^-(T,\bar{e};w,w^*)} c(e)(w^*(\bar{e}) - w(e))^p.$$

Based on (12.10), we present a polynomial time algorithm for (*PInvMST*) under the unit l_1 norm, that is, $c \equiv 1$. Let us consider the cardinality of $K^-(T, \bar{e}; w, w^*)$. If $|K^-(T, \bar{e}; w, w^*)| = 0$, (12.10) is transformed into

$$\|w^* - w\|_1 = w(\bar{e}) - w^*(\bar{e}).$$

This implies that the difference between the two weight vector is only at $\bar{e}$ and the optimal value $opt = w(\bar{e}) - w^*(\bar{e}) \leq l(\bar{e})$.

If $|K^-(T, \bar{e}; w, w^*)| > 2$, we can conclude that $w(\bar{e}) - w^*(\bar{e}) = l(\bar{e})$. Otherwise, if $w(\bar{e}) - w^*(\bar{e}) < l(\bar{e})$, then we can continue to decrease the weight of $\bar{e}$ by $\varepsilon > 0$ while reducing the weight of each edge in $K^-(T, \bar{e}; w, w^*)$ by ε, that is, make the weights of all these edges equal to $w^*(\bar{e}) - \varepsilon$. Clearly, the new weight vector is also feasible, and the objective value is equal to

$$\begin{aligned} &(w(\bar{e}) - (w^*(\bar{e}) - \varepsilon)) + \sum_{e \in K^-(T,\bar{e};w,w^*)} ((w^*(\bar{e}) - \varepsilon)) - w(e)) \\ &= \|w^* - w\|_1 - (|K^-(T, \bar{e}; w, w^*)| - 1)\varepsilon. \\ &< \|w^* - w\|_1 \end{aligned} \quad (12.11)$$

This contradicts with optimality of w^*.

For the rest case $|K^-(T, \bar{e}; w, w^*)| = 1$, as the same discussion in case $|K^-(T, \bar{e}; w, w^*)| > 2$, if $opt > l(\bar{e})$, then there exists an optimal solution such that $w(\bar{e}) - w^*(\bar{e}) = l(\bar{e})$, and if $opt \leq l(\bar{e})$, then there exists an optimal solution that is similar to the optimal solution in case $|K^-(T, \bar{e}; w, w^*)| = 0$.

In summary, when $opt \leq l(\bar{e})$, there exists an optimal solution that only changes the weight of $\bar{e}$, and the current problem is equivalent to (*PInvMST*$_1^-$). On the other hand, when $opt > l(\bar{e})$, the weight of $\bar{e}$ will be changed to $w(\bar{e}) - l(\bar{e})$, and the problem at this point is equivalent to $(PInvMST_1^+)$. Hence, we present the following algorithm and theorem:

Algorithm 12.6 An algorithm to solve $(PInvMST_1)$ with $|F| = 1$ and $c \equiv 1$ [87, Algorithm 3]

Input: An instance $I = (G, w, c, \bar{e}, l, u)$ of $(PInvMST_1)$ with $\bar{e} = st$ and $c \equiv 1$.
Output: An optimal solution w^* or claim that I has no feasible solutions.

```
Execute Algorithm 12.1 on I = (G, w, c, ē, l' ≡ ∞, u' ≡ 0) to obtain solution w_1.
if w(ē) − w_1(ē) ≤ l(ē) then
    Set w* := w_1.
else
    Set w' := w.
    Set w'(ē) := w(ē) − l(ē).
    Execute Algorithm 12.3 on I' = (G, w', c, ē, l' ≡ 0, u) to obtain w_2.
    if w_2 is "No Feasible Solutions!" then
        return "No Feasible Solutions!"
    else
        Set w* := w_2.
    end if
end if
return w*.
```

Theorem 12.13 *The problem $(PInvMST_1)$ can be solved in time $CUT(|V|, |E|)$ when $|F| = 1$ and c is a unit cost vector.*

Remark 12.8 Unfortunately, Algorithm 12.6 cannot work for *(PInvMST)* with $|F| = 1$ under the unit l_p norm. Figure 12.2 illustrates an example of it. In that figure, (a) is an instance I of *(PInvMST)* under the unit l_2 norm with $l = u \equiv \infty$. The partial solution is the dashed edge, and the number on each edge is its weight. (b) is the output of Algorithm 12.6. The weights on edges constitute a feasible solution w^* of I, and the dark edges form a feasible tree with respect to w^*. (c) is another feasible solution w' of I, and the dark edges form a feasible tree

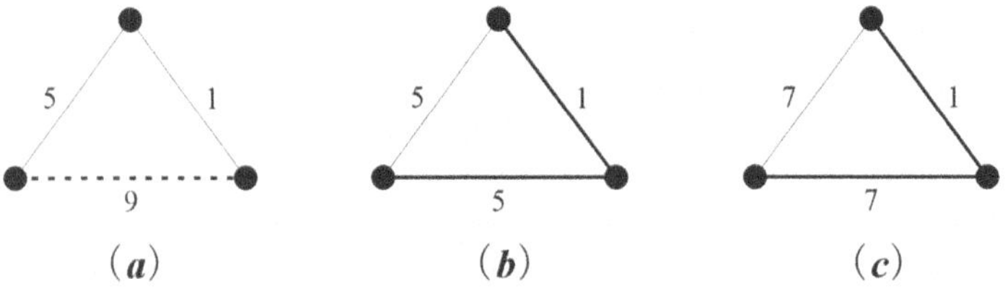

Fig. 12.2 An instance of *(PInvMST)* with $|F| = 1$ under the unit l_2 norm and its different solutions

with respect to w'. Clearly, $\|w' - w\|_2 = \sqrt{(9-5)^2} = 4$ and $\|w^* - w\|_2 = \sqrt{(9-7)^2 + (7-5)^2} = 2\sqrt{2} < 4$. Thus, it is necessary to find a new method to solve this problem. A natural idea is to make several different guesses about the weight of $\bar{e}$, then input the modified weight into Algorithm 12.3 to obtain several different solutions, and select the optimal one. For details, readers can refer to Theorem 6 and Algorithm 4 in [87].

12.2.7 A Polynomial Time Algorithm of ($PInvMST_{bH}$)

For bottleneck norm such as bottleneck Hamming distance and weighted l_∞ norm, the common algorithm design method is to first provide an algorithm for decision version and then use binary search method to obtain the optimal solution. Recall that for any two weight vectors w and w', the bottleneck Hamming distance between w and w' is

$$\|w' - w\|_{bH} = \max_{e \in E(G)} c(e) H(w'(e), w(e)).$$

Let $I = (G, w, c, F, l, u)$ be an instance of ($PInvMST_{bH}$) and $\bar{w}$ be a feasible solution of I with $\|\bar{w} - w\|_{bH} \le K$. Define a new weight vector w_K as follows:

$$w_K(e) = \begin{cases} w(e), & c(e) > K; \\ w(e) + u(e), & c(e) \le K \text{ and } e \notin F; \\ w(e) - l(e), & c(e) \le K \text{ and } e \in F. \end{cases} \tag{12.12}$$

It is not difficult to see that for any $e \in F$, $w_K(e) \le \bar{w}(e)$; for any $e \notin F$, $w_K(e) \ge \bar{w}(e)$; and for any $e \in E$,

$$w(e) - l(e) \le w_K(e) \le w(e) + u(e).$$

Let T be a feasible tree with respect to $\bar{w}$. For each edge $\bar{e} \in F$ and each edge $e \in K(T, \bar{e}) \setminus \{\bar{e}\}$, $w_K(\bar{e}) \le \bar{w}(\bar{e}) \le \bar{w}(e) \le w_K(e)$. Moreover,

$$w_K(\bar{e}) = \min\{w_K(e) : e \in K(T, \bar{e})\}.$$

Hence, by Theorem 12.1, w_K is also a feasible solution of I with $\|w_K - w\|_{bH} \le K$. Based on the above results, we can obtain the following important conclusion on decision version of ($PInvMST_{bH}$):

Lemma 12.5 *Let $I = (G, w, c, F, l, u)$ be an instance of decision version of ($PInvMST_{bH}$) with parameter $K \ge 0$. The answer of I is "True" if and only if w_K is a feasible solution of I.*

From Lemma 12.5, we can present the following algorithm to solve the decision version of $(PInvMST_{bH})$.

Algorithm 12.7 Algorithm for the decision version of $(PInvMST_{bH})$

Input: An instance $I = (G, w, F, l, u, c)$ of $(PInvMST_{bH})$ and a real number $K \geq 0$.
Output: "YES" or "NO."
1: Compute w_K by (12.12).
2: Find a minimum spanning tree T of G with respect to w_K, and choose the edges in F firstly to break the tie.
3: **if** $F \subseteq E(T)$ **then**
4: **return** "YES."
5: **else**
6: **return** "NO."
7: **end if**

By the definition of bottleneck Hamming distance, if the answer of instance I is "True" for parameter $K \geq 0$, then the answer of instance I is also "True" for any parameter $K' \geq K$. Furthermore, the answer of instance I is also "True" for parameter $c(e_K)$, where $e_K = \arg\max\{c(e) \mid c(e) \leq K\}$. Therefore, we can obtain the following theorem and algorithm:

Theorem 12.14 *Let $I = (G, w, c, F, l, u)$ be an instance of $(PInvMST_{bH})$ and $c_M = \max\{c(e) \mid e \in E\}$. I has feasible solutions if and only if w_{c_M} is a feasible solution of I. Moreover, if I has feasible solutions, then w_{c_K} is an optimal solution of I, where c_K is the minimum edge cost that makes w_{c_K} a feasible solution of I.*

Algorithm 12.8 Algorithm for $(PInvMST_{bH})$

Input: An instance $I = (G, w, F, l, u, c)$ of $(PInvMST_{bH})$.
Output: An optimal solution w^* or claim that I has no feasible solutions.
1: Set $c_M := \max\{c(e) \mid e \in E\}$.
2: **if** the output of Algorithm 12.7 on I with parameter c_M is "NO" **then**
3: **return** "No Feasible Solutions."
4: **end if**
5: Order the edges in E as $c(e_1) \leq c(e_2) \leq \cdots \leq c(e_m)$.
6: Use binary search method to find the minimum index j such that the output of Algorithm 12.7 on I with parameter c_{e_j} is "YES."
7: Set $c_j := c_{e_j}$ and compute w_{c_j} by (12.12).
8: **return** $w^* = w_{c_j}$.

Remark 12.9 The running time of Algorithm 12.7 is $O(|E| + |V| \log |V|)$. For Algorithm 12.8, line 1 to line 4 will cost $O(|E| + |V| \log |V|)$ time, line 5 will cost $O(|E| \log |E|)$ time, and line 6 will run up to $\log |E|$ iterations; each iteration will cost $O(|E| + |V| \log |V|)$ time. Thus, the running time of Algorithm 12.8 is $O(\log |V| \cdot (|E| + |V| \log |V|))$ time.

Theorem 12.15 *($PInvMST_{bH}$) can be solved in time* $O(\log |V| \cdot (|E| + |V| \log |V|))$.

12.2.8 A Polynomial Time Algorithm of ($PInvMST_\infty$)

Recall that for any two weight vector w and w', the weighted l_∞ norm between w and w' is

$$\|w' - w\|_\infty = \max_{e \in E(G)} c(e)|w'(e) - w(e)|.$$

Let $I = (G, w, c, F, l, u)$ be an instance of ($PInvMST_\infty$). Define a new weight vector w_K for a real number $K \geq 0$ as follows:

$$w_K(e) = \begin{cases} w(e) - \min\{l(e), K/c(e)\}, & e \in F; \\ w(e) + \min\{u(e), K/c(e)\}, & e \notin F. \end{cases} \tag{12.13}$$

Through a discussion similar to ($PInvMST_{bH}$), we can obtain the following important conclusion on decision version of ($PInvMST_\infty$).

Lemma 12.6 *Let $I = (G, w, c, F, l, u)$ be an instance of decision version of ($PInvMST_\infty$) with parameter $K \geq 0$. The answer of I is "True" if and only if w_K as (12.13) is a feasible solution of I.*

From Lemma 12.6, we can present the following algorithm to solve the decision version of ($PInvMST_\infty$):

Algorithm 12.9 Algorithm for the decision version of ($PInvMST_\infty$)

Input: An instance $I = (G, w, F, c, l, u)$ of ($PInvMST_\infty$) and a real number $K \geq 0$.
Output: "YES" or "NO."
1: Compute w_K by (12.13).
2: Find a minimum spanning tree T of G with respect to w_K, and choose the edges in F firstly to break the tie.
3: **if** $F \subseteq E(T)$ **then return** "YES."
4: **else return** "NO."
5: **end if**

If we directly combine Algorithm 12.9 with binary search method, we can obtain an algorithm to solve (*PInvMST*$_\infty$) with arbitrarily small error. However, there are two weak points in this algorithm. Firstly, if the weight vector is not limited to be integer, the output of the algorithm may have error. Secondly, the algorithm is not strong polynomial, because the number of iterations of binary search method relies on the values of input and prescribed error. Therefore, we hope to make the above algorithm strong polynomial by obtaining a candidate set of optimal values with polynomial scale.

In the following, we firstly study (*PInvMST*) under unit l_∞ norm, a special case of (*PInvMST*$_\infty$). Then, we generalize all results to (*PInvMST*$_\infty$). Due to the fact that the optimal solution to (*PInvMST*) is often not unique, in order to further investigate the properties of the optimal value, we provide the following definitions: Let $I = (G, w, c \equiv 1, F, l, u)$ be *an instance of unit* (*PInvMST*). Let w^* be an optimal solution and *opt* be the optimal value. With respect to w^*, an edge e is called an *optimal edge* if $|w^*(e) - w(e)| = opt$, and denote $E_{opt}(w^*)$ as the set of all optimal edges. An optimal solution w^* is called an *extreme optimal solution* of I if it satisfies the following two conditions:

1. $|E_{opt}(w^*)|$ is minimum.
2. Under condition (1), the unit l_1 norm between w and w^* is minimum.

The following result shows that extreme optimal solution satisfies *separable property*:

Lemma 12.7 ([85, Lemma 1]) *Let* $I = (G, w, c \equiv 1, F, l, u)$ *be an instance of unit (PInvMST) and* w^* *be an extreme optimal solution of* I*. Then* w^* *satisfies separable property.*

Based on *separable property*, the following two lemmas give more in-depth properties of extreme optimal solutions:

Lemma 12.8 ([85, Lemma 2]) *Let* w^* *be an extreme optimal solution of* I *and* T *be a feasible tree with respect to* w^**. For any edge* $\bar{e} \in F$*, if its weight decreases in* w^**, then there is an edge* $e \in K(T, \bar{e}) \setminus \{\bar{e}\}$*, such that* $w^*(e) = w^*(\bar{e})$*. Furthermore,*

$$w^*(\bar{e}) = \min\{w^*(e) : e \in K(T, \bar{e}) \setminus \{\bar{e}\}\}. \tag{12.14}$$

Lemma 12.9 ([85, Lemma 3]) *Let* w^* *be an extreme optimal solution and* T *be a feasible tree with respect to* w^**. For any edge* e *not included in* F*, if its weight increases in* w^**, then there is an edge* $\bar{e} \in C(T, e) \cap F$ *with* $w^*(\bar{e}) = w^*(e)$*. Furthermore, we have*

$$w^*(e) = \max\{w^*(\bar{e}) : \bar{e} \in C(T, e) \cap F\}. \tag{12.15}$$

Let w' be any weight vector on E. We call e an *extreme edge* with respect to w' if $w'(e) - w(e) = u(e)$ or $w(e) - w'(e) = l(e)$. Let $E_{ex+}(w') = \{e \mid w'(e) - w(e) = u(e)\}$ and $E_{ex-}(w') = \{e \mid w(e) - w'(e) = l(e)\}$. Furthermore, if w' satisfies *separate property*, $E_{ex-}(w') \subseteq F$ and $E_{ex+}(w') \subseteq E \setminus F$. Without ambiguity, we abbreviate $E_{opt}(w')$, $E_{ex+}(w')$, and $E_{ex-}(w')$ as E_{opt}, E_{ex+}, and E_{ex-} in the following:

Now, suppose that w^* is an extreme optimal solution with $opt > 0$ and $\bar{e} \in F \cap E_{opt}$. Then, by Lemma 12.8, there is an edge e_m that belongs to $K(T, \bar{e}) \setminus \{\bar{e}\}$ with

$$w^*(\bar{e}) = w^*(e_m) = \min\{w^*(e) : e \in K(T, \bar{e}) \setminus \{\bar{e}\}\}.$$

If $e_m \notin E_{opt} \cup E_{ex+}$, that is,

$$\delta(e_m) \triangleq w^*(e_m) - w(e_m) < \min\{opt, u(e_m)\}, \tag{12.16}$$

it implies that the weight of e_m still has room to rise. Thus, we can rise the weights of $\bar{e}$ and e_m a little ($\varepsilon > 0$) to ensure that their new weights are still equal and also meet the capacitated constraints. Figure 12.3 depicts the above operation. Furthermore, if each edge e_m satisfies (12.16), we can obtain a new function w' with $\|w' - w\|_\infty < \|w^* - w\|_\infty$ or $\|w' - w\|_\infty = \|w^* - w\|_\infty$ and $E_{opt}(w') \subsetneq E_{opt}(w^*)$. This will contradict with the extreme optimality of w^*.

Lemma 12.10 presents an accurate statement about above description.

Lemma 12.10 ([85, Lemma 4]) *Let w^* be an extreme optimal solution of I and T be a feasible tree with respect to w^*. If $opt > 0$ and $\bar{e} \in F \cap E_{opt}(w^*)$, then there is an edge $e \in K(T, \bar{e}) \setminus \{\bar{e}\}$ with $w^*(\bar{e}) = w^*(e)$ such that $e \in E_{opt}(w^*) \cup E_{ex+}(w^*)$.*

Lemma 12.11 ([85, Lemma 5]) *Let w^* be an extreme optimal solution of I, and T be a feasible tree with respect to w^*. If $opt > 0$ and $e \in (E \setminus F) \cap E_{opt}(w^*)$,*

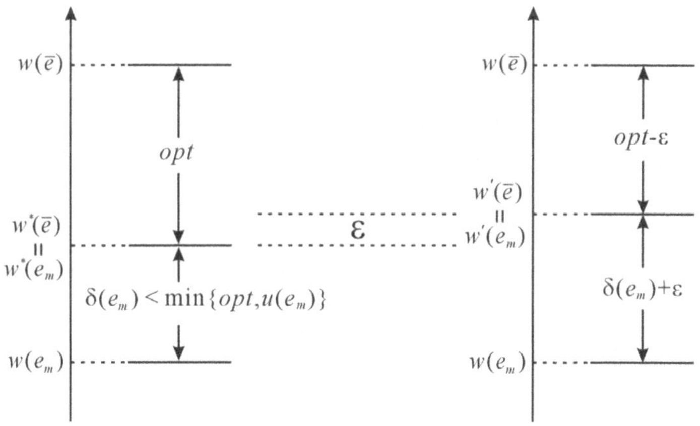

Fig. 12.3 An illustration of the increasing operation

then $e \notin E(T)$, *and there is an edge* $\bar{e} \in C(T, e) \cap F$ *with* $w^*(e) = w^*(\bar{e})$ *such that* $\bar{e} \in E_{opt}(w^*) \cup E_{ex-}(w^*)$.

Theorem 12.16 ([85, Theorem 4]) *Let* $I = (G, w, c \equiv 1, F, l, u)$ *be an instance of unit (PInvMST) and* $opt > 0$ *be the optimal value of* I. *Then, there exist two edges* $\bar{e} \in F$ *and* $e \notin F$ *such that*

$$opt = \max\left\{\frac{1}{2}\delta(\bar{e}, e), \delta(\bar{e}, e) - u(e), \delta(\bar{e}, e) - l(\bar{e})\right\}, \tag{12.17}$$

where $\delta(\bar{e}, e) = w(\bar{e}) - w(e)$.

Let

$$\Delta(\bar{e}, e) = \max\left\{\frac{1}{2}\delta(\bar{e}, e), \delta(\bar{e}, e) - u(e), \delta(\bar{e}, e) - l(\bar{e})\right\} \tag{12.18}$$

and $\Delta(I)$ be the set of all possible value $\Delta(\bar{e}, e)$ of an instance I, that is,

$$\Delta(I) = \{\Delta(\bar{e}, e) \mid \bar{e} \in F \text{ and } e \notin F\}.$$

From Theorem 12.16, we can see that $opt \in \Delta(I)$ and $|\Delta(I)| = |F||E \setminus F| = O(|V||E|)$. Hence, combining Algorithm 12.9 and binary search method on $\Delta(I)$, we can obtain Algorithm 12.10 to solve unit (*PInvMST*) with strong polynomial time. It should be noted that if $w(\bar{e}) \leq w(e)$, $\Delta(\bar{e}, e) \leq 0$, and it cannot equal to opt. Similarly, if $w(\bar{e}) - l(\bar{e}) > w(e) + u(e)$, by (12.1), e cannot be contained in the fundamental cut with respect to $\bar{e}$ in any feasible tree. Therefore, the above two cases is deleted in line 10 of Algorithm 12.10.

Algorithm 12.10 Algorithm for unit ($PInvMST_\infty$) [85, Algorithm 2]

Input: An instance $I = (G, w, c \equiv 1, F, l, u)$ of unit ($PInvMST_\infty$).
Output: An optimal solution w^* or claim that I has no feasible solutions.

1: **if** the output of Algorithm 12.9 on I with parameter 0 is "YES" **then**
2:     **return** $w^* = w$.
3: **end if**
4: Set $l_{\max} := \max_{e \in E} l(e)$ and $u_{\max} := \max_{e \in E} u(e)$.
5: **if** the output of Algorithm 12.9 on I with parameter $\max\{l_{\max}, u_{\max}\}$ is "NO" **then**
6:     **return** "No Feasible Solutions!"
7: **end if**
8: Set $\Delta(I) := \emptyset$.
9: **for** $\bar{e} \in F$ and $e \notin F$ **do**
10:     **if** $w(\bar{e}) > w(e)$ and $w(\bar{e}) - l(\bar{e}) \leq w(e) + u(e)$ **then**
11:         Calculate $\Delta(\bar{e}, e)$ by (12.18). Set $\Delta(I) := \Delta(I) \cup \Delta(\bar{e}, e)$.
12:     **end if**

13: **end for**
14: Order $\Delta(I)$.
15: Use binary search method on $\Delta(I)$ to find the minimum number *opt* satisfying the output of Algorithm 12.9 on I with parameter *opt* is "YES."
16: Compute w_{opt} by (12.13).
17: **return** $w^* = w_{opt}$.

Remark 12.10 The running time of Algorithm 12.10 is $O(|V||E|\log|V|)$. In detail, the running time of lines 1 to 3 and lines 4 to 7 is $O(|E| + |V|\log|V|)$, lines 8 to 13 is $O(|V||E|)$, and ordering $\Delta(I)$ is $O(|V||E|\log|V|)$. For the binary search stage, it should be at most $O(\log|V||E|) = O(\log|V|)$ iterations, and every iteration costs $O(|E| + |V|\log|V|)$ time. Thus, the running time of binary search stage is $O(\log|V| \cdot (|E| + |V|\log|V|))$. Therefore, the whole running time is $O(|V||E|\log|V| + \log|V| \cdot (|E| + |V|\log|V|)) = O(|V||E|\log|V|)$.

Finally, we generalize above results to (*PInvMST*$_\infty$). Let $I = (G, w, c, F, l, u)$ be an instance of (*PInvMST*$_\infty$) and w^* be an optimal solution of I. Update the definition of *optimal edge* to $c(e)|w^*(e) - w(e)| = opt$ while keeping the definition of *extreme optimal solution* unchanged. It is not difficult to verify that any extreme optimal solution of (*PInvMST*$_\infty$) also satisfies *separable property*. And Lemmas 12.8 and 12.9 also hold. Furthermore, we can update Lemmas 12.10 and 12.11 for (*PInvMST*$_\infty$) as follows:

Lemma 12.12 *[85, Lemma 6] Let w^* be an extreme optimal solution of I and T be a feasible tree with respect to w^*. If $\bar{e} \in F \cap E_{opt}$, then there is an edge $e \in K(T, \bar{e}) \setminus \{\bar{e}\}$ with $w^*(\bar{e}) = w^*(e)$ such that $e \in E_{opt} \cup E_{ex+}$. If $e \in (E \setminus F) \cap E_{opt}$, then $e \notin E(T)$, and there is an edge $\bar{e} \in C(T, e) \cap F$ with $w^*(e) = w^*(\bar{e})$ such that $\bar{e} \in E_{opt} \cup E_{ex-}$.*

Based on Lemma 12.12, Theorem 12.16 can be updated as follows:

Theorem 12.17 ([85, Theorem 5]) *Let I be an instance of (PInvMST$_\infty$) and opt be the optimal value of I. Then, there are two edges $\bar{e} \in F$ and $e \notin F$ such that*

$$opt = \max\left\{\frac{c(\bar{e})c(e)\delta(\bar{e}, e)}{c(\bar{e}) + c(e)}, c(\bar{e})\big(\delta(\bar{e}, e) - u(e)\big), c(e)\big(\delta(\bar{e}, e) - l(\bar{e})\big)\right\},$$

where $\delta(\bar{e}, e) = w(\bar{e}) - w(e)$.

Based on Theorem 12.17, we can generalize Algorithm 12.10 to solve (*PInvMST*$_\infty$) by updating $\Delta(\bar{e}, e)$ in line 11 defined below:

$$\Delta(\bar{e}, e) = \max\left\{\frac{c(\bar{e})c(e)\delta(\bar{e}, e)}{c(\bar{e}) + c(e)}, c(\bar{e})\big(\delta(\bar{e}, e) - u(e)\big), c(e)\big(\delta(\bar{e}, e) - l(\bar{e})\big)\right\}.$$

12.3 Approximation Algorithms on (*PInvMST*)

In this section, we shall discuss the approximation algorithms on (*PInvMST*$^+$) and (*PInvMST*) under the weighted l_1 norm or the weighted sum Hamming distance. Since these problems have been proven to be $\mathcal{NP}$-hard, it is necessary to study their approximation algorithms.

12.3.1 *Fitted Cut*

Let G be a graph with an edge subset F. Given any edge $e = st \in F$, we call an s-t cut *fitted for* F, if it does not contain any edges in F except e, and also call it a *fitted cut* for F. The motivation for this definition is that given any instance $I = (G, w, c, F, l, u)$ of (*PInvMST*), any feasible solution $\bar{w}$ of I and any feasible tree T with respect to $\bar{w}$, for each edge $\bar{e} \in F$, the fundamental cut $K(T, \bar{e})$ is *fitted for* F. It is not difficult to verify that if F is acyclic, fitted cut always exists. Given an edge-weighted graph G with weight vector w, an edge subset F, and an edge $e = st \in F$, a minimum fitted s-t cut for F can be calculated by Algorithm 12.11. It is easy to see that if $w'(K^*) < \infty$, K^* is a minimum fitted cut, and $w(K^*) = w'(K^*)$; otherwise, the minimum fitted cut does not exist.

Algorithm 12.11 Algorithm for minimum fitted cut [87, Algorithm 1]

Input: An instance $I = (G, w, F, s, t)$ of minimum fitted cut problem.
Output: A minimum fitted cut K^*.

```
1: Set w' := w.
2: for ē ∈ F and ē ≠ st do
3:     Set w'(ē) := ∞.
4: end for
5: Use a min-cut algorithm on (G, s, t, w') to find a minimum cut K*.
6: return K*.
```

The following Lemma shows a property of a cut restricted on subgraphs:

Lemma 12.13 ([87, Lemma 3]) *Let $G = (V, E)$ be a graph with $\bar{e} = st \in E$ and K be an s-t cut of G. Then for any subgraph $H = (V(H), E(H))$ of G containing $\bar{e}$, $K \cap E(H)$ is an s-t cut of H. Furthermore, if K is fitted for some edge subset F, then $K \cap E(H)$ is also fitted for $F \cap E(H)$.*

12.3.2 Approximation Algorithms on (*PInvMST*$^+$)

We first study the approximation algorithm on (*PInvMST*$_1^+$), and then we will extend the obtained approximation algorithm to (*PInvMST*$^+$) under other norms. From Sect. 12.2.3, although (*PInvMST*$_1^+$) is $\mathcal{NP}$-hard for $|F| \geq 2$, when $|F| = 1$, it has a polynomial time algorithm. Based on this key point, the design idea of the approximation algorithm is to decompose the multi-edge case into several single-edge case. It should be noted that since each fundamental cut is fitted for F, when dealing with single-edge subproblem at each stage, the minimum fitted cut algorithm needs to be called instead of the classical minimum cut algorithm. The pseudo-code of the algorithm is as follows:

Algorithm 12.12 Approximation algorithm for ($PInvMST_1^+$ [87, Algorithm 2])

Input: An instance $I = (G, w, c, F, l \equiv 0, u)$ of (*PInvMST*$_1^+$).
Output: A feasible solution w_f or claim that I has no feasible solutions.

Reorder the edges in F with $w(e_1) \geq w(e_2) \geq \ldots \geq w(e_k)$.
Set $G_1 := G$ and $w_f := w$.
for $j = 1 \rightarrow k$ **do**
  Set $\bar{e} := e_j$ and denote the two endpoints as s_j and t_j.
  Let H be the component of G_j containing $\bar{e}$.
  Calculate the new weight vector w' by (12.7).
  Call Algorithm 12.11 on $(H, w', F \cap E(H), s_j, t_j)$ to find a minimum fitted cut K_j^*.
  if $w'(K_j^*) = \infty$ **then return** "No Feasible Solutions!"
  else set $G_{j+1} := G_j - K_j^*$ and $w_f(e) := \max\{w(e), w(\bar{e})\}$ for $e \in K_j^*$.
  end if
end for
return w_f.

In order to understand the algorithm more clearly, we give an example to show the process of the algorithm. Figure 12.4 gives an instance $I = (G, w, c \equiv 1, F, l \equiv 0, u)$ of (*PInvMST*$_1^+$). In this instance, the two dash edges form the partial solution F. For any edge $e \in E$, there is an ordered pair $(w(e), u(e))$ on it, such as $w(uv) = 2$ and $u(uv) = 4$.

Figure 12.5 illustrates the whole process of using Algorithm 12.12 to solve above instance I. Since $w(e_1) \geq w(e_2)$, Algorithm 12.12 processes edge e_1 firstly. Figure 12.5a shows the new weight vector w' by (12.7) with respect to e_1 and $w'(e) = \infty$ for any dash-dot edge e. It is not hard to verify that the minimum fitted s_1-t_1 cut $K_1^* = \{e_1, s_2t_1, t_1t_2, t_2y\}$. Figure 12.5b shows the new graph $G_2 = G_1 - K_1^*$, and it is easy to see that e_2 is contained in H_2, the left component of G_2. The new weight vector w' with respect to e_2 is shown in Fig. 12.5c, and the minimum fitted s_2-t_2 cut in H_2 is $K_2^* = \{e_2, s_1s_2, vs_2\}$. Finally, Fig. 12.5d shows the

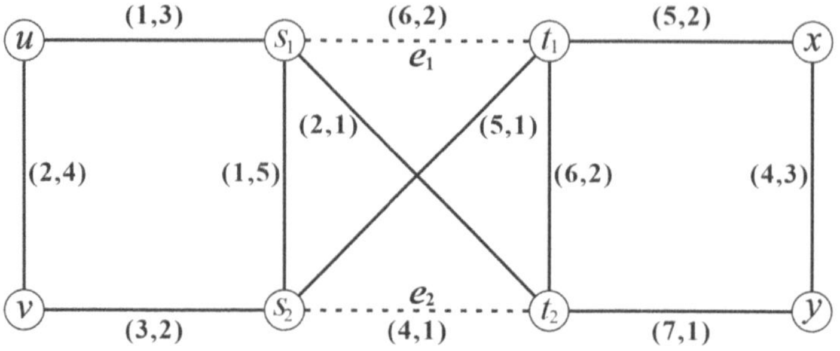

Fig. 12.4 An instance $I = (G, w, c \equiv 1, F, l \equiv 0, u)$ of $(PInvMST_1^+)$

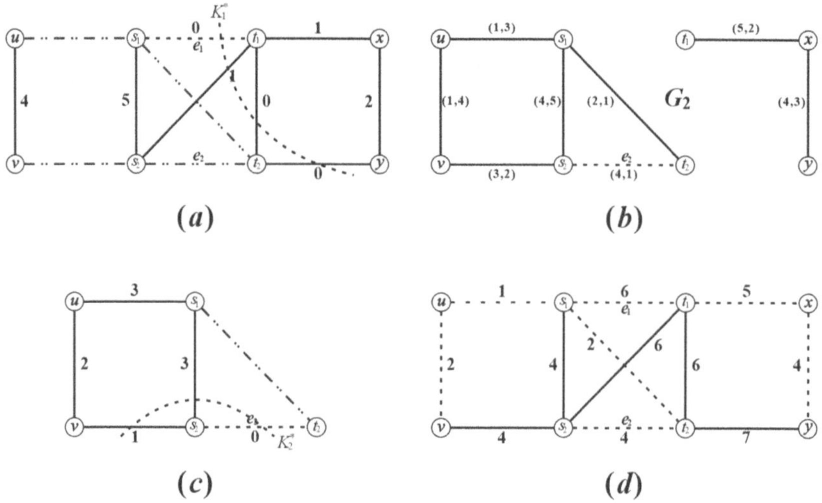

Fig. 12.5 The process of Algorithm 12.12

output result w_f of Algorithm 12.12. It is not difficulty to verify that the difference between w_f and w is only in the edges of the two fitted cut K_1^* and K_2^* and the dash edges form a feasible tree with respect to w_f.

Using Theorem 12.2 and a similar proof method as Theorem 12.8, it is not difficult to obtain the correctness of Algorithm 12.12.

Theorem 12.18 ([87, Theorem 3]) *If the output of Algorithm 12.12 is "No Feasible Solutions!", the instance I has no feasible solutions. Else, w_f is a feasible solution of I.*

In the following, we shall analyze the approximation ratio of Algorithm 12.12. Firstly, we give a lower bound of the optimal value of instance I. Let w^* be an optimal solution of I and T be a feasible tree with respect to w^*. Since $w^* \leq w$, for any integer j with $1 \leq j \leq k$,

$$\| w^* - w \|_1 = \sum_{e \in E(G)} c(e)(w^*(e) - w(e)) \geq \sum_{e \in K(T,e_j)} c(e)(w^*(e) - w(e)). \tag{12.19}$$

Moreover, for any edge $e \in K(T, e_j)$, $w(e) + u(e) \geq w^*(e) \geq w^*(e_j) \geq w(e_j)$, and

$$w^*(e) - w(e) \geq \max\{0, c(e)(w(e_j) - w(e))\}. \tag{12.20}$$

Combining (12.20) with (12.19), we have

$$\| w^* - w \|_1 \geq \sum_{e \in K(T,e_j)} c(e) \max\{0, w(e_j) - w(e)\}. \tag{12.21}$$

On the other hand, let the weight vector w' obtained in the j-th round of the for-loop be w_j. Since for any edge $e \in K(T, e_j)$, $w(e_j) \leq w(e) + u(e)$, by (12.7),

$$w_j(e) = c(e) \max\{0, w(e) - w(e_j)\}. \tag{12.22}$$

Hence, combining (12.21) with (12.22), we can obtain that

$$\| w^* - w \|_1 \geq \sum_{e \in K(T,e_j)} w_j(e) \geq w_j(K(T, e_j)). \tag{12.23}$$

Since $K(T, e_j)$ is a *fitted cut* for F, we can get the following lemma directly:

Lemma 12.14 ([87, Lemma 4]) *Let w^* be an optimal solution of I and K_j be a minimum fitted s_j-t_j cut with respect to w_j. Then,*

$$\| w^* - w \|_1 \geq \max_{1 \leq j \leq k} \{w_j(K_j)\}. \tag{12.24}$$

If the instance I has feasible solutions, the following lemma gives an upper bound of difference between w_f and w:

Lemma 12.15 ([87, Lemma 5]) *Let w_f be the output of Algorithm 12.12. Then,*

$$\| w_f - w \|_1 \leq \sum_{j=1}^{k} w_j(K_j).$$

By Lemmas 12.14 and 12.15, the approximation ratio can be obtained.

Theorem 12.19 *Algorithm 12.12 is a k-approximation algorithm for (PInvMST$_1^+$) with running time $O(k \cdot CUT(|V|, |E|))$, where $k = |F|$.*

Remark 12.11 Algorithm 12.12 can be modified to solve (*PInvMST*$^+_{wsH}$) directly. The adapted algorithm only needs to change the definition of w' in line 6 by (12.9), and the approximation ratio is also k:

Remark 12.12 Algorithm 12.12 can also be generalized to solve (*PInvMST*^+_p) directly. The adapted algorithm only needs to change the definition of w' in line 6 by (12.25), and the approximation ratio is $\sqrt[p]{k}$.

$$w'(e) = \begin{cases} 0, & w(e) \geq w(\bar{e}); \\ c(e)(w(\bar{e}) - w(e))^p, & w(e) < w(\bar{e}) \leq w(e) + u(e); \\ \infty, & w(\bar{e}) > w(e) + u(e). \end{cases} \tag{12.25}$$

Remark 12.13 Since $p \to \infty$, $\sqrt[p]{k} \to 1$. This means that although the algorithm cannot work when $p = \infty$ (because it needs to calculate $(w(e) - w(e_j))^p$), it can theoretically verify that (*PInvMST*$^+_\infty$) can be solved in strongly polynomial time, which is the main result in Sect. 12.2.2.

12.3.3 Approximation Algorithms on (PInvMST)

As the same discussion in Sect. 12.2.6, by *separable property*, when the new weights of edges in F are determined, the current problem is equivalent to (*PInvMST*$^+$). In this way, existing approximation algorithms of (*PInvMST*$^+$) can be used to solve the current problem. In this subsection, we firstly consider unit (*PInvMST*$_1$) and then generalize obtained results to (*PInvMST*) under other norms.

Let $I = (G, w, c \equiv 1, F, l, u)$ be an instance of unit (*PInvMST*$_1$). According to *separation property*, the weight of edges in F will not increase. Combining with the characteristics of the optimal solution of (*PInvMST*$_1$) with $|F| = 1$, a natural idea of determining the new weights of F is to use Algorithm 12.1 on $I' = (G, w, c, F, l' \equiv \infty, u' \equiv 0)$ of (*PInvMST*$^-$). Let w_1 be the output of Algorithm 12.1. It should be noted that if $w(\bar{e}) - w_1(\bar{e}) > l(\bar{e})$ for some edge $\bar{e} \in F$, it does not mean that there is no feasible solution for instance I, but rather that the weight of $\bar{e}$ can only be reduced by $l(\bar{e})$. Based on the above analysis, we propose the following algorithm:

Algorithm 12.13 Approximation algorithm for unit $(PInvMST_1)$

Input: An instance $I = (G, w, c \equiv 1, F, l, u)$ of unit $(PInvMST_1)$.
Output: A feasible solution w_f or claim that I has no feasible solutions.
1: Execute Algorithm 12.1 on $I = (G, w, c, F, l' \equiv \infty, u' \equiv 0)$ to obtain solution w_1.
2: **for** each edge $\bar{e} \in F$ **do**
3:     **if** $w(\bar{e}) - w_1(\bar{e}) > l(\bar{e})$ **then**
4:         Set $w_1(\bar{e}) = w(\bar{e}) - l(\bar{e})$.
5:     **end if**
6: **end for**
7: Execute Algorithm 12.12 on $I = (G, w_1, c, F, l' \equiv 0, u)$ to obtain solution w_f.
8: **return** w_f.

Remark 12.14 Compared with Algorithm 5 in [87], Algorithm 12.13 significantly simplifies the process of determining the new weights of edges in F. In fact, the results obtained by the two algorithms are exactly the same.

Based on the same analysis method in Sect. 12.3.2, we can obtain the accuracy and approximation ratio of Algorithm 12.13.

Theorem 12.20 ([87, Theorem 8]) *If I has no feasible solutions, Algorithm 12.13 returns "No Feasible Solutions!". Else, Algorithm 12.13 returns a feasible solution w_f.*

Theorem 12.21 ([87, Theorem 9]) *Algorithm 12.13 is a k-approximation algorithm for unit $(PInvMST_1)$ with running time $O(k \cdot CUT(|V|, |E|))$, where $k = |F|$.*

Remark 12.15 Algorithm 12.13 can be modified to solve $(PInvMST_{wsH})$ directly. The adapted algorithm only needs to use Algorithm 12.5 to determine the new weights of edges in F, and the approximation ratio is also k.

Remark 12.16 Algorithm 12.13 can also be modified to solve unit $(PInvMST_p)$ directly. The adapted algorithm only needs to use the method mentioned in Remark 12.8 to determine the new weights of edges in F, and the approximation ratio is $\sqrt[p]{k} + \varepsilon$. For details, readers can refer to Algorithm 5 and Theorem 9 in [87].

12.4 Conclusions

In this chapter, the computational complexities and algorithms of partial inverse minimum spanning tree problem (*PInvMST*) are introduced, and the final results are listed in Table 12.1. Note that the symbol * means that this algorithm is suitable for all monotone norms, where $n = |V|$, $m = |E|$, and $k = |F|$.

Table 12.1 The computational complexities and algorithms or (*PInvMST*$^-$), (*PInvMST*$^+$), and (*PInvMST*)

	Unit l_1 norm	Weighted l_∞ norm	wsH	bH
PIMST$^-$	$^*O(m + n\log n + k^2)$ [23]			
PIMST$^+$	$\mathcal{NP}$-Hard [86] k-approx. [87]	$O(m + n\log n + (m-n)k)$ [81]	$\mathcal{NP}$-Hard [84] k-approx. [87]	$O(\log n(m + \log n))$ [84]
PIMST	$\mathcal{NP}$-Hard [86] k-approx. [87]	$O(mn\log n)$ [85]	$\mathcal{NP}$-Hard [84] k-approx. [87]	$O(\log n(m + \log n))$ [84]

A major research direction for future work is to study partial inverse bottleneck spanning tree problems, abbreviated as (*PInvBST*). Up to now, Tayyebi and Sepasian [126] and Dong et al. [36] have respectively studied (*PInvBST*) under the weighted l_1 norm and bottleneck Hamming distance and proposed corresponding polynomial time algorithms. In addition, designing better exact and approximation algorithms on (*PInvMST*) and studying its variants are also very meaningful works.

Part IV
Generalized Inverse Center Location Problems

Location problem is one of the classic problems in operations research, which is widely used in production and life, logistics and military. The inverse center location problems play a pivotal role, focusing on the adjustment of weights within a given network to optimize a specific vertex as the most optimal center. This study is particularly relevant in scenarios requiring the enhancement of network efficiency, such as emergency services, logistics, and transportation systems.

This part delves into two distinctive types of inverse center location problems, the inverse vertex obnoxious 1-center (IVO1C) location problem and the inverse quickest 1-center (IQ1C) location problem, providing a detailed analysis of their mathematical models, algorithmic solutions, and computational experiments.

In the problem (IVO1C), the objective is to adjust edge weights within predefined bounds to ensure a specified vertex becomes the most obnoxious center, thereby maximizing the minimum distance to all other vertices. We propose an algorithm with a time complexity of $O(n^3)$ for solving the problem (IVO1C) under the weighted l_∞ norm, where n represents the number of vertices in the graph. Additionally, a binary search method is introduced for the bottleneck Hamming distance case, demonstrating a time complexity of $O(n^2 \log n)$.

The problem (IQ1C) is divided into two subproblems by regarding a given point as a vertex quickest 1-center or an absolute quickest 1-center, which are denoted by (IVQ1C) and (IAQ1C), respectively. We initially analyze the optimality conditions under the weighted ∞-norm and introduce an effective modification strategy based on a greedy approach. Subsequently, we craft a binary search algorithm with a time complexity of $O(n^2 \log n)$ to precisely determine the optimal value. Under weighted l_1 norm, we study the inverse vertex and absolute quickest 1-centers in a unified framework by introducing a maximum transmission time balance (MTTB) problem. Then, we transform them into corresponding MTTB problems, which render two $O(n^3)$ algorithms. These research outcomes not only enrich the theory of inverse combinatorial optimization but also offer viable solutions to related practical issues.

We list the research results on the inverse 1-center problems (Table IV), where the meaning of symbols are listed below:

$\uparrow$: only increase the edge weight
$\downarrow$: only decrease the edge weight
$\updownarrow$: increase or decrease the edge weight

We highlight the importance of inverse optimization in addressing practical challenges in network design and management. We also suggest avenues for further exploration, such as extending the algorithm to handle more complex network topologies and varying norm conditions.

Table IV Research results on the inverse 1-center problems

Problem type	Norm	Time complexity	Reference
Inv-OA1C	(Bounded, weighted l_1 norm)	$O(m)$	[10]
Inv-OV1C	(Bounded, weighted l_1 norm)	$\uparrow$: $O(n)$	[13]
		$\downarrow$: $O(n^3)$	
		$\updownarrow$: $O(n^6)$	
	(Bounded, weighted l_∞ norm)	$O(n^3)$	Sect. 13.3.4
	(Bounded, bottleneck HD)	$O(n^2 \log n)$	Sect. 13.3.5
Inverse VQ1C ontrees	(Bounded, weighted l_∞ norm)	$O(n^2 \log n)$	Sect. 14.2.1 [110]
	(Bounded, weighted l_1 norm)	$O(n^3)$	Sect. 14.3.2 [109]
Inverse AQ1C ontrees	(Bounded, weighted l_∞ norm)	$O(n^2 \log n)$	Sect. 14.2.2 [111]
	(Bounded, weighted l_1 norm)	$O(n^3)$	[109]
VQ1C	Original problem	$O(n \log n)$	[110]
AQ1C	Original problem	$O(n \log n)$	[111]

Chapter 13
Inverse Vertex Obnoxious 1-Center Location Problems

Abstract In this chapter, we first give a brief introduction to the inverse center and median location problems under different norms. Then we mainly consider the inverse vertex obnoxious 1-center location problem (IVO1C) on a general graph G. We aim to adjust the edge weights satisfying upper and lower bounds with the least cost, so that a given vertex s becomes the obnoxious 1-center of graph G. We construct their mathematical models and prove some properties under the weighted l_∞ norm and bottleneck Hamming distance. We design an $O(n^3)$ time algorithm to solve the problem (IVO1C$_\infty$) by solving the transcendence point of the cost function in each iteration, where n is the number of vertices in the graph G. We also propose a binary search method for the problem (IVO1C$_{bH}$) with time complexity $O(n^2 \log n)$. Finally, we show some computational experiments to verify the effectiveness of the algorithms.

Keywords Inverse center location problems · Inverse median location problems · Inverse vertex obnoxious 1-center location problems · Intersection points of two polyline functions

13.1 Inverse Center Location Problems

13.1.1 A Mathematical Model of Inverse Center Location Problems

Given a graph $G = (V, E)$, for each edge $e \in E$, there is an associated lead time $l(e) > 0$ that represents the time to traverse the edge e. For any two vertices $v, t \in V$, let P_{vt} be the path connecting vertices v and t. Define $l(P_{vt}) = \sum_{e \in P_{vt}} l(e)$. Let $\mathcal{P}_{vt}$ denote the set of path connecting v and t and the lead time of v and t be defined as $\min_{P_{vt} \in \mathcal{P}_{vt}} l(P_{vt})$. The vertex 1-center problem aims at finding a vertex $v^* \in V$ such that the maximum lead time from v^* to other vertices is minimum, which can

Chen Wang and Xiucui Guan have equally contributed to this chapter.

X. Guan et al., *Inverse Combinatorial Optimization Problems*, Springer Optimization and Its Applications 225, https://doi.org/10.1007/978-3-031-91175-0_13

be mathematically described below:

$$\textbf{(V1C)} \quad \min_{v \in V} \max_{t \in V} \min_{P_{vt} \in \mathcal{P}_{vt}} l(P_{vt}). \tag{13.1}$$

If the location of the established facility can only be placed on the vertex, it is called the vertex location problem. If it can be placed on both vertices and points on edges, it is called the absolute location problem [9].

The corresponding inverse vertex 1-center problem is to modify the lead time vector with minimum cost under some norm such that a given vertex v^* becomes a vertex 1-center of G, which can be mathematically described below:

$$\begin{aligned}
&\min\ f(\tilde{l}) = \|l - \tilde{l}\| \\
&s.t.\ \max_{t \in V} \min_{P_{v^*t} \in \mathcal{P}_{v^*t}} l(P_{v^*t}) \le \max_{t \in V} \min_{P_{vt} \in \mathcal{P}_{vt}} l(P_{vt}), \forall v \in V, \\
&\quad\ \ \tilde{l}(e) = l(e) + x(e) - y(e), e \in E, \\
&\quad\ \ 0 \le x(e) \le u(e), e \in E, \\
&\quad\ \ 0 \le y(e) \le d(e), e \in E.
\end{aligned}$$

13.1.2 Research Results of Inverse Center Location Problems

The Inverse Vertex 1-Center Problem

Cai et al. [25] first proved that even if the original problem is solvable in polynomial time, the corresponding inverse vertex 1-center problem on general networks is $\mathcal{NP}$-hard. Yang and Zhang [134] transformed the inverse vertex 1-center problem on trees into a series of minimum cost flow problems and provided an algorithm with a time complexity of $O(n^2 \log n)$.

The Inverse Absolute 1-Center Problem

Alizadeh et al. [11] studied the inverse absolute 1-center problem on trees under weighted l_1 norm, and for the case where only the edge weights are increased, they provided an algorithm with a time complexity of $O(n \log n)$. Subsequently, Alizadeh and Burkard [9] balanced the heights of two subtrees to solve the inverse problem of the 1-center location problem and improved the time complexity to $O(n^2)$. Nguyen [98] provided optimality conditions for the vertex and absolute 1-center on circles and studied the inverse problems of these two types of centers under the unit l_1 norm, providing an algorithm with a time complexity of $O(n^2 \log n)$. Jana et al. [64] provided a linear-time algorithm for the inverse 1-center problem in trapezoidal graphs. Recently, Tayyebi and Sepasian [127] studied the inverse 1-center problem on trees under the convex piecewise linear cost function. They transformed the problem into a series of parameterized minimum cost flow problems and provided an approximate algorithm based on binary search.

The Inverse Obnoxious Center Problem

Alizadeh and Burkard [10] provided an $O(m)$ algorithm for the inverse problem of the obnoxious center location on general graphs under the l_1 norm. Alizadeh and Etemad [12] provided $O(n)$ algorithms for the inverse obnoxious center location problem on general networks under the l_1 norm and weighted Hamming distance. In particular, when the adjustment amount must be an integer, this type of problem can also be solved in linear time. Etemad and Alizadeh [40] designed an $O(n \log n)$ algorithm for the inverse obnoxious center location problem on circles under the weighted l_1 norm and improved the time complexity of the algorithm to $O(n)$ under unit cost.

13.2 Inverse Median Location Problems

13.2.1 A Mathematical Model of Inverse Median Location Problem

The vertex 1-median problem aims at finding a vertex $v^* \in V$ such that the sum of lead time from v^* to other vertices is minimum, which can be mathematically described below:

$$\textbf{(V1M)} \quad \min_{v \in V} \sum_{t \in V} \min_{P_{vt} \in \mathcal{P}_{vt}} l(P_{vt}). \tag{13.2}$$

The corresponding inverse vertex 1-median problem is to modify the lead time vector with minimum cost under some norm such that a given vertex v^* becomes a vertex 1-median of G, which can be mathematically described below:

$$\begin{aligned}
&\min f(\tilde{l}) = \|l - \tilde{l}\| \\
&s.t. \sum_{t \in V} \min_{P_{v^*t} \in \mathcal{P}_{v^*t}} l(P_{v^*t}) \le \sum_{t \in V} \min_{P_{vt} \in \mathcal{P}_{vt}} l(P_{vt}), \forall v \in V, \\
&\quad \tilde{l}(e) = l(e) + x(e) - y(e), e \in E, \\
&\quad 0 \le x(e) \le u(e), e \in E, \\
&\quad 0 \le y(e) \le d(e), e \in E.
\end{aligned}$$

13.2.2 Research Results of Inverse Median Location Problems

The Inverse 1-Median Problem

Burkard et al. [21] were the first to propose an $O(n \log n)$ algorithm for solving the inverse 1-median problem. Guan et al. [55] proved that the inverse 1-median problem on trees under weighted sum Hamming distance is $\mathcal{NP}$-hard and provided an $O(n)$ algorithm for solving the problem under weighted bottleneck Hamming distance. Nguyen [97] utilized the convexity of the objective function to provide an $O(n \log n)$ algorithm for the inverse 1-median problem on block graphs. Afrashteh et al. [1] considered the inverse 1-median problem on trees under maximum profit conditions. For the weighted l_1 norm and weighted l_∞ norm measures, they provided linear-time algorithms under the condition of only decreasing weights. Alizadeh and Afrashteh [8] studied the inverse 1-median problem with cost constraints, proving that the problem is $\mathcal{NP}$-hard when the objective function is the bottleneck Hamming distance and the cost constraint is the sum Hamming distance. Soltanpour et al. [117] studied the inverse 1-median problem with risk constraints under the l_1 norm and provided an $O(n \log n)$ algorithm, where the weight vector is a random variable. Nguyen et al. [100] introduced the concept of group 1-median and characterized the optimality conditions of this median. In addition, they designed an $O(n^2 \log n)$ algorithm for the inverse problem of group 1-median on trees under the weighted l_1 norm. Nguyen-Thu et al. [102] studied the inverse 1-median problem on trees under the weighted l_1 norm combined with the weighted l_∞ norm and provided an $O(n \log n)$ algorithm using the monotonicity of the objective function.

The Inverse Median Location Problem

Burkard et al. [18] offered an $O(n \log n)$ algorithm for solving the inverse Fermat-Weber problem based on a greedy strategy. Nguyen et al. [99] designed an $O(n^2 \log n)$ algorithm for the inverse problem of convex ordered median location on trees under weighted l_∞ and weighted bottleneck Hamming distances. Afrashteh et al. [2] provided an $O(n \log n)$ algorithm for solving the inverse selective median location problem. Taghikhani and Baroughi [120] considered the inverse median location problem on trees with fuzzy weights and provided an $O(n)$ algorithm. Furthermore, Taghikhani et al. [121] proved that the inverse p-median problem on general networks under the weighted l_1 norm, where both edge weights and vertex weights are fuzzy variables, is $\mathcal{NP}$-hard and proposed a new hybrid improved particle swarm optimization algorithm.

The Inverse Obnoxious p-Median Problem

In addition, Gassner [45] considered the solution of the inverse obnoxious 1-median problem, proving that the problem is $\mathcal{NP}$-hard in general networks and transformed the inverse problem on trees into a minimum cost circulation flow problem, providing an $O(n \log n)$ algorithm based on this. Nguyen and Vui [101] transformed the inverse obnoxious p-median problem on trees into $O(p^2)$ inverse obnoxious 2-median problems. For the l_1 norm case, the authors transformed the inverse problem into a linear programming problem under the l_∞ norm and provided an $O(n \log n)$ algorithm. Alizadeh et al. [7] provided $O(p^2 n^2 \log^2 n)$, $O(p^2 n)$ algorithms for the inverse obnoxious p-median problem on trees under weighted l_1 norm and l_∞ norm, respectively. Afrashteh et al. [3] explored the inverse obnoxious p-median problem under the sum Hamming distance in extended star graphs, providing an $O(p^2 n \log n)$ algorithm when the adjustment amount is an integer. Mohammadi et al. [96] studied the inverse obnoxious p-median problem on trees under the bottleneck Hamming distance with the adjustment amount as an integer constraint and designed a linear-time algorithm.

13.3 Inverse Vertex Obnoxious 1-Center Location Problems

We first construct a mathematical model of inverse vertex obnoxious 1-center location problems (IVO1C) under the weighted l_∞ norm and bottleneck Hamming distance, followed by proving some properties. Then we describe two polyline functions and propose an algorithm to solve the intersection point of the above functions. Next, we design an $O(n^3)$ time algorithm to solve the problem (IVO1C$_\infty$) under the weighted l_∞ norm. We also propose a binary search method in $O(n^2 \log n)$ time for the problem (IVO1C$_{bH}$) under the bottleneck Hamming distance. Finally, we perform some numerical experiments to compare the results of these algorithms.

13.3.1 An Introduction to the Problem (IVO1C)

Location problems have been studied in recent years. The aim of the classic location problem is to select the location of one or more facilities in a system such that these facilities serve existing customer in an optimal way. Facilities are mainly divided into two categories: One type is desirable facilities, to which all the customers want to be as close as possible, such as fire stations, supermarkets, schools, etc. The other type is obnoxious facilities [10], from which the existing customers want to be as far away as possible, such as airports, prisons, garbage disposal centers, etc.

Letting $G = (V, E, l)$ be a weighted undirected graph, we define $P_{s,t}$ as a path from s to t and $d_l(P_{s,t}) = \sum_{e \in P_{s,t}} l(e)$ as length of path $P_{s,t}$ under weight l. Let $\mathcal{P}_{s,t}$ be the set of all $s - t$ paths in G. Find a path $P^*_{s,t} \in \mathcal{P}_{s,t}$ in G such that $d_l(P^*_{s,t}) = \min_{P_{s,t} \in \mathcal{P}_{s,t}} d_l(P_{s,t})$. The path $P^*_{s,t}$ is said to be a shortest path under edge weight l.

There are two main types of obnoxious location problems [45]: The first is the obnoxious p-median location problem (OpM). Select p facilities $s_1, s_2, \cdots, s_p$ to maximize the sum of the distances between the demand vertices $v_i (i = 1, \cdots, n)$

and its nearest facility s_k under the weight l. The model is as follows:

$$(OpM) \quad \max \sum_{i=1}^{n} \min_{k=1,\cdots,p} d_l(P^*_{v_i,s_k})$$

The second is the obnoxious p-center location problem (OpC). Select p facilities $s_1, s_2, \cdots, s_p$ to maximize the minimum of the distance between the demand vertices $v_i (i = 1, \cdots, n)$ and its nearest facility s_k under the weight l. The model is as follows:

$$(OpC) \quad \max \min_{v_i \in V} \min_{k=1,\cdots,p} d_l(P^*_{v_i,s_k})$$

Especially, when $p = 1$, it is the obnoxious 1-center location problem $(O1C)$. Select one facility s such that the minimum distance from s to each vertex $v_i (i = 1, \cdots, n)$ is the maximum under the weight l. The model is below:

$$(O1C) \quad \max \min_{v_i \in V} d_l(P^*_{v_i,s})$$

Some researchers considered the inverse obnoxious p-center location problem $(IOpC)$. Adjust the edge weights from l to $\tilde{l}$, such that some given facilities $s_1, \cdots, s_p$ become the obnoxious p-center under the new weights $\tilde{l}$. We aim to adjust the edge weights with the least cost under a certain norm. Its model is below:

$$\begin{aligned} & \min \ \|l - \tilde{l}\| \\ (IOpC) \quad & s.t. \ s_1, \cdots, s_p \text{ are obnoxious } p\text{-center under } \tilde{l}, \\ & \underline{l} \le \tilde{l} \le \bar{l}. \end{aligned} \tag{13.3}$$

In 2013, Alizadeh and Burkard [10] considered the inverse absolute obnoxious 1-center location problem on general graphs and proposed an $O(m)$ time algorithm under weighted l_1 norm, where m is the number of edges in the graph. In 2018, Alizadeh and Etemad [13] studied the inverse vertex obnoxious 1-center location problem under weighted l_1 norm on general graphs. They divided the problem into three cases. When edge weights can only be increased, they proposed an $O(n)$ time algorithm. When the edge weight can only be reduced, they transformed the problem into a minimum edge coverage problem and solved it by finding a maximum matching in the graph in $O(n^3)$ time. When the edge weight can both be increased and decreased, the problem can be solved in $O(n^6)$ time.

13.3.2 A Mathematical Model of the Problem (IVO1C)

This section mainly studies the problem (IVO1C) under the weight l_∞ norm and the bottleneck Hamming distance. Let $G = (V, E)$ be an undirected connected graph, where $V = \{s, v_1, v_2, \cdots, v_{n-1}\}$ is the vertex set of G and $E = \{e_1, e_2, \cdots, e_m\}$ is the edge set. A given vertex s represents an established facility. The set $V_0 = \{v_1, v_2, \cdots, v_{n-1}\}$ represents the existing customers. Each edge $e \in E$ has a positive edge weight $l(e)$. Let $x(e)$ and $y(e)$ represent the increments and decrements of the edge e, respectively, which satisfy an upper bound constraints $x(e) \leq u^x(e)$ and $y(e) \leq u^y(e) \leq l(e)$, where $u^x(e), u^y(e) > 0$. Define $A(v)$ as the edge set adjacent to vertex v in the graph G. Let $D_l(v)$ be the smallest edge weight in $A(v)$. Obviously, $D_l(v) = \min_{e \in A(v)}\{l(e)\} = \min_{v' \in V \setminus \{v\}} d_l(P_{v,v'})$ is the shortest distance from vertex v to other vertices. The problem (IVO1C) aims to minimize the cost of adjusting edge weights from l to $\tilde{l}$, thereby making vertex s the obnoxious 1-center of graph G under the edge weight $\tilde{l}$.

The mathematical model of problem (IVO1C$_\infty$) under weight l_∞ norm can be stated as follows:

$$
\begin{aligned}
&\min \max_{e \in E}\{c(e)x(e), d(e)y(e)\} \\
(IVO1C_\infty)\ \ &s.t.\ \ \tilde{l}(e) = l(e) + x(e) - y(e),\ \forall e \in E, \\
&\qquad D_{\tilde{l}}(s) \geq D_{\tilde{l}}(v_i),\ \forall v_i \in V_0, \\
&\qquad 0 \leq x(e) \leq u^x(e),\ \forall e \in E, \\
&\qquad 0 \leq y(e) \leq u^y(e),\ \forall e \in E.
\end{aligned}
\tag{13.4}
$$

The mathematical model of problem (IVO1C$_{bH}$) under bottleneck Hamming distance can be stated as follows:

$$
\begin{aligned}
&\min \max_{e \in E}\{c(e)H(x(e), 0), d(e)H(y(e), 0)\} \\
&s.t.\ \ \tilde{l}(e) = l(e) + x(e) - y(e),\ \forall e \in E, \\
&\qquad D_{\tilde{l}}(s) \geq D_{\tilde{l}}(v_i),\ \forall v_i \in V_0, \\
&\qquad 0 \leq x(e) \leq u^x(e),\ \forall e \in E, \\
&\qquad 0 \leq y(e) \leq u^y(e),\ \forall e \in E.
\end{aligned}
\tag{13.5}
$$

13.3.3 An Algorithm to Solve the Intersection Point of Two Polyline Functions

In this subsection, we first present some important definitions. Then we describe the two polyline functions in detail and propose an algorithm to solve the intersection point of the above functions.

Definition 13.1 In a general graph G, we call the maximum value of the shortest adjacent edge weight of all vertices $v_i \in V_0$ as the longest obnoxious distance of the graph G, denoted as D_l^{max}; then $D_l^{max} = \max_{v_i \in V_0}\{D_l(v_i)\}$.

First, we consider the case which the edge weight adjustment has no upper and lower bound constraints. If we want the shortest adjacent edge weight of vertex s under the edge weight $\tilde{l}$ to be equal to $\rho(\geq D_l(s))$, letting $f_e(\rho) = c(e) \cdot (\rho - l(e))$, then the corresponding cost function is $F_s(\rho) = \max_{e \in A(s)} f_e(\rho)$.

Lemma 13.1 *The cost function* $F_s(\rho) = \max_{e \in A(s)} f_e(\rho)$ *corresponding to the shortest adjacent edge weight* ρ *of the vertex* s *is a strictly monotonically increasing function on* $\rho \in [D_l(s), +\infty)$.

Proof Let any ρ_1, ρ_2 satisfy $D_l(s) \leq \rho_1 < \rho_2$, and the cost function $F_s(\rho)$ is considered in two cases below.

Suppose that $F_s(\rho_1) and F_s(\rho_2)$ are both obtained on the edge $e^* \in A(s)$. Since $f_{e^*}(\rho) = c(e^*) \cdot (\rho - l(e^*))$ is a strictly monotonically increasing function, so $f_{e^*}(\rho_1) < f_{e^*}(\rho_2)$, that is, $F_s(\rho_1) = f_{e^*}(rho_1) < f_{e^*}(\rho_2) = F_s(\rho_2)$.

Suppose that $F_s(\rho_1) and F_s(\rho_2)$ are obtained on the edge $e^*, e^{**} \in A(s)(e^* \neq e^{**})$, respectively. Obviously, we have $f_{e^*}(\rho_2) < f_{e^{**}}(\rho_2)$. Since $f_{e^*}(\rho)$ is a strictly monotonically increasing function, we have $f_{e^*}(\rho_1) < f_{e^*}(\rho_2) < f_{e^{**}}(\rho_2)$, that is, $F_s(\rho_1) = f_{e^*}(\rho_1) < f_{e^{**}}(\rho_2) = F_s(\rho_2)$.

To sum up, $F_s(\rho)$ is a strictly monotonically increasing function on $\rho \in [D_l(s), +\infty)$: □

The following is an algorithm to solve the cost function $F_s(\rho)$ corresponding to vertex s on $\rho \in [D_l(s), D_l^{max}]$.

1. Sort the edge $e \in A(s)$ according to the value of $l(e)$ nondecreasingly by

$$D_l(s) = l(e_{j_1}) \leq l(e_{j_2}) \leq \cdots \leq l(e_{j_{|A(s)|}}). \tag{13.6}$$

   For each edge $e \in A(s)$, define the cost function $f_{e_{j_k}}(\rho) = c(e_{j_k}) \cdot (\rho - l(e_{j_k}))$.
2. Initialize $c_{max} := c(e_{j_1})$, $J^1(\rho) = f_{e_{j_1}}(\rho)$.
3. If $c(e_{j_2}) \leq c_{max}$, then the function $f_{e_{j_2}}(\rho)$ will never exceed the function $J^1(\rho)$ on $\rho \in [D_l(s), D_l^{max}]$, so let $J^2(\rho) = J^1(\rho)$. If $c(e_{j_2}) > c_{max}$, let $c_{max} = c(e_{j_2})$. At this time, it is possible that the two functions will have an intersection point, so we need to determine the function value when $\rho = D_l^{max}$. If $f_{e_{j_2}}(D_l^{max}) \leq J^1(D_l^{max})$, then $f_{e_{j_2}}(\rho)$ will never exceed the function $J^1(\rho)$

on $\rho \in [D_l(s), D_l^{max}]$, thus letting $J^2(\rho) = J^1(\rho)$. If $f_{e_{j_2}}(D_l^{max}) > J^1(D_l^{max})$, then the two functions must produce the intersection point. We define the abscissa of this intersection point as $\bar{\rho}_1$, then letting

$$J^2(\rho) = \begin{cases} J^1(\rho), & if\ D_l(s) \le \rho \le \bar{\rho}_1; \\ f_{e_{j_2}}(\rho), & if\ \bar{\rho}_1 < \rho \le D_l^{max}. \end{cases}$$

The first iteration ends and the next iteration begins.

4. Repeat the above steps until the end of the $|A(s)| - 1$ iteration, and finally get the cost function $F_s(\rho) = J^{|A(s)|}(\rho)$.

Based on the above analysis, Algorithm13.1 is given.

Algorithm 13.1 An algorithm for the polyline function $F_s(\rho)$ on $\rho \in [D_l(s), D_l^{max}]$. $(z_s, k_s, b_s) :=$ **Increase**$(D_l(s), D_l^{max}, A(s), l, c)$

Input: $D_l(s), D_l^{max}, A(s), l, c$.
Output: The turning point abscissa set z_s, the corresponding line segment slope set k_s, and the intercept set b_s.

Sort the edge $e \in A(s)$ as shown in (13.6).
Initialize $c_{max} := c(e_{j_1}), z_s := \{D_l(s)\}, k_s := \{c(e_{j_1})\}, b_s := \{-c(e_{j_1}) \cdot l(e_{j_1})\}$.
if $|A(s)| > 1$ **then**
  for $k = 1 : |A(s)| - 1$ **do**
    if $c(e_{j_{k+1}}) > c_{max}$ **then** $c_{max} := c(e_{j_{k+1}})$. $f_{e_{j_{k+1}}}(D_l^{max}) := c(e_{j_{k+1}}) \cdot (D_l^{max} - l(e_{j_{k+1}}))$. $J^k(D_l^{max}) := k_s^{|z_s|} \cdot D_l^{max} + b_s^{|z_s|}$.
      if $f_{e_{j_{k+1}}}(D_l^{max}) > J^k(D_l^{max})$ **then**
        if $|z_s| = 1$ **then** $\bar{\rho}_k := \frac{b_s^1 + c(e_{j_{k+1}}) \cdot l(e_{j_{k+1}})}{c(e_{j_{k+1}}) - k_s^1}$.
          if $\bar{\rho}_k = z_s^1$ **then** $k_s^1 := c(e_{j_{k+1}}), b_s^1 := -c(e_{j_{k+1}}) \cdot l(e_{j_{k+1}})$.
          else $z_s := z_s \cup \{\bar{\rho}_k\}, k_s := k_s \cup \{c(e_{j_{k+1}})\}, b_s := b_s \cup \{-c(e_{j_{k+1}}) \cdot l(e_{j_{k+1}})\}$.
          end if
        else
          for $p = 1 : |z_s| - 1$ **do**
            $\bar{\rho}_k := \frac{b_s^p + c(e_{j_{k+1}}) \cdot l(e_{j_{k+1}})}{c(e_{j_{k+1}}) - k_s^p}$.
            if $z_s^p < \bar{\rho}_k \le z_s^{p+1}$ **then** $z_s := \{z_s^1, \cdots, z_s^p\} \cup \{\bar{\rho}_k\}$. $k_s := \{k_s^1, \cdots, k_s^p\} \cup \{c(e_{j_{k+1}})\}$. $b_s := \{b_s^1, \cdots, b_s^p\} \cup \{-c(e_{j_{k+1}}) \cdot l(e_{j_{k+1}})\}$. **Break.**
            else if $p = |z_s| - 1$ **then** $\bar{\rho}_k := \frac{b_s^{|z_s|} + c(e_{j_{k+1}}) \cdot l(e_{j_{k+1}})}{c(e_{j_{k+1}}) - k_s^{|z_s|}}$. $z_s := z_s \cup \{\bar{\rho}_k\}, k_s := k_s \cup \{c(e_{j_{k+1}})\}, b_s := b_s \cup \{-c(e_{j_{k+1}}) \cdot l(e_{j_{k+1}})\}$.
            end if
          end for
        end if

```
            end if
        end if
    end for
end if
return (z_s, k_s, b_s).
```

Theorem 13.1 *The time complexity of Algorithm 13.1 is $O(n^2)$.*

For vertex $v_i \in V_0^>$, we also consider the case which the edge weight adjustment has no upper and lower bound constraints. If we want the shortest adjacent edge weight of vertex v_i under the edge weight $\tilde{l}$ to be equal to $\rho(\leq D_l^{max})$, we let $g_e(\rho) = d(e) \cdot (l(e) - \rho)$, and then the corresponding cost function is $G_{v_i}(\rho) = \min_{e \in A(v_i)} \max\{g_e(\rho), 0\}$. Similarly, we can conclude that

Lemma 13.2 *The cost function $G_{v_i}(\rho) = \min_{e \in A(v_i)} \max\{g_e(\rho), 0\}$ corresponding to the shortest adjacent edge weight ρ of the vertex $v_i \in V_0^>$ is a strictly monotonically decreasing function on $\rho \in [D_l(s), D_l(v_i)]$ and $G_{v_i}(\rho) = 0$ on $\rho \in [D_l(v_i), +\infty)$.*

Proof Let any ρ_1, ρ_2 satisfy $D_l(s) \leq \rho_1 < \rho_2 \leq D_l(v_i)$, and the cost function $G_{v_i}(\rho)$ is considered in two cases below:

Suppose that $G_{v_i}(\rho_1) and G_{v_i}(\rho_2)$ are both obtained on the edge $e^* \in A(v_i)$. Since $g_{e^*}(\rho) = d(e^*) \cdot (l(e^*) - \rho)$ is a strictly monotonically decreasing function, so $g_{e^*}(\rho_1) > g_{e^*}(\rho_2)$, that is, $G_{v_i}(\rho_1) = g_{e^*}(\rho_1) > g_{e^*}(\rho_2) = G_{v_i}(\rho_2)$.

Suppose that $G_{v_i}(\rho_1) and G_{v_i}(\rho_2)$ are obtained on the edge $e^*, e^{**} \in A(v_i)(e^* \neq e^{**})$, respectively. Obviously, we have $g_{e^*}(\rho_2) > g_{e^{**}}(\rho_2)$. Since $g_{e^*}(\rho)$ is a strictly monotonically decreasing function, we have $g_{e^*}(\rho_1) > g_{e^*}(\rho_2) > g_{e^{**}}(\rho_2)$, that is, $G_{v_i}(\rho_1) = g_{e^*}(\rho_1) > g_{e^{**}}(\rho_2) = G_{v_i}(\rho_2)$.

To sum up, $G_{v_i}(\rho)$ is a strictly monotonically decreasing function on $\rho \in [D_l(s), D_l(v_i)]$, and $G_{v_i}(\rho) = 0$ on $\rho \in [D_l(v_i), +\infty)$. □

Then design an algorithm to solve the cost function $G_{v_i}(\rho)$ corresponding to vertex v_i on $\rho \in [D_l(v_i), +\infty)$:

1. Sort the edge $e \in A(v_i)$ according to the value of $l(e)$ nondecreasingly by

$$l(e_{j_1}) \leq l(e_{j_2}) \leq \cdots \leq l(e_{j_{|A(v_i)|}}). \tag{13.7}$$

For each edge $e \in A(v_i)$, define the cost function $g_{e_{j_k}}(\rho) = d(e_{j_k}) \cdot (l(e_{j_k}) - \rho)$.

2. Initialize $d_{min} = d(e_{j_1})$,

$$J^1(\rho) = \begin{cases} g_{e_{j_1}}(\rho), & if\ D_l(s) \leq \rho < l(e_{j_1}); \\ 0, & if\ l(e_{j_1}) \leq \rho \leq D_l^{max}. \end{cases}$$

3. If $d(e_{j_2}) \geq d_{min}$, then the function $g_{e_{j_2}}(\rho)$ will never be lower than the function $J^1(\rho)$ on $\rho \in [D_l(s), D_l^{max}]$, so let $J^2(\rho) = J^1(\rho)$. If $d(e_{j_2}) < d_{min}$, let

$d_{min} = d(e_{j2})$. At this time, it is possible that the two functions will have an intersection point, so we need to determine the function value when $\rho = D_l(s)$. If $g_{e_{j_2}}(D_l(s)) \geq J^1(D_l(s))$, also the function $g_{e_{j_2}}(\rho)$ will never be lower than the function $J^1(\rho)$ on $\rho \in [D_l(s), D_l^{max}]$, thus letting $J^2(\rho) = J^1(\rho)$. If $g_{e_{j_2}}(D_l(s)) < J^1(D_l(s))$, then the two functions must produce the intersection point. We define the abscissa of this intersection point as $\underline{\rho}_1$ and then let

$$J^2(\rho) = \begin{cases} g_{e_{j_2}}(\rho), & \text{if } D_l(s) \leq \rho \leq \underline{\rho}_1; \\ J^1(\rho), & \text{if } \underline{\rho}_1 < \rho \leq l(e_{j_1}); \\ 0, & \text{if } l(e_{j_1}) < \rho \leq D_l^{max}. \end{cases}$$

The first iteration ends and the next iteration begins.

4. Repeat the above steps until the end of the $|A(v_i)| - 1$ iteration, and finally get the cost function $G_{v_i}(\rho) = J^{|A(v_i)|}(\rho)$.

Based on the above analysis, Algorithm13.2 is given.

Algorithm 13.2 An algorithm for the polyline function $G_{v_i}(\rho)$ on $\rho \in [D_l(s), D_l^{max}]$. $(z_{v_i}, k_{v_i}, b_{v_i}) :=$ **Decrease**$(D_l(s), D_l^{max}, A(v_i), l, d)$

Input: $D_l(s), D_l^{max}, A(v_i), l, d$.
Output: The turning point abscissa set z_{v_i}, the corresponding line segment slope set k_{v_i}, and the intercept set b_{v_i}.

Sort the edge $e \in A(v_i)$ as shown in (13.7).
Initialize $d_{min} := d(e_{j_1}), z_{v_i} := \{l(e_{j_1})\}, k_{v_i} := \{-d(e_{j_1})\}, b_{v_i} := \{d(e_{j_1}) \cdot l(e_{j_1})\}$.
if $|A(v_i)| > 1$ **then**
  for $k = 1 : |A(v_i)| - 1$ **do**
    if $d(e_{j_{k+1}}) < d_{min}$ **then** $d_{min} := d(e_{j_{k+1}})$. $g_{e_{j_{k+1}}}(D_l(s)) := d(e_{j_{k+1}}) \cdot (l(e_{j_{k+1}}) - D_l(s))$, $J^k(D_l(s)) := k_{v_i}^1 \cdot D_l(s) + b_{v_i}^1$.
      if $g_{e_{j_{k+1}}}(D_l(s)) < J^k(D_l(s))$ **then**
        if $|z_{v_i}| = 1$ **then** $\underline{\rho}_k := \frac{b_{v_i}^1 - d(e_{j_{k+1}}) \cdot l(e_{j_{k+1}})}{-d(e_{j_{k+1}}) - k_{v_i}^1}$.
          if $\underline{\rho}_k = z_{v_i}^1$ **then** $k_{v_i}^1 := -d(e_{j_{k+1}}), b_{v_i}^1 := d(e_{j_{k+1}}) \cdot l(e_{j_{k+1}})$.
          else $z_{v_i} := \{\underline{\rho}\} \cup z_{v_i}, k_{v_i} := \{-d(e_{j_{k+1}})\} \cup k_{v_i}, b_{v_i} := \{d(e_{j_{k+1}}) \cdot l(e_{j_{k+1}})\} \cup b_{v_i}$.
          end if
        else
          for $p = |z_{v_i}| : -1 : 2$ **do**
            $\underline{\rho}_k := \frac{b_{v_i}^p - d(e_{j_{k+1}}) \cdot l(e_{j_{k+1}})}{-d(e_{j_{k+1}}) - k_{v_i}^p}$.
            if $z_{v_i}^{p-1} \leq \underline{\rho}_k < z_{v_i}^p$ **then** $z_{v_i} := \{\underline{\rho}_k\} \cup \{z_{v_i}^p, \cdots, z_{v_i}^{|z_{v_i}|}\}$.

$k_{v_i} := \{-d(e_{j_{k+1}})\} \cup \{k_{v_i}^p, \cdots, k_{v_i}^{|k_{v_i}|}\}$. $b_{v_i} := \{d(e_{j_{k+1}}) \cdot l(e_{j_{k+1}})\} \cup \{b_{v_i}^p, \cdots, b_{v_i}^{|b_{v_i}|}\}$. **Break.**
else if $p = 2$ **then** $\underline{\rho}_k := \frac{b_s^1 - d(e_{j_{k+1}}) \cdot l(e_{j_{k+1}})}{-d(e_{j_{k+1}}) - k_{v_i}^1}$. $z_{v_i} := \underline{\rho}_k \cup z_{v_i}$. $k_{v_i} := \{-d(e_{j_{k+1}})\} \cup k_{v_i}$. $b_{v_i} := \{d(e_{j_{k+1}}) \cdot l(e_{j_{k+1}})\} \cup b_{v_i}$.
end if
end for
end if
end if
end if
end for
end if
return $(z_{v_i}, k_{v_i}, b_{v_i})$, $z_{v_i} := \{D_l(s)\} \cup z_{v_i}$, $k_{v_i} := k_{v_i} \cup \{0\}$, $b_{v_i} := b_{v_i} \cup \{0\}$.

Theorem 13.2 *The time complexity of Algorithm 13.2 is $O(n^2)$.*

From the above Lemmas 13.1 and 13.2, we have

Theorem 13.3 *If the cost function $F_s(\rho)$ and $G_{v_i}(\rho)$ have intersection, then there must be a unique intersection point.*

Next, we solve the intersection point $(\rho(v_i), C(v_i))$ based on the binary search method:

1. Reorder the values in the turning point set $z_s \cup z_{v_i}$ increasingly, and put them into the set H; then

$$D_l(s) = H_1 < H_2 \cdots < H_{N_1} (H_{N_1} \geq D_l(v_i)). \tag{13.8}$$

2. Initialize $a := 1, b := N_1$, and when $b - a > 1$, perform a while loop. Let $k := \lfloor \frac{a+b}{2} \rfloor$. If $F_s(H_k) < G_{v_i}(H_k)$, then the intersection point of $F_s(\rho)$ and $G_{v_i}(\rho)$ is on $(H_k, H_{N_1}]$, so let $a := k$; if $F_s(H_k) \geq G_{v_i}(H_k)$, then the intersection point of $F_s(\rho)$ and $G_{v_i}(\rho)$ is on $(H_1, H_k]$, so let $b := k$. Then repeat the above steps until $b - a = 1$, and get the interval $(H_a, H_b]$ where the intersection point is.
3. Define z_s^j as the abscissa of the jth vertex in the function $F_s(\rho)$, and $k_s^j and b_s^j$ are the slope and intercept of the corresponding line segment to the right of the jth vertex in the function $F_s(\rho)$, respectively. Let $z_{v_i}^j$ be the abscissa of the jth vertex in the function $G_{v_i}(\rho)$ and $k_{v_i}^j and b_{v_i}^j$ be the slope and intercept of the corresponding line segment to the right of the jth vertex in the function $G_{v_i}(\rho)$, respectively. Let $j_s := \arg\max_{1 \leq j \leq |z_s|} \{z_s^j \leq H_a\}$, $j_{v_i} := \arg\max_{1 \leq j \leq |z_{v_i}|} \{z_{v_i}^j \leq H_a\}$. Thus, the slope and intercept of the corresponding line segment of the function $F_s(\rho)$ on $(H_a, H_b]$ are $k_s^{j_s}, b_s^{j_s}$, respectively. The slope and intercept of the corresponding line segment of the function $G_{v_i}(\rho)$ on $(H_a, H_b]$ are

$k_{v_i}^{j_{v_i}} and b_{v_i}^{j_{v_i}}$, respectively. Finally, the intersection point $(\rho(v_i), C(v_i))$ can be obtained by the formula of the intersection point of the two lines.

Based on the above analysis, Algorithm13.3 is given.

Algorithm 13.3 An algorithm to find the intersection point $(\rho(v_i), C(v_i))$ of the function $F_s(\rho)$ and $G_{v_i}(\rho)$. $(\rho(v_i), C(v_i)) :=$ **Intersection**$(z_s, k_s, b_s, z_{v_i}, k_{v_i}, b_{v_i})$

Input: $z_s, k_s, b_s, z_{v_i}, k_{v_i}, b_{v_i}$.
Output: the intersection point $(\rho(v_i), C(v_i))$.
1: Reorder the values in $z_s \cup z_{v_i}$ into the set H, as shown in (13.8).
2: Initialize $a := 1, b := N_1$.
3: **while** $b - a > 1$ **do**
4: $\quad k := \lfloor \frac{a+b}{2} \rfloor$.
5: $\quad$**if** $F_s(H_k) < G_{v_i}(H_k)$ **then** $a := k$.
6: $\quad$**else** $b := k$.
7: $\quad$**end if**
8: **end while**
9: Let $j_s := \arg\max_{1 \le j \le |z_s|}\{z_s^j \le H_a\}$, $j_{v_i} := \arg\max_{1 \le j \le |z_{v_i}|}\{z_{v_i}^j \le H_a\}$.
10: Let $\rho(v_i) := \frac{b_{v_i}^{j_s} - b_s^{j_s}}{k_s^{j_s} - k_{v_i}^{j_s}}$, $C(v_i) := k_s^{j_s} \cdot \rho(v_i) + b_s^{j_s}$.
11: **return** $(\rho(v_i), C(v_i))$.

13.3.4 (Inv, VO1C, Bounded, Weighted l_∞ Norm)

In this subsection, we define the concept of the transcendence point and analyze some important properties. We design an $O(n^3)$ algorithm for the inverse vertex obnoxious 1-center location problem under the weighted l_∞ norm and prove its optimality.

For the problem $(IVO1C_\infty)$, we first need to judge whether the condition $D_l(s) \ge D_l^{max}$ holds. If so, then the vertex s is already an obnoxious 1-center of graph G, and no further adjustment is needed. Otherwise, since the adjustment amount of the increase or decrease of the edge weight corresponds to an upper bound $u^x(e), u^y(e)$, there may be some edges that cannot reach the required value. As a result, the cost functions $F_s(\rho)$ and $G_{v_i}(\rho)$ of the adjacent edges of vertex s and vertex v_i, $v_i \in V_0^>$ will change.

For edge $e \in E$, record the upper and lower bounds after edge weight adjustment as $l^x(e), l^y(e)$; then $l^x(e) = l(e) + u^x(e), l^y(e) = l(e) - u^y(e)$.

For vertex $v_i \in V_0^>$, put the edges satisfying $l^y(e) \le \rho$ in its adjacent edges into the set $A_{v_i}^{\le}(\rho)$, that is, $A_{v_i}^{\le}(\rho) = \{e \in A(v_i) | l^y(e) \le \rho\}$.

Let

$$\rho_{max} = \min\left\{\min_{e \in A(s)} \{l^x(e)\}, D_l^{max}\right\}. \tag{13.9}$$

Obviously, due to the limitation of upper and lower bounds, the edge weight of the shortest adjacent edge of the vertex s should not exceed $\min_{e\in A(s)}\{l^x(e)\}$. It follows from Lemma 13.1 that the cost function $F_s(\rho)$ corresponding to the shortest adjacent edge weight ρ of the vertex s on $\rho \in [D_l(s), \min_{e\in A(s)}\{l^x(e)\}]$ is a strictly monotonically increasing function.

Similarly, for $v_i \in V_0^>$, it is possible that the original minimum cost is obtained on the edge $e^* \in A(v_i)$, but when $\rho < l(e^*) - u^y(e^*)$, the edge weight of e^* cannot reduce to ρ, and then the minimum cost is obtained on the edge $e^{**} \in A(v_i)$, $e^* \neq e^{**}$. Therefore, the cost function $G_{\overline{v}_i}^{\leq}(\rho) = \min_{e\in A_{\overline{v}_i}^{\leq}(rho)} \max\{g_e(\rho), 0\}$ may be a piecewise function with breakpoints, and breakpoints can only exist in the set $\{l^y(e)|e \in A(v_i)\}$.

Lemma 13.3 *The cost function* $G_{\overline{v}_i}^{\leq}(\rho) = \min_{e\in A_{\overline{v}_i}^{\leq}(rho)} \max\{g_e(\rho), 0\}$ *corresponding to the shortest adjacent edge weight* ρ *of the vertex* $v_i \in V_0^>$ *is a strictly monotonically decreasing function on* $\rho \in [\min_{e\in A(v_i)}\{l^y(e)\}, D_l(v_i)]$, *and* $G_{\overline{v}_i}^{\leq}(\rho) = 0$ *on* $\rho \in [D_l(v_i), +\infty)$.

Proof Obviously, when $\rho \geq D_l(v_i)$, $G_{\overline{v}_i}^{\leq}(\rho) = 0$. Next, we prove that the cost function $G_{\overline{v}_i}^{\leq}(\rho)$ corresponding to the shortest adjacent edge weight ρ of the vertex $v_i \in V_0^>$ is a strictly monotonically decreasing function on $\rho \in [\min_{e\in A(v_i)}\{l^y(e)\}, D_l(v_i)]$.

Putting $l^y(e), e \in A(v_i)$ that satisfy $\min_{e\in A(v_i)}\{l^y(e)\} < l^y(e) < D_l(v_i)$ and $\min_{e\in A(v_i)}\{l^y(e)\}, D_l(v_i)$ into the set LY in an ascending order, then

$$\min_{e\in A(v_i)}\{l^y(e)\} = l_1^y < l_2^y < \cdots < l_{N_2}^y = D_l(v_i).$$

For any $\min_{e\in A(v_i)}\{l^y(e)\} \leq \rho_1 < \rho_2 = D_l(v_i)$, obviously, there is $G_{\overline{v}_i}^{\leq}(\rho_1) > 0 = G_{\overline{v}_i}^{\leq}(\rho_2)$. For any $\min_{e\in A(v_i)}\{l^y(e)\} \leq \rho_1 < \rho_2 < D_l(v_i)$, the following can be divided into three categories:

Case 1. $|LY| = 2$.
Then $LY = \{\min_{e\in A(v_i)}\{l^y(e)\}, D_l(v_i)\}$. At this time, $G_{\overline{v}_i}^{\leq}(\rho)$ is continuous on $\rho \in [\min_{e\in A(v_i)}\{l^y(e)\}, D_l(v_i))$, so $G_{\overline{v}_i}^{\leq}(\rho) = G_{v_i}(\rho)$; according to Lemma 13.2, we have $G_{\overline{v}_i}^{\leq}(\rho_1) > G_{\overline{v}_i}^{\leq}(\rho_2)$.

Case 2. $|LY| = 3$. There are two subcases:

2.1. While $\rho_1, \rho_2 \in [l_1^y, l_2^y)$ or $\rho_1, \rho_2 \in [l_2^y, l_3^y)$, then the edge set $A_{\overline{v}_i}^{\leq}(\rho_1)$ is the same as the edge set $A_{\overline{v}_i}^{\leq}(\rho_2)$; similar to the proof of Lemma13.2, it can be seen that $G_{\overline{v}_i}^{\leq}(\rho_1) > G_{\overline{v}_i}^{\leq}(\rho_2)$.

2.2. While $\rho_1 \in [l_1^y, l_2^y)$ and $\rho_2 \in [l_2^y, l_3^y)$, it can be divided into two situations:

2.2.1. If the cost function $G_{\overline{v}_i}^{\leq}(\rho)$ is continuous on $\rho \in [l_1^y, l_3^y)$, similar to the proof of Lemma 13.2, $G_{\overline{v}_i}^{\leq}(\rho)$ is a strictly monotonically decreasing function on $\rho \in [l_1^y, l_2^y)$ and $\rho \in [l_2^y, l_3^y)$, respectively. And since the cost function $G_{\overline{v}_i}^{\leq}(\rho)$ is continuous on $\rho \in [l_1^y, l_3^y)$, then $G_{\overline{v}_i}^{\leq}(\rho_1) > G_{\overline{v}_i}^{\leq}(l_2^y) \geq G_{\overline{v}_i}^{\leq}(\rho_2)$.

2.2.2. If the cost function $G^{\leq}_{\overline{v}_i}(\rho)$ is discontinuous on $\rho \in [l^y_1, l^y_3)$, suppose the cost function $G^{\leq}_{\overline{v}_i}(\rho)$ on $\rho \in [l^y_2 - \epsilon, l^y_2)$ is obtained on the edge $e^* \in A^{\leq}_{\overline{v}_i}(\rho_1)$. Since the cost function $G^{\leq}_{\overline{v}_i}(\rho)$ has a breakpoint at $\rho = l^y_2$, there must be an edge $e^{**} \in A^{\leq}_{\overline{v}_i}(\rho_2) \setminus A^{\leq}_{\overline{v}_i}(\rho_1)$ such that $\max\{g_{e^*}(l^y_2), 0\} > \max\{g_{e^{**}}(l^y_2), 0\}$. Hence, the cost function $G^{\leq}_{\overline{v}_i}(\rho)$ is a strictly monotonically decreasing function on $\rho \in [l^y_1, l^y_2)$ and $\rho \in [l^y_2, l^y_3)$, respectively. So $G^{\leq}_{\overline{v}_i}(\rho_1) > \max\{g_{e^*}(l^y_2), 0\} > \max\{g_{e^{**}}(l^y_2), 0\} \geq G^{\leq}_{\overline{v}_i}(\rho_2)$. Thus, in this situation, we have $G^{\leq}_{\overline{v}_i}(\rho_1) > G^{\leq}_{\overline{v}_i}(\rho_2)$.

Case 3. $|LY| > 3$. There are three subcases.

3.1. While $\rho_1, \rho_2 \in [l^y_j, l^y_{j+1})$, similar to case 2.1, we have $G^{\leq}_{\overline{v}_i}(\rho_1) > G^{\leq}_{\overline{v}_i}(\rho_2)$.

3.2. While $\rho_1 \in [l^y_j, l^y_{j+1})$ and $\rho_2 \in [l^y_{j+1}, l^y_{j+2})$, it can be seen from case 2.2 that $G^{\leq}_{\overline{v}_i}(\rho_1) > G^{\leq}_{\overline{v}_i}(\rho_2)$.

3.3. While $\rho_1 \in [l^y_j, l^y_{j+1})$ and $\rho_2 \in [l^y_k, l^y_{k+1}), k > j+1$, according to case 3.2, we have $G^{\leq}_{\overline{v}_i}(\rho_1) > G^{\leq}_{\overline{v}_i}(l^y_{j+1}) > G^{\leq}_{\overline{v}_i}(l^y_{j+2}) > \cdots > G^{\leq}_{\overline{v}_i}(l^y_k) \geq G^{\leq}_{\overline{v}_i}(\rho_2)$. Thus, in this case, we have $G^{\leq}_{\overline{v}_i}(\rho_1) > G^{\leq}_{\overline{v}_i}(\rho_2)$.

As a result, the cost function $G^{\leq}_{\overline{v}_i}(\rho)$ corresponding to the shortest adjacent edge weight ρ of the vertex $v_i \in V^{>}_0$ is a strictly monotonically decreasing function on $\rho \in [\min_{e \in A(v_i)}\{l^y(e)\}, D_l(v_i)]$, and $G^{\leq}_{\overline{v}_i}(\rho) = 0$ on $\rho \in [D_l(v_i), +\infty)$. □

Definition 13.2 If $F_s(\rho^1) \geq G^{\leq}_{\overline{v}_i}(\rho^1)$ is satisfied and does not exist $\rho^2 < \rho^1$, such that $F_s(\rho^2) \geq G^{\leq}_{\overline{v}_i}(\rho^2)$, then we call the point $(\rho^1, F_s(\rho^1))$ a transcendence point.

Based on the above analysis, the cost function $G^{\leq}_{\overline{v}_i}(\rho)$ may be a piecewise function with breakpoints. Therefore, we need to solve the transcendence point of the cost function $F_s(\rho)$ corresponding to vertex s and the cost function $G^{\leq}_{\overline{v}_i}(\rho)$ corresponding to vertex $v_i \in V^{>}_0$.

Now, supposing that we have determined that the transcendence point is on $(a, b]$ and there is no value in $\{l^y(e), e \in A(v_i)\}$ on (a, b), then we have $G^{\leq}_{\overline{v}_i}(b) \leq F_s(b)$, and the transcendence point will be specifically solved below:

1. Calculate the corresponding cost $F_s(b)$ when the shortest adjacent edge weight of vertex s is b.
2. For edge $e \in A^{\leq}_{\overline{v}_i}(b)$, sort the values of $G_e(b) = \max\{g_e(b), 0\}$ nondecreasingly by

$$G_{e_{q_1}}(b) \leq G_{e_{q_2}}(b) \leq \cdots \leq G_{e_{q_\tau}}(b) \tag{13.10}$$

   where $\tau = |A^{\leq}_{\overline{v}_i}(b)|$.
3. While $G_{e_{q_1}}(b) = F_s(b)$, the transcendence point is $(b, F_s(b))$. Otherwise, we consider two cases. If $G_e(b) < F_s(b)$ is satisfied for any $e \in A^{\leq}_{\overline{v}_i}(b)$, then define the edge set $A^{<}_{\overline{v}_i}(b) = \{e \in A^{\leq}_{\overline{v}_i}(b) | l^y(e) < b\}$. If there is an edge e_{q_k} such that

$G_{e_{q_k}}(b) \geq F_s(b)$ ($k \geq 2$ at this time; otherwise, it contradicts $G_{e_{q_1}}(b) < F_s(b)$), then define the edge set $A_{v_i}^{<}(b) = \{e \in \{e_{q_1}, \cdots, e_{q_{k-1}}\} | l^y(e) < b\}$.

4. While $A_{v_i}^{<}(b)$ is an empty set, the transcendence point is $(b, F_s(b))$; otherwise, the function $F_s(\rho)$ and $G_{v_i}^{\leq}(\rho)$ must have a unique intersection point on (a, b).
5. Call Algorithms 13.2 and 13.3 successively to solve the intersection point $(\rho(v_i), C(v_i))$ of the two cost functions, which is the transcendence point at this time; the algorithm is summarized as follows:

Algorithm 13.4 An algorithm to solve the transcendence point $(\rho(v_i), C(v_i))$ on $(a, b]$: $(\rho(v_i), C(v_i))$:= **Trans − Point**$(z_s, k_s, b_s, a, b, l^y, d, l, D_l(s), \rho_{max})$

Input: $z_s, k_s, b_s, a, b, l^y, d, l, D_l(s), \rho_{max}$.
Output: the transcendence point $(\rho(v_i), C(v_i))$.

$A_{v_i}^{\leq}(b) := \{e \in A(v_i) | l^y(e) \leq b\}$, $G_e(b) := \max\{g_e(b), 0\}$, $e \in A_{v_i}^{\leq}(b)$.
Sort edge $e \in A_{v_i}^{\leq}(b)$ as shown by (13.10).
if $G_{e_{q_1}}(b) = F_s(b)$ **then** $\rho(v_i) := b$, $C(v_i) := F_s(b)$.
else
  if $G_{e_{q_\tau}}(b) < F_s(b)$ **then** $A_{v_i}^{<}(b) := \{e \in A_{v_i}^{\leq}(b) | l^y(e) < b\}$.
  else $\hat{k} := \arg\min_{1 \leq k \leq \tau}\{G_{e_{q_k}}(b) \geq F_s(b)\}$.
    $A_{v_i}^{<}(b) := \{e \in \{e_{q_1}, \cdots, e_{q_{\hat{k}-1}}\} | l^y(e) < b\}$.
  end if
  if $A_{v_i}^{<}(b) = \emptyset$ **then return** $\rho(v_i) := b$, $C(v_i) := F_s(b)$.
  else
    Call Alg. 13.2 $(z_{v_i}^{<}, k_{v_i}^{<}, b_{v_i}^{<})$=**Decrease**$(D_l(s), \rho_{max}, A_{v_i}^{<}(b), l, d)$ to solve the cost function $G_{v_i}^{<}(\rho)$ corresponding to vertex v_i on $\rho \in [D_l(s), \rho_{max}]$.
    Call Alg. 13.3 $(\rho(v_i), C(v_i))$=**Intersection**$(z_s, k_s, b_s, z_{v_i}^{<}, k_{v_i}^{<}, b_{v_i}^{<})$ to solve the intersection point of the functions $F_s(\rho)$ and $G_{v_i}^{<}(\rho)$.
  end if
end if
return $(\rho(v_i), C(v_i))$.

Theorem 13.4 *If there is a vertex $v_i \in V_0^{>}$ that satisfies $\min_{e \in A(v_i)}\{l^y(e)\} > \min_{e \in A(s)}\{l^x(e)\}$, then the problem (IVO1C$_\infty$) is infeasible.*

Proof Due to $\min_{e \in A(v_i)}\{l^y(e)\} > \min_{e \in A(s)}\{l^x(e)\}$, no matter how much the edge weight is adjusted, the shortest adjacent edge weight $D_{\bar{l}}(s)$ of vertex s is always smaller than the shortest adjacent edge weight $D_{\bar{l}}(v_i)$ of vertex v_i. At this time, $D_{\bar{l}}(s) \geq D_{\bar{l}}(v_j)$, $v_j \in V_0$ does not hold, so the problem (IVO1C$_\infty$) is infeasible. □

Theorem 13.5 *Adjusting the edge weight at the optimal value C^* in the problem (IVO1C$_\infty$), then the shortest adjacent edge weight of vertex s cannot be greater than $\rho_{max} = \min\left\{\min_{e \in A(s)}\{l^x(e)\}, D_l^{max}\right\}$.*

Proof While $\min_{e \in A(s)}\{l^x(e)\} \leq D_l^{max}$, since the upper bound constraint of increasing the edge weight, it is obvious that the shortest adjacent edge weight of vertex s will not exceed $\rho_{max} = \min_{e \in A(s)}\{l^x(e)\}$. While $\min_{e \in A(s)}\{l^x(e)\} > D_l^{max}$, if the edge weights of $e \in \{A(s)|l(e) < D_l^{max}\}$ are increased to D_l^{max} and other edge weights remain unchanged, then vertex s will become an obnoxious 1-center, so it is a feasible solution. The cost is $F_s(D_l^{max})$ at this time. Next, we prove it by contradiction. Assume that the edge weight is adjusted at the optimal value C^*. Then the shortest adjacent edge weight of vertex s is $\rho^* > D_l^{max}$, according to Lemma 13.1. It can be seen that $C^* \geq F_s(\rho^*) > F_s(D_l^{max})$, which contradicts the optimal value of C^*. In summary, the shortest adjacent edge weight of vertex s cannot be greater than $\rho_{max} = \min\left\{\min_{e \in A(s)}\{l^x(e)\}, D_l^{max}\right\}$. □

When the shortest adjacent edge weight of vertex s is ρ_{max}, we can directly call Algorithm 13.1 to solve the corresponding cost $F_s(\rho_{max})$. When the shortest adjacent edge weight of vertex v_i is ρ_{max}, the corresponding cost is $G_{\overline{v}_i}^{\leq}(\rho_{max}) = \min_{e \in A_{\overline{v}_i}^{\leq}(\rho_{max})} \max\{g_e(\rho_{max}), 0\}$. If $G_{\overline{v}_i}^{\leq}(\rho_{max}) > F_s(\rho_{max})$ is satisfied, let $(\rho(v_i), C(v_i)) = (\rho_{max}, G_{\overline{v}_i}^{\leq}(\rho_{max}))$. Otherwise, it means that there must be a transcendental point on $\rho \in (D_l(s), \rho_{max}]$, and the following is divided into three cases to solve the transcendence point:

1. If for vertex $v_i \in V_0^>$ $\max_{e \in A(v_i)}\{l^y(e)\} \leq D_l(s)$ is satisfied, then $G_{\overline{v}_i}^{\leq}(\rho) = G_{v_i}(\rho)$ on $\rho \in [D_l(s), \rho_{max}]$. At this time, directly call Algorithm 13.2 to solve the cost function $G_{\overline{v}_i}^{\leq}(\rho)$ corresponding to vertex v_i on $\rho \in [D_l(s), \rho_{max}]$. Obviously, $F_s(D_l(s)) = 0 < G_{v_i}(D_l(s))$; otherwise, it will contradict $D_l(v_i) > D_l(s)$. It follows from Lemmas 13.1 and 13.3 that the cost function $F_s(\rho)$ is a strictly monotonically increasing function on $\rho \in [D_l(s), \min\{\rho_{max}, D_l(v_i)\}]$, and the cost function $G_{\overline{v}_i}^{\leq}(\rho)$ is a strictly monotonically decreasing function on $\rho \in [D_l(s), \min\{\rho_{max}, D_l(v_i)\}]$. So there must be a unique intersection point between the function $F_s(\rho)$ and $G_{\overline{v}_i}^{\leq}(\rho)$. Then directly call Algorithm 13.3 to find the intersection point $(\rho(v_i), C(v_i))$ of the two functions, which is the transcendence point at this time.
2. If the vertex $v_i \in V_0^>$ satisfies $\max_{e \in A(v_i)}\{l^y(e)\} > D_l(s)$ and $\{e \in A(v_i)|D_l(s) < l^y(e) \leq \rho_{max}\}$ is an empty set, then the edge set $A(v_i)$ must be composed of two parts: $A_{\overline{v}_i}^{\leq}(D_l(s))$ and $\{e \in A(v_i)|l^y(e) > \rho_{max}\}$. From Theorem 13.5, we just need to consider the edge set $A_{\overline{v}_i}^{\leq}(D_l(s))$ and use it as an input parameter to call Algorithm 13.2 to solve the cost function $G_{\overline{v}_i}^{\leq}(\rho)$ corresponding to the vertex v_i on $\rho \in [D_l(s), \rho_{max}]$. At this time, the cost function $G_{\overline{v}_i}^{\leq}(\rho)$ is continuous. Similar to case 1, directly call Algorithm 13.3 to find the unique intersection point $(\rho(v_i), C(v_i))$ of the functions $F_s(\rho)$ and $G_{\overline{v}_i}^{\leq}(\rho)$, which is the transcendence point.
3. If $\{e \in A(v_i)|D_l(s) < l^y(e) \leq \rho_{max}\}$ is not an empty set for vertex $v_i \in V_0^>$, then the cost function $G_{\overline{v}_i}^{\leq}(\rho)$ on $\rho \in [D_l(s), \rho_{max}]$ may be a piecewise linear function with breakpoints. Next, we need to determine the interval where the transcendence point is located and ensure that the value in $\{l^y(e), e \in A(v_i)\}$ does

not exist on the open interval. Reorder the edge weight $\{l^y(e)|e \in A(v_i), D_l(s) < l^y(e) \leq \rho_{max}\}$ increasingly by

$$l_1^y < l_2^y < \cdots < l_{N_3}^y. \tag{13.11}$$

Calculate the corresponding cost $F_s(l_k^y)$ and $G_{\bar{v}_i}^{\leq}(l_k^y)$ when the shortest adjacent edge weight of vertices s and v_i is l_k^y, respectively. From Lemmas 13.1 and 13.3, we need to find the smallest k^* so that $G_{\bar{v}_i}^{\leq}(l_{k^*}^y) \leq F_s(l_{k^*}^y)$ is satisfied. If we cannot find $1 \leq k^* \leq N_3$ such that $G_{\bar{v}_i}^{\leq}(l_{k^*}^y) \leq F_s(l_{k^*}^y)$ is satisfied, then there must be $l_{N_3}^y < \rho_{max}$. Otherwise, it contradicts $G_{\bar{v}_i}^{\leq}(\rho_{max}) \leq F_s(\rho_{max})$. Therefore, the transcendence point is on $(l_{N_3}^y, \rho_{max}]$. If we find $k^* = 1$, then the transcendence point is on $(D_l(s), l_1^y]$; otherwise, the transcendence point is on $(l_{k^*-1}^y, l_{k^*}^y]$. After determining the interval where the transcendence point is located, call Algorithm 13.4 to solve the transcendence point $(\rho(v_i), C(v_i))$ on this interval.

Theorem 13.6 *The optimal value of (IVO1C$_\infty$) is* $C^* = \max_{v_i \in V_0^>}\{C(v_i)\}$.

Proof For any $v_i \in V_0^>$, the obtained $(\rho(v_i), C(v_i))$ must belong to one of the following two cases:

1. While $G_{\bar{v}_i}^{\leq}(\rho_{max}) > F_s(\rho_{max})$, if $\rho_{max} = D_l^{max}(< \min_{e \in A(s)}\{l^x(e)\})$, then the cost function $G_{\bar{v}_i}^{\leq}(\rho_{max}) = 0$; obviously, $F_s(\rho_{max}) > 0$; it contradicts the premise $G_{\bar{v}_i}^{\leq}(\rho_{max}) > F_s(\rho_{max})$, so $\rho_{max} = \min_{e \in A(s)}\{l^x(e)\}(< D_l^{max})$ can be obtained. Then the shortest adjacent edge weight of vertex s can no longer be increased. According to Lemmas 13.1 and 13.3, $C(v_i) = G_{\bar{v}_i}^{\leq}(\rho_{max})$ is the minimum cost that can satisfy $D_{\tilde{l}}(s) \geq D_{\tilde{l}}(v_i)$, so if we want the vertex s to become the obnoxious 1-center of the graph G , then it must satisfy $C^* \geq C(v_i)$.
2. If $(\rho(v_i), C(v_i))$ is a transcendental point, we prove by contradiction. Assuming that the edge weights are adjusted at the optimal value $C^* < C(v_i)$, then the vertex s can become an obnoxious 1-center. It follows from Lemma 13.1 that the cost function $F_s(\rho)$ is a strictly monotonically increasing function on $\rho \in [D_l(s), \min\{\rho_{max}, D_l(v_i)\}]$; then we get $D_{l^*}(s) < \rho(v_i)$. From Lemma 13.3, the cost function $G_{\bar{v}_i}^{\leq}(\rho)$ on $\rho \in [\min_{e \in A(v_i)}\{l^y(e)\}, \min\{\rho_{max}, D_l(v_i)\}]$ is a strictly monotonically decreasing function. Then we have $D_{l^*}(v_i) > \rho(v_i) > D_{l^*}(s)$. Obviously, $D_{l^*}(s) \geq D_{l^*}(v_j), v_j \in V_0$ does not hold, which contradicts the original assumption. Hence, if the vertex s becomes the obnoxious 1-center of the graph G, then it needs to satisfy $C^* \geq C(v_i)$.

Therefore, we have $C^* \geq \max_{v_i \in V_0^>}\{C(v_i)\}$, and it is feasible to solve problem (IVO1C$_\infty$) at the cost $\max_{v_i \in V_0^>}\{C(v_i)\}$, which is just the optimal value. □

For the problem $(IVO1C_\infty)$, when the given vertex s is not the obnoxious 1-center of the graph G, we first need to judge whether the problem is feasible. If it is feasible, according to the above analysis, for each $v_i \in V_0^>$, solve $(\rho(v_i), C(v_i))$, and take the largest $C(v_i)$ as the optimal value of the problem $(IVO1C_\infty)$. The algorithm is summarized as follows:

Algorithm 13.5 An algorithm to solve the problem $(\mathbf{IVO1C}_\infty)$:
$(C^*, x^*, y^*) = \mathbf{IVO1C}_\infty(G, l, c, d, u^x, u^y)$

Input: $G(V, E), l, c, d, u^x, u^y$.
Output: The optimal value C^* and the optimal solution x^*, y^*.

Initialize $C^* := 0,\ x^* := 0,\ y^* := 0$.
$A(v) := \{e | e = (v, t) \in E\}$, $D_l(v) := \min_{e \in A(v)}\{l(e)\}$, $v \in V$.
$D_l^{max} := \max_{v_i \in V_0}\{D_l(v_i)\}$.
if $D_l(s) \geq D_l^{max}$ **then**
  Output: "The vertex s is already the obnoxious 1-center of the graph G and no further adjustment is needed."
  return (C^*, x^*, y^*).
else
  Let $V_0^> := \{v_i \in V_0 | D_l(v_i) > D_l(s)\}$.
  Let $l^x(e) := l(e) + u^x(e), l^y(e) := l(e) - u^y(e), e \in E$.
  if $\{v_i \in V_0^> \,|\, \min_{e \in A(v_i)}\{l^y(e)\} > \min_{e \in A(s)}\{l^x(e)\}\} \neq \emptyset$ **then**
    Output: " The instance is infeasible."
    Return (C^*, x^*, y^*) with $C^* := +\infty,\ x^* := \emptyset,\ y^* := \emptyset$.
  end if
  Let $\rho_{max} := \min\left\{\min_{e \in A(s)}\{l^x(e)\}, D_l^{max}\right\}$.
  Let $l_v := l(A(v)), v \in V, c_s := c(A(s)), d_{v_i} := d(A(v_i)), v_i \in V_0$.
  Call Algorithm 13.1(z_s, k_s, b_s)=**Increase**$(D_l(s), \rho_{max}, A(s), l_s, c_s)$ to solve the cost function $F_s(\rho)$ corresponding to vertex s on $\rho \in [D_l(s), \rho_{max}]$.
  for $v_i \in V_0^>$ **do**
    Let $A_{v_i}^{\leq}(\rho) := \{e \in A(v_i) | l^y(e) \leq \rho\}$.
    Let $F_s(\rho_{max}) := k_s^{|z_s|} \cdot \rho_{max} + b_s^{|z_s|}$.
    Let $G_{v_i}^{\leq}(\rho_{max}) := \min\limits_{e \in A_{v_i}^{\leq}(\rho max)} \max\{g_e(\rho_{max}), 0\}$.
    if $G_{v_i}^{\leq}(\rho_{max}) > F_s(\rho_{max})$ **then**
      Let $\rho(v_i) := \rho_{max}, C(v_i) := G_{v_i}^{\leq}(\rho_{max})$.
    else if $\max_{e \in A(v_i)}\{l^y(e)\} \leq D_l(s)$ **then**
      Call Alg. 13.2$(z_{v_i}^{\leq}, k_{v_i}^{\leq}, b_{v_i}^{\leq})$:=**Decrease**$(D_l(s), \rho_{max}, A(v_i), l_{v_i}, d_{v_i})$ to solve the function $G_{v_i}^{\leq}(\rho)$ corresponding to vertex v_i on $\rho \in [D_l(s), \rho_{max}]$.
      Call Alg. 13.3 $(\rho(v_i), C(v_i))$:=**Intersection**$(z_s, k_s, b_s, z_{v_i}^{\leq}, k_{v_i}^{\leq}, b_{v_i}^{\leq})$ to solve the intersection point of the functions $F_s(\rho)$ and $G_{v_i}^{\leq}(\rho)$.
    else if $\{e \in A(v_i) | D_l(s) < l^y(e) \leq \rho_{max}\} = \emptyset$ **then**
      Call Alg. 13.2$(z_{v_i}^{\leq}, k_{v_i}^{\leq}, b_{v_i}^{\leq})$:=**Decrease**$(D_l(s), \rho_{max}, A_{v_i}^{\leq}(D_l(s)), l_{v_i}, d_{v_i})$ to solve the cost function $G_{v_i}^{\leq}(\rho)$ corresponding to vertex v_i on $\rho \in [D_l(s), \rho_{max}]$.
      Call Alg. 13.3 $(\rho(v_i), C(v_i))$:=**Intersection**$(z_s, k_s, b_s, z_{v_i}^{\leq}, k_{v_i}^{\leq}, b_{v_i}^{\leq})$ to solve the intersection point of the functions $F_s(\rho)$ and $G_{v_i}^{\leq}(\rho)$.

else reorder the values in $\{l^y(e)\}, e \in \{A(v_i)|D_l(s) < l^y(e) \leq \rho_{max}\}$, as shown in (13.11). Set $k^* := 0$.
for $k = 1 : N_3$ **do**
Let $j_s := \arg\max_{1\leq j\leq |z_s|}\{z_s^j \leq l_k^y\}$, $F_s(l_k^y) := k_s^{j_s} \cdot l_k^y + b_s^{j_s}$.
Let $G_{\bar{v}_i}^{\leq}(l_k^y) := \min_{e\in A_{\bar{v}_i}^{\leq}(l_k^y)} \max\{g_e(l_k^y), 0\}$.
if $G_{\bar{v}_i}^{\leq}(l_k^y) \leq F_s(l_k^y)$ **then** let $k^* := k$, **break.**
end if
end for
if $k^* = 0$ **then**
call Alg. 13.4 $(\rho(v_i), C(v_i))$=**Trans-Point**$(z_s, k_s, b_s, l_{N_3}^y, \rho_{max}, l^y, d_{v_i}, l_{v_i}, D_l(s), \rho_{max})$ to solve the transcendental point on $(l_{N_3}^y, \rho_{max}]$.
else if $k^* = 1$ **then**
call Alg. 13.4 $(\rho(v_i), C(v_i))$=**Trans-Point**$(z_s, k_s, b_s, D_l(s), l_1^y, l^y, d_{v_i}, l_{v_i}, D_l(s), \rho_{max})$ to solve the transcendental point on $(D_l(s), l_1^y]$.
else
call Alg. 13.4 $(\rho(v_i), C(v_i))$=**Trans-Point**$(z_s, k_s, b_s, l_{k^*-1}^y, l_{k^*}^y, l^y, d_{v_i}, l_{v_i}, D_l(s), \rho_{max})$ to solve the transcendental point on $(l_{k^*-1}^y, l_{k^*}^y]$.
end if
end if
if $C(v_i) > C^*$ **then** $\rho^* := \rho(v_i), C^* := C(v_i)$.
end if
end for

$$x^*(e) := \begin{cases} \rho^* - l(e), & if\ e \in A(s), l(e) < \rho^*; \\ 0, & otherwise. \end{cases}$$

$$y^*(e) := \begin{cases} \min\{\frac{C^*}{d(e)}, l(e) - \rho^*, u^y(e)\}, & if\ e \in A(s), l(e) > \rho^*; \\ \min\{\frac{C^*}{d(e)}, u^y(e)\}, & if\ e \notin A(s), l(e) > \rho^*; \\ 0, & otherwise. \end{cases}$$

Return (C^*, x^*, y^*).
end if

Theorem 13.7 *The time complexity of Algorithm 13.5 is $O(n^3)$.*

13.3.5 (Inv, VO1C, Bounded, Bottleneck Hamming Distance)

In this subsection, we first analyze some properties of the inverse vertex obnoxious 1-center location problem under the bottleneck Hamming distance and then design an $O(n^2 \log n)$ algorithm based on a binary search method.

13.3.5.1 Some Properties of the Problem (IVO1C$_{bH}$)

Firstly, we need to judge whether $D_l(s) \geq D_l^{max}$ is true. If it is true, then the vertex s is already the obnoxious 1-center of the graph G. Otherwise, we need to judge the feasibility of the problem (IVO1C$_{bH}$). If the problem is feasible, then the optimal value C^* must be a certain component in the cost vector c or d.

Theorem 13.8 *Adjust the edge weight at the cost* $\max_{e\in E}\{c(e), d(e)\}$*; if exists,* $v_{i^*} \in \{V_0 | D_l(v_i) > \min_{e\in A(s)}\{l^x(e)\}\}$ *so that* $\min_{e\in A(v_{i^*})}\{l^y(e)\} > \min_{e\in A(s)}\{l^x(e)\}$*; then the problem (IVO1C$_{bH}$) is infeasible; otherwise, (IVO1C$_{bH}$) is feasible.*

Proof Obviously, for the edge $e \in A(s)$, we get $c(e) \leq \max_{e\in E}\{c(e), d(e)\}$. Since there is an upper bound for increasing the amount of edge weights, the edge weight of edge $e \in \{A(s) | l(e) < \min_{e\in A(s)}\{l^x(e)\}\}$ can be adjusted to $\min_{e\in A(s)}\{l^x(e)\}$, so $D_{\tilde{l}}(s) = \min_{e\in A(s)}\{l^x(e)\}$. If exists, $v_{i^*} \in \{V_0 | D_l(v_i) > \min_{e\in A(s)}\{l^x(e)\}\}$, so that $\min_{e\in A(v_{i^*})}\{l^y(e)\} > \min_{e\in A(s)}\{l^x(e)\}$. Even if the edge weight of edge $e \in A(v_{i^*})$ is adjusted to the lower bound, then $D_{\tilde{l}}(v_{i^*}) = \min_{e\in A(v_{i^*})}\{l^y(e)\}$. At this time, we get $D_{\tilde{l}}(v_{i^*}) > D_{\tilde{l}}(s)$. Therefore, the vertex s cannot become the obnoxious 1-center of G under the edge weight $\tilde{l}$, and the problem (IVO1C$_{bH}$) is infeasible.

If for any $v_i \in \{V_0 | D_l(v_i) > \min_{e\in A(s)}\{l^x(e)\}\}$, $l^y(e_i^*) = \min_{e\in A(v_i)}\{l^y(e)\} \leq \min_{e\in A(s)}\{l^x(e)\}$ holds. Obviously, $d(e_i^*) \leq \max_{e\in E}\{c(e), d(e)\}$, and then the edge e_i^* can be adjusted to the lower bound of the edge weight. At this time, we get $D_{\tilde{l}}(v_i) = l^y(e_i^*) \leq \min_{e\in A(s)}\{l^x(e)\} = D_{\tilde{l}}(s)$. Therefore, the vertex s is the obnoxious 1-center of G under the edge weight $\tilde{l}$, and the problem(IVO1C$_{bH}$) is feasible.

If the vertex set $\{v_i \in V_0 | D_l(v_i) > \min_{e\in A(s)}\{l^x(e)\}\}$ is an empty set, then we have $D_l(v_i) \leq \min_{e\in A(s)}\{l^x(e)\} = D_{\tilde{l}}(s)$ for any $v_i \in V_0$. Therefore, the problem (IVO1C$_{bH}$) is feasible. □

When problem (IVO1C$_{bH}$) is feasible, we put all components in the cost vector c, d into the cost set C in ascending order, and then

$$C_1 < C_2 < \cdots < C_{N_5} \tag{13.12}$$

Next, we need to find the minimum cost in the set C that makes vertex s an obnoxious 1-center of the graph G.

Increase the edge weight of edge $e \in A(s)$ as much as possible at the cost $C_k (k = 1, 2, \cdots, N_5)$, and record it as

$$l_k(e) = \begin{cases} l^x(e), & if\ c(e) \leq C_k; \\ l(e), & otherwise. \end{cases}$$

For edge $e \in A(s)$, denoting the smallest edge weight $l_k(e)$ as ρ_k, then $\rho_k = \min_{e\in A(s)}\{l_k(e)\}$.

For vertex $v_i \in V_0$, putting the vertices satisfying $D_l(v_i) > \rho_k$ into set V_k, then $V_k = \{v_i \in V_0 | D_l(v_i) > \rho_k\}, k = 1, 2, \cdots, N_5$.

For edge $e \in A(v_i)$, putting the edge satisfying $l^y(e) \le \rho_k$ and $d(e) \le C_k$ into the set E_i^k, then $E_i^k = \{e \in A(v_i) | l^y(e) \le \rho_k, d(e) \le C_k\}, k = 1, 2, \cdots, N_5$.

For vertex $v_i \in V_k (k = 1, 2, \cdots, N_5)$, putting the vertices that satisfy $E_i^k = \emptyset$ into set $\hat{V}_k$, then $\hat{V}_k = \{v_i \in V_k | E_i^k = \emptyset\}$.

Theorem 13.9 *When the vertex set V_1 is an empty set, then the optimal value of problem (IVO1C$_{bH}$) is C_1, and*

$$\tilde{x}(e) = \begin{cases} \rho_1 - l(e), & if\ e \in A(s), l(e) < \rho_1; \\ 0, & otherwise. \end{cases} \quad \tilde{y} = 0. \tag{13.13}$$

is an optimal solution.

Proof By the definition of $l_1(e)$, ρ_1, if the edge $e \in \{A(s) | l(e) < \rho_1\}$, then $c(e) \le C_1$ and $l^x(e) \ge \rho_1$, so $\rho_1 - l(e) \le u^x(e)$ and $\tilde{l}(e) = l(e) + \tilde{x}(e) = l(e) + \rho_1 - l(e) = \rho_1$. For the edge $e \in \{A(s) | l(e) \ge \rho_1\}$, $\tilde{l}(e) = l(e) \ge \rho_1$, then $D_{\tilde{l}}(s) = \rho_1$. Since the vertex set V_1 is an empty set, then $D_l(v_i) \le \rho_1$ for any $v_i \in V_0$. Hence, $D_{\tilde{l}}(s) \ge D_{\tilde{l}}(v_j)$, $v_j \in V_0$ holds, and $\tilde{x}$ given by (13.13) is a feasible solution under the cost C_1. Since $C_1 = \min\{C_k, k = 1, 2, \cdots, N_5\}$, C_1 is the optimal value of the problem (IVO1C$_{bH}$), and $\tilde{x}$ given by (13.13) is an optimal solution. □

Theorem 13.10 *When vertex set V_1 is not an empty set and if vertex set $\hat{V}_1$ is an empty set, then the optimal value of problem (IVO1C$_{bH}$) is C_1, and*

$$\tilde{x}(e) = \begin{cases} \rho_1 - l(e), & if\ e \in A(s), l(e) < \rho_1; \\ 0, & otherwise. \end{cases} \tag{13.14}$$

$$\tilde{y}(e) := \begin{cases} \min\{u^y(e), l(e) - \rho_1\}, & if\ e \in E, l(e) > \rho_1, d(e) \le C_1; \\ 0, & otherwise. \end{cases} \tag{13.15}$$

is an optimal solution. Otherwise, $C^* > C_1$.

Proof When the vertex set $V_1 \ne \emptyset$ and $\hat{V}_1 = \emptyset$, then $E_i^1 \ne \emptyset$ for any $v_i \in V_1$. Thus, there is $e_i^* \in A(v_i)$, such that $l^y(e_i^*) \le \rho_1$ and $d(e_i^*) \le C_1$ holds. Due to $v_i \in V_1$, $l(e_i^*) > \rho_1$, we have $\tilde{l}(e_i^*) = l(e_i^*) - \tilde{y}(e_i^*) = l(e_i^*) - \min\{u^y(e_i^*), l(e_i^*) - \rho_1\} = l(e_i^*) - (l(e_i^*) - \rho_1) = \rho_1$. Therefore, $D_{\tilde{l}}(v_i) \le \rho_1$. Adjusting the edge weights at the cost C_1 for the adjacent edges of vertex s, we have

$$l_1(e) = \begin{cases} l^x(e), & if\ c(e) \le C_1; \\ l(e), & otherwise. \end{cases}$$

According to the definition of ρ_1, for the edge $e \in \{A(s) | l(e) < \rho_1\}$, we have $c(e) \le C_1$ and $l^x(e) \ge \rho_1$, so $\rho_1 - l(e) \le u^x(e)$, and $\tilde{l}(e) = l(e) + \tilde{x}(e) = l(e) + \rho_1 -$

$l(e) = \rho_1$. For edge $e \in \{A(s)|l(e) \geq \rho_1\}$, we have $\tilde{l}(e) = l(e) \geq \rho_1$. At this time, $D_{\tilde{l}}(s) = \rho_1 \geq D_{\tilde{l}}(v_i)$, $v_i \in V_1$. For $v_i \in V_0 \setminus V_1$, we have $D_{\tilde{l}}(s) = \rho_1 \geq D_{\tilde{l}}(v_i)$. So $D_{\tilde{l}}(s) \geq D_{\tilde{l}}(v_j)$, $v_j \in V_0$.

Thus, $(\tilde{x}, \tilde{y})$ given by (13.14) and (13.15) is a feasible solution with the cost C_1, and $C_1 = \min\{C_k, k = 1, 2, \cdots, N_5\}$. Therefore, C_1 is the optimal value of the problem (IVO1C$_{bH}$); $(\tilde{x}, \tilde{y})$ is an optimal solution.

Obviously, no matter how much edge weights are added to edge $e \in A(s)$, $D_{\tilde{l}}(s)$ will never exceed ρ_1. If $v_{i^*} \in \hat{V}_1$ exists, such that $E^1_{i^*} = \emptyset$. Hence, for any $e \in A(v_{i^*})$, $l^y(e) \leq \rho_1$ and $d(e) \leq C_1$ cannot hold; then the adjacent edge of vertex v_{i^*} either belongs to $E_1 = \{e \in A(v_{i^*})|l^y(e) > \rho_1, d(e) \leq C_1\}$ or belongs to $E_2 = \{e \in A(v_{i^*})|d(e) > C_1\}$. For edge $e \in E_1$, since there is an upper bound on the adjustment amount of edge weight reduction, no matter how much edge weights are reduced, we still have $\tilde{l}(e) > \rho_1$. For edge $e \in E_2$, obviously $l(e) > \rho_1$; otherwise, it contradicts $v_{i^*} \in V_1$.

Due to $d(e) > C_1$, the side weight $l(e)$ cannot be reduced, so $\tilde{l}(e) = l(e) > \rho_1$. In summary, $D_{\tilde{l}}(v_{i^*}) > \rho_1 \geq D_{\tilde{l}}(s)$. $D_{\tilde{l}}(s) \geq D_{\tilde{l}}(v_j)$, $v_j \in V_0$ does not hold, so the problem (IVO1C$_{bH}$) is infeasible by adjusting the edge weight at the cost C_1; therefore, $C^* > C_1$. □

13.3.5.2 An $O(n^2 \log n)$ Algorithm for the Problem (IVO1C$_{bH}$)

If the edge weights are adjusted at the cost C_1, the problem (IVO1C$_{bH}$) is infeasible, and then the binary search method is used to solve the problem.

Initialize $a := 1$, $b := N_5$; when $b - a > 1$, perform while loop. Let $k := \left\lfloor \frac{a+b}{2} \right\rfloor$. If the vertex set V_k is an empty set, then similar to the proof of the Theorem 13.9, the problem (IVO1C$_{bH}$) is feasible by adjusting the edge weights at the cost C_k. At this time, the optimal value $C^* \leq C_k$, let $b := k$. When the vertex set V_k is not an empty set, then similar to the proof of the Theorem 13.10, if the vertex set $\hat{V}_k$ is empty, then the problem (IVO1C$_{bH}$) is also feasible with the cost C_k, and let $b := k$. If there is a vertex $v_{i^*} \in \hat{V}_k$, so that the edge set $E^k_{i^*}$ is an empty set, the problem (IVO1C$_{bH}$) is infeasible and the optimal value $C^* > C_k$. Let $a := k$.

Next, repeat the above steps until $b - a = 1$, and get the interval $(C_a, C_b]$ where the optimal value is located. Hence, C_b is the minimum cost in the cost set C that can make vertex s an obnoxious 1-center of graph G. Then construct an optimal solution according to different situations. If the vertex set V_b is an empty set, then

$$\tilde{x}(e) = \begin{cases} \rho_b - l(e), & if\ e \in A(s), l(e) < \rho_b; \\ 0, & otherwise. \end{cases} \quad \tilde{y} = 0. \tag{13.16}$$

is an optimal solution; otherwise,

$$\tilde{x}(e) = \begin{cases} \rho_b - l(e), \ if \ e \in A(s), l(e) < \rho_b; \\ 0, \qquad otherwise. \end{cases} \tag{13.17}$$

$$\tilde{y}(e) := \begin{cases} \min\{u^y(e), l(e) - \rho_b\}, \ if \ e \in E, l(e) > \rho_b, d(e) \le C_b; \\ 0, \qquad otherwise. \end{cases} \tag{13.18}$$

is an optimal solution.

The algorithm is summarized as follows:

Algorithm 13.6 An algorithm to solve the problem $(\mathbf{IVO1C}_{bH}) : (C^*, x^*, y^*) = \mathbf{IVO1C}_{bH}(G, l, c, d, u^x, u^y)$

Input: $G(V, E), l, c, d, u^x, u^y$.
Output: The optimal value C^* and the optimal solution x^*, y^*.

Initialize $C^* := 0,\ x^* := 0,\ y^* := 0$.
Let $A(v) := \{e|e = (v, t) \in E\}$, $D_l(v) := \min_{e \in A(v)}\{l(e)\}$, $v \in V$
Let $D_l^{max} := \max_{v_i \in V_0}\{D_l(v_i)\}$.
if $D_l(s) \ge D_l^{max}$ **then**
  Output: "The vertex s is already the obnoxious 1-center of G." **Return**(C^*, x^*, y^*).
else
  Let $l^x(e) := l(e) + u^x(e), l^y(e) := l(e) - u^y(e), e \in E$.
  for $v_i \in V_0$ **do**
    if $D_l(v_i) > \min_{e \in A(s)}\{l^x(e)\}$, $\min_{e \in A(v_i)}\{l^y(e)\} > \min_{e \in A(s)}\{l^x(e)\}$ **then** Output: "The instance is infeasible." **Return**(C^*, x^*, y^*) with $C^* := +\infty,\ x^* := \emptyset,\ y^* := \emptyset$.
    end if
  end for
  All components in the cost c, d are sorted, as shown in (13.12).
  Let $l_1(e) := \begin{cases} l^x(e), \ if \ c(e) \le C_1; \\ l(e), \ otherwise. \end{cases}$, for $e \in A(s)$.
  Let $\rho_1 := \min_{e \in A(s)}\{l_1(e)\}$, $V_1 := \{v_i \in V_0 | D_l(v_i) > \rho_1\}$.
  if $V_1 = \emptyset$ **then** $C^* := C_1, x^*, y^*$ as shown in (13.13).
  else $E_i^1 := \{e \in A(v_i) | l^y(e) \le \rho_1, d(e) \le C_1\}$, $v_i \in V_1$, $\hat{V}_1 := \{v_i \in V_1 | E_i^1 = \emptyset\}$.
    if $\hat{V}_1 = \emptyset$ **then** $C^* := C_1$,x^*, y^* as shown in (13.14) and (13.15).
    else Initialize $a := 1, b := N_5$ (13.12).
      while $b - a > 1$ **do**
        $k := \lfloor \frac{a+b}{2} \rfloor$.
        $l_k(e) := \begin{cases} l^x(e), \ if \ c(e) \le C_k; \\ l(e), \ otherwise. \end{cases}$, $e \in A(s)$.
        $\rho_k := \min_{e \in A(s)}\{l_k(e)\}$, $V_k := \{v_i \in V_0 | D_l(v_i) > \rho_k\}$.

        if $V_k = \emptyset$ **then** $b := k$.
        else $E_i^k := \{e \in A(v_i) | l^y(e) \le \rho_k, d(e) \le C_k\}.,v_i \in V_k$.
          $\hat{V}_k := \{v_i \in V_k | E_i^k = \emptyset\}$.
          if $\hat{V}_k = \emptyset$ **then** $b := k$.
          else $a := k$.
          end if
        end if
      end while
      if $b = N_5$ **then** $\rho_{N_5} := \min_{e \in A(s)} l^x(e)$, $V_{N_5} := \{v_i \in V_0 | D_l(v_i) > \rho_{N_5}\}$.
      end if
      if $V_b = \emptyset$ **then** $C^* := C_b$, x^*, y^* assignment as in (13.16).
      else $C^* := C_b$, x^*, y^* assignment as in (13.17) and (13.18).
      end if
    end if
  end if
  return (C^*, x^*, y^*).
end if

Theorem 13.11 *The time complexity of Algorithm 13.6 is* $O(n^2 \log n)$.

13.3.6 Computational Experiments

Now, we will show the results of numerical experiments for Algorithms 13.5 and 13.6. The programs were coded in MATLAB 9.1.0.441655 (R2016b) and run on a PC Intel(R) Core(TM) i5-6200U CPU @2.30GHz 2.40GHz under Windows 10. The two algorithms are tested on six classes of random graphs, where the number of nodes n varies from 200 to 10,000 and the number of edges m varies from 500 to 30,000. For each case, we randomly generate 300 instances. In each instance, five vectors l, c, d, u^x, and u^y are randomly generated, where l is evenly distributed on the interval [50,100], c and d are evenly distributed on the interval [30, 50], and u^x and u^y are evenly distributed on the interval [10, 20]. For each randomly generated graph, denote Atime1, min_time1, and max_time1, respectively, the average, minimum, and maximum values of the CPU when solving problem (IVO1C$_\infty$) by Algorithm 13.5. Denote Atime2, min_time2 and max_time2, respectively, the average, minimum, and maximum values of the CPU when solving problem (IVO1C$_{bH}$) by Algorithm 13.6.

As can be seen in Table 13.1, for each type of random graph, Algorithm 13.6 for the problem (IVO1C$_{bH}$) with time complexity $O(n^2 \log n)$ operates better than Algorithm 13.5 for the problem (IVO1C$_\infty$) running in $O(n^3)$ time. Therefore, the effectiveness of algorithms can be verified.

Table 13.1 The experimental results of Algorithms 13.5 and 13.6

n	200	500	1000	2000	5000	10,000
m	500	800	5000	8000	18,000	30,000
Atime1	0.1406	0.7863	3.7170	10.8583	81.9267	281.0573
min_time1	0.0972	0.6139	2.4520	10.4305	72.5414	271.0632
max_time1	0.5240	1.3968	4.5408	13.4030	126.4194	297.3597
Atime2	0.1243	0.7710	3.3646	10.8240	65.8044	273.6985
min_time2	0.0952	0.6915	2.5071	10.1718	63.3744	252.0563
max_time2	0.4046	1.0800	3.7015	16.2282	70.5623	302.0707

13.4 Conclusions and Further Research

In this chapter, we consider the inverse vertex obnoxious 1-center location problem under the weighted l_∞ norm and the bottleneck Hamming distance. For the problem (IVO1C$_\infty$), we design an algorithm with time complexity $O(n^3)$ by describing the adjustment cost functions corresponding to the adjacent edge of the facility point and the customer points and solving the transcendental point of the above functions. For the problem (IVO1C$_{bH}$), if this problem is feasible, we propose an algorithm with time complexity $O(n^2 \log n)$ based on the idea of binary search method. In each iteration, for a given specific cost, determine whether the problem is feasible at this cost. If feasible, seek a feasible solution with a smaller target value in the left half of the search interval. Otherwise, the right half of the search interval is found. The half part continues to search for the optimal value, and the above process iteratively terminates until there is only one value in the search interval, and this value is the optimal value. Otherwise, continue to search for the optimal value in the right half of the search interval, and the above process iteratively terminates until there is only one value in the search interval, and this value is the optimal value. Finally, numerical experiments are carried out to verify the effectiveness of the corresponding algorithms.

In the future, we still need to conduct in-depth research on the inverse vertex obnoxious 1-center location problem under the weighted sum Hamming distance. For the inverse vertex obnoxious 1-center location problem under the weighted l_1 norm, it is challenging to improve the time complexity of the known algorithm in $O(n^6)$ time [13]. And the inverse and the improvement problem of obnoxious p-center location problems under different norms are topics worthy of follow-up research.

Chapter 14
Inverse Quickest 1-Center Location Problem on Trees

Abstract This chapter investigates the inverse quickest 1-center (IQ1C) location problem on trees under different norms, focusing on modifying network capacities to optimize emergency response times. We first introduce optimality conditions for vertex and absolute quickest 1-center problems. Then we propose an $O(n^2 \log n)$ algorithm for the inverse problem (IQ1C) under the weighted l_∞ norm and an $O(n^3)$ algorithm under the weighted l_1 norm. The algorithms ensure a designated vertex becomes the quickest 1-center, enhancing network performance for critical applications.

Keywords Inverse quickest 1-center · Vertex and absolute quickest 1-center problems · Maximum transmission time balance problem

14.1 Vertex and Absolute Quickest 1-Center Problems

Let $T = (V, E)$ be a tree, where V and E are the sets of vertices and edges, respectively, and $n = |V|$. For each edge $e \in E$, there is an associated lead time $l(e) > 0$ that represents the time to traverse the edge e. For any two vertices $v, t \in V$, let P_{vt} be the only path connecting vertices v and t. Define $l(P_{vt}) = \sum_{e \in P_{vt}} l(e)$. For every edge $e \in E$, let $c(e) > 0$ represent the capacity of e, which denotes the upper bound that can flow on the edge. Define $c(P_{vt}) = \min_{e \in P_{vt}} c(e)$. Given a value σ, a transmission time to send σ units data from v to t is defined as $Q(v, t, \sigma) = l(P_{vt}) + \frac{\sigma}{c(P_{vt})}$. This leads to a vertex quickest 1-center (**VQ1C**) location problem. The problem (**VQ1C**) on trees is to determine a vertex $v^* \in V$ with the property that the maximum transmission time to transmit σ units data from v^* to the other vertices is minimum, which can mathematically be described below.

$$(\textbf{VQ1C}) \quad \min_{v \in V} \max_{t \in V} \{Q(v, t, \sigma)\}. \tag{14.1}$$

Xinqiang Qian and Xiucui Guan have equally contributed to this chapter.

X. Guan et al., *Inverse Combinatorial Optimization Problems*, Springer Optimization and Its Applications 225, https://doi.org/10.1007/978-3-031-91175-0_14

For any point $s \in T$, let $e_s = (v_s^1, v_s)$ be the edge going through s and $l(s, v_s) = \theta_s l(e_s)$, $l(v_s^1, s) = (1-\theta_s)l(e_s)$, $0 \leq \theta_s \leq 1$. For any $t \in V$ with $v_s^1 \in P_{st}$, define $l(P_{st}) = l(s, v_s^1) + l(P_{v_s^1 t})$ and $c(P_{st}) = c(P_{v_s t})$. The transmission time to send σ units data from a point s to t is defined as $Q(s, t, \sigma) = l(P_{st}) + \frac{\sigma}{c(P_{st})}$. The absolute quickest 1-center (**AQ1C**) problem on trees is to determine a point $s^* \in T$ such that the maximum transmission time to send σ units data is minimum, which can mathematically be written below:

$$\textbf{(AQ1C)} \quad \min_{s\in T} \max_{t\in V} Q(s, t, \sigma). \tag{14.2}$$

Keshtkar and Ghiyasvand [72] first proposed the problem (*AQ1C*) on trees, providing the necessary and sufficient optimality conditions of this center, but did not provide an effective algorithm to determine such a center. In 2022, they [46] designed an $O(r|D|(m + n\log n))$ algorithm for solving the problem (*AQ1C*) on general graphs, where $n = |V|$, $m = |E|$, and r and $|D|$ represent the number of different capacity values and the number of demand points, respectively.

Definition 14.1 ([110]) Let v_i be a vertex in $V \backslash Z(T)$ and assume that $deg(v_i) = s$, where $Z(T)$ is the set of leaves. We partition the tree T into nontrivial subtrees $T_{v_i}^1, \cdots, T_{v_i}^s$ with

$$T = \bigcup_{k=1}^{s} T_{v_i}^k,\; T_{v_i}^k \bigcap T_{v_i}^j = \{v_i\}, \text{ for any } k \neq j.$$

Let $t_{v_i}^*$ be the farthest leaf from v_i with $Q(v_i, t_{v_i}^*, \sigma) = \max_{t\in Z(T)}\{Q(v_i, t, \sigma)\}$. Without loss of generality, we assume that $P_{v_i t_{v_i}^*} \in T_{v_i}^1$, $T_{v_i}' = T_{v_i}^1$, $T_{v_i}'' = \bigcup_{k=2}^{s} T_{v_i}^k$. We define the **gate** of v_i as the first vertex v_{i^*} from v_i on the path $P_{v_i t_{v_i}^*}$. We obviously have $Q(v_i, t_{v_i}^*, \sigma) = \max_{t\in Z(T_{v_i}')}\{Q(v_i, t, \sigma)\}$.

We can conclude the following sufficient and necessary condition of the vertex quickest 1-center.

Theorem 14.1 ([110]) *A vertex $v_i \in V$ is a vertex quickest 1-center of T if and only if the following inequality holds:*

$$Q(v_i, t_{v_i}^*, \sigma) \leq \max_{t\in Z(T_{v_i}'')}\{Q(v_{i^*}, t, \sigma)\}. \tag{14.3}$$

Assumption 14.1 Assume that vertex s^* is not a vertex quickest 1-center for T under the original capacity vector c. According to Theorem 14.1, it can be deduced that

$$Q_c' > Q_c^{v_{s^*}} > Q_c''. \tag{14.4}$$

Similar to Definition 14.1, we present the definition of partition trees with respect to the absolute quickest 1-center in order to characterize the sufficient and necessary optimality conditions.

Definition 14.2 ([110]) Let s be a point in T. Let $deg(s)$ be the degree of s if s is a vertex and $deg(s) = 2$ if s is an interior point on an edge. We partition the tree T into nontrivial subtrees $T_s^1, \cdots, T_s^{deg(s)}$ with

$$T = \bigcup_{k=1}^{deg(s)} T_s^k, \quad T_s^k \bigcap T_s^j = \{s\}, \text{ for any } k \neq j.$$

Let t_s be the farthest leaf from s satisfying $Q(s, t_s, \sigma) = \max_{t \in Z(T)} Q(s, t, \sigma)$. Without loss of generality, we assume that the path $P_{st_s} \in T_s^1$, $T_s' = T_s^1$, $T_s'' = \bigcup_{k=2}^{deg(s)} T_s^k$. Let v_s be the first vertex from s on the path P_{st_s} and $e_s = (v_s^1, v_s)$ be the edge traversing s with $v_s^1 \in V$. Notice that $v_s^1 = s$ when $s \in V$. For convenience, we call e_s the **critical edge** under c and v_s the **gate** to enter the path P_{st_s}.

Condition 14.1 For a point $s \in T$, there is $t_s'' \in Z(T_s'')$ with $Q(s, t_s, \sigma) = Q(s, t_s'', \sigma)$.

Condition 14.2 For a point $s \in T$, there is $t_s'' \in Z(T_s'')$ with $Q(s, t_s, \sigma) \leq l(P_{st_s''}) + \frac{\sigma}{c(e_s)}$.

Theorem 14.2 ([72]) *A point $s \in T$ is an absolute quickest 1-center of T if and only if s satisfies Conditions 14.1 or 14.2.*

14.2 Inverse Vertex and Absolute Quickest 1-Center Problems Under Weighted l_∞ Norm

14.2.1 (Inv, VQ1C, Bounded, Weighted l_∞ Norm)

In this subsection, we investigate the inverse vertex quickest 1-center problem ($IVQ1C_\infty$) under weighted l_∞ norm and present an $O(n^2 \log n)$ time algorithm.

The problem ($IVQ1C_\infty$) is defined as follows: We aim to modify edge capacities from c to $\tilde{c}$ under bounded constraints to make a given vertex $v^* \in V$ a vertex quickest 1-center of T. The objective function is to minimize the maximum modification cost. The positive values $q^+(e)$ and $q^-(e)$ represent the unit costs of increasing and decreasing the edge capacity of $e \in E$, respectively. The value $Q_{\tilde{c}}(v^*, t, \sigma)$ denotes the transmission time from v^* to t under $\tilde{c}$. The variables $x(e)$ and $y(e)$ are the increments and decrements of the edge capacity of e, respectively.

The values $u(e)$ and $d(e)$ are the upper bounds for the increments and decrements of the edge capacity of e, respectively, and $d(e) < c(e)$. The problem ($IVQ1C_\infty$) can be mathematically described below:

$$\min f(\tilde{c}) = \max_{e \in E}\{q^+(e)x(e), q^-(e)y(e)\}$$

$$s.t. \max_{t \in Z(T)} \{Q_{\tilde{c}}(v^*, t, \sigma)\} \leq \max_{t \in Z(T)} \{Q_{\tilde{c}}(v, t, \sigma)\}, \forall v \in V, \tag{14.5}$$

$$\tilde{c}(e) = c(e) + x(e) - y(e), e \in E, \tag{14.6}$$

$$0 \leq x(e) \leq u(e), e \in E, \tag{14.7}$$

$$0 \leq y(e) \leq d(e), e \in E. \tag{14.8}$$

We first introduce some symbols. For any capacity vector $\tilde{c}$, we partition the tree T from v^* into two subtrees $T'_{\tilde{c}}$ and $T''_{\tilde{c}}$ as described in Definition 14.1, where $P_{v^*t_{\tilde{c}}} \in T'_{\tilde{c}}$ with $Q_{\tilde{c}}(v^*, t_{\tilde{c}}, \sigma) = \max_{t \in Z(T)}\{Q_{\tilde{c}}(v^*, t, \sigma)\}$. For simplicity, denote $T' = T'_c$, $T'' = T''_c$, v_{i^*} the **gate** of v^*, the edge $e^* = (v^*, v_{i^*})$ a **gate edge** of T, $E' = E(T')$, $E'' = E(T'')$, $Z' = Z(T')$, and $Z'' = Z(T'')$. It is obvious that any path P_{v^*t} with $t \in Z'$ or $P_{v_{i^*}t}$ with $t \in Z''$ must traverse e^*. Furthermore, let $Q'_{\tilde{c}} = \max_{t \in Z'}\{Q_{\tilde{c}}(v^*, t, \sigma)\}$, $Q''_{\tilde{c}} = \max_{t \in Z''}\{Q_{\tilde{c}}(v^*, t, \sigma)\}$, and $Q^{i^*}_{\tilde{c}} = \max_{t \in Z''}\{Q_{\tilde{c}}(v_{i^*}, t, \sigma)\}$ under $\tilde{c}$. It is obvious that $Q^{i^*}_{\tilde{c}} > Q''_{\tilde{c}}$. Let $S_E = \{l, c, u, d, q^+, q^-, \sigma\}$ be the set of vectors defined on E and the value σ.

Lemma 14.1 ([110]) *Let Q be a given value with $\max_{t \in Z'}\{l(P_{v^*t})\} < Q < Q'_c$ and $\tilde{c}$ be a new capacity vector after increasing the capacities $c(e)$ of $e \in E'$ satisfying $Q'_{\tilde{c}} = Q$. Then, for every $e \in E'$ with $c(e) < \xi_e$, we have $\tilde{c}(e) = \xi_e$ and $\tilde{c}(e) = c(e)$ for other edges, where $\xi_e = \max\{\frac{\sigma}{Q - l(P_{v^*t})} \mid t \in Z', e \in P_{v^*t}\}$.*

Lemma 14.2 ([110]) *Let Q be a transmission time satisfying $Q > Q''_c$ and $\tilde{c}$ be a new capacity vector after decreasing the capacities $c(e)$ of $e \in E''$ satisfying $Q''_{\tilde{c}} = Q$. Then, for a given edge $\tilde{e} \in E''$, we have $\tilde{c}(\tilde{e}) = \gamma_{\tilde{e}}$ and $\tilde{c}(e) = c(e)$ for other edges, where $\gamma_{\tilde{e}} = \max\{\frac{\sigma}{Q - l(P_{v^*t})} \mid t \in Z'', \tilde{e} \in P_{v^*t}\}$.*

Lemma 14.3 ([110]) *Suppose ($IVQ1C_\infty$) is feasible with optimal objective value D^*. If there is a capacity vector $\tilde{c}$ satisfying (14.6)–(14.8) with*

$$\max_{t \in Z(T)} \{Q_{\tilde{c}}(v^*, t, \sigma)\} > \max_{t \in Z'}\{Q_{\tilde{c}}(v^*, t, \sigma)\}, \tag{14.9}$$

then we have $f(\tilde{c}) > D^*$.

Corollary 14.1 ([110]) *If there is an optimal capacity vector c^* of ($IVQ1C_\infty$), then we have $T' = T'_{c^*}$, $T'' = T''_{c^*}$.*

Based on Lemma 14.3 and Corollary 14.1, we next always consider the capacity vector $\tilde{c}$ with $T' = T'_{\tilde{c}}$ and $T'' = T''_{\tilde{c}}$, which implies that $\max_{t \in Z(T)}$ $\{Q_{\tilde{c}}(v^*, t, \sigma)\} = \max_{t \in Z'}\{Q_{\tilde{c}}(v^*, t, \sigma)\}$, v_{i^*} is always the **gate**, and $e^* = (v^*, v_{i^*})$

is always the **gate edge** of T. Together with Theorem 14.1, we can reformulate the mathematical model (14.5)–(14.8) of *(IVQ1C$_\infty$)* as follows:

$$\min f(\tilde{c}) = \max_{e\in E}\{q^+(e)x(e), q^-(e)y(e)\}$$

$$s.t. \max_{t\in Z'}\{Q_{\tilde{c}}(v^*, t, \sigma)\} \leq \max_{t\in Z''}\{Q_{\tilde{c}}(v_{i^*}, t, \sigma)\}, \tag{14.10}$$

$$\max_{t\in Z'}\{Q_{\tilde{c}}(v^*, t, \sigma)\} = \max_{t\in Z(T)}\left\{Q_{\tilde{c}}(v^*, t, \sigma)\right\}, \tag{14.11}$$

$$\tilde{c}(e) = c(e) + x(e) - y(e), e \in E, \tag{14.12}$$

$$0 \leq x(e) \leq u(e), e \in E, \tag{14.13}$$

$$0 \leq y(e) \leq d(e), e \in E. \tag{14.14}$$

Define

$$l(P_{v^*t_l^1}) = \max_{t\in Z'}\{l(P_{v^*t})\}, \quad l(P_{v_{i^*}t_l^2}) = \max_{t\in Z''}\{l(P_{v_{i^*}t})\}. \tag{14.15}$$

Lemma 14.4 ([110]) *If $l(P_{v^*t_l^1}) > l(P_{v_{i^*}t_l^2})$, then we have $c^*(P_{v_{i^*}t_i^*}) < c^*(e^*)$, where c^* is an optimal solution of (IVQ1C$_\infty$) and $t_i^* \in Z''$ satisfying $Q_{c^*}(v_{i^*}, t_i^*, \sigma) = Q_{c^*}^{i^*}$.*

Lemma 14.5 ([110]) *If $l(P_{v^*t_l^1}) \leq l(P_{v_{i^*}t_l^2})$, then we have $c(P_{v^*t_{v^*}}) < c(e^*)$, where $t_{v^*} \in Z'$ satisfying $Q_c(v^*, t_{v^*}, \sigma) = Q'_{c}$.*

Lemma 14.6 ([110]) *If c^* is an optimal solution of (IVQ1C$_\infty$), then we have*

$$\max_{t\in Z'}\{Q_{c^*}(v^*, t, \sigma)\} \leq \max_{t\in Z'}\{Q_c(v^*, t, \sigma)\}. \tag{14.16}$$

Corollary 14.2 ([110]) *If c^* is an optimal solution of (IVQ1C$_\infty$), then we have*

$$\max_{t\in Z''}\{Q_{c^*}(v^*, t, \sigma)\} \geq \max_{t\in Z''}\{Q_c(v^*, t, \sigma)\}.$$

Theorem 14.3 ([110]) *Suppose the problem **IVQ1C$_\infty$** is feasible with an optimal value D^*. If $l(P_{v^*t_l^1}) > l(P_{v_{i^*}t_l^2})$, then there is an optimal solution c^* with $c^*(e^*) \geq c(e^*)$.*

Corollary 14.3 ([110]) *If the problem **IVQ1C$_\infty$** is feasible and $l(P_{v^*t_l^1}) \leq l(P_{v_{i^*}t_l^2})$, then there is an optimal solution c^* with $c^*(e^*) \leq c(e^*)$.*

We partition the edge set E into two disjoint subsets E^+ and E^-, where the edge capacities of $e \in E^+$ should be increased and the capacities of $e \in E^-$ should be decreased:

$$\begin{cases} E^+ = E', & E^- = E'', \quad \text{if } l(P_{v^*t_l^1}) > l(P_{v_{i^*}t_l^2}), \\ E^+ = E' \setminus \{e^*\}, & E^- = E'' \cup \{e^*\}, \quad \text{if } l(P_{v^*t_l^1}) \le l(P_{v_{i^*}t_l^2}). \end{cases} \tag{14.17}$$

Definition 14.3 ([110]) For a given value D, define an **effective modification** c_D under D as follows:

$$c_D(e) = \begin{cases} c(e) + \min\left\{\dfrac{D}{q^+(e)}, u(e)\right\}, & e \in E^+, \\ c(e) - \min\left\{\dfrac{D}{q^-(e)}, d(e)\right\}, & e \in E^-. \end{cases} \tag{14.18}$$

Furthermore, we call c_D an **admissible solution** of $(IVQ1C_\infty)$ if

$$\max_{t \in Z'}\{Q_{c_D}(v^*, t, \sigma)\} \le \max_{t \in Z''}\{Q_{c_D}(v_{i^*}, t, \sigma)\}. \tag{14.19}$$

Theorem 14.4 ([110]) *Supposing the problem $(IVQ1C_\infty)$ is feasible with an optimal value D^*, then $D \ge D^*$ if and only if c_D defined as in (14.18) is an admissible solution.*

Lemma 14.7 ([110]) *The problem $(IVQ1C_\infty)$ is feasible if and only if $\bar{c}$ is an admissible solution, where*

$$\bar{c}(e) = \begin{cases} c(e) + u(e), & e \in E^+, \\ c(e) - d(e), & e \in E^-. \end{cases} \tag{14.20}$$

Define four significant subsets of E as follows:

$$E_u^+ = \{e \in E^+ \mid q^+(e)u(e) < D^*\},\ E_{D^*}^+ = \{e \in E^+ \mid q^+(e)u(e) \ge D^*\},$$
$$E_d^- = \{e \in E^- \mid q^-(e)d(e) < D^*\},\ E_{D^*}^- = \{e \in E^- \mid q^-(e)d(e) \ge D^*\}.$$

Define

$$\bar{D} = \min\left\{\min_{e \in E_{D^*}^+}\{q^+(e)u(e)\},\ \min_{e \in E_{D^*}^-}\{q^-(e)d(e)\}\right\}, \tag{14.21}$$

$$\underline{D} = \max\left\{\max_{e \in E_u^+}\{q^+(e)u(e)\},\ \max_{e \in E_d^-}\{q^-(e)d(e)\}\right\}. \tag{14.22}$$

For any $D \in [\underline{D}, D^*]$, the equation (14.18) can be equivalently transformed into the following form:

$$c_D(e) = \begin{cases} c(e) + u(e), & e \in E_u^+, \\ c(e) + \dfrac{D}{q^+(e)}, & e \in E_{D^*}^+, \\ c(e) - d(e), & e \in E_d^-, \\ c(e) - \dfrac{D}{q^-(e)}, & e \in E_{D^*}^-. \end{cases} \tag{14.23}$$

Let $Y_u^+, Y_D^+, Y_d^-, Y_D^-$ denote the set of linear functions defined on $D \in [\underline{D}, \bar{D}]$:

$$Y_u^+ = \left\{ y(D) = c(e) + u(e) \mid e \in E_u^+ \right\},$$

$$Y_D^+ = \left\{ y(D) = c(e) + \frac{D}{q^+(e)} \mid e \in E_{D^*}^+ \right\},$$

$$Y_d^- = \left\{ y(D) = c(e) - d(e) \mid e \in E_d^- \right\},$$

$$Y_D^- = \left\{ y(D) = c(e) - \frac{D}{q^-(e)} \mid e \in E_{D^*}^- \right\}.$$

When $l_{e^*}^1 > l_{e^*}^2$, let $I' = I_{11} \cup I_{12}$ be the set of intersection points of any two functions in $Y_u^+ \cup Y_D^+$. If $e^* \in E_u^+$, let $I'' = I_{21} \cup I_{22} \cup \{D_i \mid e_i \in E^-\}$ be the set of intersection points of any two functions in $Y_d^- \cup Y_D^- \cup \{y(D) = c(e^*) + u(e^*)\}$; if $e^* \in E_{D^*}^+$, let $I'' = I_{21} \cup I_{22} \cup \{D_i \mid e_i \in E^-\}$ be the set of intersection points of any two functions in $Y_d^- \cup Y_D^- \cup \left\{ y(D) = c(e^*) + \frac{D}{q^+(e^*)} \right\}$, where $I_{11}, I_{12}, I_{21}, I_{22}, and D_i$ are calculated as follows:

$$I_{11} = \left\{ D_{ij} = q^+(e_j)(c(e_i) + u(e_i) - c(e_j)) \mid e_i \in E_u^+, e_j \in E_{D^*}^+ \right\}, \tag{14.24}$$

$$I_{12} = \left\{ D_{ij} = \frac{q^+(e_i)q^+(e_j)(c(e_j) - c(e_i))}{q^+(e_j) - q^+(e_i)} \mid e_i, e_j \in E_{D^*}^+, q^+(e_i) \neq q^+(e_j) \right\}, \tag{14.25}$$

$$I_{21} = \left\{ D_{ij} = q^-(e_j)(c(e_j) + d(e_i) - c(e_i)) \mid e_i \in E_d^-, e_j \in E_{D^*}^- \right\}, \tag{14.26}$$

$$I_{22} = \left\{ D_{ij} = \frac{q^-(e_i)q^-(e_j)(c(e_j) - c(e_i))}{q^-(e_i) - q^-(e_j)} \mid e_i, e_j \in E_{D^*}^-, q^-(e_i) \neq q^-(e_j) \right\}, \tag{14.27}$$

$$D_i = \begin{cases} (c(e_i) - c(e^*) - u(e^*))q^-(e_i), & e_i \in E_{D^*}^-, e^* \in E_u^+, \\ (c(e_i) - c(e^*) - d(e_i))q^+(e^*), & e_i \in E_d^-, e^* \in E_{D^*}^+, \\ \frac{q^+(e^*)q^-(e_i)(c(e_i) - c(e^*))}{q^+(e^*) + q^-(e_i)}, & e_i \in E_{D^*}^-, e^* \in E_{D^*}^+. \end{cases} \tag{14.28}$$

It is obvious that the **gate edge** $e^* \in E_u^+$ or $E_{D^*}^+$ if $l(P_{v^*t_l^1}) > l(P_{v_{i^*}t_l^2})$ and $e^* \in E_d^-$ or $E_{D^*}^-$ if $l(P_{v^*t_l^1}) \le l(P_{v_{i^*}t_l^2})$.

Next, we present **Algorithm 14.1** to find the four sets $(E_u^+, E_{D^*}^+, E_d^-, E_{D^*}^-)$ based on Theorem 14.4.

Let $\Lambda^+ = \{D = q^+(e)u(e) \mid e \in E^+\}$. For every $\Lambda_i^+ \in \Lambda^+$, calculate $c_{\Lambda_i^+}$ by (14.18). We call $(E_u^+, E_{D^*}^+)=$ **Findsets** $(T', T'', E^+, S_{E^+}, \Lambda^+, c_{\Lambda^+}, Q'_{c_{\Lambda^+}}, Q^{i^*}_{c_{\Lambda^+}})$ to find the sets $E_u^+, E_{D^*}^+$.

Let $\Lambda^- = \{D = q^-(e)d(e) \mid e \in E^-\}$. For every $\Lambda_i^- \in \Lambda^-$, calculate $c_{\Lambda_i^-}$ by (14.18). We call $(E_d^-, E_{D^*}^-) =$ **Findsets**$(T', T'', E^-, S_{E^-}, \Lambda^-, c_{\Lambda^-}, Q'_{c_{\Lambda^-}}, Q^{i^*}_{c_{\Lambda^-}})$ to find the sets $E_d^-, E_{D^*}^-$.

Algorithm 14.1 $[E_{11}, E_{12}] =$ **Findsets**$(T', T'', E_1, S_{E_1}, \Lambda, c_\Lambda, Q'_{c_\Lambda}, Q^{i^*}_{c_\Lambda})$ [110]

Input: Two subtrees T' and T'' as described in Definition 14.1, a set E_1 of edges, the set S_{E_1} of parameters defined on E_1, a set Λ of values, a set c_Λ of vectors defined on the set Λ, two functions $Q'_{c_\Lambda}, Q^{i^*}_{c_\Lambda}$ with respect to c_Λ, where c_{Λ_i} is defined as in (14.18).
Output: Two subsets E_{11},E_{12} of E_1.

Initialize $E_{11} := \varnothing$, $E_{12} := \varnothing$.
for $i = 1 : |E_1|$ **do**
  if $Q'_{c_{\Lambda_i}} > Q^{i^*}_{c_{\Lambda_i}}$ **then**
    $E_{11} := E_{11} \cup \{e_i\}$
  else
    $E_{12} := E_{12} \cup \{e_i\}$
  end if
end for
Return E_{11}, E_{12}.

Theorem 14.5 ([110]) *If $(IVQ1C_\infty)$ is feasible, then we have* $\max_{t \in Z'}\{Q_{c^*}(v^*, t, \sigma)\} = \max_{t \in Z''}\{Q_{c^*}(v_{i^*}, t, \sigma)\}$, *where c^* is defined below:*

$$c^*(e) = \begin{cases} c(e) + u(e), & e \in E_u^+, \\ c(e) + \dfrac{D^*}{q^+(e)}, & e \in E_{D^*}^+, \\ c(e) - d(e), & e \in E_d^-, \\ c(e) - \dfrac{D^*}{q^-(e)}, & e \in E_{D^*}^-. \end{cases} \tag{14.29}$$

Corollary 14.4 ([110]) *If $(IVQ1C_\infty)$ is feasible, then c^* defined as in (14.29) is an optimal solution.*

Remark 14.1 Based on Corollary 14.4, one can first solve the equation $Q^1_{c_D} = Q^{i^*}_{c_D}$ to obtain the optimal value D^*; then construct the optimal solution c^* using equation (14.29). To solve the equation $Q^1_{c_D} = Q^{i^*}_{c_D}$, it is necessary to determine the analytical expressions for $Q^1_{c_D}$ and $Q^{i^*}_{c_D}$.

Algorithm 14.2 $[D_{\tau_1}, D_{\tau_2}] =$ **Binary Search**$(T', T'', S_E, \Lambda, c_\Lambda, Q'_{c_\Lambda}, Q^{i^*}_{c_\Lambda})$ [110]

Input: Two subtrees T' and T'' as described in Definition 14.1, the set S_E of parameters, a set Λ of values, a set c_Λ of vectors defined on the set Λ, two functions $Q'_{c_\Lambda}, Q^{i^*}_{c_\Lambda}$ with respect to c_Λ.
Output: The interval $(D_{\tau_1}, D_{\tau_2}]$ that the optimal value D^* lies in.

```
Sort the different values in Λ nondecreasingly. Denote the sorted set still by Λ.
Initialize a := 1, b := |Λ|.
if b ≤ 2 then return D_τ1 := min Λ, D_τ2 := max Λ.
end if
while b − a > 1 do
    Let k := ⌊(a + b)/2⌋.
    if Q'_{cΛk} ≤ Q^{i*}_{cΛk} then
        update b := k.
    else
        update a := k.
    end if
end while
Return D_τ1 := Λ_a, D_τ2 := Λ_b.
```

Let H^+ and H^- represent the sets of curves of the transmission time from v^* to $t \in Z'$ and from v_{i^*} to $t \in Z''$ defined on $D \in [D_{\tau_1}, D_{\tau_2}]$, respectively. We define H^+, H^- by (14.30) if $l(P_{v^*t^1_l}) > l(P_{v_{i^*}t^2_l})$ and by (14.31) if $l(P_{v^*t^1_l}) \leq l(P_{v_{i^*}t^2_l})$:

$$\begin{cases} H^+ = \left\{z(D) = l(P_{v^*t}) + \frac{\sigma}{c_D(P_{v^*t})} \mid t \in Z'\right\}, \\ H^- = \left\{z(D) = l(P_{v_{i^*}t}) + \frac{\sigma}{c_D(P_{v_{i^*}t})} \mid t \in Z'', c_D(P_{v_{i^*}t}) < c_D(e^*)\right\}. \end{cases} \tag{14.30}$$

$$\begin{cases} H^+ = \left\{z(D) = l(P_{v^*t}) + \frac{\sigma}{c_D(P_{v^*t})} \mid t \in Z', c_D(P_{v^*t}) < c_D(e^*)\right\}, \\ H^- = \left\{z(D) = l(P_{v_{i^*}t}) + \frac{\sigma}{c_D(P_{v_{i^*}t})} \mid t \in Z''\right\}. \end{cases} \tag{14.31}$$

Lemma 14.8 ([110]) *For every curve $z(D) \in H^+ \cup H^-$, $z(D)$ is a continuous function on $[D_{\tau_1}, D_{\tau_2}]$.*

Denote $I^+_H = \{D_{ij} \mid t_i, t_j \in Z'\}$ the set of intersections of every two curves of H^+. For any $t_i, t_j \in Z'$ and $D \in [D_{\tau_1}, D_{\tau_2}]$ with $c_D(\hat{e}_i) = c_D(P_{v^*t_i})$, $c_D(\hat{e}_j) =$

$c_D(P_{v^*t_j})$, let

$$D_{ij} = \begin{cases} \Big(\frac{\sigma}{m_1+\frac{\sigma}{c(\hat{e}_i)+u(\hat{e}_i)}} - c(\hat{e}_j)\Big)q^+(\hat{e}_j), \ \ \hat{e}_i \in E_u^+, \ \ \hat{e}_j \in E_{D^*}^+, \\ \frac{q^+(\hat{e}_i)q^+(\hat{e}_j)(c(\hat{e}_j)-c(\hat{e}_i))}{q^+(\hat{e}_j)-q^+(\hat{e}_i)}, \hat{e}_i, \hat{e}_j \in E_{D^*}^+, l(P_{v^*t_i}) = l(P_{v^*t_j}), q^+(\hat{e}_j) \neq q^+(\hat{e}_i), \\ d_{ij} \in (D_{\tau_1}, D_{\tau_2}], \ \ \hat{e}_i, \ \ \hat{e}_j \in E_{D^*}^+, \ \ l(P_{v^*t_i}) \neq l(P_{v^*t_j}), \end{cases} \tag{14.32}$$

where $d_{ij} = \frac{-m_6 \pm \sqrt{(m_6)^2 - 4m_1(m_1m_2m_3 + m_3m_4 - m_2m_5)}}{2m_1}$, $m_1 = l(P_{v^*t_i}) - l(P_{v^*t_j})$, $m_2 = c(\hat{e}_i)q^+(\hat{e}_i)$, $m_3 = c(\hat{e}_j)q^+(\hat{e}_j)$, $m_4 = \sigma q^+(\hat{e}_i)$, $m_5 = \sigma q^+(\hat{e}_j)$, and $m_6 = m_1(m_2 + m_3) + m_4 - m_5$.

Denote $I_H^- = \{D_{ij} \mid t_i, t_j \in Z''\}$ the set of intersections of every two curves of H^-. For any $t_i, t_j \in Z''$ and $D \in [D_{\tau_1}, D_{\tau_2}]$ with $c_D(\hat{e}_i) = c_D(P_{v_{i^*}t_i})$, $c_D(\hat{e}_j) = c_D(P_{v_{i^*}t_j})$, let

$$D_{ij} = \begin{cases} \Big(c(\hat{e}_j) - \frac{\sigma}{m_1+\frac{\sigma}{c(\hat{e}_i)-d(\hat{e}_i)}}\Big)q^-(\hat{e}_j), \ \ \hat{e}_i \in E_d^-, \ \ \hat{e}_j \in E_{D^*}^-, \\ \frac{q^-(\hat{e}_i)q^-(\hat{e}_j)(c(\hat{e}_j)-c(\hat{e}_i))}{q^-(\hat{e}_i)-q^-(\hat{e}_j)}, \hat{e}_i, \hat{e}_j \in E_{D^*}^-, l(P_{v_{i^*}t_i}) = l(P_{v_{i^*}t_j}), q^-(\hat{e}_i) \neq q^-(\hat{e}_j), \\ d_{ij} \in (D_{\tau_1}, D_{\tau_2}], \ \ \hat{e}_i, \ \ \hat{e}_j \in E_{D^*}^-, \ \ l(P_{v_{i^*}t_i}) \neq l(P_{v_{i^*}t_j}), \end{cases} \tag{14.33}$$

where $d_{ij} = \frac{-m_6 \pm \sqrt{(m_6)^2 - 4m_1(m_1m_2m_3 + m_3m_4 - m_2m_5)}}{2m_1}$, $m_1 = l(P_{v_{i^*}t_i}) - l\ (P_{v_{i^*}t_j})$, $m_2 = c(\hat{e}_i)q^-(\hat{e}_i)$, $m_3 = c(\hat{e}_j)q^-(\hat{e}_j)$, $m_4 = \sigma q^-(\hat{e}_i)$, $m_5 = \sigma q^-(\hat{e}_j)$, and $m_6 = m_1(-m_2 - m_3) - m_4 + m_5$.

Let $I^* = I_H^+ \cup I_H^- \cup \{D_{\tau_1}, D_{\tau_2}\} \cap [D_{\tau_1}, D_{\tau_2}]$. For every $I_i^* \in I^*$, $c_{I_i^*}$ is defined as in (14.23). We call **Binary Search**($T', T'', S_E, I^*, c_{I^*}, Q'_{c_{I^*}}, Q^{i^*}_{c_{I^*}}$) again to determine a more exact interval $(D^*_{\tau_1}, D^*_{\tau_2}]$ including D^*. For any $z_1(D) = Q_{c_D}(v^*, t^1, \sigma)$, $z_2(D) = Q_{c_D}(v^*, t^2, \sigma) \in H^+$, we can easily determine the quantitative relationship between $z_1(D)$ and $z_2(D)$. More exactly, let $\hat{D}^* = \frac{D^*_{\tau_1} + D^*_{\tau_2}}{2}$. If $z_1(\hat{D}^*) < z_2(\hat{D}^*)$, then $z_1(D) \leq z_2(D)$ holds for any $D \in [D^*_{\tau_1}, D^*_{\tau_2}]$. Similarly, for any $z_1(D) = Q_{c_D}(v_{i^*}, t^1, \sigma)$, $z_2(D) = Q_{c_D}(v_{i^*}, t^2, \sigma) \in H^-$, we can easily determine the quantitative relationship between $z_1(D)$ and $z_2(D)$. Therefore, let

$$Q_{c_{\hat{D}^*}}(v^*, t', \sigma) = \max_{t \in Z'}\{Q_{c_{\hat{D}^*}}(v^*, t, \sigma)\}, c_{\hat{D}^*}(P_{v^*t'}) = c_{\hat{D}^*}(e'), \tag{14.34}$$

$$Q_{c_{\hat{D}^*}}(v_{i^*}, t'', \sigma) = \max_{t \in Z''}\{Q_{c_{\hat{D}^*}}(v_{i^*}, t, \sigma)\}, c_{\hat{D}^*}(P_{v_{i^*}t''}) = c_{\hat{D}^*}(e''). \tag{14.35}$$

Then, for any $D \in [D^*_{\tau_1}, D^*_{\tau_2}]$, we have

$$Q_{c_D}(v^*, t', \sigma) = \max_{t \in Z'}\{Q_{c_D}(v^*, t, \sigma)\}, \quad c_D(P_{v^*t'}) = c_D(e'),$$
$$Q_{c_D}(v_{i^*}, t'', \sigma) = \max_{t \in Z''}\{Q_{c_D}(v_{i^*}, t, \sigma)\}, \; c_D(P_{v_{i^*}t''}) = c_D(e'').$$

Now, we present the following lemma to describe that $q^+(e')u(e') < D^*$ and $q^-(e'')d(e'') < D^*$ cannot hold simultaneously:

Lemma 14.9 ([110]) *For the edges e' and e'', we have that $e' \in E^+_u$ and $e'' \in E^-_d$ cannot hold simultaneously.*

From Corollary14.4 and Remark14.1, it is known that to determine the optimal value D^*, one only needs to solve $Q_{c_D}(v^*, t', \sigma) = Q_{c_D}(v_{i^*}, t'', \sigma)$, which is the equation described below:

$$l(P_{v^*t'}) + \frac{\sigma}{c_D(e')} = l(P_{v_{i^*}t''}) + \frac{\sigma}{c_D(e'')}. \tag{14.36}$$

We can calculate the optimal value D^* in the following four cases in (14.37) based on Lemma 14.9. In the first three cases, the formula (14.36) involves a linear equation with one unknown, and then D^* can be calculated directly. However, in the fourth case, the formula (14.36) involves a quadratic equation with one unknown, then there are two solutions, and we need the one $d^* \in (D^*_{\tau_1}, D^*_{\tau_2}]$:

$$D^* = \begin{cases} q^-(e'')\Big(c(e'') - \frac{\sigma}{\frac{\sigma}{c(e')+u(e')}-k_1}\Big), & e' \in E^+_u, \;\; e'' \in E^-_{D^*}, \\ q^+(e')\Big(\frac{\sigma}{k_1+\frac{\sigma}{c(e'')-d(e'')}} - c(e')\Big), & e' \in E^+_{D^*}, \;\; e'' \in E^-_d, \\ \frac{(c(e'')-c(e'))q^+(e')q^-(e'')}{q^+(e')+q^-(e'')}, & e' \in E^+_{D^*}, e'' \in E^-_{D^*}, l(P_{v^*t'}) = l(P_{v_{i^*}t''}), \\ d^*, d^* \in (D^*_{\tau_1}, D^*_{\tau_2}], & e' \in E^+_{D^*}, e'' \in E^-_{D^*}, l(P_{v^*t'}) \neq l(P_{v_{i^*}t''}), \end{cases} \tag{14.37}$$

where $d^* = \frac{k_6 \pm \sqrt{(k_6)^2 - 4k_1(k_3k_4 - k_2k_5 - k_1k_2k_3)}}{2k_1}$, $k_1 = l(P_{v_{i^*}t''}) - l(P_{v^*t'})$, $k_2 = c(e')q^+(e')$, $k_3 = c(e'')q^-(e'')$, $k_4 = \sigma q^+(e')$, and $k_5 = \sigma q^-(e'')$, $k_6 = k_1(k_3 - k_2) + k_4 + k_5$.

Algorithm 14.3 $[c^*, D^*] = (IVQ1C_\infty)(T, l, c, u, d, q^+, q^-, \sigma, v^*)$ [110]

Input: A tree $T = (V, E)$, a lead time vector l, a capacity vector c, an upper bound vector u, a lower bound vector d, two cost vectors q^+ and q^-, a value σ, a vertex v^*.

Output: The optimal solution c^* and optimal value D^* of $(IVQ1C_\infty)$.

1: Partition the tree into T' and T'' as described in Definition 14.1. Determine the **gate edge** $e^* := (v^*, v_{i^*})$ and the leaves t^1_l and t^2_l by (14.15). Calculatc Q'_c and $Q^{i^*}_c$.

if $Q'_c \leq Q^{i^*}_c$ **then return** $c^* := c$, $D^* := 0$.
else if $l(P_{v^*t_l^1}) > l(P_{v_{i^*}t_l^2})$ **then** let $E^+ := E'$, $E^- := E''$.
else let $E^+ := E' \setminus \{e^*\}$, $E^- := E'' \cup \{e^*\}$.
end if
Construct $\bar{c}$ as described in (14.20). Calculate $Q'_{\bar{c}} := \max_{t \in Z'} \{Q_{\bar{c}}(v^*, t, \sigma)\}$ and $Q^{i^*}_{\bar{c}} := \max_{t \in Z''}\{Q_{\bar{c}}(v_{i^*}, t, \sigma)\}$, respectively.
if $Q'_{\bar{c}} > Q^{i^*}_{\bar{c}}$ **then return** c^* an empty vector and $D^* := +\infty$.
else
  Let $\Lambda^+ := \{D_i = q^+(e_i)u(e_i) \mid e_i \in E^+\}$ and c_{D_i} be calculated by (14.18). Call $[E^+_u, E^+_{D^*}] :=$. **Findsets**$(T', T'', E^+, S_{E^+}, \Lambda^+, c_{\Lambda^+}, Q'_{c_{\Lambda^+}}, Q^{i^*}_{c_{\Lambda^+}})$.
  Let $\Lambda^- := \{D_i = q^-(e_i)d(e_i) \mid e_i \in E^-\}$ and c_{D_i} be calculated by (14.18). Call $[E^-_d, E^-_{D^*}] :=$ **Findsets**$(T', T'', E^-, S_{E^-}, \Lambda^-, c_{\Lambda^-}, Q'_{c_{\Lambda^-}}, Q^{i^*}_{c_{\Lambda^-}})$.
  Calculate $\bar{D}$ and $\underline{D}$ by (14.21) and (14.22), respectively.
  Calculate I' by (14.24), (14.25), (14.28), I'' by (14.26)–(14.28), respectively. Let $I := I' \cup I'' \cup \{\underline{D}, \bar{D}\} \cap [\underline{D}, \bar{D}]$.
  Call $[D_{\tau_1}, D_{\tau_2}] :=$ **BinarySearch**$(T', T'', S_E, I, c_I, Q'_{c_I}, Q^{i^*}_{c_I})$
  Calculate I^+_H, I^-_H by (14.32), (14.33), respectively and let $I^* := I^+_H \cup I^-_H \cup \{D_{\tau_1}, D_{\tau_2}\} \cap [D_{\tau_1}, D_{\tau_2}]$.
  Call $[D^*_{\tau_1}, D^*_{\tau_2}] :=$ **Binary Search**$(T', T'', S_E, I^*, c_{I^*}, Q'_{c_{I^*}}, Q^{i^*}_{c_{I^*}})$.
  Determine the paths $P_{v^*t'}$ and $P_{v_{i^*}t''}$ and the edges e' and e'' by (14.34) and (14.35), respectively.
  Calculate the optimal value D^* and optimal solution c^* by (14.37) and (14.29), respectively.
end if
Return c^* and D^*.

Theorem 14.6 ([110]) *The time complexity of Algorithm 14.3 is $O(n^2 \log n)$.*

14.2.2 (Inv, AQ1C, Bounded, Weighted l_∞ Norm)

In this subsection, we introduce the mathematical model of the inverse absolute quickest 1-center (**IAQ1C**$_\infty$) problem under weighted l_∞ norm and some notations.

For the problem (*IAQ1C*$_\infty$), we are intended to modify the capacity vector from c to $\tilde{c}$ with minimum cost under weighted l_∞ norm such that a given point $s^* \in T$ becomes an absolute quickest 1-center. Denote the positive unit costs of increasing and decreasing the capacity $c(e)$ by $q^+(e)$ and $q^-(e)$, respectively. The nonnegative variables $x(e)$ and $y(e)$ represent the values of increments and decrements of $c(e)$, respectively. The nonnegative values $u(e)$ and $d(e)$ are the upper bounds for the increments and decrements of $c(e)$ and $d(e) < c(e)$. Let $Q_{\tilde{c}}(s^*, t, \sigma)$ be the transmission time from s^* to t under a capacity vector $\tilde{c}$. The problem $(IAQ1C_\infty)$ can be mathematically described below:

$$\min\ f(\tilde{c}) = \max_{e\in E}\{q^+(e)x(e), q^-(e)y(e)\}$$

$$(\mathbf{IAQ1C_\infty})\ s.t.\ \max_{t\in Z(T)}\{Q_{\tilde{c}}(s^*,t,\sigma)\} \le \max_{t\in Z(T)}\{Q_{\tilde{c}}(s,t,\sigma)\}, \forall s\in T, \quad (14.38)$$

$$\tilde{c}(e) = c(e) + x(e) - y(e), \quad (14.39)$$

$$0 \le x(e) \le u(e), \quad (14.40)$$

$$0 \le y(e) \le d(e). \quad (14.41)$$

For any capacity vector $\tilde{c}$, we can partition the tree T from s^* to $T'_{\tilde{c}}$ and $T''_{\tilde{c}}$ by Definition 14.2. For convenience, denote $T' = T'_c$, $T'' = T''_c$, $E' = E(T')$, $E'' = E(T'')$, $Z' = Z(T')$, and $Z'' = Z(T'')$. Let $\lambda_i^+ = q^+(e_i)u(e_i)$ and $\lambda_i^- = q^-(e_i)d(e_i)$ for each $e_i \in E$. Denote the critical edge by $e^* = (v^1_{s^*}, v_{s^*})$ under the original capacity vector c. Let $Q'_{\tilde{c}} = \max_{t\in Z'} Q_{\tilde{c}}(s^*,t,\sigma)$ and $Q''_{\tilde{c}} = \max_{t\in Z''} Q_{\tilde{c}}(s^*,t,\sigma)$.

Notably, the given point s^* may be located on a vertex or the interior of the critical edge e^*. When s^* is located on a vertex, it is obvious that the critical edge $e^* \in E'$. However, when $s^* \in Int(e^*)$, the point s^* divides e^* into two sub-edges $e''^* = (v^1_{s^*}, s^*) \in E''$ and $e'^* = (s^*, v_{s^*}) \in E'$. What is especially important is that we should assure $\tilde{c}(e''^*) = \tilde{c}(e'^*) = \tilde{c}(e^*)$ under any capacity vector $\tilde{c}$.

According to Theorem 14.2, we have to modify the original capacity vector c to make the given point s^* satisfy *Condition* 1 or 2. We denote the problem $(IAQ1C_\infty)$ satisfying *Condition* 1 and 2 by $(IAQ1C1_\infty)$ and $(IAQ1C2_\infty)$, respectively. In the subsequent, we always assume that s^* does not satisfy both *Conditions* 1 and 2 and then $Q'_c > Q''_c$ and $Q'_c > \max_{t\in Z''}\{l(P_{s^*t}) + \frac{\sigma}{\tilde{c}(e^*)}\}$.

Remark 14.2 Even if s^* is located on a vertex, the problem $(IAQ1C_\infty)$ is different with the inverse vertex quickest 1-center problem because we should assure the inequality (14.38) holds for any $s \in T$, which is stricter than $\max_{t\in Z} Q_{\tilde{c}}(s^*,t,\sigma) \le \max_{t\in Z} Q_{\tilde{c}}(v,t,\sigma)$ for any $v \in V$ in the $(IVQ1C)$ problem.

Due to the complexity of the algorithms in the literature, we only introduce the main idea here. For detailed algorithms, readers can refer to the paper [111]. We independently solve the two problems $(IAQ1C1_\infty)$ and $(IAQ1C2_\infty)$, respectively. We mainly focus on the case that the point s^* is located on a vertex and then demonstrate that the case $s^* \in Int(e^*)$ can be similarly solved. In each subproblem, we determine how to modify the capacity of every edge and make use of a binary search method twice to find an exact interval where the optimal value lies in. Then, we calculate the optimal solution and the optimal value of the corresponding subproblem. Finally, we choose the one that has the smaller optimal objective value.

14.3 Inverse Vertex and Absolute Quickest 1-Center Problems Under Weighted l_1 Norm

In order to deal with the two inverse problems in a unified framework, we first deal with a maximum transmission time balance **(MTTB)** problem in this subsection. We investigate **(MTTB)** by transforming it into a series of minimum value problems with one variable and choose the one whose total cost is minimum. Then we transform the two inverse problems into two **(MTTB)** problems with different parameters.

14.3.1 Maximum Transmission Time Balance Problem

Given two trees $T_1(V_1, E_1)$, $T_2(V_2, E_2)$ rooted at vertices $s_1 ands_2$, respectively, assume that $E_1 \cap E_2 = \emptyset$. Let $l(e)$ and $c(e)$ represent the lead time and capacity for every edge $e \in E_1 \cup E_2$. Denote the values of increments and decrements of $c(e)$ by $x(e)$ and $y(e)$, which are upper bounded by $u(e)$ and $d(e)$ with $d(e) < c(e)$. Let the positive values $q^+(e), q^-(e)$ represent the unit costs of increasing and decreasing the capacity $c(e)$. Let $Z(T_i)$ be the set of leaves of T_i and $Q^i_{\tilde{c}}(s_i, t, \sigma)$ be the transmission time to transmit σ units data from s_i to $t \in Z(T_i)$ under a capacity vector $\tilde{c}$ with $i = 1, 2$.

The problem (*MTTB*) consists in modifying capacity vector from c to $\tilde{c}$ with minimum cost under weighted l_1 norm such that $\max\limits_{t\in Z(T_1)} \{Q^1_{\tilde{c}}(s_1, t, \sigma)\} = \max\limits_{t\in Z(T_2)} \{Q^2_{\tilde{c}}(s_2, t, \sigma)\}$, which can be mathematically described below:

$$\min g(\tilde{c}) = \sum_{e\in E_1\cup E_2} (q^+(e)x(e) + q^-(e)y(e))$$

$$\textbf{(MTTB)}\ s.t.\ \max_{t\in Z(T_1)} \{Q^1_{\tilde{c}}(s_1, t, \sigma)\} = \max_{t\in Z(T_2)} \{Q^2_{\tilde{c}}(s_2, t, \sigma)\}, \tag{14.42}$$

$$\tilde{c}(e) = c(e) + x(e) - y(e), e \in E_1 \cup E_2, \tag{14.43}$$

$$0 \le x(e) \le u(e), e \in E_1 \cup E_2, \tag{14.44}$$

$$0 \le y(e) \le d(e), e \in E_1 \cup E_2. \tag{14.45}$$

For convenience, let $Q^i_{\tilde{c}} = \max_{t\in Z(T_i)} Q^i_{\tilde{c}}(s_i, t, \sigma)$, $i = 1, 2$ under a capacity vector $\tilde{c}$, and we call $\tilde{c}$ a feasible solution of the problem (*MTTB*) if $\tilde{c}$ satisfies constraints (14.42)–(14.45).

It is a trivial case that $Q^1_c = Q^2_c$. We next assume that $Q^1_c > Q^2_c$. Let Q be a given value. Next, we first find the minimum cost of modifications of capacities to satisfy $Q^1_{\tilde{c}} = Q$. Then we find the cost $g^2_k(Q)$ of decreasing capacity of one edge $e_k \in E_2$ to satisfy $Q^2_{\tilde{c}} = Q$. Finally, we propose an algorithm to solve the problem

(*MTTB*) by calculating a series of subproblems ($MTTB_{e_k}$), which can be solved by a minimum value problem with one variable.

Let $\bar{c}(e) = c(e) + u(e)$ for $e \in E_1$. Define $L_i(e) = \{t \in Z(T_i) \mid e \in P_{s_i t}\}$ be the set of leaves controlled by $e \in E_i$, $i = 1, 2$.

Lemma 14.10 ([109]) *Let c^* be an optimal solution of the problem (MTTB). Then the maximum transmission time $Q^1_{c^*}$ is in the interval $[Q^1_{\bar{c}}, Q^1_c]$.*

Let $G^1_Q(\tilde{c})$ be the minimum cost of increasing capacities of $e \in E_1$ such that the maximum transmission time $Q^1_{\tilde{c}}$ in T_1 is Q. More exactly, for $Q \in [Q^1_{\bar{c}}, Q^1_c]$, $G^1_Q(\tilde{c})$ is the optimal value of the following optimization problem:

$$G^1_Q(\tilde{c}) = \min \sum_{e \in E_1} q^+(e)x(e) \tag{14.46}$$

$$s.t. \quad \max_{t \in Z(T_1)} \{Q^1_{\tilde{c}}(s_1, t, \sigma)\} = Q, \tag{14.47}$$

$$\tilde{c}(e) = c(e) + x(e), e \in E_1, \tag{14.48}$$

$$0 \le x(e) \le u(e), e \in E_1. \tag{14.49}$$

We next try to present the analytical form of $G^1_Q(\tilde{c})$. For every edge $e \in E_1$, define $l^1_e = \max\{l(P_{s_1 t}) \mid t \in L_1(e)\}$ and $\xi^1_e = \frac{\sigma}{Q - l^1_e}$. Define $S_Q = \{Q^1_e = l^1_e + \frac{\sigma}{c(e)} \mid e \in E_1, Q^1_e \ge Q^1_{\bar{c}}\}$, and sort the elements in S_Q such that $Q^1_{e_{j_1}} \le Q^1_{e_{j_2}} \le \cdots \le Q^1_{e_{jm}} = Q^1_c$. For convenience, let $Q^1_{e_{j_0}} = Q^1_{\bar{c}}$ and $g^1_{j_h}(Q) = q^+(e_{j_h})(\xi^1_{e_{j_h}} - c(e_{j_h}))$.

Theorem 14.7 ([109]) *Let Q be a value in $[Q^1_{e_{j_i}}, Q^1_{e_{j_{i+1}}})$ with $i = 0, 1, \cdots, m-1$. For $e \in E_1$, let $\tilde{c}^1(e) = \xi^1_e$ if $c(e) < \xi^1_e$ and $\tilde{c}^1(e) = c(e)$ for other edges. Then $\tilde{c}^1$ is an optimal solution of (14.46)–(14.49) and the minimum cost $G^1_Q(\tilde{c}^1)$ can be calculated by*

$$G^1_Q(\tilde{c}^1) = \sum_{h=i+1}^{m} g^1_{j_h}(Q). \tag{14.50}$$

We next consider decreasing capacity of exactly one edge such that $Q^2_{\tilde{c}} = Q$. For any $e \in E_2$, define $l^2_e = \max\{l(P_{s_2 t}) \mid t \in L_2(e)\}$ and $\xi^2_e = \frac{\sigma}{Q - l^2_e}$.

Lemma 14.11 ([109]) *Let Q be a value with $Q > Q^2_c$. Define $\tilde{c}^2(e_k) = \xi^2_{e_k}$ for a given edge $e_k \in E_2$ and $\tilde{c}^2(e) = c(e)$ for $e \in E_2 \setminus \{e_k\}$. Then we have $Q^2_{\tilde{c}^2} = Q$.*

Lemma 14.12 ([109]) *Let c^* be an optimal solution of the problem (MTTB). Then we have $Q^2_{c^*} \ge Q^2_c$. Correspondingly, we have $c^*(e) \le c(e)$ for $e \in E_2$. Particularly, there is at most one edge with $c^*(e) < c(e)$.*

For every edge $e_k \in E_2$, define $Q^2_{e_k} = l^2_{e_k} + \frac{\sigma}{c(e_k)-d(e_k)}$ and $E^{\geq}_2 = \{e_k \in E_2 \mid Q^2_{e_k} \geq Q^2_{\tilde{c}}\}$. According to Lemma 14.12, we only consider the edge $e_k \in E^{\geq}_2$. For every $e_k \in E^{\geq}_2$, let $g^2_k(Q) = q^-(e_k)(c(e_k) - \xi^2_{e_k})$ with $Q \in [Q^2_c, Q^2_{e_k}]$ represent the total cost of decreasing the capacity of e_k such that the maximum transmission time in T_2 is Q.

For every edge $e_k \in E^{\geq}_2$, we consider a subproblem ($MTTB_{e_k}$), which consists in increasing the capacity of $e \in E_1$ and decreasing the capacity of e_k such that $Q^1_{\tilde{c}} = Q^2_{\tilde{c}} = l^2_{e_k} + \frac{\sigma}{\tilde{c}(e_k)}$. It can be mathematically stated below:

$$
\begin{aligned}
&\min\ g_k(Q) = \sum_{e \in E_1} q^+(e)x(e) + q^-(e_k)(c(e_k) - \tilde{c}(e_k))\\
(\mathbf{MTTB}_{e_k})\ \ &s.t.\ \max_{t \in Z(T_1)} \{Q_{\tilde{c}}(s_1, t, \sigma)\} = l^2_{e_k} + \frac{\sigma}{\tilde{c}(e_k)} = Q,\\
&\tilde{c}(e) = c(e) + x(e), e \in E_1,\\
&0 \leq x(e) \leq u(e), e \in E_1,\\
&0 \leq c(e_k) - \tilde{c}(e_k) \leq d(e_k).
\end{aligned}
$$

Lemma 14.13 ([109]) *If $Q^2_{e_k} < Q^1_{\tilde{c}}$, then the subproblem ($MTTB_{e_k}$) is infeasible.*

If $Q^2_{e_k} \geq Q^1_{\tilde{c}}$, let $[\underline{Q}, \bar{Q}^k] = [Q^1_{\tilde{c}}, Q^1_c] \cap [Q^2_c, Q^2_{e_k}]$. Define $Q^1_{e_{j_{i*}}} = \min\{Q^1_e \in S_Q \mid Q^1_e \geq \underline{Q}\}$ and $Q^1_{e_{j_{p*}}} = \min\{Q^1_e \in S_Q \mid Q^1_e \geq \bar{Q}^k\}$. For convenience, let $Q_{e_{j_{i^0}}} = \underline{Q}$ and $Q_{e_{j_{p^0}}} = \bar{Q}^k$. Then according to Theorem 14.7 and Lemma 14.12, we present the analytical form of $g_k(Q)$ with $Q \in [\underline{Q}, \bar{Q}^k]$ as follows:

Corollary 14.5 ([109]) *The total cost $g_k(Q)$ with $Q \in [\underline{Q}, \bar{Q}^k]$ can be represented as $g_k(Q) = \sum_{h=i+1}^m g^1_{j_h}(Q) + g^2_k(Q)$, $Q \in [Q^1_{e_{j_i}}, Q^1_{e_{j_{i+1}}})$, $i = i^0, \cdots, p^0 - 1$ if $i^* < p^*$, or $g_k(Q) = \sum_{h=p^*}^m g^1_{j_h}(Q) + g^2_k(Q)$, $Q \in [\underline{Q}, \bar{Q}^k]$ when $i^* = p^*$.*

Notice that for a given index i, $g_k(Q)$ is in the form of

$$
g_k(Q) = \sum_{h=i+1}^{m} q^+(e_{j_h}) \left(\frac{\sigma}{Q - l^1_{e_{j_h}}} - c(e) \right) + q^-(e_k) \left(c(e_k) - \frac{\sigma}{Q - l^2_{e_k}} \right). \tag{14.51}
$$

Obviously, the function $g_k(Q)$ is continuous in $[\underline{Q}, \bar{Q}^k]$. Then $g_k(Q)$ can attain a minimum value $g_k(Q_k)$ according to Weierstrass extreme value theorem. Furthermore, we can calculate the optimal value $g_k(Q_k)$ of the problem ($MTTB_{e_k}$) for every $e_k \in E^{\geq}_2$ and choose the one e_{k^*} whose optimal value $g_{k^*}(Q_{k^*})$ is minimum. Therefore, the optimal value of the problem ($MTTB$) is $g_{k^*}(Q_{k^*})$.

We first present an algorithm of solving the subproblem $(MTTB_{e_k})$.

Algorithm 14.4 $[Q_k, g_k(Q_k)] =$ $\mathbf{MTTB}_{e_k}(T_1, T_2, Q_l^1, Q_u^1, Q_l^2, Q_u^2)$ [109]

Input: Two trees $T_1(V_1, E_1)$ and $T_2(V_2, E_2)$ rooted at s_1, s_2, respectively, with vectors $l, c, u, d, q^+, and q^-$, a value σ, and four values $Q_l^1, Q_u^1, Q_l^2, and Q_u^2$.
Output: The transmission time Q_k and the optimal value $g_k(Q_k)$ of $(MTTB_{e_k})$.

if $Q_l^1 > Q_u^2$ **then**
  The problem $(MTTB_{e_k})$ is infeasible. **Return** $Q_k := +\infty$, $g_k(Q_k) := +\infty$.
else
  Let $[\underline{Q}, \bar{Q}^k] := [Q_l^1, Q_u^1] \cap [Q_l^2, Q_u^2]$, $S_Q := \{Q_e^1 = l_e^1 + \frac{\sigma}{c(e)} \mid e \in E_1, Q_e^1 \geq \underline{Q}\}$. Sort the values in S_Q nondecreasingly, and denote the sorted set still by S_Q.
  Let $Q^1_{e_{j_{i^*}}} := \min\{Q_e^1 \in S_Q \mid Q_e^1 \geq \underline{Q}\}$, $Q^1_{e_{j_{p^*}}} := \min\{Q_e^1 \in S_Q \mid Q_e^1 \geq \bar{Q}^k\}$.
  if $i^* < p^*$ **then**
    $SP := \{Q_e^1 \mid Q_e^1 \in S_Q, Q_e^1 \leq \bar{Q}^k\} \cup \{\underline{Q}, \bar{Q}^k\}$, $Q_{e_{j_{i^0}}} := \underline{Q}$, $Q_{e_{j_{p^0}}} := \bar{Q}^k$.
    for $i = i^0 : p^0 - 1$ **do**
      Let $g_k(Q) := \sum_{h=i+1}^{m} g^1_{j_h}(Q) + g_k^2(Q)$.
      Let $g_k'(Q) := -\sum_{h=i+1}^{m} q^+(e_{j_h})\frac{\sigma}{(Q-l^1_{e_{j_h}})^2} + q^-(e_k)\frac{\sigma}{(Q-l^2_{e_k})^2}$.
      Let $SP := SP \cup \{Q \in [Q^1_{e_{j_i}}, Q^1_{e_{j_{i+1}}}) \mid g_k'(Q) = 0\}$.
    end for
  else
    Let $g_k(Q) := \sum_{h=p^*}^{m} g^1_{j_h}(Q) + g_k^2(Q)$, $Q \in [\underline{Q}, \bar{Q}^k]$.
    Let $g_k'(Q) := -\sum_{h=p^*}^{m} q^+(e_{j_h})\frac{\sigma}{(Q-l^1_{e_{j_h}})^2} + q^-(e_k)\frac{\sigma}{(Q-l^2_{e_k})^2}$.
    Let $SP := \{Q \in [\underline{Q}, \bar{Q}^k] \mid g_k'(Q) = 0\} \cup \{\underline{Q}, \bar{Q}^k\}$.
  end if
  Return $Q_k := \arg\min\{g_k(Q) \mid Q \in SP\}$ and $g_k(Q_k)$.
end if

Suppose $|V_1| = n_1$ and we have the following theorem:

Theorem 14.8 ([109]) *The time complexity of Algorithm 14.4 is* $O(n_1^2)$.

For any $e_k \in E_{\bar{2}}^{\geq}$, we call $[Q_k, g_k(Q_k)]$ =**MTTB**$_{e_k}(T_1, T_2, Q_{\bar{c}}^1, Q_c^1, Q_{\bar{c}}^2, Q_{e_k}^2)$. Let $g_{k^*}(Q_{k^*}) = \min\{g_k(Q_k) \mid e_k \in E_{\bar{2}}^{\geq}\}$ be achieved at e_{k^*}. Correspondingly, the optimal solution c^* of (*MTTB*) is given below with $\xi_e^i = \frac{\sigma}{Q_{k^*} - l_e^i}, i = 1, 2$:

$$c^*(e) = \begin{cases} \xi_e^1, & e \in E_1, c(e) < \xi_e^1, \\ c(e), & e \in E_1, c(e) \geq \xi_e^1, \\ \xi_e^2, & e = e_{k^*}, \\ c(e), & e \in E_2 \setminus \{e_{k^*}\}. \end{cases} \tag{14.52}$$

Algorithm 14.5 $[c^*, g(c^*)] = MTTB(T_1, T_2, E_{\bar{2}}^{\geq}, Q_l^1, Q_u^1, Q_{\bar{l}}^2)$ [109]

Input: Two trees $T_1(V_1, E_1)$ and $T_2(V_2, E_2)$ rooted at s_1 and s_2, respectively, with vectors $l, c, u, d, q^+, andq^-$, a value σ, a set $E_{\bar{2}}^{\geq}$ of edges, and three values $Q_l^1, Q_u^1, andQ_{\bar{l}}^2$.

Output: An optimal solution c^* and the optimal value $g(c^*)$ of (*MTTB*).

1: Initialize $Q_{k^*} := +\infty$, $g_{k^*} := +\infty$, $e_{k^*} := \emptyset$.
2: **for** $e_k \in E_{\bar{2}}^{\geq}$ **do**
3: Calculate $Q_{e_k}^2 := l_{e_k}^2 + \frac{\sigma}{c(e_k) - d(e_k)}$.
4: Call $[Q_k, g_k(Q_k)]$:=**MTTB**$_{e_k}(T_1, T_2, Q_l^1, Q_u^1, Q_{\bar{l}}^2, Q_{e_k}^2)$.
5: **if** $g_k(Q_k) < g_{k^*}$ **then**
6: Update $Q_{k^*} := Q_k$, $g_{k^*} := g_k(Q_k)$, $e_{k^*} := e_k$.
7: **end if**
8: **end for**
9: **if** $g_{k^*} = +\infty$ **then return** c^* an empty vector and $g(c^*) := +\infty$.
10: **else return** an optimal solution c^* by (14.52) and the value $g(c^*) := g_{k^*}$.
11: **end if**

Suppose $|V_2| = n_2$ and we have the following theorem:

Theorem 14.9 ([109]) *The time complexity of Algorithm 14.5 is* $O(n_1^2 n_2)$.

14.3.2 (Inv, VQ1C, Bounded, Weighted l_1 Norm)

In this subsection, we consider the problem ($IVQ1C_1$). We first characterize a necessary optimality condition and then deal with the problem from two cases. In each case, we transform the problem into a maximum transmission time balance problem.

Given a vertex s^*, the problem ($IVQ1C_1$) consists in modifying capacity vector from c to $\check{c}$ with minimum cost under weighted l_1 norm such that s^* becomes a vertex quickest 1-center, which can be mathematically stated below:

$$\min f(\check{c}) = \sum_{e \in E} (q^+(e)x(e) + q^-(e)y(e)) \tag{14.53}$$

$$(\mathbf{IVQ1C}_1) \; s.t. \; \max_{t \in Z(T)} \{Q_{\check{c}}(s^*, t, \sigma)\} \le \max_{t \in Z(T)} \{Q_{\check{c}}(s, t, \sigma)\}, \forall s \in V, \tag{14.54}$$

$$\check{c}(e) = c(e) + x(e) - y(e), e \in E, \tag{14.55}$$

$$0 \le x(e) \le u(e), e \in E, \tag{14.56}$$

$$0 \le y(e) \le d(e), e \in E. \tag{14.57}$$

We first present a profound property which demonstrates that the subtrees T' and T'' remain unchanged under an optimal capacity vector.

Define $L'(e) = \{t \in Z(T') \mid e \in P_{s^*t}\}$, $L''(e) = \{t \in Z(T'') \mid e \in P_{s^*t}\}$, $l'_e = \max\{l(P_{s^*t}) \mid t \in L'(e)\}$, and $l''_e = \max\{l(P_{s^*t}) \mid t \in L''(e)\}$.

Lemma 14.14 ([109]) *Let D^* be the optimal objective value of the problem ($IVQ1C_1$). If there is a capacity vector $\check{c}$ satisfying (14.55)–(14.57) with*

$$\max_{t \in Z(T)} \{Q_{\check{c}}(s^*, t, \sigma)\} > \max_{t \in Z(T')} \{Q_{\check{c}}(s^*, t, \sigma)\}, \tag{14.58}$$

then we have $f(\check{c}) > D^*$.

According to Lemma 14.14, we can infer that $\check{c}$ could not be an optimal capacity vector unless $T'_{\check{c}} = T'_c = T'$. Therefore, we next always consider the capacity vector $\check{c}$ such that $\max_{t \in Z(T')}\{Q_{\check{c}}(s^*, t, \sigma)\} = \max_{t \in Z(T)}\{Q_{\check{c}}(s^*, t, \sigma)\}$, which implies that $T'_{\check{c}} = T'$, v_{s^*} is also the **gate** and $e^* = (s^*, v_{s^*})$ is also the **gate edge** under $\check{c}$. Furthermore, according to Theorem 14.1, we can reformulate the mathematical model (14.53)–(14.57) of the problem $\mathbf{IVQ1C}_1$ below:

$$\min f(\check{c}) = \sum_{e \in E} (q^+(e)x(e) + q^-(e)y(e))$$

$$(\mathbf{IVQ1C}_1) \; s.t. \; \max_{t \in Z(T')} \{Q_{\check{c}}(s^*, t, \sigma)\} \le \max_{t \in Z(T'')} \{Q_{\check{c}}(v_{s^*}, t, \sigma)\}, \tag{14.59}$$

$$\max_{t \in Z(T')} \{Q_{\check{c}}(s^*, t, \sigma)\} = \max_{t \in Z(T)} \{Q_{\check{c}}(s^*, t, \sigma)\}, \tag{14.60}$$

$$\check{c}(e) = c(e) + x(e) - y(e), e \in E, \tag{14.61}$$

$$0 \le x(e) \le u(e), e \in E, \tag{14.62}$$

$$0 \le y(e) \le d(e), e \in E. \tag{14.63}$$

For convenience, we call $\check{c}$ a feasible solution if $\check{c}$ satisfies (14.59)–(14.63).

In order to transform the problem $(IVQ1C_1)$ into a maximum transmission time balance problem, we define some symbols:

$T_1 = T'$ rooted at s^*, $T_2 = T'' \cup \{e^*\}$ rooted at v_{s^*},
$Z(T_1) = Z(T')$, $Z(T_2) = Z(T'')$,
$E_1 = E(T')$, $E_2 = E(T'') \cup \{e^*\}$,
$L_1(e) = \{t \in Z(T_1) \mid e \in P_{s^*t}\}, e \in E_1$, $L_2(e) = \{t \in Z(T_2) \mid e \in P_{v_{s^*}t}\}, e \in E_2$,
$l_e^1 = \max\{l(P_{s^*t}) \mid t \in L_1(e)\}, e \in E_1$, $l_e^2 = \max\{l(P_{v_{s^*}t}) \mid t \in L_2(e)\}, e \in E_2$,
$Q_{\check{c}}^1 = \max_{t \in Z(T_1)}\{Q_{\check{c}}(s^*, t, \sigma)\}$, $Q_{\check{c}}^2 = \max_{t \in Z(T_2)}\{Q_{\check{c}}(v_{s^*}, t, \sigma)\}$,
$\xi_e^1 = \frac{\sigma}{Q - l_e^1}$, $\xi_e^2 = \frac{\sigma}{Q - l_e^2}$.

Obviously, we have $l_e^1 = l'_e$ for $e \in E_1$, $l_e^2 = l(e^*) + l''_e$ for $e \in E_2 \setminus \{e^*\}$, $l_{e^*}^1 = \max_{t \in Z(T_1)} l(P_{s^*t})$, $l_{e^*}^2 = \max_{t \in Z(T_2)} l(P_{v_{s^*}t})$, $Q_{\check{c}}^1 = Q'_{\check{c}}$, and $Q_{\check{c}}^2 = Q_{\check{c}}^{v_{s^*}}$.

Due to inequalities (14.4) and (14.59), a straightforward idea is to reduce Q_c^1 and augment Q_c^2 by increasing and decreasing the capacities of $e \in E_1$ and $e \in E_2$. However, it is difficult to determine whether we increase or decrease the capacity of the **gate edge** e^* since $E_1 \cap E_2 = \{e^*\}$. Next, we solve the dilemma by checking the numerical relationship between $l_{e^*}^1$ and $l_{e^*}^2$. We show that $c(e^*)$ should not be decreased when $l_{e^*}^1 > l_{e^*}^2$ and should not be increased otherwise.

For convenience, denote the problem by $(IVQ1C_1^{>})$ and $(IVQ1C_1^{\leq})$ when $l_{e^*}^1 > l_{e^*}^2$ and $l_{e^*}^1 \leq l_{e^*}^2$, respectively.

Lemma 14.15 ([109]) *Let c^* be an optimal solution of $(IVQ1C_1^{>})$. Then $c^*(P_{v_{s^*}t_2^*}) < c^*(e^*)$, where $t_2^* \in Z(T_2)$ satisfies $Q_{c^*}(v_{s^*}, t_2^*, \sigma) = Q_{c^*}^2$.*

Lemma 14.16 ([109]) *Let c^* be an optimal solution of the problem $(IVQ1C_1^{>})$. Then we have $Q_{c^*}^1 \leq Q_c^1$. Correspondingly, we have $c^*(e) \geq c(e)$ for $e \in E_1$.*

Corollary 14.6 ([109]) *The transmission time $Q_{c^*}^1$ with respect to an optimal solution c^* of the problem $(IVQ1C_1^{>})$ is in the interval $[Q_{\hat{c}}^1, Q_c^1]$.*

Lemma 14.17 ([109]) *Let c^* be an optimal solution of the problem $(IVQ1C_1^{>})$. Then we have $Q''_{c^*} \geq Q''_c$. Correspondingly, we have $c^*(e) \leq c(e)$ for $e \in E(T'')$. Particularly, there is at most one edge $e \in E(T'')$ with $c^*(e) < c(e)$.*

Theorem 14.10 ([109]) *If c^* is an optimal solution of $(IVQ1C_1^{>})$, then $Q_{c^*}^1 = Q_{c^*}^2$.*

According to Theorem 14.10, we transform the problem $(IVQ1C_1^{>})$ into a maximum transmission time balance problem **MTTB**^{1}.

Lemma 14.18 ([109]) *The two problems $(IVQ1C_1^{>})$ and **MTTB**^{1} when $T_1 = T'$, $T_2 = T'' \cup \{e^*\}$, $s_1 = s^*$, $s_2 = v_{s^*}$ are equivalent with a same optimal solution.*

Let $Q_l^2 = l(e^*) + Q''_c$. According to Lemmas 14.15 and 14.17, there is at most one edge $e \in E(T'')$ with $c^*(e) < c(e)$ under an optimal solution c^* and $Q_{c^*}^2 = l(e^*) + Q''_{c^*} \geq Q_l^2$. Therefore, let $Q_{e_k}^2 = l_{e_k}^2 + \frac{\sigma}{c(e_k) - d(e_k)}$ and $E_2^{\geq} = \{e_k \in E(T'') \mid$

$Q_{e_k}^2 \geq Q_l^2\}$. Then, we call $[c^*, f(c^*)]$ =$MTTB(T_1, T_2, E_2^{\geq}, Q_{\bar{c}}^1, Q_c^1, Q_l^2)$ to find an optimal solution c^* with optimal value $f(c^*)$.

Lemma 14.19 ([109]) *Let $t_c \in Z(T_1)$ satisfy $Q_c(s^*, t_c, \sigma) = Q_c^1$. Then we have $c(P_{s^*t_c}) < c(e^*)$ under the original capacity vector c.*

Partition the edge set E_1 into $E_1^< = \{e \in E_1 \mid c(e) < c(e^*)\}$ and $E_1^{\geq} = E_1 \setminus E_1^<$. Similar to Lemma 14.16, we have the following conclusion:

Lemma 14.20 ([109]) *Let c^* be an optimal solution of the problem $(IVQ1C_1^{\leq})$. Then we have $Q_{c^*}^1 \leq Q_c^1$. Especially, we have $c^*(e) \geq c(e)$ for $e \in E_1^<$, $c^*(e) = c(e)$ for $e \in E_1^{\geq} \setminus \{e^*\}$ and $c^*(e^*) \leq c(e^*)$.*

For $e \in E_1$, define

$$\bar{c}(e) = \begin{cases} c(e) + \min\{c(e^*) - c(e), u(e)\}, & e \in E_1^<, \\ c(e), & e \in E_1^{\geq}. \end{cases} \tag{14.64}$$

Lemma 14.21 ([109]) *The transmission time $Q_{c^*}^1$ with respect to an optimal solution c^* of the problem $(IVQ1C_1^{\leq})$ is in the interval $[Q_{\bar{c}}^1, Q_c^1]$.*

Lemma 14.22 ([109]) *Let c^* be an optimal solution of $(IVQ1C_1^{\leq})$. Then we have*

Similar to Theorem 14.10, we have the following corollary:

Corollary 14.7 ([109]) *If c^* is an optimal solution of $(IVQ1C_1^{\leq})$, then $Q_{c^*}^1 = Q_{c^*}^2$.*

According to Lemma 14.22, there is at most one edge $e \in E_2$ with $c^*(e) < c(e)$ under an optimal solution c^* and $Q_{c^*}^2 \geq Q_c^2$. Therefore, define $Q_{e_k}^2 = l_{e_k}^2 + \frac{\sigma}{c(e_k) - d(e_k)}$ for $e_k \in E_2$ and $E_2^{\geq} = \{e_k \in E_2 \mid Q_{e_k}^2 \geq Q_c^2\}$. Then, we transform the problem $(IVQ1C_1^{\leq})$ into the problem **MTTB**^{2} and call $[c^*, f(c^*)]$ =$(MTTB)$ $(T_1, T_2, E_2^{\geq}, Q_{\bar{c}}^1, Q_c^1, Q_c^2)$ to find an optimal solution c^* with optimal value $f(c^*)$.

Algorithm 14.6 $[c^*, f(c^*)] =$ **IVQ1C**$_1(T, s^*)$ [109]

Input: A tree $T(V, E)$ with vectors $l, c, u, d, q^+, and q^-$, a value σ, and a vertex s^*.

Output: An optimal solution c^* and the optimal value $f(c^*)$ of $(IVQ1C_1)$.

Calculate $\max_{t \in Z(T)}\{Q_c(s^*, t, \sigma)\}$ and partition the tree at s^* into T', T''. Determine the **gate** v_{s^*} and the **gate edge** $e^* := (s^*, v_{s^*})$ under c. Let $Q_c' := \max_{t \in Z(T)}\{Q_c(s^*, t, \sigma)\}$ and calculate $Q_c'' := \max_{t \in Z(T'')}\{Q_c(s^*, t, \sigma)\}$, $Q_c^{v_{s^*}} := \max_{t \in Z(T'')}\{Q_c(v_{s^*}, t, \sigma)\}$.

if $Q_c' \leq Q_c^{v_{s^*}}$ **then**

  s^* is a vertex quickest 1-center of T under c, **return** c^*:=c and $f(c^*)$:=0 .

else

  Initialize $T_1 := T'$ and $T_2 := T'' \cup \{e^*\}$ rooted at s^* and v_{s^*}, respectively, with $E_1 := E(T')$, $E_2 := E(T'') \cup \{e^*\}$. Let $Q_c^1 := Q_c'$, $Q_c^2 := Q_c^{v_{s^*}}$.

if $l_{e^*}^1 > l_{e^*}^2$ **then**
  Let $\bar{c}(e) := c(e) + u(e)$ for $e \in E_1$ and calculate $Q_{\bar{c}}^1$.
  Let $Q_l^2 := l(e^*) + Q_c''$, $E_2^{\geq} := \{e_k \in E_2 \setminus \{e^*\} \mid Q_{e_k}^2 \geq Q_l^2\}$.
else
  Let $E_1^{<} := \{e \in E_1 \mid c(e) < c(e^*)\}$, $E_1^{\geq} := E_1 \setminus E_1^{<}$, $\bar{c}(e) := c(e) + \min\{c(e^*) - c(e), u(e)\}$ if $e \in E_1^{<}$, $\bar{c}(e) := c(e)$, if $e \in E_1^{\geq}$ and calculate $Q_{\bar{c}}^1$.
  Let $Q_l^2 := Q_c^2 = Q_c^{v_s*}$, $E_2^{\geq} := \{e_k \in E_2 \mid Q_{e_k}^2 \geq Q_l^2\}$.
end if
Call $[c^*, g(c^*)] = MTTB(T_1, T_2, E_2^{\geq}, Q_{\bar{c}}^1, Q_c^1, Q_l^2)$.
Return an optimal solution c^* with optimal value $f(c^*) := g(c^*)$.
end if

Theorem 14.11 ([109]) *The time complexity of Algorithm 14.6 is* $O(n^3)$.

For the inverse absolute quickest 1-center problem under weighted l_1 norm, the main idea is similar to that of **IVQ1C**$_1$. However, there is an **MTTB** which is not entirely equivalent to the corresponding inverse problem. The readers can refer to the related papers for more details.

14.4 Conclusions

In this chapter, we introduce the main results of the classic inverse center and median location problems. In addition, we focus on the inverse quickest 1-center problem under weighted l_∞ and l_1 norm.

For the inverse quickest 1-center problem under weighted l_∞ norm, we characterize the optimality conditions for inverse vertex and absolute 1-center problems to find that the partition of the tree T under the optimal solution is unchanged. Then, we determine how the capacity of each edge should be modified. Further, we propose an effective modification by means of a greedy strategy and design a corresponding binary algorithm to determine the interval where the optimal value lies. Finally, we present two $O(n^2 \log n)$ algorithms.

For the inverse quickest 1-center problem under weighted l_1 norm, we study the inverse vertex and absolute quickest 1-centers in a unified framework by introducing a maximum transmission time balance problem. Then, we analyze the optimality conditions of the two inverse problems and transform them into corresponding maximum transmission time balance problem. Finally, we design two $O(n^3)$ algorithms.

References

1. Afrashteh, E., Alizadeh, B., Baroughi, F.: Linear time optimal approaches for max–profit inverse 1–median location problems. AP JOR. **35**(5), 1–22 (2018)
2. Afrashteh, E., Alizadeh, B., Baroughi ,F.: Optimal algorithms for selective variants of the classical and inverse median location problems on trees. Optim. Methods Softw. **34**(6), 1213–1230 (2019)
3. Afrashteh, E., Alizadeh, B., Baroughi, F.: Optimal algorithms for integer inverse undesirable p–median location problems on weighted extended star networks. J. Oper. Res. Soc. China. **9**(1), 99–117 (2021)
4. Ahuja, R.K., Magnanti, T.L., Orlin, J.B.: Network Flows, Theory, Algorithms and Applications. Prentice Hall, Englewood Cliffs, NJ (1993)
5. Ahuja, R.K., Orlin, J.B.: Inverse Optimization. Oper. Res. **49**(5), 771–783 (2001)
6. Ahuja, R.K., Orlin, J.B.: A faster algorithm for the inverse spanning tree problem. J. Algorithms **34**, 177–193 (2000)
7. Alizadeh, B., Afrashteh, E., Baroughi, F.: Inverse obnoxious p-median location problems on trees with edge length modifications under different norms. Theor. Comput. Sci, **772**(1), 73–87 (2019)
8. Alizadeh, B., Afrashteh, E.: Budget constrained inverse median facility location problem on tree networks. Appl. Math. Comput. **375**(12), 50–78 (2020)
9. Alizadeh, B., Burkard, R.E.: Combinatorial algorithms for inverse absolute and vertex 1-center location problems on trees. Networks **58**(3), 190–200 (2011)
10. Alizadeh, B., Burkard, R.: A linear time algorithm for inverse obnoxious center location problems on networks. Cent. Eur. J. Oper. Res. **21**(3), 585–594 (2013)
11. Alizadeh, B., Burkard, R., Pferschy, U.: Inverse 1-center location problems with edge length augmentation on trees. Computing **86**(4), 331–343 (2009)
12. Alizadeh, B., Etemad, R.: Linear time optimal approaches for reverse obnoxious center location problems on networks. Optimization **65**(11), 2025–2036 (2016)
13. Alizadeh, B., Etemad, R.: Optimal algorithms for inverse vertex obnoxious center location problems on graphs. Theor. Comput. Sci. **707**(1), 36–45 (2018)
14. Ayer, T.: Inverse optimization for assessing emerging technologies in breast cancer screening. Ann. Oper. Res. **230**(1), 57–85 (2015)
15. Bazgan, C., Nichterlein, A., et al.: A refined complexity analysis of finding the most vital edges for undirected shortest paths: algorithms and complexity. LNCS **9079**, 47–60 (2015)
16. Bondy, J.A., Murty, U.S.R.: Graph Theory. Springer, New York (2008)
17. Bulut, A., Ralphs, T.K.: On the complexity of inverse mixed integer linear optimization. SIAM J. Optim. **31**(4), 3014–3043 (2021)

X. Guan et al., *Inverse Combinatorial Optimization Problems*, Springer Optimization and Its Applications 225, https://doi.org/10.1007/978-3-031-91175-0

18. Burkard, R.E., Galavii, M., Gassner, E.: The inverse Fermat–Weber problem. Eur. J. Oper. Res. **206**(1), 11–17 (2010)
19. Burkard, R., Gassner, E., Hatzl, J.: A linear time algorithm for the reverse 1-median problem on a cycle. Networks **48**(1), 16–23 (2006)
20. Burkard, R., Lin, Y., Zhang, J.Z.: Weight reduction problems with certain bottleneck objectives. Eur. J. Oper. Res. **153**, 191–199 (2004)
21. Burkard, R.E., Pleschiutschnig, C., Zhang, J.Z.: Inverse median problems. Discrete Optim. **1**(1), 23–39 (2004)
22. Burton, D., Toint, P.L.: On an instance of the inverse shortest paths problem. Math. Program. **53**(1–3), 45–61 (1992)
23. Cai, M.C., Duin, C.W., Yang, X.G., Zhang, J.Z.: The partial inverse minimum spanning tree problem when weight increasing is forbidden. Eur. J. Oper. Res. **188**, 348–353 (2008)
24. Cai, M., Yang, X., Li, Y.: Inverse problems of submodular functions on digraphs. J. Optim. Theory App. **104**(3), 559–575 (2000)
25. Cai, M.C., Yang, X.G., Zhang, J.Z.: The complexity analysis of the inverse center location problem. J. Global Optim. **15**(2), 213–218 (1999)
26. Chan, T.C., Mahmood, R., Zhu, I.Y.: Inverse optimization: Theory and applications. Oper. Res. (2023). https://doi.org/10.1287/opre.2022.0382
27. Christofides, N.: Graph Theory: An Algorithmic Approach. Academic Press, London (1975)
28. Chung, Y.: Inverse bin-packing number problems: hardness and approximation results. Manag. Sci. Financ. Eng. **18**(2), 19–22 (2012)
29. Chung, Y.: Inverse bin-packing number problems: polynomially solvable cases. Manag. Sci. Financ. Eng. **19**(1), 25–28 (2013)
30. Chung, Y., Demange, M.: The 0-1 inverse maximum stable set problem. Discret. Appl. Math. **156**(13), 2501–2516 (2008)
31. Chung, Y., Demange, M.: On inverse traveling salesman problems. 4OR-Q. J. Oper. Res. **10**, 193–209 (2012)
32. Chung, Y., Park, M.J.: Notes on inverse bin-packing problems. Inform. Process. Lett. **115**(1), 60–68 (2015)
33. Cormen, T.H., Leiserson, C.E., Rivest, R.L.: Introduction to Algorithms, 4th edn. MIT Press, Cambridge (2022)
34. Dahlhaus, E., Johnson, D.S., Papadimitriou, C.H., Seymour, P.D., Yannakakis, M.: The complexity of multiterminal cuts. SIAM J. Comput. **23**, 864–894 (1994)
35. Deaconu, A.M., Tayyebi, J.: Inverse maximum capacity path problems under sum-type and max-type distances and their practical application to transportation networks. IEEE Access **8**, 225957–225966 (2020)
36. Dong, Q.Z., Li, X.Y., Yang, Y.: Partial inverse min-max spanning tree problem under the weighted bottleneck hamming distance. J. Comb. Optim. **46**(4), 27 (2023). https://doi.org/10.1007/s10878-023-01093-8
37. Duin, C.W., Volgenant, W.: Some inverse optimization problems under the Hamming distance. Eur. J. Oper. Res. **170**, 887–899 (2006)
38. Edmons, J., Giles, R.: A min–max relation for submodular functions on graphs. Ann. Discrete Math. **1**, 185–204 (1977)
39. Ertugrl, A., Gokhan, O., Cevriye, T.G.: Determining the most vital arcs on the shortest path for fire trucks in terrorist actions that will cause fire. Commun. Fac. Sci. Univ. Ank. Ser. A1 Math. Stat. **68**(1), 441–450 (2019)
40. Etemad, R., Alizadeh, B.: Combinatorial algorithms for reverse selective undesirable center location problems on cycle graphs. J. Oper. Res. Soc. China **5**(3), 347–361 (2017)
41. Fathali, J.: A row generation method for inverse continuous facility location problem. Comput. Ind. Eng. **171**, 108482 (2022)
42. Frank, A.: An algorithm for submodular functions on graphs. Ann. Discrete Math. **16**, 97–120 (1982)
43. Fredman, M.L., Tarjan, R.E.: Fibonacci heaps and their uses in improved network optimization algorithms. J. ACM. **34**(3), 596–615 (1987)

44. Garey, M.R., Johnson, D.S.: Computers and Intractability – A Guide to the Theory of NP-Completeness. Freeman, San Francisco (1979)
45. Gassner, E.: The inverse 1–maxian problem with edge length modification. J. Comb. Optim. **16**(1), 50–67 (2008)
46. Ghiyasvand, M., Keshtkar, I.: Solving the absolute 1–center problem in the quickest path case. B. Iran. Math. Soc. **48**(2), 643–671 (2022)
47. Goldberg, A.V., Tarjan, R.E.: Solving minimum cost flow problem by successive approximations. Math. Oper. Res. **15**, 430–466 (1990)
48. Grötschel, M., Lovász, L., Schrijver,A.: Geometric Algorithms and Combinatorial Optimization. Algorithms Combin. Springer, New York (1993)
49. Guan, X.C., Cao, Y.B.: Constrained and bicriteria inverse bottleneck optimization problems under weighted Hamming distance. Optimization **61**(2), 129–142 (2012)
50. Guan, X.C., He, X.Y., Pardalos, P.M., Zhang, B.W.: Inverse max + sum spanning tree problem under Hamming distance by modifying the sum-cost vector. J. Glob. Optim. **69**, 911–925 (2017)
51. Guan, X.C., Pardalos, P.M., Zhang, B.W.: Inverse max+sum spanning tree problem under weighted l_1 norm by modifying the sum-cost vector. Optim. Lett. **5**, 1–13 (2017)
52. Guan X.C., Pardalos PM, Zhang B.W.: Inverse max+sum spanning tree problem under weighted l_1 norm by modifying the sum-cost vector. Optim. Lett. **12**(5), 1065–1077 (2018)
53. Guan, X.C., Pardalos, P.M., Zuo, X.: Inverse max + sum spanning tree problem by modifying the sum-cost vector under weighted l_∞ Norm. J. Glob. Optim. **61**(1), 165–182 (2015)
54. Guan, X.C., Su, J.: Further research on node based bottleneck improvement problem for multicut. Lec. Notes Comput. Sci. **4113**, 895–900 (2006)
55. Guan, X.C., Zhang, B.W.: Inverse 1–median problem on trees under weighted Hamming distance. J. Glob. Optim. **54**(1), 75–82 (2012)
56. Guan, X.C., Zhang, J.Z.: Inverse bottleneck optimization problems on networks. Lec. Notes Comput. Sci. **4041**, 220–230 (2006)
57. Guan, X.C., Zhang, J.Z.: A class of node based bottleneck improvement problems. Euro. J. Oper. Res. **174**(3), 1540–1552 (2006)
58. Guan, X.C., Zhang, J.Z.: Inverse constrained bottleneck problems under weighted l_∞ norm. Comput. Oper. Res. **34**(11), 3243–3254 (2007)
59. Guan, X.C., Zhang, J.Z.: Improving multicut in directed trees by upgrading nodes. Euro. J. Oper. Res. **183**(3), 971–980 (2007)
60. Hambrusch, S.E., Tu, H.Y.: Edge weight reduction problems in directed acyclic graphs. J. Alg. **24**(1), 66–93 (1997)
61. Hasanzadeh, M., Alizadeh, B., Baroughi, F.: The cardinality constrained inverse center location problems on tree networks with edge length augmentation. Theor. Comput. Sci. **865**, 12–33 (2021)
62. He, Y., Zhang, B.W., Yao, E.Y.: Weighted inverse minimum spanning tree problems under Hamming distance. J. Comb. Optim. **9**, 91–100 (2005)
63. Huang, S.M., Liu, Z.H.: On the inverse problem of linear programming and its application to minimum weight perfect k-matching. Eur. J. Oper. Res. **112**, 421–426 (1999)
64. Jana, B., Mondal, S., Pal, M.: Computation of inverse 1–center location problem on the weighted trapezoid graphs. Missouri J. Math. Sci. **31**(1), 14–35 (2019)
65. Jia, J.H., Guan, X.C., Wang, H., Zhang, B.W., Pardalos, P.M.: Combinatorial algorithms for solving the restricted bounded inverse optimal value problem on minimum spanning tree under weighted l_∞ norm. J. Comput. Appl. Math. **419**, 114754 (2023)
66. Jia, J.H., Guan, X.C., Zhang, B.W., Qian, X.Q., Pardalos, P. M.: Combinatorial algorithms for restricted inverse optimal value problems on minimum spanning tree under weighted l_1 norm. J. Comput. Appl. Math. **451**, 116110 (2024)
67. Jia, J.H., Guan, X.C., Zhang, Q., Qian, X.Q., Pardalos, P.M.: Inverse max+sum spanning tree problem under weighted l_∞ norm by modifying max weight vector. J. Glob. Optim. **84**(3), 715–738 (2022)

68. Jiang, Y., Lin, W.F., Liu L.C., Peng, A.Z.: Constrained inverse minimum flow problems under the weighted Hamming distance. Theor. Comput. Sci. **883**, 59–68 (2021)
69. Jiang, Y., Xiao, X.T., Zhang, L.W., Zhang, J.Z.: A perturbation approach for a type of inverse linear programming problems. Int. J. Comput. Math. **88**(3), 508–516 (2011)
70. Karger, D.R., Koller, D., Phillips, S.J.: Finding the hidden path: time bounds for all-pairs shortest paths. SIAM J. Comput. **22**(6), 1199–1217 (1993)
71. Karimi, M., Aman, M., Dolati, A.: Inverse generalized minimum cost flow problem under the hamming distances. J. Oper. Res. Soc. China **7**(2), 355–364 (2019)
72. Keshtkar, I., Ghiyasvand, M.: Inverse quickest center location problem on a tree. Disc. Appl. Math., **260**, 188–202 (2019)
73. Khachiyan, L., Boros, E., Borys, K., Elbassioni, K., Gurvich, V., Rudolf, G., Zhao J.: On short paths interdiction problems: total and node-wise limited interdiction. Theor. Comput. Syst. **43**(2), 204–233 (2008)
74. Korte, B.H., Vygen, J., Korte, B.: Combinatorial Optimization. Springer, Berlin (2011)
75. Krumke, S.O., Marathe, M.V., Noltemeier, H., Ravi, R., Ravi, S.S.: Approximation algorithms for certain network improvement problems. J. Comb. Optim. **2**, 257–288 (1998)
76. Krumke, S.O., Marathe, M.V., Noltemeier, H., Ravi, R., Ravi, S.S., Sundaram, R., Wirth, H.C.: Improving minimum cost spanning trees by upgrading nodes. J. Alg. **33**, 92–111 (1999)
77. Krumke, S.O., Marathe, M.V., Noltemeier, H., Ravi, S.S., Wirth, H.C.: Upgrading bottleneck constrained forests. Disc. Appl. Math. **108**, 129–142 (2001)
78. Nardelli, E., Proietti, G., Widmyer, P.: A faster computation of the most vital edge of a shortest path between two nodes. Inf. Process Lett. **79**(2), 81–85 (2001)
79. Krumke, S.O., Noltemeier, H., Wirth, H.C., Marathe, M.V., Ravi, R., Ravi, S.S., Sundaram, R.: Improving spanning trees by upgrading nodes. Theor. Comput. Sci. **221**, 139–155 (1999)
80. Lawler, E.L.: Combinatorial Optimization: Networks and Matroids. Holt, Rinehart Winston (1976)
81. Li, S.S., Zhang, Z., Lai, H.J.: Algorithm for constraint partial inverse matroid problem with weight increase forbidden. Theoret. Comput. Sci. **640**, 119–124 (2016)
82. Li X., Guan, X.C., Jia, J.H., Pardalos, P.M.: Double interdiction problem on trees on the sum of root-leaf distances by upgrading edges. J. Global Optim. (2025). https://doi.org/10.1007/s10898-025-01490-9
83. Li X., Guan, X.C., Zhang Q., Yin, X.Y., Pardalos, P.M.: The sum of root-leaf distance interdiction problem with cardinality constraint by upgrading edges on trees. J. Comb. Optim. **48**(39), (2024). https://doi.org/10.1007/s10878-024-01230-x.
84. Li, X.Y., Shu, X.C., Huang, H.J., Bai, J.J.: Capacitated partial inverse maximum spanning tree under the weighted Hamming distance. J. Comb. Optim. **38**, 1005–1018 (2019)
85. Li, X.Y., Yang, R.W., Zhang, H.P., Zhang, Z.: Partial inverse maximum spanning tree problem under the Chebyshev norm. J. Comb. Optim. **44**, 3331–3350 (2022)
86. Li, X.Y., Zhang, Z., Du, D.Z.: Partial inverse maximum spanning tree in which weight can only be decreased under l_p-norm. J. Global Optim. **70**(3), 677–685 (2018)
87. Li, X.Y., Zhang, Z., Yang, R.W., Zhang, H.P., Du, D.Z.: Approximation algorithms for capacitated partial inverse maximum spanning tree problem. J. Global Optim. **77**(2), 319–340 (2020)
88. Libura, M.: On the adjustment problem for linear programs. Euro. J. Oper. Res. **183**(1), 125–134 (2007)
89. Liu, L.C., Yao, E.Y.: Inverse min–max spanning tree problem under the weighted sum-type Hamming distance. Theor. Comput. Sci. **396**, 28–34 (2008)
90. Lei, Y., Shao, H., Wu, T.: An accelerating algorithm for maximum shortest path interdiction problem by upgrading edges on trees under unit Hamming distance. Optim. Lett. **17**, 453–469 (2023)
91. Lov*á*sz, L.: Flats in Matroids and Geometric Graphs, Combinatorial Surveys, Edited by P.J. Cameron, pp. 45–86. Academic Press, New York, NY (1977)
92. Magnouche, Y., Martin, S., Most vital vertices for the shortest s–t path problem: complexity and Branch-and-Cut algorithm. Optim. Lett. **14**(2), 2039–2053 (2020)

93. Martello, S., Toth, P.: Knapsack Problems: Algorithms and Computer Implementations. Wiley, Chichester, UK (1990)
94. McCormick,S.T., Smallwood, S.R., Spieksma, F.C.R.: A polynomial algorithm for multiprocessor scheduling with two job lengths. Math. Oper. Res. **26**, 31–49 (2001)
95. Mohammadi, A., Javad, T.: Maximum capacity path interdiction problem with fixed costs. Asia Pac. J. Oper. Res. **36**(4), 1–21 (2019)
96. Mohammadi, S., Alizadeh, B., Afrashteh, E.: Optimal algorithms for integer inverse obnoxious p-median location problems on tree networks. Comput. Appl. Math. **42**(23), 1–28 (2023)
97. Nguyen, K.T.: Inverse 1-median problem on block graphs with variable vertex weights. J. Optim. Theory Appl. **168**(3), 944–957 (2016)
98. Nguyen, K.T.: The inverse 1-center problem on cycles with variable edge lengths. Cent. Eur. J. Oper. Res. **27**(1), 263–274 (2019)
99. Nguyen, K.T., Chassein, A.: The inverse convex ordered 1-median problem on trees under Chebyshev norm and Hamming distance. Eur. J. Oper. Res. **247**(3), 774–781 (2015)
100. Nguyen, K.T., Hieu, V.N.M., Pham, V.: Inverse group 1-median problem on trees. J. Ind. Manage. Optim. **17**(1), 221–232 (2021)
101. Nguyen, K.T., Vui, P.T.: The inverse p-maxian problem on trees with variable edge lengths. Taiwan. J. Math. **20**(6), 1437–1449 (2016)
102. Nguyen–Thu, H., Nguyen, K.T., Toan, N.T.: The max–sum inverse median location problem on trees with budget constraint. Appl. Math. Comput. **460**, 128296 (2024)
103. Orlin, J.B.: A faster strongly polynomial minimum cost flow algorithm. Oper. Res. **41**(2), 338–350 (1993)
104. Orlin, J.B.: Max flows in $O(nm)$ time, or better. In Proceedings of the forty-fifth annual ACM symposium on Theory of Computing (STOC '13), pp. 765–774. Association for Computing Machinery, New York, NY, USA (2013)
105. Papadimitriou, C.H.: Computational Complexity. Addison-Wesley, Reading, MA (1994)
106. Papadimitriou, C.H., Steiglitz, K.: Combinatorial Optimization: Algorithms and Complexity, 2nd edn. Prentice Hall, Hoboken (1998)
107. Pettie, S.: A new approach to all-pairs shortest paths on real-weighted graphs. Theor. Comput. Sci. **312**(1), 47–74 (2004)
108. Punnen, A.P.: A linear time algorithm for the maximum capacity path problem. Eur. J. Oper. Res. **53**(3), 402–404 (1991)
109. Qian, X.Q., Guan, X.C., Jia J.H., Pardalos, P.M.: Inverse Vertex/Absolute quickest 1-center location problem on a tree under weighted l_1 norm. J. Optim. Theory Appl. **200**(2), 524–554, (2024)
110. Qian, X.Q., Guan, X.C., Jia J.H., Zhang Q., Pardalos, P.M.: Vertex quickest 1-center location problem on trees and its inverse problem under weighted l_∞ norm. J. Global Optim. **85**(2), 461–485 (2023)
111. Qian, X.Q., Guan, X.C., Jia J.H., Zhang Q., Pardalos, P.M.:The absolute quickest 1-center location problem on trees and its inverse problem under weighted l_∞ norm. Optimization (2023). Submitted
112. Radzik, T.: Parametric flow, weighted means of cuts, and fractional combinatorial optimization. In: P. Pardalos (Ed.), Complexity in Numerical Optimization, pp. 351–386. World Scientific Publishing, Singapore (1993)
113. Raz, R., Safra, S.: A sub-constant error-probability low-degree test, and sub-constant error-probability PCP characterization of NP. In: Proc. 29th Ann. ACM Symp. on Theory of Computing, pp. 475–484. ACM (1997)
114. Roland, J., Figueira, J.R., De Smet, Y.: The inverse {0, 1}-knapsack problem: Theory, algorithms and computational experiments. Discrete Optim. **10**, 181–192 (2013)
115. Schaefer, A.J.: Inverse integer programming. Optim. Lett. **3**(4), 483–489 (2009)
116. Sokkalingam, P.T., Ahuja, R.K., Orlin, J.B.: Solving inverse spanning tree problems through network flow techniques. Oper. Res. **47**(2), 291–298 (1999)
117. Soltanpour, A., Baroughi, F., Alizadeh, B.: The inverse 1-median location problem on uncertain tree networks with tail value at risk criterion. Inform. Sci. **506**(1), 383–394 (2020)

118. Stackelberg, H.: The Theory of the Market Economy. William Hodge and Co., London (1952)
119. Stoer, M., Wagner, F.: A simple min-cut algorithm. J. ACM. **44**, 585–591 (1997)
120. Taghikhani, S., Baroughi, F.: Fuzzy random classical and inverse median location problems. Soft Comput. **27**(13), 8821–8839 (2023)
121. Taghikhani, S., Baroughi, F., Alizadeh, B.: A hybrid modified PSO algorithm for the inverse p-median location problem in fuzzy random environment. Theor. Comput. Sci. **1000**, 114–574 (2024)
122. Tarantola, A.: Inverse Problem Theory: Methods for Data Fitting and Model Parameter Estimation. Elsevier, Amsterdam (1987)
123. Tardos, $\acute{E}$.: A strongly polynomial algorithm to solve combinatorial linear programs. Oper. Res. **34**, 250–256 (1986)
124. Tayyebi, J., Aman, M.: On inverse linear programming problems under the bottleneck-type weighted Hamming distance. Discrete Appl. Math. **240**, 92–101 (2018)
125. Tayyebi, J., Mitra, A., Sefair, J.A.: The continuous maximum capacity path interdiction problem. Eur. J. Oper. Res. **305**(1), 38–52 (2023)
126. Tayyebi, J., Sepasian, A.R.: Partial inverse min-max spanning tree problem. J. Comb. Optim. **40**, 1075–1091 (2020)
127. Tayyebi, J., Sepasian, A.R.: Reverse 1-centre problem on trees under convex piecewise–linear cost function. Optimization **72**(3), 843–860 (2021)
128. Wang, H., Guan, X.C., Zhang, Q., Zhang, B.W.: Capacitated inverse optimal value problem on minimum spanning tree under bottleneck Hamming distance, J. Comb. Optim. **41**(4), 861–887 (2021)
129. Wang, L.Z.: Cutting plane algorithms for the inverse mixed integer linear programming problem. Oper. Res. Lett. **37**(2), 114–116 (2009)
130. Wang, Q., Yang, T., Wu, L.: General restricted inverse assignment problems under l_1 and l_∞ norms. J. Comb. Optim. **44**, 2040–2055 (2020)
131. Yang, C., Zhang, J.Z.: Two general methods for inverse optimization problems. Appl. Math. Lett. **12**(2), 69–72 (1999)
132. Yang C., Zhang J.Z.: Inverse maximum capacity problems. OR Spectr. **20**(2), 97–100 (1998)
133. Yang, X.G., Zhang, J.Z.: Some inverse min-max network problems under weighted l_1 and l_∞ norms with bound constraints on changes. J. Comb. Optim. **13**, 123–135 (2007)
134. Yang, X.G., Zhang, J.Z.: Inverse center location problem on a tree. J. Syst. Sci. Complex. **21**(4), 651–664 (2008)
135. Zhang, B.W., Guan, X.C., Pardalos, P.M., He, C.Y.: An algorithm for the shortest path improvement problem on rooted trees under the unit Hamming distance. J. Optim. Theory and App. **178**(2), 538–559 (2018)
136. Zhang, B.W., Guan, X.C., Pardalos, P.M., Wang, H., Zhang, Q., Liu, Y., Chen, S.: The lower bounded inverse optimal value problem on minimum spanning tree under unit l_∞ norm. J. Glob. Optim. **79**(3), 757–777 (2021)
137. Zhang, B.W., Guan, X.C., Wang, Q., He, C.Y., Samson, H.: The complexity analysis of the shortest path improvement problem under the Hamming distance. Pac. J. Optim. **11**(4), 605–608 (2015)
138. Zhang, B.W., Guan, X.C., Zhang, Q.: Inverse optimal value problem on minimum spanning tree under unit l_∞ norm. Optim. Lett. **14**(8), 2301–2322 (2020)
139. Zhang, B.W., Zhang, J.Z., He, Y.: Constrained inverse minimum spanning tree problems under bottleneck-type Hamming distance. J. Comb. Optim. **34**, 467–474 (2006)
140. Zhang, B.W., Zhang, J.Z., Qi, L.Q.: The shortest path improvement problem under Hamming distance. J. Comb. Optim. **12**, 351–361 (2006)
141. Zhang, H.L., Xu, Y.F., Wen, X.G.: Optimal shortest path set problem in undirected graphs, J. Comb. Optim. **29**(3), 511–530 (2015)
142. Zhang, J.Z., Lin, Y.X.: Computation of the reverse shortest path problem. J. Global Optim. **25**, 243–261 (2003)
143. Zhang, J.Z., Liu, Z.H.: Calculating some inverse linear programming problem. J. Comput. Appl. Math. **72**, 261–273 (1996)

144. Zhang, J.Z., Liu, Z.H.: A further study on inverse linear programming problems. J. Comput. Appl. Math. **106**(2), 345–359 (1999)
145. Zhang, J.Z., Liu, Z.H.: An oracle strongly polynomial algorithm for bottleneck expansion problems. Optim. Method Softw. **17**, 61–75 (2002)
146. Zhang, J.Z., Liu, Z.H.: A general model of some inverse combinatorial optimization problems and its solution method under l_∞ norm. J. Comb. Optim. **6**(2), 207–227 (2002)
147. Zhang, J.Z., Liu, Z.H., Ma, Z.F.: Some reverse location problems. Eur. J. Oper. Res. **124**(1), 77–88 (2000)
148. Zhang, J.Z., Ma, Z.F.: A network method for solving some inverse combinatorial optimization problems. Optimization **37**, 59–72 (1996)
149. Zhang, J.Z., Ma, Z.F., Yang, C.: A column generation method for inverse shortest path problems. ZOR **41**(3), 347–358 (1995)
150. Zhang, J.Z., Yang, C., Lin, Y.: A class of bottleneck expansion problems. Comput. Oper. Res. **28**, 505–519 (2001)
151. Zhang, Q., Guan, X.C., Jia, J.H., Qian, X.Q.: The sum of root-leaf distance interdiction problem by upgrading edges/nodes on trees. J. Comb. Optim. **44**, 74–93 (2022)
152. Zhang, Q., Guan, X.C., Jia, J.H., Qian, X.Q., Pardalos, P.M.: The restricted inverse optimal value problem on shortest path under l_1 norm on trees. J. Glob. Optim. **86**(1), 251–284 (2023)
153. Zhang, Q., Guan, X.C., Pardalos, P.M: Maximum shortest path interdiction problem by upgrading edges on trees under weighted l_1 norm. J. Glob. Optim. **79**(4), 959–987 (2021)
154. Zhang, Q., Guan, X.C., Wang, H., Pardalos, P.M.: Maximum shortest path interdiction problem by upgrading edges on trees under Hamming distance. Optim. Lett. **15**(8), 2661–2680 (2021)
155. Zhang, Z., Li, S.S., Lai, H.J., Du, D.Z.: Algorithms for the partial inverse matroid problem in which weights can only be increased. J. Global Optim. **65**(4), 801–811 (2016)